Cognitive Informatics for Revealing Human Cognition:

Knowledge Manipulations in Natural Intelligence

Yingxu Wang
University of Calgary, Canada

Information Science
REFERENCE

Managing Director:	Lindsay Johnston
Editorial Director:	Joel Gamon
Book Production Manager:	Jennifer Romanchak
Publishing Systems Analyst:	Adrienne Freeland
Assistant Acquisitions Editor:	Kayla Wolfe
Typesetter:	Nicole Sparano
Cover Design:	Nick Newcomer

Published in the United States of America by
Information Science Reference (an imprint of IGI Global)
701 E. Chocolate Avenue
Hershey PA 17033
Tel: 717-533-8845
Fax: 717-533-8661
E-mail: cust@igi-global.com
Web site: http://www.igi-global.com

Library of Congress Cataloging-in-Publication Data

Cognitive informatics for revealing human cognition: knowledge manipulations in natural intelligence / Yingxu Wang, editor.
    p. cm.
  Includes bibliographical references and index.
  Summary: "This book presents indepth research that builds a link between natural and life sciences with informatics and computer science for investigating cognitive mechanisms and the human information processes"-- Provided by publisher.
  ISBN 978-1-4666-2476-4 (hardcover) -- ISBN 978-1-4666-2477-1 (ebook) -- ISBN 978-1-4666-2478-8 (print & perpetual access) 1. Neural computers. 2. Cognitive science. I. Wang, Yingxu.
  QA76.87.C662 2013
  006.3'2--dc23
                            2012023350

British Cataloguing in Publication Data
A Cataloguing in Publication record for this book is available from the British Library.

The views expressed in this book are those of the authors, but not necessarily of the publisher.

Table of Contents

Section 1
Cognitive Informatics

Section 2
Cognitive Computing

Section 3
Denotational Mathematics

Section 4
Computational Intelligence

Section 5
Applications of Cognitive Informatics and Cognitive Computing

Detailed Table of Contents

Section 1
Cognitive Informatics

Chapter 1

Yingxu Wang, University of Calgary, Canada

The contemporary wonder of sciences and engineering recently refocused on the starting point: how the brain processes internal and external information autonomously rather than imperatively as those of conventional computers? This paper explores the interplay and synergy of cognitive informatics, neural informatics, abstract intelligence, denotational mathematics, brain informatics, and computational intelligence. A key notion recognized in recent studies in cognitive informatics is that the root and profound objective in natural, abstract, and artificial intelligence, and in cognitive informatics and cognitive computing, is to seek suitable mathematical means for their special needs. A layered reference model of the brain and a set of cognitive processes of the mind are systematically developed towards the exploration of the theoretical framework of cognitive informatics. A wide range of applications of cognitive informatics and denotational mathematics are recognized in the development of highly intelligent systems such as cognitive computers, cognitive knowledge search engines, autonomous learning machines, and cognitive robots.

Chapter 2

Yingxu Wang, University of Calgary, Canada
Bernard Widrow, Stanford University, USA
Bo Zhang, Tsinghua University, China
Witold Kinsner, University of Manitoba, Canada
Kenji Sugawara, Chiba Institute of Technology, Japan
Fuchun Sun, Tsinghua University, China
Jianhua Lu, Tsinghua University, China
Thomas Weise, University of Science and Technology of China, China
Du Zhang, California State University, USA

The contemporary wonder of sciences and engineering has recently refocused on the beginning point of: how the brain processes internal and external information autonomously and cognitively rather than imperatively like conventional computers. Cognitive Informatics (CI) is a transdisciplinary enquiry of computer science, information sciences, cognitive science, and intelligence science that investigates the internal information processing mechanisms and processes of the brain and natural intelligence, as well as their engineering applications in cognitive computing. This paper reports a set of eight position statements presented in the plenary panel of IEEE ICCI'10 on Cognitive Informatics and Its Future Development contributed from invited panelists who are part of the world's renowned researchers and scholars in the field of cognitive informatics and cognitive computing.

Chapter 3

With increased understanding of cognitive informatics and the advance of computer technologies, it is becoming clear that human-computer interaction (HCI) is an interaction between two kinds of intelligences, i.e., natural intelligence and artificial intelligence. This paper attempts to clarify interaction-related terminologies through step-by-step definitions, and discusses the nature of HCI, arguing that shared models are the most important aspect of HCI. This paper also proposes that a role-based interaction can be taken as an appropriate shared model for HCI, i.e., Role-Based HCI.

Chapter 4

In many fields including digital image processing and artificial retina design, they always confront a balance issue among real-time, accuracy, computing load, power consumption, and other factors. It is difficult to achieve an optimal balance among these conflicting requirements. However, human retina can balance these conflicting requirements very well. It can efficiently and economically accomplish almost all the visual tasks. This paper presents a bio-inspired model of the retina, not only to simulate various types of retina cells but also to simulate complex structure of retina. The model covers main information processing pathways of retina so that it is much closer to the real retina. In this paper, the authors did some research on various characteristics of retina via large-scale statistical experiments, and further analyzed the relationship between retina's structure and functions. The model can be used in bionic chip design, physiological assumptions verification, image processing and computer vision.

Chapter 5

Language comprises a central component of a complex that is sometimes called "the human capacity." This complex seems to have crystallized fairly recently among a small group in East Africa of whom people are all descendants. Common descent has been important in the evolution of the brain, such that avian and mammalian brains may be largely homologous, particularly in the case of brain regions involved in auditory perception, vocalization and auditory memory. There has been convergent evolution of the capacity for auditory-vocal learning, and possibly for structuring of external vocalizations, such that apes lack the abilities that are shared between songbirds and humans. Language's recent evolutionary origin suggests that the computational machinery underlying syntax arose via the introduction of a single, simple, combinatorial operation. Further, the relation of a simple combinatorial syntax to the sensory-motor and thought systems reveals language to be asymmetric in design: while it precisely matches the representations required for inner mental thought, acting as the "glue" that binds together other internal cognitive and sensory modalities, at the same time it poses computational difficulties for externalization, that is, parsing and speech or signed production. Despite this mismatch, language syntax leads directly to the rich cognitive array that marks us as a symbolic species.

Section 2
Cognitive Computing

Chapter 6

Sébastien Dourlens, Université de Versailles Saint Quentin, France
Amar Ramdane-Cherif, Université de Versailles Saint Quentin, France

Since 1960, AI researchers have worked on intelligent and reactive architectures capable of managing multiple events and acts in the environment. This issue is part of the Robotics domain. An extraction of meaning at different levels of abstraction and the decision process must be implemented in the robot brain to accomplish the multimodal interaction with humans in a human environment. This paper presents a semantic agents architecture giving the robot the ability to understand what is happening and thus provide more robust responses. Intelligence and knowledge about objects like behaviours in the environment are stored in two ontologies linked to an inference engine. To store and exchange information, an event knowledge representation language is used by semantic agents. This architecture brings other advantages: pervasive, cooperating, redundant, automatically adaptable, and interoperable. It is independent of platforms.

Chapter 7

Jason McLaughlin, Indiana University-Purdue University, USA
Shiaofen Fang, Indiana University-Purdue University, USA
Sandra W. Jacobson, Wayne State University, USA, & University of Cape Town, South Africa
H. Eugene Hoyme, Sanford School of Medicine, USA
Luther Robinson, State University of New York, USA
Tatiana Foroud, Indiana University, USA

A new visual approach to the surface shape analysis and classification of 3D facial images is presented. It allows the users to visually explore the natural patterns and geometric features of 3D facial scans to provide decision-making information for face classification which can be used for the diagnosis of diseases that exhibit facial characteristics. Using surface feature analysis under a digital geometry analysis framework, the method employs an interactive feature visualization technique that allows interactive definition, modification and exploration of facial features to provide the best discriminatory power for a given classification problem. OpenGL based surface shading and interactive lighting are employed to generate visual maps of discriminatory features to visually represent the salient differences between labeled classes. This technique will be applied to a medical diagnosis application for Fetal Alcohol Syndrome (FAS) which is known to exhibit certain facial patterns.

The cognitive mechanisms of knowledge representation, memory establishment, and learning are fundamental issues in understanding the brain. A basic approach to studying these mental processes is to observe and simulate how knowledge is memorized by little children. This paper presents a simulation tool for knowledge acquisition and memory development for young children of two to five years old. The cognitive mechanisms of memory, the mathematical model of concepts and knowledge, and the fundamental elements of internal knowledge representation are explored. The cognitive processes of children's memory and knowledge development are described based on concept algebra and the object-attribute-relation (OAR) model. The design of the simulation tool for children's knowledge acquisition and memory development is presented with the graphical representor of memory and the dynamic concept network of knowledge. Applications of the simulation tool are described by case studies on children's knowledge acquisition about family members, relatives, and transportation. This work is a part of the development of cognitive computers that mimic human knowledge processing and autonomous learning.

Emotion recognition is a very hot topic, which is related with computer science, psychology, artificial intelligence, etc. It is always performed on facial or audio information with classical method such as ANN, fuzzy set, SVM, HMM, etc. Ensemble learning theory is a novelty in machine learning and ensemble method is proved an effective pattern recognition method. In this paper, a novel ensemble learning method is proposed, which is based on selective ensemble feature selection and rough set theory. This method can meet the tradeoff between accuracy and diversity of base classifiers. Moreover, the proposed method is taken as an emotion recognition method and proved to be effective according to the simulation experiments.

Cognitive Informatics (CI) is a transdisciplinary enquiry of computer science, information sciences, cognitive science, and intelligence science that investigates into the internal information processing mechanisms and processes of the brain and natural intelligence, as well as their engineering applications in cognitive computing. The latest advances in CI leads to the establishment of cognitive computing theories and methodologies, as well as the development of Cognitive Computers (CogC) that perceive, infer, and learn. This paper reports a set of nine position statements presented in the plenary panel of IEEE ICCI*CC'11 on Cognitive Informatics in Year 10 and Beyond contributed from invited panelists who are part of the world's renowned researchers and scholars in the field of cognitive informatics and cognitive computing.

Section 3
Denotational Mathematics

Chapter 11

Inference as the basic mechanism of thought is one of the gifted abilities of human beings. It is recognized that a coherent theory and mathematical means are needed for dealing with formal causal inferences. This paper presents a novel denotational mathematical means for formal inferences known as Inference Algebra (IA). IA is structured as a set of algebraic operators on a set of formal causations. The taxonomy and framework of formal causal inferences of IA are explored in three categories: (a) Logical inferences on Boolean, fuzzy, and general logic causations; (b) Analytic inferences on general functional, correlative, linear regression, and nonlinear regression causations; and (c) Hybrid inferences on qualification and quantification causations. IA introduces a calculus of discrete causal differential and formal models of causations; based on them nine algebraic inference operators of IA are created for manipulating the formal causations. IA is one of the basic studies towards the next generation of intelligent computers known as cognitive computers. A wide range of applications of IA are identified and demonstrated in cognitive informatics and computational intelligence towards novel theories and technologies for machine-enabled inferences and reasoning.

Chapter 12

In spite of their striking diversity, numerous tasks and architectures of intelligent systems such as those permeating multivariable data analysis, decision-making processes along with their underlying models, recommender systems and others exhibit two evident commonalities. They promote (a) human centricity and (b) vigorously engage perceptions (rather than plain numeric entities) in the realization of the systems and their further usage. Information granules play a pivotal role in such settings. Granular Computing delivers a cohesive framework supporting a formation of information granules and facilitating their processing. The author exploits two essential concepts of Granular Computing. The first one deals with the construction of information granules. The second one helps endow constructs of intelligent systems with a much needed conceptual and modeling flexibility. The study elaborates in detail on the three representative studies. In the first study being focused on the Analytic Hierarchy Process (AHP) used in decision-making, the author shows how an optimal allocation of granularity helps improve the quality of the solution and facilitate collaborative activities in models of group decision-making. The second study is concerned with a granular interpretation of temporal data where the role of information granularity is profoundly visible when effectively supporting human centric description of relationships existing in data. The third study concerns a formation of granular logic descriptors on a basis of a family of logic descriptors.

Yingxu Wang, University of Calgary, Canada
Yousheng Tian, University of Calgary, Canada
Kendal Hu, University of Calgary, Canada

Towards the formalization of ontological methodologies for dynamic machine learning and semantic analyses, a new form of denotational mathematics known as concept algebra is introduced. Concept Algebra (CA) is a denotational mathematical structure for formal knowledge representation and manipulation in machine learning and cognitive computing. CA provides a rigorous knowledge modeling and processing tool, which extends the informal, static, and application-specific ontological technologies to a formal, dynamic, and general mathematical means. An operational semantics for the calculus of CA is formally elaborated using a set of computational processes in real-time process algebra (RTPA). A case study is presented on how machines, cognitive robots, and software agents may mimic the key ability of human beings to autonomously manipulate knowledge in generic learning using CA. This work demonstrates the expressive power and a wide range of applications of CA for both humans and machines in cognitive computing, semantic computing, machine learning, and computational intelligence.

Jun Zhang, Shanghai University, China
Xiangfeng Luo, Shanghai University, China
Xiang He, Shanghai University, China
Chuanliang Cai, Shanghai University, China

Dealing with the large-scale text knowledge on the Web has become increasingly important with the development of the Web, yet it confronts with several challenges, one of which is to find out as much semantics as possible to represent text knowledge. As the text semantic mining process is also the knowledge representation process of text, this paper proposes a text knowledge representation model called text semantic mining model (TSMM) based on the algebra of human concept learning, which both carries rich semantics and is constructed automatically with a lower complexity. Herein, the algebra of human concept learning is introduced, which enables TSMM containing rich semantics. Then the formalization and the construction process of TSMM are discussed. Moreover, three types of reasoning rules based on TSMM are proposed. Lastly, experiments and the comparison with current text representation models show that the given model performs better than others.

Bing Zhou, University of Regina, Canada
Yiyu Yao, University of Regina, Canada

Decision-Theoretic Rough Set (DTRS) model provides a three-way decision approach to classification problems, which allows a classifier to make a deferment decision on suspicious examples, rather than being forced to make an immediate determination. The deferred cases must be reexamined by collecting further information. Although the formulation of DTRS is intuitively appealing, a fundamental question that remains is how to determine the class of the deferment examples. In this paper, the authors introduce an adaptive learning method that automatically deals with the deferred examples by searching for effective granulization. A decision tree is constructed for classification. At each level, the authors sequentially choose the attributes that provide the most effective granulization. A subtree is added recursively if the conditional probability lies in between of the two given thresholds. A branch reaches its leaf node when the conditional probability is above or equal to the first threshold, or is below or equal to the second threshold, or the granule meets certain conditions. This learning process is illustrated by an example.

Section 4
Computational Intelligence

Chapter 16

The main topics covered in this paper address the following four issues: (1) Distinction between how adaptation and cognition are viewed with respect to each other, (2) With human cognition viewed as the framework for cognition, the following cognitive processes are identified: the perception-action cycle, memory, attention, intelligence, and language. With language being outside the scope of the paper, detailed accounts of the other four cognitive processes are discussed, (3) Cognitive radar is singled out as an example application of cognitive dynamic systems that "mimics" the visual brain; experimental results on tracking are presented using simulations, which clearly demonstrate the information-processing power of cognition, and (4) Two other example applications of cognitive dynamic systems, namely, cognitive radio and cognitive control, are briefly described.

Chapter 17

This paper presents a new cryptosystem based on chaotic continuous-interval cellular automata (CCA) to increase data protection as demonstrated by their flexibility to encrypt and decrypt information from distinct sources. Enhancements to cryptosystems are also presented including (i) a model based on a new chaotic CCA attractor, (ii) the dynamical integration of modules containing dynamical systems to generate complex sequences, and (iii) an enhancement for symmetric cryptosystems by allowing them to generate an unlimited number of keys. This paper also presents a process of mixing chaotic sequences obtained from cellular automata, instead of using differential equations, as a basis to achieve higher security and higher speed for the encryption and decryption processes, as compared to other recent approaches. The complexity of the mixed sequences is measured using the variance fractal dimension trajectory to compare them to the unmixed chaotic sequences to verify that the former are more complex. This type of polyscale measure and evaluation has never been done in the past outside this research group.

Chapter 18

This paper examines the inherited persistent behavior of particle swarm optimization and its implications to cognitive machines. The performance of the algorithm is studied through an average particle's trajectory through the parameter space of the Sphere and Rastrigin function. The trajectories are decomposed into position and velocity along each dimension optimized. A threshold is defined to separate the transient period, where the particle is moving towards a solution using information about the position of its best neighbors, from the steady state reached when the particles explore the local area surrounding the solution to the system. Using a combination of time and frequency domain techniques, the inherited long-term dependencies that drive the algorithm are discerned. Experimental results show the particles balance exploration of the parameter space with the correlated goal oriented trajectory driven by their social interactions. The information learned from this analysis can be used to extract complexity measures to classify the behavior and control of particle swarm optimization, and make proper decisions on what to do next. This novel analysis of a particle trajectory in the time and frequency domains presents clear advantages of particle swarm optimization and inherent properties that make this optimization algorithm a suitable choice for use in cognitive machines.

Chapter 19

From the point of view of an autonomous agent the world consists of high-dimensional dynamic sensorimotor data. Interface algorithms translate this data into symbols that are easier to handle for cognitive processes. Symbol grounding is about whether these systems can, based on this data, construct symbols that serve as a vehicle for higher symbol-oriented cognitive processes. Machine learning and data mining techniques are geared towards finding structures and input-output relations in this data by providing appropriate interface algorithms that translate raw data into symbols. This work formulates the interface design as global optimization problem with the objective to maximize the success of the overlying symbolic algorithm. For its implementation various known algorithms from data mining and machine learning turn out to be adequate methods that do not only exploit the intrinsic structure of the subsymbolic data, but that also allow to flexibly adapt to the objectives of the symbolic process. Furthermore, this work discusses the optimization formulation as a functional perspective on symbol grounding that does not hurt the zero semantical commitment condition. A case study illustrates technical details of the machine symbol grounding approach.

Chapter 20

In the research of Web content-based image retrieval, how to reduce more of the image dimensions without losing the main features of the image is highlighted. Many features of dimensional reduction schemes are determined by the breaking of higher dimensional general covariance associated with the selection of a particular subset of coordinates. This paper starts with analysis of commonly used methods for the dimension reduction of Web images, followed by a new algorithm for nonlinear dimensionality reduction based on the HSV image features. The approach obtains intrinsic dimension estimation by similarity calculation of two images. Finally, some improvements were made on the Parallel Genetic Algorithm (APGA) by use of the image similarity function as the self-adaptive judgment function to improve the genetic operators, thus achieving a Web image dimensionality reduction and similarity retrieval. Experimental results illustrate the validity of the algorithm.

Section 5
Applications of Cognitive Informatics and Cognitive Computing

Chapter 21

Rong-Hua Li, The Hong Kong Polytechnic University, Hong Kong

Shuang Liang, The Hong Kong Polytechnic University, Hong Kong

George Baciu, The Hong Kong Polytechnic University, Hong Kong

Eddie Chan, The Hong Kong Polytechnic University, Hong Kong

Singularity problems of scatter matrices in Linear Discriminant Analysis (LDA) are challenging and have obtained attention during the last decade. Linear Discriminant Analysis via QR decomposition (LDA/QR) and Direct Linear Discriminant analysis (DLDA) are two popular algorithms to solve the singularity problem. This paper establishes the equivalent relationship between LDA/QR and DLDA. They can be regarded as special cases of pseudo-inverse LDA. Similar to LDA/QR algorithm, DLDA can also be considered as a two-stage LDA method. Interestingly, the first stage of DLDA can act as a dimension reduction algorithm. The experiment compares LDA/QR and DLDA algorithms in terms of classification accuracy, computational complexity on several benchmark datasets and compares their first stages. The results confirm the established equivalent relationship and verify their capabilities in dimension reduction.

Chapter 22

Jun Peng, Chongqing University of Science and Technology, China

Du Zhang, California State University, USA

Xiaofeng Liao, Chongqing University, China

This paper proposes a novel image block encryption algorithm based on three-dimensional Chen chaotic dynamical system. The algorithm works on 32-bit image blocks with a 192-bit secret key. The idea is that the key is employed to drive the Chen's system to generate a chaotic sequence that is inputted to a specially designed function G, in which we use new 8x8 S-boxes generated by chaotic maps (Tang, 2005). In order to improve the robustness against differential cryptanalysis and produce desirable avalanche effect, the function G is iteratively performed several times and its last outputs serve as the key-streams to encrypt the original image block. The design of the encryption algorithm is described along with security analyses. The results from key space analysis, differential attack analysis, and information entropy analysis, correlation analysis of two adjacent pixels prove that the proposed algorithm can resist cryptanalytic, statistical and brute force attacks, and achieve a higher level of security. The algorithm can be employed to realize the security cryptosystems over the Internet.

Chapter 23

Mingming Li, Beijing University of Post and Telecommunication, China

Jiaru Lin, Beijing University of Post and Telecommunication, China

Fazhong Liu, College of Computing & Communication Engineering, GUCAS, China

Dongxu Wang, China Unicom, China

Li Guo, Beijing University of Post and Telecommunication, China

The authors consider a cognitive radio network in which a set of cognitive users make opportunistic spectrum access to one primary channel by time-division multiplexing technologies. Multiple Input Multiple Output techniques (MIMO) are similarly considered to enhance the stable throughput for cognitive links while they should guarantee co-channel interference constraints to the primary link. Here, two different cases are considered: one is that cognitive radio network is distributed; the other is centrally-controlled that cognitive radio network has a cognitive base station. In the first case, how to choose one fixed cognitive user and power control for each transmission antenna at the cognitive base station are considered to maximize the cognitive link's stable throughput. In the second case, a scheme to choose a group of cognitive users and a Zero-Forcing method to pre-white co-channel interference to the primary user, are also proposed in order to maximize cognitive base station's sum-rate. The algorithm can be employed to realize opportunistic spectrum transmission over the wireless fading channels.

Applications of switched reluctance motor (SRM) to direct drive robot are increasingly popular because of its valuable advantages. However, the greatest potential defect is its torque ripple owing to the significant nonlinearities. In this paper, a fuzzy neural network (FNN) is applied to control the SRM torque at the goal of the torque-ripple minimization. The desired current provided by FNN model compensates the nonlinearities and uncertainties of SRM. On the basis of FNN-based current closed-loop system, the trajectory tracking controller is designed by using the dynamic model of the manipulator, where the torque control method cancels the nonlinearities and cross-coupling terms. A single link robot manipulator directly driven by a four-phase 8/6-pole SRM operates in a sinusoidal trajectory tracking rotation. The simulated results verify the proposed control method and a fast convergence that the robot manipulator follows the desired trajectory in a 0.9-s time interval.

Preface

Cognitive informatics (CI) is a new discipline that studies the natural intelligence and internal information processing mechanisms of the brain, as well as the processes involved in perception and cognition. CI was initiated by Yingxu Wang and his colleagues in 2002. The development and the cross fertilization among computer science, information science, cognitive science, brain science, and intelligence science have led to a whole range of extremely interesting new research fields known as CI, which investigates the internal information processing mechanisms and processes of the natural intelligence – human brains and minds – and their engineering applications in computational intelligence.

The theories of informatics and their perceptions on the object of information have evolved from the classic information theory, modern informatics, to cognitive informatics in the last six decades. The *classic information theories*, particularly Shannon's information theory, are the first-generation informatics that study signals and channel behaviors based on statistics and probability theory. The *modern informatics* studies information as properties or attributes of the natural world that can be distinctly elicited, generally abstracted, quantitatively represented, and mentally processed. The first- and second-generation informatics put emphases on external information processing, which are yet to be extended to observe the fundamental fact that human brains are the original sources and final destinations of information. Any information must be cognized by human beings before it is understood, comprehended, and consumed. The aforementioned observations have led to the establishment of the third-generation informatics, *cognitive informatics*, a term coined by Yingxu Wang in a keynote to the First IEEE International Conference on Cognitive Informatics in 2002. It is recognized in CI that *information* is the third essence of the natural world supplementing to matter and energy, which is any property or attribute of the natural world that can be distinctly elicited, generally abstracted, quantitatively represented, and mentally processed. On the basis of the evolvement of intension and extension of the term information, *informatics* is the science of information that studies the nature of information, its processing, and ways of transformation between information, matter and energy.

In many disciplines of human knowledge, almost all of the hard problems yet to be solved share a common root in the understanding of the mechanisms of natural intelligence and the cognitive processes of the brain. Therefore, CI is a discipline that forges links between a number of natural science and life science disciplines with informatics and computing science. CI provides a coherent set of fundamental theories, and contemporary mathematics, which form the foundation for most information and knowledge based science and engineering disciplines.

This book, entitled *Cognitive Informatics for Revealing Human Cognition: Knowledge Manipulations in Natural Intelligence,* is the fifth volume in the IGI series of Advances in Cognitive Informatics and Natural Intelligence. The book encompasses 24 chapters of expert contributions selected from the International Journal of Cognitive Informatics and Natural Intelligence during 2011. The book is organized in five sections on: (1) Cognitive informatics; (2) Cognitive computing; (3) Denotational mathematics; (4) Computational intelligence; and (5) Applications in cognitive informatics and cognitive computing.

SECTION 1: COGNITIVE INFORMATICS

Cognitive Informatics (CI) is a transdisciplinary enquiry of computer science, information science, cognitive science, and intelligence science that investigates into the internal information processing mechanisms and processes of the brain and natural intelligence, as well as their engineering applications in cognitive computing. Fundamental theories developed in CI covers the Information-Matter-Energy-Intelligence (IME-I) model, the Layered Reference Model of the Brain (LRMB), the Object-Attribute-Relation (OAR) model of internal information representation in the brain, the cognitive informatics model of the brain, natural intelligence (NI), abstract intelligence (αI), neuroinformatics (NeI), denotational mathematics (DM), and cognitive systems. Recent studies on LRMB in cognitive informatics reveal an entire set of cognitive functions of the brain and their cognitive process models, which explain the functional mechanisms and cognitive processes of the natural intelligence with 43 cognitive processes at seven layers known as the sensation, memory, perception, action, meta-cognitive, meta-inference, and higher cognitive layers.

According to CI, natural intelligence, in the narrow sense, is a human or a system ability that transforms information into behaviors; while in the broad sense, it is any human or system ability that autonomously transfers the forms of abstract information between data, information, knowledge, and behaviors in the brain. The history of human quest to understand the brain and natural intelligence is certainly as long as human history itself. It is recognized that artificial intelligence is a subset of natural intelligence. Therefore, the understanding of natural intelligence is a foundation for investigating into artificial, machinable, and computational intelligence.

The section on cognitive informatics encompasses the following five chapters:

- **Chapter 1:** Towards the Synergy of Cognitive Informatics, Neural Informatics, Brain Informatics, and Cognitive Computing
- **Chapter 2:** Perspectives on the Field of Cognitive Informatics and Its Future Development
- **Chapter 3:** Role-Based Human-Computer Interaction
- **Chapter 4:** Main Retina Information Processing Pathways Modeling
- **Chapter 5:** Songs to Syntax: Cognition, Combinatorial Computation, and the Origin of Language

Chapter 1, *Towards the Synergy of Cognitive Informatics, Neural Informatics, Brain Informatics, and Cognitive Computing,* by Yingxu Wang, recognizes that the contemporary wonder of sciences and engineering has recently refocused on the starting point of them: how the brain processes internal and external information autonomously rather than imperatively as those of conventional computers? The interplay and synergy of cognitive informatics, neural informatics, abstract intelligence, denotational mathematics, brain informatics, and computational intelligence are explored in this chapter. A key notion recognized in recent studies in cognitive informatics is that the root and profound objective in natural, abstract, and artificial intelligence in general, and in cognitive informatics and cognitive computing in particular, is to seek suitable mathematical means for their special needs that were missing. A layered reference model of the brain and a set of cognitive processes of the mind are systematically developed towards the exploration of the theoretical framework of cognitive informatics. A wide range of applications of cognitive informatics and denotational mathematics are recognized in the development of highly intelligent systems such as cognitive computers, cognitive knowledge search engines, autonomous learning machines, and cognitive robots.

Chapter 2, *Perspectives on the Field of Cognitive Informatics and its Future Development*, by Yingxu Wang, Bernard Widrow, Bo Zhang, Witold Kinsner, Kenji Sugawara, Fuchun Sun, Jianhua Lu, Thomas Weise, and Du Zhang, presents Cognitive Informatics (CI) as a transdisciplinary enquiry of computer science, information sciences, cognitive science, and intelligence science that investigates into the internal information processing mechanisms and processes of the brain and natural intelligence, as well as their engineering applications in cognitive computing. This chapter reports a set of eight position statements presented in the plenary panel of IEEE ICCI'10 on *Cognitive Informatics and its Future Development* contributed from invited panelists who are part of the world's renowned researchers and scholars in the field of cognitive informatics and cognitive computing.

Chapter 3, *Role-Based Human-Computer Interaction*, by Haibin Zhu and Min Hou, presents that with the increased understanding of cognitive informatics and the advance of computer technologies, it is becoming clear that human-computer interaction (HCI) is an interaction between two kinds of intelligences, i.e., natural intelligence and artificial intelligence. This chapter attempts to clarify interaction-related terminologies through step-by-step definitions, and discusses the nature of HCI, arguing that shared models are the most important aspect of HCI. This chapter also proposes that a role-based interaction can be taken as an appropriate shared model for HCI, i.e., role-based HCI.

Chapter 4, *Main Retina Information Processing Pathways Modeling*, by Hui Wei, Qingsong Zuo, and Xudong Guan, identifies there always confront a balance issue among real-time, accuracy, computing load, power consumption and other factors in digital image processing and artificial retina design. It is difficult to achieve an optimal balance among these conflicting requirements. However, human retina can balance these conflicting requirements very well. It can efficiently and economically accomplish almost all the visual tasks. This chapter presents a bio-inspired model of the retina, not only to simulate various types of retina cells but also to simulate complex structure of retina. The model covers main information processing pathways of retina so that it is much closer to the real retina. In this chapter, the authors researched various characteristics of retina via large-scale statistical experiments, and further analyzed the relationship between retina's structure and functions. The model can be used in bionic chip design, physiological assumptions verification, image processing, and computer vision.

Chapter 5, *Songs to Syntax: Cognition, Combinatorial Computation, and the Origin of Language*, by Robert C. Berwick, reveals that language comprises a central component of what the co-founder of modern evolutionary theory, Alfred Russell Wallace, called "man's intellectual and moral nature" – the human cognitive capacities for creative imagination, language and symbolism generally, a complex that is sometimes simply called "the human capacity." This complex seems to have crystallized fairly recently among a small group in East Africa of whom we are all descendants, distinguishing contemporary humans sharply from all other animals, with enormous consequences for the whole of the biological world, as well as for the study of computational cognition. How can we explain this evolutionary leap? On the one hand, common descent has been important in the evolution of the brain, such that avian and mammalian brains may be largely homologous, particularly in the case of brain regions involved in auditory perception, vocalization and auditory memory. On the other hand, there has been convergent evolution of the capacity for auditory-vocal learning, and possibly for structuring of external vocalizations, such that apes lack the abilities that are shared between songbirds and humans. Language's recent evolutionary origin suggests that the computational machinery underlying syntax arose via the introduction of a single, simple, combinatorial operation. Further, the relation of a simple combinatorial syntax to the sensory-motor and thought systems reveals language to be asymmetric in design: while it precisely matches the representations required for inner mental thought, acting as the "glue" that binds together other internal cognitive and sensory modalities, at the same time it poses computational difficulties for externalization, that is, parsing and speech or signed production. Despite this mismatch, language syntax leads directly to the rich cognitive array that marks us as a symbolic species, including mathematics, music, and much more.

SECTION 2: COGNITIVE COMPUTING

Computing systems and technologies can be classified into the categories of *imperative, autonomic,* and *cognitive* computing from the bottom up. The imperative computers are a passive system based on stored-program controlled behaviors for data processing. The autonomic computers are goal-driven and self-decision-driven machines that do not rely on instructive and procedural information. Cognitive computers are more intelligent computers beyond the imperative and autonomic computers, which embody major natural intelligence behaviors of the brain such as thinking, inference, and learning.

Cognitive Computing (CC) is a novel paradigm of intelligent computing methodologies and systems based on CI that implements computational intelligence by autonomous inferences and perceptions mimicking the mechanisms of the brain. CC is emerged and developed based on the multidisciplinary research in CI. The latest advances in CI and CC, as well as denotational mathematics, enable a systematic solution for the future generation of intelligent computers known as *cognitive computers* (CogCs) that think, perceive, learn, and reason. A CogC is an intelligent computer for knowledge processing as that of a conventional von Neumann computer for data processing. CogCs are designed to embody *machinable intelligence* such as computational inferences, causal analyses, knowledge manipulation, machine learning, and autonomous problem solving.

The section on cognitive computing encompasses the following five chapters:

- **Chapter 6:** Cognitive Memory for Semantic Agents Architecture in Robotic Interaction
- **Chapter 7:** Interactive Feature Visualization and Detection for 3D Face Classification
- **Chapter 8:** A Computational Simulation of the Cognitive Process of Children Knowledge Acquisition and Memory Development
- **Chapter 9:** A Novel Emotion Recognition Method Based on Ensemble Learning and Rough Set Theory
- **Chapter 10:** Cognitive Informatics and Cognitive Computing in Year 10 and Beyond

Chapter 6, *Cognitive Memory for Semantic Agents Architecture in Robotic Interaction*, by Sébastien Dourlens and Amar Ramdane-Cherif, presents that since 1960, lots of AI researchers work on intelligent and reactive architectures able to manage multiple events and act in the environment. This issue is also part of robotics domain. An extraction of meaning at different levels of abstraction and the decision process must be implemented in the robot brain to accomplish the multimodal interaction with human in human environment. In this chapter, the authors present a semantic agents architecture giving the robot the ability to well understand what is happening and thus provide more robust responses. They describe their agent component. Intelligence and knowledge about objects like behaviours in the environment are stored in two ontologies linked to an inference engine. To store and exchange information, an event knowledge representation language is used by semantic agents. This architecture brings other advantages: pervasive, cooperating, redundant, automatically adaptable and interoperable. It is independent of platforms.

Chapter 7, *Interactive Feature Visualization and Detection for 3D Face Classification*, by Jason Mclaughlin, Shiaofen Fang, Luther Robinson, Sandra Jacobson, Tatiana Foroud, and H Eugene Hoyme, presents a new visual approach to the surface shape analysis and classification of 3D facial images. It aims to allow the users to visually explore the natural patterns and geometric features of 3D facial scans to provide decision-making information for face classification which can be used for the diagnosis of diseases that exhibit facial characteristics. Using surface feature analysis under a digital geometry analysis framework, the authors employ an interactive feature visualization technique that allows interactive definition, modification and exploration of facial features to provide the best discriminatory power for a given classification problem. OpenGL based surface shading and interactive lighting are employed to generate visual maps of discriminatory features to visually represent the salient differences between labeled classes. This technique will be applied to a medical diagnosis application for Fetal Alcohol Syndrome (FAS) which is known to exhibit certain facial patterns.

Chapter 8, *A Computational Simulation of the Cognitive Process of Children Knowledge Acquisition and Memory Development*, by Jeff Bancroft and Yingxu Wang, identifies that the cognitive mechanisms of knowledge representation, memory establishment, and learning are fundamental issues in understanding the brain. A basic approach to study these mental processes is to observe and simulate how knowledge is memorized by little children. This chapter presents a simulation tool for knowledge acquisition and memory development for young children of 2 to 5 year-old. The cognitive mechanisms of memory, the mathematical model of concepts and knowledge, and the fundamental elements of internal knowledge representation are explored. The cognitive processes of children memory and knowledge development are described based on concept algebra and the object-attribute-relation (OAR) model. The design of the simulation tool for children knowledge acquisition and memory development is presented with the graphical representor of memory and the dynamic concept network of knowledge. Applications of the simulation tool are described by case studies on children knowledge acquisition about family members, relatives, and transportation. This work is a part of the development of cognitive computers that mimic human knowledge processing and autonomous learning.

Chapter 9, *A Novel Emotion Recognition Method Based on Ensemble Learning and Rough Set Theory*, by Yong Yang and Guoyin Wang, presents an emotion recognition technology related to computer science, psychology, and artificial intelligence. It is always performed on facial or audio information with classical method such as ANN, fuzzy set, SVM, HMM, et cetera. Ensemble learning theory is a novelty in machine learning and ensemble method is proved an effective pattern recognition method. In this chapter, a novel ensemble learning method is proposed, which is based on selective ensemble feature selection and rough set theory. This method can meet the tradeoff between accuracy and diversity of base classifiers. Moreover, the proposed method is taken as an emotion recognition method and proved to be effective according to the simulation experiments.

Chapter 10, *Cognitive Informatics and Cognitive Computing in Year 10 and Beyond*, by Yingxu Wang, Robert C. Berwick, Simon Haykin, Witold Pedrycz, Witold Kinsner, George Baciu, Du Zhang, Virendrakumar C. Bhavsar, and Marina Gavrilova, presents Cognitive Informatics (CI) as a transdisciplinary enquiry of computer science, information sciences, cognitive science, and intelligence science that investigates into the internal information processing mechanisms and processes of the brain and natural intelligence, as well as their engineering applications in cognitive computing. The latest advances in CI leads to the establishment of cognitive computing theories and methodologies, as well as the development of Cognitive Computers (CogC) that perceive, infer, and learn. This chapter reports a set of nine position statements presented in the plenary panel of IEEE ICCI*CC'11 on *Cognitive Informatics in Year 10 and Beyond* contributed from invited panelists who are part of the world's renowned researchers and scholars in the field of cognitive informatics and cognitive computing.

SECTION 3: DENOTATIONAL MATHEMATICS

The needs for complex and long-series of causal inferences in cognitive computing, αI, computational intelligence, software engineering, and knowledge engineering have led to new forms of mathematics collectively known as denotational mathematics. *Denotational Mathematics* (DM) is a category of expressive mathematical structures that deals with high-level mathematical entities beyond numbers and sets, such as abstract objects, complex relations, perceptual information, abstract concepts, knowledge, intelligent behaviors, behavioral processes, and systems.

It is recognized that the maturity of any scientific discipline is characterized by the maturity of its mathematical means, because the nature of mathematics is a generic meta-methodological science. In recognizing mathematics as the *metamethodology* for all sciences and engineering disciplines, a set of DMs has been created and applied in CI, αI, AI, CC, CogC, soft computing, computational intelligence, and computational linguistics. Typical paradigms of DM are such as *concept algebra* (Wang, 2008), *system algebra* (Wang, 2008), *real-time process algebra* (Wang, 2002), *granular algebra* (Wang, 2009), *visual semantic algebra* (Wang, 2009), and *inference algebra* (Wang, 2011). DM provides a coherent set of contemporary mathematical means and explicit expressive power for cognitive informatics, cognitive computing, artificial intelligence, and computational intelligence.

The section on denotational mathematics encompasses the following five chapters:

- **Chapter 11:** Inference Algebra (IA): A Denotational Mathematics for Cognitive Computing and Machine Reasoning (I)
- **Chapter 12:** Human Centricity and Perception-Based Perspective and Their Centrality to the Agenda of Granular Computing
- **Chapter 13:** Semantic Manipulations and Formal Ontology for Machine Learning Based on Concept Algebra
- **Chapter 14:** Text Semantic Mining Model Based on the Algebra of Human Concept Learning
- **Chapter 15:** In Search of Effective Granulization with DTRS for Ternary Classification

Chapter 11, *Inference Algebra (IA): A Denotational Mathematics for Cognitive Computing and Machine Reasoning (I)*, by Yingxu Wang, presents that inference as the basic mechanism of thought is one of the gifted abilities of human beings, which is a cognitive process that creates rational causations between a pair of cause and effect based on empirical arguments, formal reasoning, and/or statistical norms. It is recognized that a coherent theory and mathematical means are needed for dealing with formal causal inferences. This chapter presents a novel denotational mathematical means for formal inferences known as *Inference Algebra* (IA). IA is structured as a set of algebraic operators on a set of formal causations. The taxonomy and framework of formal causal inferences of IA are explored in three categories: a) *Logical inferences* on Boolean, fuzzy, and general logic causations; b) *Analytic inferences* on general functional, correlative, linear regression, and nonlinear regression causations; and c) *Hybrid inferences* on qualification and quantification causations. IA introduces a calculus of *discrete causal differential* and formal models of causations, based on them nine algebraic inference operators of IA are created for manipulating the formal causations. IA elicits and formalizes the common and empirical reasoning processes in a rigorous form, which enable artificial intelligence and computational intelligent systems to mimic human inference abilities by cognitive computing. IA is one of the basic studies towards the next generation of intelligent computers known as *cognitive computers*. A wide range of applications of IA are identified and demonstrated in cognitive informatics and computational intelligence towards novel theories and technologies for machine-enabled inferences and reasoning. This work is presented in two parts due to its excessive length. The structure of formal inference, the framework of IA, and the mathematical models of formal causations are described in this chapter; while the inference operators of IA as well as their extensions and applications are elaborated in *IJCINI, 6*(1).

Chapter 12, *Human Centricity and Perception-Based Perspective and Their Centrality to the Agenda of Granular Computing*, by Witold Pedrycz, presents that in spite of their striking diversity, numerous tasks and architectures of intelligent systems such as those permeating multivariable data analysis (e.g., time series, spatio-temporal, and spatial dependencies), decision-making processes along with their underlying models, recommender systems and others exhibit two evident commonalities. They promote (a) human centricity and (b) vigorously engage perceptions (rather than plain numeric entities) in the realization of the systems and their further usage. Information granules play a pivotal role in such settings. Granular Computing delivers a cohesive framework supporting a formation of information granules and facilitating their processing. The chapter exploits two essential concepts of Granular Computing. The first one, formed with the aid of a principle of justifiable granularity, deals with the construction of information granules. The second one, based on an idea of an optimal allocation of information granularity, helps endow constructs of intelligent systems with a much needed conceptual and modeling flexibility. The study elaborates in detail on the three representative studies. In the first study being focused on the Analytic Hierarchy Process (AHP) used in decision-making, the author shows how an optimal allocation of granularity helps improve the quality of the solution and facilitate collaborative activities (e.g., consensus building) in models of group decision-making. The second study is concerned with a granular interpretation of temporal data where the role of information granularity is profoundly visible when effectively supporting human centric description of relationships existing in data. The third study concerns a formation of granular logic descriptors on a basis of a family of logic descriptors.

Chapter 13, *Semantic Manipulations and Formal Ontology for Machine Learning Based on Concept Algebra*, by Yingxu Wang, Yousheng Tian, and Kendall Hu, presents a new form of denotational mathematics known as concept algebra towards the formalization of ontological methodologies for dynamic machine learning and semantic analyses. *Concept Algebra* (CA) is a denotational mathematical structure for formal knowledge representation and manipulation in machine learning and cognitive computing. CA provides a rigorous knowledge modeling and processing tool, which extends the informal, static, and application-specific ontological technologies to a formal, dynamic, and general mathematical means. An operational semantics for the calculus of CA is formally elaborated using a set of computational processes in real-time process algebra (RTPA). A case study is presented on how machines, cognitive robots, and software agents may mimic the key ability of human beings to autonomously manipulate knowledge in generic learning using CA. This work demonstrates the expressive power and a wide range of applications of CA for both humans and machines in cognitive computing, semantic computing, machine learning, and computational intelligence.

Chapter 14, *Text Semantic Mining Model Based on the Algebra of Human Concept Learning*, by Jun Zhang, Xiangfeng Luo, Xiang He, and Chuanliang Cai, presents that dealing with the large-scale text knowledge on the Web has become increasingly important with the development of the Web, yet it confronts with several challenges, one of which is to find out as much semantics as possible to represent text knowledge. As the text semantic mining process is also the knowledge representation process of text, this chapter proposes a text knowledge representation model called text semantic mining model (TSMM) based on the algebra of human concept learning, which both carries rich semantics and is constructed automatically with a lower complexity. Herein, the algebra of human concept learning is introduced, which enables TSMM containing rich semantics. Then the formalization and construction processes of TSMM are discussed. Moreover, three types of reasoning rules based on TSMM are proposed. Lastly, experiments and the comparison with current text representation models show that our model performs better than others.

Chapter 15, *In Search of Effective Granulization with DTRS for Ternary Classification*, by Bin Zhou and Yiyu Yao, presents a decision-Theoretic Rough Set (DTRS) model. DTRS provides a three-way decision approach to classification problems, which allows a classifier to make a deferment decision on suspicious examples, rather than being forced to make an immediate determination. The deferred cases must be reexamined by collecting further information. Although the formulation of DTRS is intuitively appealing, a fundamental question that remains is how to determine the class of the deferment examples. In this chapter, we introduce an adaptive learning method that automatically deals with the deferred examples by searching for effective granulization. A decision tree is constructed for classification. At each level, we sequentially choose the attributes that provide the most effective granulization. A subtree is added recursively if the conditional probability lies in between of the two given thresholds. A branch reaches its leaf node when the conditional probability is above or equal to the first threshold, or is below or equal to the second threshold, or the granule meets certain conditions. This learning process is illustrated by an example.

SECTION 4: COMPUTATIONAL INTELLIGENCE

Intelligence science studies theories and models of the brain at all levels, and the relationship between the concrete physiological brain and the abstract soft mind. Intelligence science is a new frontier with the fertilization of biology, psychology, neuroscience, cognitive science, cognitive informatics, philosophy, information science, computer science, anthropology, and linguistics. A fundamental view developed in software and intelligence sciences is known as *abstract intelligence* (αI), which provides a unified foundation for the studies of all forms and paradigms of intelligence such as natural, artificial, machinable, and computational intelligence. αI is an enquiry of both natural and artificial intelligence at the neural, cognitive, functional, and logical levels from the bottom up. In the narrow sense, αI is a human or a system ability that transforms information into behaviors. However, in the broad sense, αI is any human or system ability that autonomously transfers the forms of abstract information between data, information, knowledge, and behaviors in the brain or intelligent systems.

Computational intelligence (CoI) is an embodying form of abstract intelligence (αI) that implements intelligent mechanisms and behaviors by computational methodologies and software systems, such as expert systems, fuzzy systems, cognitive computers, cognitive robots, software agent systems, genetic/evolutionary systems, and autonomous learning systems. The theoretical foundations of computational intelligence root in cognitive informatics, software science, and denotational mathematics.

The section on computational intelligence encompasses the following five chapters:

- **Chapter 16:** Cognitive Dynamic Systems
- **Chapter 17:** A Modular Dynamical Cryptosystem Based On Continuous Cellular Automata
- **Chapter 18:** Time and Frequency Analysis of Particle Swarm Trajectories for Cognitive Machines
- **Chapter 19:** On Machine Symbol Grounding and Optimization
- **Chapter 20:** Image Dimensionality Reduction Based on the Intrinsic Dimension and Parallel Genetic Algorithm

Chapter 16, *Cognitive Dynamic Systems*, by Simon Haykin, address the following four issues: a) Distinction between how adaptation and cognition are viewed with respect to each other; b) With human cognition viewed as the framework for cognition, the following cognitive processes are identified: the perception-action cycle, memory, attention, intelligence, and language. With language being outside the scope of the chapter, detailed accounts of the other four cognitive processes are discussed; c) Cognitive radar is singled out as an example application of cognitive dynamic systems that "mimics" the visual brain; experimental results on tracking are presented using simulations, which clearly demonstrate the information-processing power of cognition; and d) Two other example applications of cognitive dynamic systems, namely, cognitive radio and cognitive control, are briefly described.

Chapter 17, *A Modular Dynamical Cryptosystem Based on Continuous Cellular Automata*, by Jesus D.T. Gonzalez and Witold Kinsner, presents a new cryptosystem based on chaotic continuous-interval cellular automata (CCA) to increase data protection as demonstrated by their flexibility to encrypt and decrypt information from distinct sources. Enhancements to cryptosystems are also presented including (i) a model based on a new chaotic CCA attractor, (ii) the dynamical integration of modules containing dynamical systems to generate complex sequences, and (iii) an enhancement for symmetric cryptosystems by allowing them to generate an unlimited number of keys. This chapter also presents a process of mixing chaotic sequences obtained from cellular automata, instead of using differential equations, as a basis to achieve higher security and higher speed for the encryption and decryption processes, as compared to other recent approaches. The complexity of the mixed sequences is measured using the variance fractal dimension trajectory to compare them to the unmixed chaotic sequences to verify that the former are more complex. This type of polyscale measure and evaluation has never been done in the past outside this research group.

Chapter 18, *Time and Frequency Analysis of Particle Swarm Trajectories for Cognitive Machines*, by Dario Schor and Witold Kinsner, examines the inherited persistent behavior of particle swarm optimization and its implications to cognitive machines. The performance of the algorithm is studied through an average particle's trajectory through the parameter space of the Sphere and Rastrigin function. The trajectories are decomposed into position and velocity along each dimension being optimized. Then, a threshold is defined to separate the transient period, where the particle is moving towards a solution using information about the position of its best neighbors, from the steady state reached when the particles explore the local area surrounding the solution to the system. Using a combination of time and frequency domain techniques, the inherited long-term dependencies that drive the algorithm are discerned. Experimental results show the particles balance exploration of the parameter space with the correlated goal oriented trajectory driven by their social interactions. The information learned from this analysis can be used further to extract complexity measures in order to classify the behavior and control of particle swarm optimization, and make proper quick decisions on what to do next. Thus, this novel analysis of a particle trajectory in the time and frequency domains presents clear advantages of particle swarm optimization and inherent properties that make this optimization algorithm a suitable choice for use in cognitive machines.

Chapter 19, *On Machine Symbol Grounding and Optimization*, by Oliver Kramer, perceives that the world consists of high-dimensional dynamic sensorimotor data from the point of view of an autonomous agent. Interface algorithms translate this data into symbols that are easier to handle for cognitive processes. Symbol grounding is about whether these systems can, based on this data, construct symbols that serve as a vehicle for higher symbol-oriented cognitive processes. Machine learning and data mining techniques are geared towards finding structures and input-output relations in this data by providing appropriate interface algorithms that translate raw data into symbols. This work formulates the interface design as global optimization problem with the objective to maximize the success of the overlying symbolic algorithm. For its implementation various known algorithms from data mining and machine learning turn out to be adequate methods that do not only exploit the intrinsic structure of the subsymbolic data, but that also allow to flexibly adapt to the objectives of the symbolic process. Furthermore, this work discusses the optimization formulation as a functional perspective on symbol grounding that does not hurt the zero semantical commitment condition. A case study illustrates technical details of the machine symbol grounding approach.

Chapter 20, *Image Dimensionality Reduction Based on the Intrinsic Dimension and Parallel Genetic Algorithm*, by Liang Lei, TongQing Wang, Jun Peng, and Bo Yang, presents that without losing the main features of the image, how to reduce more of the image dimensions is highlighted in the research of web content-based image retrieval. Many features of dimensional reduction schemes are determined by the breaking of higher dimensional general covariance associated with the selection of a particular subset of coordinates. This chapter started by analysis of commonly used methods for the dimension reduction of Web images, followed a new algorithm for nonlinear dimensionality reduction based on the HSV image features. The approach obtains intrinsic dimension estimation by similarity calculation of two images. Finally, some improvements were made on the Parallel Genetic Algorithm (APGA) by use of the image similarity function as the self-adaptive judgment function to improve the genetic operators, thus achieving a Web image dimensionality reduction and similarity retrieval. Experimental results illustrate the validity of the authors' algorithm.

SECTION 5: APPLICATIONS OF COGNITIVE INFORMATICS AND COGNITIVE COMPUTING

A series of fundamental breakthroughs have been recognized and a wide range of applications has been developed in cognitive informatics and cognitive computing in the last decade. This section reviews applications of theories, models, methodologies, mathematical means, and techniques of CI and CC toward the exploration of the natural intelligence and the brain, as well novel cognitive computers. The key application areas of CI can be divided into two categories. The first category of applications uses informatics and computing techniques to investigate cognitive science problems, such as memory, learning, and reasoning. The second category adopts cognitive theories to investigate problems in informatics, computing, and software/knowledge engineering. CI focuses on the nature of information processing in the brain, such as information acquisition, representation, memory, retrieve, generation, and communication. Through the interdisciplinary approach and with the support of modern information and neuroscience technologies, mechanisms of the brain and the mind may be systematically explored.

The section on applications of cognitive informatics and cognitive computing encompasses the following four chapters:

- **Chapter 21:** Equivalence between LDA/QR and Direct LDA
- **Chapter 22:** A Novel Algorithm for Block Encryption of Digital Image Based on Chaos
- **Chapter 23:** Cognitive MIMO Radio: Performance Analysis and Precoding Strategy
- **Chapter 24:** Fuzzy Neural Network Control for Robot Manipulator Directly Driven by Switched Reluctance Motor

Chapter 21, *Equivalence between LDA/QR and Direct LDA*, by Rong-Hua Li, Shuang Liang, George Baciu, and Eddie Chan, presents the singularity problem of scatter matrices in Linear Discriminant Analysis (LDA), which is very challenging and has obtained a lot of attentions during the last decade. Linear Discriminant Analysis via QR decomposition (LDA/QR) and Direct Linear Discriminant analysis (DLDA) are two popular algorithms to solve the singularity problem. In this chapter, we establish the equivalent relationship between LDA/QR and DLDA. We show that they can be regarded as special cases of pseudo-inverse LDA. Similar to LDA/QR algorithm, DLDA can also be considered as a two-stage LDA method. Interestingly, we find out that the first stage of DLDA can act as a dimension reduction algorithm. In our experiment, we compare LDA/QR and DLDA algorithms in terms of classification accuracy, computational complexity on several benchmark datasets. We also conduct experiments to compare their first stages on these datasets. Our results confirm the established equivalent relationship and verify their capabilities in dimension reduction.

Chapter 22, *A Novel Algorithm for Block Encryption of Digital Image Based on Chaos*, by Jun Peng, Du Zhang, and Xiaofeng Liao, presents a novel image block encryption algorithm based on three-dimensional chaotic dynamical system. The algorithm works on 32-bit image blocks with a 192-bit secret key. The main idea is that the key is employed to drive the Chen's system to generate a chaotic sequence that is inputted to a specially designed function G, in which we use new 8x8 S-boxes generated by chaotic maps in (Tang, 2005). In order to improve the robustness against difference cryptanalysis and produce desirable avalanche effect, the function G is iteratively performed several times and its last outputs serve as the keystreams to encrypt the original image block. The design of the encryption algorithm is described in detail, along with security analyses. The results from key space analysis, differential attack analysis, information entropy analysis, correlation analysis of two adjacent pixels have proven that the proposed algorithm can resist cryptanalytic, statistical and brute force attacks, and achieve higher level of security. Moreover, the algorithm can also be easily employed to realize the security cryptosystems over the Internet.

Chapter 23, *Cognitive MIMO Radio: Performance Analysis and Precoding Strategy*, by Mingming Li, Jiaru Lin, Fazhong Liu, Dongxu Wang, and Li Guo, presents a cognitive radio network in which a set of cognitive users make opportunistic spectrum access to one primary channel by time-division multiplexing technologies. Multiple Input Multiple Output techniques (MIMO) are similarly considered to enhance the stable throughput for cognitive links while they should guarantee co-channel interference constraints to the primary link. Here, we consider two different cases: one is that cognitive radio network is distributed; the other is centrally-controlled that cognitive radio network has a cognitive base station. In the first case, how to choose one fixed cognitive user and power control for each transmission antenna at the cognitive base station are considered to maximize the cognitive link's stable throughput. In the second case, a scheme how to choose a group of cognitive users and a Zero-Forcing method how to pre-white co-channel interference to the primary user are also proposed in order to maximize cognitive base station's sum-rate. The algorithm can be employed to realize opportunistic spectrum transmission over the wireless fading channels.

Chapter 24, *Fuzzy Neural Network Control for Robot Manipulator Directly Driven by Switched Reluctance Motor*, by Baoming Ge and Aníbal T. de Almeida, finds that applications of switched reluctance motor (SRM) to direct drive robot are increasingly popular because of its valuable advantages. However, a greatest potential defect is its torque ripple owing to the significant nonlinearities. In this chapter, a fuzzy neural network (FNN) is applied to control the SRM torque at the goal of the torque-ripple minimization. The desired current provided by FNN model compensates the nonlinearities and uncertainties of SRM. On the basis of FNN-based current closed-loop system, the trajectory tracking controller is designed by using the dynamic model of the manipulator, where the torque control method cancels the nonlinearities and cross-coupling terms. A single link robot manipulator directly driven by a four-phase 8/6-pole SRM operates in a sinusoidal trajectory tracking rotation. The simulated results verify the proposed control method and a fast convergence that the robot manipulator follows the desired trajectory in a 0.9-s time interval.

This book is intended to the readership of researchers, engineers, graduate students, senior-level undergraduate students, and instructors as an informative reference book in the cutting-edge fields of cognitive informatics, natural intelligence, abstract intelligence, and cognitive computing. The editor expects that readers of *Developments in Natural Intelligence Research and Knowledge Engineering* will benefit from the 24 selected chapters of this book, which represent the latest advances in research in cognitive informatics, cognitive computing, and computational intelligence as well as their engineering applications.

Yingxu Wang
University of Calgary, Canada

Acknowledgment

Many persons have contributed their dedicated work to this book and related research. The editor would like to thank all authors, the associate editors of IJCINI, the editorial board members, and invited reviewers for their great contributions to this book. I would also like to thank the IEEE Steering Committee and organizers of the series of IEEE International Conference on Cognitive Informatics and Cognitive Computing (ICCI*CC) in the last 12 years, particularly Lotfi A. Zadeh, Witold Kinsner, Witold Pedrycz, Bo Zhang, Du Zhang, George Baciu, Phillip Sheu, Jean-Claude Latombe, James A. Anderson, Robert C. Berwick, and Dilip Patel. I would like to acknowledge the publisher of this book, IGI Global, USA. I would like to thank Dr. Mehdi Khosrow-Pour, Jan Travers, Kristin M. Klinger, Erika L. Carter, and Myla Merkel, for their professional editorship.

Section 1
Cognitive Informatics

Chapter 1

Towards the Synergy of Cognitive Informatics, Neural Informatics, Brain Informatics, and Cognitive Computing

Yingxu Wang
University of Calgary, Canada

ABSTRACT

The contemporary wonder of sciences and engineering recently refocused on the starting point: how the brain processes internal and external information autonomously rather than imperatively as those of conventional computers? This paper explores the interplay and synergy of cognitive informatics, neural informatics, abstract intelligence, denotational mathematics, brain informatics, and computational intelligence. A key notion recognized in recent studies in cognitive informatics is that the root and profound objective in natural, abstract, and artificial intelligence, and in cognitive informatics and cognitive computing, is to seek suitable mathematical means for their special needs. A layered reference model of the brain and a set of cognitive processes of the mind are systematically developed towards the exploration of the theoretical framework of cognitive informatics. A wide range of applications of cognitive informatics and denotational mathematics are recognized in the development of highly intelligent systems such as cognitive computers, cognitive knowledge search engines, autonomous learning machines, and cognitive robots.

DOI: 10.4018/978-1-4666-2476-4.ch001

1. INTRODUCTION

Cognitive informatics studies the natural intelligence and the brain from a theoretical and a computational approach, which rigorously explains the mechanisms of the brain by a fundamental theory known as abstract intelligence, which formally models the brain by contemporary denotational mathematics. The contemporary wonder of sciences and engineering has recently refocused on the starting point of them: how the brain processes internal and external information autonomously and cognitively rather than imperatively as those of conventional computers? The latest advances and engineering applications of CI have led to the emergence of *cognitive computing* and the development of *cognitive computers* that perceive, learn, and reason (Wang, 2006, 2009d, 2009f, 2010a, 2011). CI has also fundamentally contributed to autonomous agent systems (Wang, 2009a) and cognitive robots (Wang, 2010a). A wide range of applications of CI are identified such as in the development of cognitive computers, cognitive robots, cognitive agent systems, cognitive search engines, cognitive learning systems, and artificial brains. The work in CI may also lead to a fundamental solution to computational linguistics, Computing with Natural Language (CNL), and Computing with Words (CWW) (Zadeh, 1975, 1999).

Cognitive Informatics is a term coined by Wang in the first IEEE International Conference on Cognitive Informatics (ICCI 2002) (Wang, 2002a). Cognitive informatics (Wang, 2002a, 2003, 2007b; Wang & Wang, 2006; Wang & Kinsner, 2006; Wang, Jonston, & Smith, 2002; Wang, Wang, Patel, & Patel, 2006; Wang, Zhang, Latombe, & Kinsner, 2008; Wang, Kinsner, & Zhang, 2009; Wang, Zhang, & Tsumoto, 2009; Wang et al., 2009, Wang & Chiew, 2010) studies the natural intelligence and the brain from a theoretical and a computational approach, which rigorously explains the mechanisms of the brain by a fundamental theory known as *abstract intel-*

ligence. Cognitive informatics formally models the brain by contemporary *denotational mathematics* such as *concept algebra* (Wang, 2008b), *real-time process algebra* (RTPA) (Wang, 2002b, 2008d), *system algebra* (Wang, 2008c; Wang, Zadeh, & Yao, 2009), and *visual semantic algebra* (VSA) (Wang, 2009e). The latest advances in CI have led to a systematic solution for explaining brain informatics and the future generation of intelligent computers.

A key notion recognized in recent studies in cognitive informatics is that the root and profound objective in natural, abstract, and artificial intelligence in general, and in cognitive informatics and cognitive computing in particular, is to seek suitable mathematical means for their special needs, which were missing in these multidisciplinary areas. This is a general requirement for searching the metamethodology in any discipline particularly those of emerging fields where no suitable mathematics has been developed or of traditional fields where persistent hard problems have been unsolved efficiently or completely (Bender, 1996; Boole, 2003; Russell, 1996; Wang, 2008a, 2010b, 2011).

This paper is an extended summary of the invited keynote lecture presented in the 2010 *International Conference on Brain Informatics* (BI 2010), which covers some of the theoretical foundations of brain informatics developed in cognitive informatics and denotational mathematics. In this paper, cognitive informatics as the science of abstract intelligence and cognitive computing is briefly described in Section 2. Neural informatics is presented to explain how intelligence and knowledge are represented in the brain. The fundamental theories and expressive tools for cognitive informatics, brain Informatics, and computational intelligence, collectively known as denotational mathematics, are introduced in Section 4. Brain informatics as an interdisciplinary field for studying brain mechanisms by computing and medical imaging technologies is introduced in Section 5. Applications of cognitive informatics and denotational mathematics in cognitive computing are elaborated in Sections 6 towards the development of the next generation of cognitive computers.

2. COGNITIVE INFORMATICS: THE SCIENCE OF ABSTRACT INTELLIGENCE AND COMPUTATIONAL INTELLIGENCE

Information is the third essence of the word supplementing energy and matter. A key discovery in information science is the basic unit of information, *bit*, abbreviated from "binary digit", which forms a shared foundation of both computer science and information science.

The science of information, *informatics*, has gone through three generations of evolution, known as the classic, modern, and cognitive informatics, since Shannon proposed the classic notion of information (Shannon, 1948). The *classical* information theory founded by Shannon (1948) defined *information* as a probabilistic measure of the variability of message that can be obtained from a message source. Along with the development in computer science and in the IT industry, the domain of informatics has been dramatically extended in the last few decades. This led to the *modern* informatics that treats information as entities of messages rather than a probabilistic measurement of the variability of messages as in that of the classic information theory. The new perception of information is found better to explain the theories in computer science and practices in the IT industry. However, both classic and modern views on information are only focused on external information. The real sources and destinations of information, the human brains, are often overlooked. This leads to the third generation of informatics, cognitive informatics, which focuses on the nature of information in the brain, such as information acquisition, memory, categorization, retrieve, generation, representation, and communication. Information

in cognitive informatics is defined as the abstract artifacts and their relations that can be modeled, processed, stored and processed by human brains. Cognitive informatics (Wang, 2002a, 2003, 2007b; Wang & Wang, 2006; Wang & Kinsner, 2006; Wang & Wang, 2006; Wang, Kinsner, & Zhang, 2009; Wang, Zhang, & Tsumoto, 2009) is emerged and developed based on the multidisciplinary research in cognitive science, computing science, information science, abstract intelligence, and denotational mathematics since the inauguration of the 1st IEEE ICCI'02 (Wang, 2002a).

Definition 1: Cognitive informatics (CI) is a transdisciplinary enquiry of computer science, information science, cognitive science, and intelligence science that investigates into the internal information processing mechanisms and processes of the brain and natural intelligence, as well as their engineering applications in cognitive computing.

CI is a cutting-edge and multidisciplinary research area that tackles the fundamental problems shared by modern informatics, computation, software engineering, AI, cybernetics, cognitive science, neuropsychology, medical science, philosophy, linguistics, brain sciences, and many others. The development and the cross fertilization among the aforementioned science and engineering disciplines have led to a whole range of extremely interesting new research areas.

The theoretical framework of CI encompasses four main areas of basic and applied research (Wang, 2007b) such as: (a) fundamental theories of natural intelligence; (b) abstract intelligence; (c) denotational mathematics; and d) cognitive computing. Fundamental theories developed in CI covers the Information-Matter-Energy (IME) model (Wang, 2003), the Layered Reference Model of the Brain (LRMB) (Wang, Wang, Patel, & Patel, 2006), the Object-Attribute-Relation (OAR) model of information/knowledge representation in the brain (Wang, 2007c), the cognitive informatics model of the brain (Wang, 2010a; Wang & Wang, 2006), Natural Intelligence (NI) (Wang, 2003; Wang & Chiew, 2010), and neural informatics (Wang, 2007c). Recent studies on LRMB in cognitive informatics reveal an entire set of cognitive functions of the brain and their cognitive process models, which explain the functional mechanisms of the natural intelligence with 43 cognitive processes at seven layers known as the sensation, memory, perception, action, metacognitive, meta-inference, and higher cognitive layers (Wang, Wang, Patel, & Patel, 2006).

It is conventionally deemed that only mankind and advanced species possess intelligence. However, the development of computers, robots, software agents, and autonomous systems indicates that intelligence may also be created or embodied by machines and man-made systems. Therefore, it is one of the key objectives in cognitive informatics and intelligence science to seek a coherent theory for explaining the nature and mechanisms of both natural and artificial intelligence.

Intelligence is an ability to acquire and use knowledge and skills, or to inference in problem solving. It is a profound human wonder on how conscious intelligence is generated as a highly complex cognitive state in human mind on the basis of biological and physiological structures. How natural intelligence functions logically and physiologically? How natural and artificial intelligence are converged on the basis of brain, software, and intelligence science?

Definition 2: Abstract Intelligence (αI) is the general form of intelligence as an abstract mathematical model that transfers information into behaviors and knowledge.

Abstract intelligence is also a discipline of human enquiries.

Definition 3: The discipline of abstract intelligence studies the foundations of intelligence science focusing the core properties of intelligence as a natural mechanism that transfers information into behaviors and knowledge.

In the *narrow sense*, αI is a human or a system ability that transforms information into behaviors. While, in the *broad sense*, αI is any human or system ability that autonomously transfers the forms of abstract information between *data, information, knowledge,* and *behaviors* in the brain or systems.

The studies on αI form a field of enquiry for both natural and artificial intelligence at the reductive levels of neural, cognitive, functional, and logical from the bottom up (Wang, 2009c). The paradigms of αI are such as natural, artificial, machinable, and computational intelligence. With the clarification of the intension and extension of the concept of αI, its paradigms or concrete forms in the real-world can be derived as summarized in Table 1.

Table 1. Taxonomy of abstract intelligence and its embodying forms

No.	Form of intelligence	Embodying means	Paradigms
1	Natural intelligence (NI)	Naturally grown biological and physiological organisms	Human brains and brains of other well developed species
2	Artificial intelligence (AI)	Cognitively-inspired artificial models and man-made systems	Intelligent systems, knowledge systems, decision-making systems, and distributed agent systems
3	Machinable intelligence (MI)	Complex machine and wired systems	Computers, robots, autonomic circuits, neural networks, and autonomic mechanical machines
4	Computational intelligence (CoI)	Computational methodologies and software systems	Expert systems, fuzzy systems, autonomous computing, intelligent agent systems, genetic/evolutionary systems, and autonomous learning systems

Box 1.

$$
\begin{aligned}
\S\alpha\mathbf{IST} &\triangleq (\mathfrak{I}_I, \mathfrak{I}_A, \mathfrak{I}_C) \\
&= \{ \quad (B_e, B_t, B_{int}) && // \ \mathrm{I_I} \text{ - Imperative intelligence} \\
&\quad\ \| (B_e, B_t, B_{int}, B_g, B_d) && // \ \mathrm{I_A} \text{ - Autonomic intelligence} \\
&\quad\ \| (B_e, B_t, B_{int}, B_g, B_d, B_p, B_{inf}) && // \ \mathrm{I_C} \text{ - Cognitive intelligence} \\
&\quad\ \}
\end{aligned} \tag{1}
$$

Definition 4. The behavioral model of αI, §αI**ST**, is an abstract logical model denoted by a set of parallel processes that encompasses the *imperative* intelligence I_I, *autonomic* intelligence I_A, and *cognitive* intelligence I_C from the bottom-up, see Equation (1) in Box 1.

According to Definition 4, the relationship among the three forms of intelligence is as follows:

$$ \mathfrak{I}_I \subseteq \mathfrak{I}_A \subseteq \mathfrak{I}_C \tag{2} $$

Both Equations (2) and (3) indicate that any lower layer intelligence and behavior is a subset of those of a higher layer. In other words, any higher layer intelligence and behavior is a natural extension of those of lower layers.

It is noteworthy that all paradigms of αI share the same cognitive informatics foundation as described in the following theorems, because they are an artificial or machine implementation or embodiment of αI.

Theorem 1: The compatible intelligent capability state that *natural intelligence* (NI), *artificial intelligence* (AI), *machinable intelligence* (MI), and *computational intelligence* (CoI), are compatible by sharing the same mechanisms of αI, i.e.:

$$ CoI \cong MI \cong AI \cong NI \cong \alpha I \tag{3} $$

On the basis of Theorem 1, the differences between NI, AI, MI, and CoI are only distinguishable by: (a) The means of their implementation; and (b) The extent of their intelligent capability. Corollary 1 indicates that AI, CoI, and MI are dominated by NI and αI. Therefore, one should not expect a computer or a software system to solve a problem where human cannot. In other words, no AI or computer systems may be designed and/or implemented for a given problem where there is no solution being known collectively by human beings as a whole. Further, Theorem 1 and Corollary 1 explain that the development and implementation of AI rely on the understanding of the mechanisms and laws of NI.

3. NEURAL INFORMATICS: HOW INTELLIGENCE AND KNOWLEDGE ARE REPRESENTED IN THE BRAIN

An important area of development in cognitive informatics is neural informatics (Wang, 2007b), which reduces cognitive informatics theories and the studies on the internal information processing mechanisms of the brain onto the neuron and physiological level.

Definition 5: Neural informatics (NeI) is a new interdisciplinary enquiry on the biological and physiological representation of information and knowledge in the brain at the neuron level as well as their abstract mathematical models.

5

In the studies of NeI, memory is recognized as the foundation, platform, as well as constraints, of any natural or machine intelligence. The cognitive models of human memory (Wang & Wang, 2006), particularly the Sensory Buffer Memory (SBM), Short-Term Memory (STM), Long-Term Memory (LTM), Action-Buffer Memory (ABM), Conscious State Memory (CSM), and their mapping onto the physiological organs of the brain, reveal the fundamental mechanisms of neural informatics.

Definition 6: The Cognitive Models of Memory (CMM) states that the architecture of human memory is parallel configured by SBM, STM, LTM, CSM, and ABM, where the ABM is newly identified in Wang and Wang (2006).

The major organ that accommodates memories in the brain is the cerebrum or the cerebral cortex. In particular, the association and premotor cortex in the frontal lobe, the temporal lobe, sensory cortex in the frontal lobe, visual cortex in the occipital lobe, primary motor cortex in the frontal lobe, supplementary motor area in the frontal lobe, and procedural memory in cerebellum (Wilson & Keil, 2001; Wang & Wang, 2006). The CMM model and the mapping of the four types of human memory onto the physiological organs in the brain reveal a set of fundamental mechanisms of neural informatics (Wang, 2007b).

Definition 7: The functional model of LTM can be described as a *Hierarchical Neural Cluster* (HNC) model with partially connected neurons via synapses.

The HNC model of LTM consists of dynamic and partially interconnected neural networks. In the HNC model, a physiological connection between a pair of neurons via a synapse represents a logical relation between two abstract objects or concepts. The hierarchical and partially connected neural clusters are the foundation for information and knowledge representation in LTM.

Definition 8: The relational model of memory is a logical memory model that states information is represented and retained in the memory by relations, which is embodied by the synaptic connections among neurons. In contrary to the conventional container metaphor, the relational metaphor indicates that the brain does not create new neurons to represent newly acquired information; instead, it generates new synapses between the existing neurons in order to represent new information.

The reconfigurable neural clusters of STM cohere and connect related objects such as images, data, and concepts, and their attributes by synapses in order to form contexts and threads of thinking. Therefore, the main function of STM may be analogized to an index memory connecting to other memories, particularly LTM. STM is the working memory of the brain. The capacity of STM is much smaller than that of LTM, but it is a hundred times greater than 7±2 digits as Miller proposed (Miller, 1956). Limited by the temporal space of STM, one has to write complicated things on paper or other types of external memories in order to compensate the required working memory space in a thinking process.

Theorem 2: The dynamic neural cluster model states that the LTM is dynamic. New neurons (to represent objects or attributes) are assigning, and new connections (to represent relations) are creating and reconfiguring all the time in the brain.

To rigorously explain the hierarchical and dynamic neural cluster model of memory at physiological level, a logical model of memory is needed, as given in Definition 9, known as the

Object-Attribute-Relation (OAR) model (Wang, 2007c).

Definition 9: The OAR model of LTM can be described as a triple, i.e.:

$$OAR \triangleq (O, A, R) \qquad (4)$$

where O is a finite set of objects identified by unique symbolic names, A is a finite set of attributes for characterizing the object, and R is a finite set of relations between two objects, two attributes, and/or an object and an attribute.

An abstract illustration of the OAR model between two objects is shown in Figure 1. The relations between objects can be established via pairs of object-object, object-attribute, and/or attribute-attribute. The connections could be highly complicated, while the mechanism is fairly simple that it can be deducted to the physiological links of neurons via synapses in LTM.

It is noteworthy as in the OAR model that the *relations* themselves represent information and knowledge in the brain. The relational metaphor is totally different from the traditional container metaphor in neuropsychology and computer science, because the latter perceives that memory and knowledge are *stored* in individual neurons and the neurons function as containers. According to the OAR model, the result of knowledge acquisition or learning can be embodied by the updating of the existing OAR in the brain.

Definition 10: Memorization is a cognitive process of the brain at the meta cognitive layer that establishes (encodes and retains) and reconstructs (retrieves and decodes) information in LTM.

Corresponding to the forms of memories in the brain, human knowledge as cognized or comprehended information can be defined in both a narrow and a broad sense.

Definition 11: Knowledge, in the narrow sense, is acquired information in LTM or acquired skills in ABM through learning. In the broad sense, knowledge is acquired information in forms of abstract knowledge, intelligence, experience, and skills through learning in LTM or ABM.

Figure 1. The OAR model of logical memory architectures

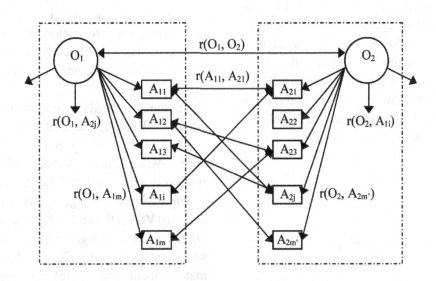

Theorem 3: The entire knowledge model maintained in the brain states that the internal memory or the representation of learning results in the form of the OAR structure, which can be updated by concept compositions between the existing OAR and the newly created sub-OAR (sOAR), that is:

$$OAR' \textbf{ST} = OAR \textbf{ ST} \uplus sOAR\textbf{ST}$$
$$= OAR \textbf{ ST} \uplus (O_s, A_s, R_s) \quad (5)$$

where **ST** is the system structure type suffix as defined in Real-Time Process Algebra (RTPA) (Wang, 2002b, 2003, 2006, 2007a, 2008d), and â denotes the concept composition operation in concept algebra (Wang, 2008b).

The theories of NeI explain a number of important questions in the study of natural intelligence such as the mechanisms and the 24-hour law of long-term memory establishment. The latest development in neural informatics has led to the determination of the magnitude of human memory and the mechanisms of internal knowledge representation, memorization, and learning. Enlightening findings in neural informatics are such as (a) LTM establishment is a subconscious process; (b) LTM is established during sleeping; (c) The major mechanism for LTM establishment is by sleeping; (d) The general acquisition cycle of LTM is equal to or longer than 24 hours; (e) The mechanism of LTM establishment is to update the entire memory of information represented as an OAR model in the brain (Wang, 2007c); and (f) Eye movement and dreams play an important role in LTM creation.

4. DENOTATIONAL MATHEMATICS: A GENERAL METAMETHODOLOGY FOR COGNITIVE INFORMATICS, BRAIN INFORMATICS, COGNITIVE COMPUTING, AND COMPUTATIONAL INTELLIGENCE

It is recognized that the maturity of a scientific discipline is characterized by the maturity of its mathematical (meta-methodological) means. A key notion recognized in recent studies in cognitive informatics and computational intelligence is that the root and profound problem in natural, abstract, and artificial intelligence in general, and in cognitive informatics and cognitive computing in particular, is to seek suitable mathematical means for their special needs. This is a general need and requirement for searching the metamethodology in any discipline particularly the emerging fields where no suitable mathematics has been developed and the traditional fields where persistent hard problems have been unsolved efficiently or completely (Bender, 1996; Boole, 2003; Kline, 1972; Russell, 1996; Wang, 2007a, 2008a).

Definition 12: Denotational mathematics (DM) is a category of expressive mathematical structures that deals with high-level mathematical entities beyond numbers and sets, such as abstract objects, complex relations, perceptual information, abstract concepts, knowledge, intelligent behaviors, behavioral processes, inferences, and systems.

A number of DMs have been created and developed (Wang, 2008a, 2009f) such as *concept algebra* (Wang, 2008b), *system algebra* (Wang, 2008c; Wang, Zadeh, & Yao, 2009), *real-time process algebra* (RTPA) (Wang, 2002b, 2008d), *granular algebra* (Wang, 2008a), *visual semantic algebra* (VSA) (Wang, 2009e), and *formal causal inference methodologies* (Wang, in press). As summarized in Figure 2 with their structures, mathematical entities, algebraic operations, and

Figure 2. Paradigms of denotational mathematics

Paradigm	Structure	Mathematical entities	Algebraic operations	Usage
Concept algebra (CA)	$CA \triangleq (C, OP, \Theta) = (\{O, A, R^c, R^i, R^o\},$ $\{\bullet_r, \bullet_c\}, \Theta_C)$	$c \triangleq (O, A, R^c, R^i, R^o)$	$\bullet_r \triangleq \{\leftrightarrow, \nleftrightarrow, \prec, \succ, =, \cong, \sim, \triangleq\}$ $\bullet_c \triangleq \{\Rightarrow, \overset{-}{\Rightarrow}, \overset{+}{\Rightarrow}, \overset{\sim}{\Rightarrow}, \uplus, \pitchfork, \Leftarrow, \vdash, \mapsto\}$	Algebraic manipulations on abstract concepts
System algebra (SA)	$SA \triangleq (S, OP, \Theta) = (\{C, R^c, R^i, R^o, B, \Omega\},$ $\{\bullet_r, \bullet_c\}, \Theta)$	$S \triangleq (C, R^c, R^i, R^o, B, \Omega, \Theta)$	$\bullet_r \triangleq \{\leftrightarrow, \nleftrightarrow, \prod, =, \sqsubseteq, \sqsupseteq\}$ $\bullet_c \triangleq \{\Rightarrow, \overset{-}{\Rightarrow}, \overset{+}{\Rightarrow}, \overset{\sim}{\Rightarrow}, \boxminus, \uplus, \pitchfork, \Leftarrow, \vdash\}$	Algebraic manipulations on abstract systems
Real-time process algebra (RTPA)	$RTPA \triangleq (\mathfrak{T}, \mathfrak{P}, \mathfrak{N})$	$\mathfrak{P} \triangleq \{:=, \blacklozenge, \Rightarrow, \Leftarrow, \nRightarrow, \triangleright, \lhd, \vert\triangleright,$ $\vert\lhd, @, \triangleq, \uparrow, \downarrow, !, \otimes, \boxtimes, \S\}$ $\mathfrak{T} \triangleq \{\mathbf{N, Z, R, S, BL, B, H, P, TI, D, DT,}$ $\mathbf{RT, ST}, @e\mathbf{S}, @t\mathbf{TM}, @int\odot, \textcircled{S}s\mathbf{BL}\}$	$\mathfrak{R} \triangleq \{\rightarrow, \curvearrowright, \vert, \vert\ldots\vert\ldots\vert\ldots, R^*, R^+, R^i,$ $\circlearrowleft, \hookrightarrow, \parallel, \oint, \parallel\parallel, \gg, \lightning, \hookrightarrow_t, \hookrightarrow_e, \hookrightarrow_i\}$	Algebraic manipulations on abstract processes
Visual semantic algebra (VSA)	$VSA \triangleq (O, \bullet_{VSA})$	$O \triangleq \{H \cup S \cup F \cup L\}$	$\bullet_{VSA} \triangleq \{\uparrow, \downarrow, \leftarrow, \rightarrow, \odot, \otimes, \boxplus, \angle,$ $@(p), @(x, y, x), \curvearrowright, \mapsto,$ $\overset{n-1N}{\underset{N=0}{R}}(A_i \mapsto A_{i+1})\}$	Algebraic manipulations on abstract visual objects/patterns
Granular algebra (GrA)	$GA \triangleq (G, \bullet_r, \bullet_p, \bullet_c)$ $= ((C, R^c, R^i, R^o, B, \Omega),$ $\bullet_r, \bullet_p, \bullet_c)$	$G \triangleq (C, R^c, R^i, R^o, B, \Omega, \Theta)$	$\bullet_r \triangleq \{\leftrightarrow, \nleftrightarrow, \prod, =, \sqsubseteq, \sqsupseteq\}$ $\bullet_p \triangleq \{\Rightarrow, \overset{-}{\Rightarrow}, \overset{+}{\Rightarrow}, \overset{\sim}{\Rightarrow}\}$ $\bullet_c \triangleq \{\uplus, \pitchfork, \Leftarrow\}$	Algebraic manipulations on abstract granules

usages, the set of DMs provide a coherent set of contemporary mathematical means and explicit expressive power for CI, αI, CC, AI, and computational intelligence.

Among the above collection of denotational mathematics, concept algebra is an abstract mathematical structure for the formal treatment of concepts as the basic unit of human reasoning and their algebraic relations, operations, and association rules for composing complex concepts. It is noteworthy that, according to concept algebra, although the semantics of words in natural languages may be ambiguity, the semantics of concept is always unique and precise in cognitive computing and computational intelligence.

Example 1: The word, "bank", is ambiguity because it may be a notion of a financial institution, a geographic location of raised ground of a river/lake, and/or a storage of something. However, the three distinguished concepts related to "bank", i.e., $b_o =$ bank(organization), $b_r =$ bank(river), and $b_s =$ bank(storage), are precisely unique, which can be formally described in concept algebra (Wang, 2008b) for CC as shown in Figure 3, where K represents the entire concepts existed in the analyser's knowledge.

All given concrete concepts share a generic framework, known as the universal abstract concept as modeled in concept algebra as given below.

Figure 3. Formal and distinguished concepts derived from the word "bank"

$$b_o\mathbf{ST} \triangleq (A, O, R^c, R^i, R^o) \qquad // \text{ bank(organization)}$$
$$= (\ b_o\mathbf{ST}.A = \{\text{organization, company, financial business,}$$
$$\text{money, deposit, withdraw, invest, exchange}\},$$
$$b_o\mathbf{ST}.O = \{\text{international bank, national bank, local bank,}$$
$$\text{investment bank, ATM}\}$$
$$b_o\mathbf{ST}.R^c = O \times A,$$
$$b_o\mathbf{ST}.R^i = \mathcal{K} \times b_o\mathbf{ST},$$
$$b_o\mathbf{ST}.R^o = b_o\mathbf{ST} \times \mathcal{K}$$
$$)$$

$$b_r\mathbf{ST} \triangleq (A, O, R^c, R^i, R^o) \qquad // \text{ bank(river)}$$
$$= (\ b_r\mathbf{ST}.A = \{\text{sides of a river, raised ground, a pile of earth, location}\},$$
$$b_r\mathbf{ST}.O = \{\text{river bank, lake bank, canal bank}\}$$
$$b_r\mathbf{ST}.R^c = O \times A,$$
$$b_r\mathbf{ST}.R^i = \mathcal{K} \times b_r\mathbf{ST},$$
$$b_r\mathbf{ST}.R^o = b_r\mathbf{ST} \times \mathcal{K}$$
$$)$$

$$b_s\mathbf{ST} \triangleq (A, O, R^c, R^i, R^o) \qquad // \text{ bank(storage)}$$
$$= (\ b_s\mathbf{ST}.A = \{\text{storage, container, place, organization}\},$$
$$b_s\mathbf{ST}.O = \{\text{information bank, human resource bank, blood bank}\}$$
$$b_s\mathbf{ST}.R^c = O \times A,$$
$$b_s\mathbf{ST}.R^i = \mathcal{K} \times b_s\mathbf{ST},$$
$$b_s\mathbf{ST}.R^o = b_s\mathbf{ST} \times \mathcal{K}$$
$$)$$

Definition 13: An abstract concept, c, is a 5-tuple, i.e.:

$$c \triangleq (O, A, R^c, R^i, R^o) \qquad (6)$$

where

- O is a nonempty set of objects of the concept, $O = \{o_1, o_2, ..., o_m\} \subseteq \mathbb{P}O$, where $\mathbb{P}O$ denotes a power set of abstract objects in the universal discourse U, $U = (O, A, R)$.
- A is a nonempty set of attributes, $A = \{a_1, a_2, ..., a_n\} \subseteq \mathbb{P}A$, where $\mathbb{P}A$ denotes a power set of attributes in U.

- $R^c = O \times A$ is a set of internal relations.
- $R^i \subseteq C' \times c$ is a set of input relations, where C' is a set of external concepts in U.
- $R^o \subseteq c \times C'$ is a set of output relations.

Concept algebra provides a set of 8 relational and 9 compositional operations on abstract concepts as summarized in Figure 2. Detailed definitions of operations defined in concept algebra may be referred to (Wang, 2008b). Additional concept operations may be introduced in order to reveal the underpinning mechanisms of learning and natural language comprehension. One of the advanced operations in concept algebra for knowledge processing is known as knowledge differential, which can be formalized as follows.

Definition 14: Knowledge differential, *dK/dt*, is an eliciting operation on a set of knowledge *K* represented by a set of concepts over time that recalls new concepts learnt during a given period t_1 through t_2, i.e.:

$$\frac{dK}{dt} \triangleq \frac{d(OAR)}{dt}$$
$$= OAR(t_2) - OAR(t_1) \qquad (7)$$
$$= OAR.C(t_2) \setminus OAR.C(t_1)$$

where the set of concepts, $OAR.C(t_1)$, are existing concepts that have already been known at time point t_1.

Example 2: As given in Example 1, assume the following concepts, $OAR.C(t_1) = \{C_o\}$, are known at t_1, and the system's learning result at t_2 is $OAR.C(t_2) = \{C_o, C_r, C_s\}$. Then, a knowledge differential can be carried out using Equation (7) as follows:

$$\frac{dK}{dt} \triangleq \frac{d(OAR)}{dt}$$
$$= OAR.C(t_2) \setminus OAR.C(t_1)$$
$$= \{C_o, C_r, C_s\} \setminus \{C_o\}$$
$$= \{C_r, C_s\}$$

Concept algebra provides a powerful denotational mathematical means for algebraic manipulations of abstract concepts. Concept algebra can be used to model, specify, and manipulate generic "*to be*" type problems, particularly system architectures, knowledge bases, and detail-level system designs, in cognitive informatics, intelligence science, computational intelligence, computing science, software science, and knowledge science. The work in this area may also lead to a fundamental solution to computational linguistics,

Computing with Natural Language (CNL), and Computing with Words (CWW) (Zadeh, 1975, 1999).

5. BRAIN INFORMATICS: EXPLORING THE IMAGES AND MECHANISMS OF THE BRAIN

The notion of brain informatics is proposed (Zhong, 2009; Wang, 2010e). A functional and logical reference model of the brain and a set of cognitive processes of the mind are systematically developed towards the exploration of the theoretical framework of brain informatics. Based on it, the current methodologies for brain studies are reviewed and their strengths and weaknesses are analyzed.

Definition 15: Brain Informatics (BI) is a joint field of brain and information sciences that studies the information processing mechanisms of the brain by computing and medical imagination technologies.

A variety of life functions and their cognitive processes have been identified in cognitive informatics, neuropsychology, cognitive science, and neurophilosophy. Based on the advances of research in cognitive informatics and related fields, the Layered Reference Model of the Brain (LRMB) (Wang, Wang, Patel, & Patel, 2006) is presented as shown in Figure 4, which explains the functional mechanisms and cognitive processes of the brain.

According to LRMB, the hierarchical life functions of the brain as a natural intelligent system (NI-Sys) can be divided into two categories: the *subconscious* and *conscious* life functions. The former known as the NI operating system (NI-OS) encompasses the layers of sensation, memory, perception, and action (Layers 1 through 4). The latter known as the NI applications (NI-App) includes the layers of meta-cognitive, meta-inference, and higher cognitive functions (Layers

Figure 4. The Layered Reference Model of the Brain (LRMB)

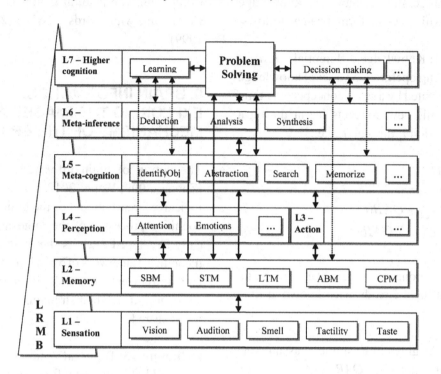

5 through 7). In the LRMB model, cross layer communications are denoted by the horizontal arrows. In LRMB, the subconscious layers of the brain (NI-OS) are inherited, fixed, and relatively mature when a person is born. Therefore, the subconscious function layers are usually neither directly controllable nor intentionally accessible by the conscious life function layers. This is why it is used to be called nonconscious life functions in psychology and cognitive science (Matlin, 1998). Contrary to the subconscious NI-OS, the conscious layers of the brain (NI-App) are acquired, highly plastic, programmable, and can be controlled intentionally based on willingness, goals, and motivations.

Detailed configurations of cognitive processes on different layers of LRMB may be referred to (Wang, Wang, Patel, & Patel, 2006). The LRMB model explains the functional mechanisms and cognitive processes of the natural and artificial brains with 43 cognitive processes at seven layers. LRMB elicits the core and highly repetitive recur-

rent cognitive processes from a huge variety of life functions, which may shed light on the study of the fundamental mechanisms and interactions of complicated mental processes as well as of cognitive systems, particularly the relationships and interactions between the inherited and the acquired life functions as well as those of the subconscious and conscious cognitive processes. Any everyday life function or behavior, such as reading or driving, is a concurrent combination of part or all of the 43 fundamental cognitive processes according to LRMB.

The basic methodologies in CI and BI are: (a) logic (formal and mathematical) modeling and reasoning; (b) empirical introspection; (c) experiments (particularly abductive observations on brain patients); and (d) using high technologies particularly brain imaging technologies. The central roles of formal logical and functional modeling for BI have been demonstrated in Sections 2 through 4 by CI, αI, NeI, and denotational mathematics.

Figure 5. Major imaging technologies in brain studies

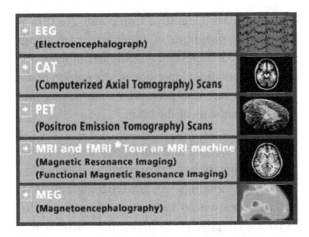

Modern brain imaging technologies such as EEG, fMRI, MEG, and PET are illustrated as shown in Figure 5. Although many promising results on cognitive functions of the brain have been derived by brain imaging studies in cognitive tests and neurobiology, they are limited to simple cognitive functions compared with the entire framework of the brain as revealed in LRMB. Moreover, there is a lack of a systematic knowledge about what roles particular types of neurons may play in complex cognitive functions such as learning and memorization, because neuroimages cannot pinpoint to detailed relationships between structures and functions in the brain.

The limitations of current brain imaging technologies such as PET and fMRI towards understanding the functions of the brain may be equivalent to the problem to examine the functions of a computer by looking at its layout and locations where they are active using imaging technologies. It is well recognized that without understanding the logical and functional models and mechanisms of the CPU as shown in Figure 6, nobody can explain the functions of it by only using fine pictures of the intricate interconnections of millions of transistors (gates). Further, it would be more confusing because the control unit (CU) and arithmetic and logic unit (ALU) of the CPU and related buses are always active for almost any different kind of operations. So do, unfortunately, brain science and neurobiology. Without a rational guide to the high-level life functions and cognitive processes as shown in the LRMB reference model, nobody may pinpoint rational functional relationship between a brain image and a specific behaviour such as an action of learning

Figure 6. A motherboard and a CPU layout of a computer

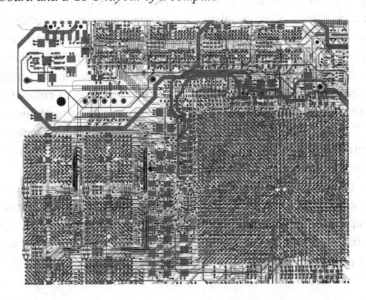

and its effect in memory, a recall of a particular knowledge retained in long-term memory, and a mapping of the same mental object from STM to LTM.

The above case study indicates that neuroscience theories and artificial intelligence technologies toward the brain have been studied at almost separate levels so far in biophysics, neurology, cognitive science, and computational/artificial intelligence. However, a synergic model as that of LRMB that maps the architectures and functions of the brain crossing individual disciplines is necessary in order to explain the complexity and underpinning mechanisms of the brain. This coherent approach will lead to the development of novel engineering applications of CI, αI, NeI, DM, CC, and BI, such as cognitive computers, artificial brains, cognitive robots, and cognitive software agents, which mimic the natural intelligence of the brain based on the theories and denotational mathematical means developed in cognitive informatics and abstract intelligence.

6. COGNITIVE COMPUTING: TOWARD THE NEXT GENERATION OF COGNITIVE COMPUTERS

The term *computing* in a narrow sense is an application of computers to solve a given problem by imperative instructions; while in a broad sense, it is a process to implement the instructive intelligence by a system that transfers a set of given information or instructions into expected intelligent behaviors. The latest advances and engineering applications of CI have led to the emergence of cognitive computing and the development of cognitive computers that perceive, reason, and learn. *Cognitive Computing* is an emerging paradigm of intelligent computing methodologies and systems based on cognitive informatics that implements computational intelligence by autonomous inferences and perceptions mimicking the

mechanisms of the brain (Wang, 2002a, 2009b; Wang et al., 2010).

Definition 16: Cognitive Computing (CC) is a novel paradigm of intelligent computing methodologies and systems that implements computational intelligence by autonomous inferences and perceptions mimicking the mechanisms of the brain.

CC is emerged and developed based on the transdisciplinary research in cognitive informatics and abstract intelligence. The term computing in a narrow sense is an application of computers to solve a given problem by imperative instructions; while in a broad sense, it is a process to implement the instructive intelligence by a system that transfers a set of given information or instructions into expected intelligent behaviors.

Computing systems and technologies can be classified into the categories of *imperative, autonomic,* and *cognitive* computing from the bottom up. The imperative computers are a traditional and passive system based on stored-program controlled behaviors for data processing (Wang, 2009b). The autonomic computers are goal-driven and self-decision-driven machines that do not rely on instructive and procedural information (Kephart & Chess, 2003; Wang, 2009f). Cognitive computers are more intelligent computers beyond the imperative and autonomic computers, which embody major natural intelligence behaviors of the brain such as thinking, inference, and learning.

The *essences* of computing are both its *data objects* and their predefined computational *operations*. From these facets, different computing paradigms may be comparatively analyzed as shown in Figure 7.

The latest advances in cognitive informatics, abstract intelligence, and denotational mathematics have led to a systematic solution for the future generation of intelligent computers known as cognitive computers (Wang, 2006, 2009d).

Figure 7. Conventional computing vs. cognitive computing

Paradigm	Data objects	Basic operations	Advanced operations
Conventional computing technology	Abstract bits Structured data based on bits	Logic operations Arithmetic operations	Functional operations
Cognitive computing technology	Concepts Words Syntax Semantics	Syntactic analysis Semantic analysis	Concept formulation Knowledge representation Comprehension Learning Inferences Causal analysis

Definition 17: A cognitive computer (cC) is an intelligent computer for knowledge processing that perceive, think, learn, and reason.

As that of a conventional von Neumann computers for *data* processing, cCs are designed to embody *machinable intelligence* such as computational inferences, causal analyses, knowledge manipulations, learning, and problem solving. According to the above analyses, a cC is driven by a *cognitive CPU* with a cognitive learning engine and formal inference engine for intelligent operations on abstract concepts as the basic unit of human knowledge. cCs are designed based on contemporary denotational mathematics (Wang, 2008a, 2009f), particularly concept algebra, as that of Boolean algebra for the conventional von Neumann architecture computers. cC is an important extension of conventional computing in both data objects modeling capabilities and their advanced operations at the abstract level of concept beyond bits. Therefore, cC is an intelligent knowledge processor that is much closer to the capability of human thinking at the level of concepts rather than bits. According to concept algebra, the basic unit of human knowledge in natural language representation is a concept rather than a word (Wang, 2008b), because the former conveys the structured semantics of the latter with its intention (attributes), extension (objects),

and relations to other concepts in the context of a knowledge network.

Recent studies in cognitive computing reveal that the computing power in computational intelligence can be classified at four levels: *data, information, knowledge,* and *intelligence* from the bottom up. Traditional von Neumann computers are designed to implement imperative data and information processing by stored-program-controlled mechanisms. However, the increasing demand for advanced computing technologies for knowledge and intelligence processing in the high-tech industry and everyday lives require novel cognitive computers for providing autonomous computing power mimicking the natural intelligence of the brain.

The above analyses indicate that cC is an important extension of conventional computing in both data objects modeling capabilities and their advanced operations at the abstract level of concept beyond bits. Therefore, cC is an intelligent knowledge processor that is much closer to the capability of human thinking at the level of concepts rather than bits. A Cognitive Learning Engine (CLE) that serves as the "CPU" of cCs is under developing on the basis of concept algebra, which implements the basic and advanced cognitive computational operations of concepts and knowledge for cCs. The work in this area may also lead to a fundamental solution to computational linguistics and Computing with Words (CWW) as Zadeh proposed (Zadeh, 1975, 1999), as well as Computing with Natural Language (CNL) (Zadeh, 1999; Wang, 2010c, 2010d).

7. CONCLUSION

The studies in CI, αI, NeI, DM, and CC lay a theoretical foundation toward revealing the basic mechanisms of different forms of intelligence. As a result, cognitive computers may be developed, which are characterized as a knowledge processor beyond those of data processors in conventional computing. Key applications in the above cutting-edge fields of CI and CC can be divided into two categories. The first category of applications uses informatics and computing techniques to investigate problems of intelligence science, cognitive science, and brain science, such as abstract intelligence, memory, learning, and reasoning. The second category of applications includes the areas that use cognitive informatics theories to investigate problems in informatics, computing, software engineering, knowledge engineering, and computational intelligence. CI focuses on the nature of information processing in the brain, such as information acquisition, representation, memory, retrieval, creation, and communication. Through the interdisciplinary approach and with the support of modern information and neuroscience technologies, mechanisms of the brain and the mind may be systematically explored based on the theories and cognitive models of CI.

Because CI and CCs provide a common and general platform for the next generation of cognitive computing, a wide range of applications of CI, αI, NeI, CC, and DM are expected toward the implementation of highly intelligent machinable thought such as formal inference, symbolic reasoning, problem solving, decision making, cognitive knowledge representation, semantic searching, and autonomous learning. Some expected innovations that will be enabled by CI and CCs are as follows, *inter alia*: a) An *inference machine* for complex and long-series of reasoning, problem solving, and decision making beyond traditional logic and if-then-rule based technologies; b) An *autonomous learning system* for cognitive knowledge acquisition and processing; c) A novel *search engine* for providing comprehendable and formulated knowledge via the Internet; d) A *cognitive medical diagnosis system* supporting evidence-based medical care and clinical practices; e) A *cognitive computing node* for the next generation of the intelligent Internet; and f) A *cognitive processor* for implementing cognitive robots and cognitive agents.

ACKNOWLEDGMENT

The author would like to acknowledge the Natural Science and Engineering Council of Canada (NSERC) for its partial support to this work. The author also thanks the anonymous reviewers for their valuable comments and suggestions.

REFERENCES

Bender, E. A. (1996). *Mathematical methods in artificial intelligence*. Washington, DC: IEEE Computer Society.

Boole, G. (2003). *The laws of thought, 1854*. Amherst, NY: Prometheus Books.

Kephart, J., & Chess, D. (2003). The vision of autonomic computing. *IEEE Computer, 26*(1), 41–50.

Kline, M. (1972). *Mathematical thought: From ancient to modern times*. New York, NY: Oxford University Press.

Matlin, M. W. (1998). *Cognition* (4th ed.). New York, NY: Harcourt Brace College.

Miller, G. A. (1956). The magical number seven, plus or minus two: Some limits of our capacity for processing information. *Psychological Review, 63*, 81–97. doi:10.1037/h0043158

Russell, B. (1996). *The principles of mathematics, 1903*. New York, NY: W.W. Norton.

Shannon, C. E. (1948). A mathematical theory of communication. *The Bell System Technical Journal, 27*, 379–423, 623–656.

Wang, Y. (2002a, August). Keynote: On cognitive informatics. In *Proceedings of the 1st IEEE International Conference on Cognitive Informatics*, Calgary, Canada, (pp. 34-42). Washington, DC: IEEE Computer Society.

Wang, Y. (2002b). The real-time process algebra (RTPA). *Annals of Software Engineering, 14*, 235–274. doi:10.1023/A:1020561826073

Wang, Y. (2003). On cognitive informatics. *Brain and Mind: A Transdisciplinary Journal of Neuroscience and Neurophilosophy, 4*(3), 151-167.

Wang, Y. (2006, July). Keynote: Cognitive informatics - towards the future generation computers that think and feel. In *Proceedings of the 5th IEEE International Conference on Cognitive Informatics*, Beijing, China (pp. 3-7). Washington, DC: IEEE Computer Society.

Wang, Y. (2007a). *Software engineering foundations: A software science perspective*. New York, NY: Auerbach.

Wang, Y. (2007b). The theoretical framework of cognitive informatics. *International Journal of Cognitive Informatics and Natural Intelligence, 1*(1), 1–27. doi:10.4018/jcini.2007010101

Wang, Y. (2007c). The OAR model of neural informatics for internal knowledge representation in the brain. *International Journal of Cognitive Informatics and Natural Intelligence, 1*(3), 64–75. doi:10.4018/jcini.2007070105

Wang, Y. (2008a). On contemporary denotational mathematics for computational intelligence. *Transactions of Computational Science, 2*, 6–29. doi:10.1007/978-3-540-87563-5_2

Wang, Y. (2008b). On concept algebra: A denotational mathematical structure for knowledge and software modeling. *International Journal of Cognitive Informatics and Natural Intelligence, 2*(2), 1–19. doi:10.4018/jcini.2008040101

Wang, Y. (2008c). On system algebra: A denotational mathematical structure for abstract system modeling. *International Journal of Cognitive Informatics and Natural Intelligence, 2*(2), 20–42. doi:10.4018/jcini.2008040102

Wang, Y. (2008d). RTPA: A denotational mathematics for manipulating intelligent and computational behaviors. *International Journal of Cognitive Informatics and Natural Intelligence, 2*(2), 44–62. doi:10.4018/jcini.2008040103

Wang, Y. (2009a). A cognitive informatics reference model of autonomous agent systems (AAS). *International Journal of Cognitive Informatics and Natural Intelligence, 3*(1), 1–16. doi:10.4018/jcini.2009010101

Wang, Y. (Ed.). (2009b). *International journal of software science and computational intelligence (Vol. 1)*. Hershey, PA: IGI Global.

Wang, Y. (2009c). On abstract intelligence: Toward a unified theory of natural, artificial, machinable, and computational intelligence. *International Journal of Software Science and Computational Intelligence, 1*(1), 1–18. doi:10.4018/jssci.2009010101

Wang, Y. (2009d). On cognitive computing. *International Journal of Software Science and Computational Intelligence, 1*(3), 1–15. doi:10.4018/jssci.2009070101

Wang, Y. (2009e). On visual semantic algebra (VSA): A denotational mathematical structure for modeling and manipulating visual objects and patterns. *International Journal of Software Science and Computational Intelligence, 1*(4), 1–15. doi:10.4018/jssci.2009062501

Wang, Y. (2009f). Paradigms of denotational mathematics for cognitive informatics and cognitive computing. *Fundamenta Informaticae, 90*(3), 282–303.

Wang, Y. (2010a). Cognitive robots: A reference model towards intelligent authentication. *IEEE Robotics and Automation, 17*(4), 54–62. doi:10.1109/MRA.2010.938842

Wang, Y. (2010b). A sociopsychological perspective on collective intelligence in metaheuristic computing. *International Journal of Applied Metaheuristic Computing, 1*(1), 110–128. doi:10.4018/jamc.2010102606

Wang, Y. (2010c). On formal and cognitive semantics for semantic computing. *International Journal of Semantic Computing, 4*(2), 203–237. doi:10.1142/S1793351X10000833

Wang, Y. (2010d). On concept algebra for computing with words (CWW). *International Journal of Semantic Computing, 4*(3), 331–356. doi:10.1142/S1793351X10001061

Wang, Y. *(2010e, August). Keynote: Cognitive informatics and denotational mathematics means for brain informatics. In* Proceedings of the 1st International Conference on Brain Informatics, *Toronto, ON, Canada (pp. 2-13).*

Wang, Y. (in press). On cognitive models of causal inferences and causation networks. International Journal of Software Science and Computational Intelligence, 3*(1).*

Wang, Y., Baciu, G., Yao, Y., Kinsner, W., Chan, K., & Zhang, B. (2010). Perspectives on cognitive informatics and cognitive computing. *International Journal of Cognitive Informatics and Natural Intelligence, 4*(1), 1–29. doi:10.4018/jcini.2010010101

Wang, Y., & Chiew, V. (2010). On the cognitive process of human problem solving. *Cognitive Systems Research: An International Journal, 11*(1), 81–92. doi:10.1016/j.cogsys.2008.08.003

Wang, Y., Johnston, R., & Smith, M. (Eds.). (2002, August). *Proceedings of the 1st IEEE International Conference,* Calgary, AB, Canada. Washington, DC: IEEE Computer Society.

Wang, Y., & Kinsner, W. (2006). Recent advances in cognitive informatics. *IEEE Transactions on Systems, Man and Cybernetics. Part C, Applications and Reviews, 36*(2), 121–123. doi:10.1109/TSMCC.2006.871120

Wang, Y., Kinsner, W., Anderson, J. A., Zhang, D., Yao, Y., & Sheu, P. (2009). A doctrine of cognitive informatics. *Fundamenta Informaticae, 90*(3), 203–228.

Wang, Y., Kinsner, W., & Zhang, D. (2009). Contemporary cybernetics and its faces of cognitive informatics and computational intelligence. *IEEE Trans. on System, Man, and Cybernetics (B), 39*(4), 1–11.

Wang, Y., & Wang, Y. (2006). Cognitive informatics models of the brain. *IEEE Transactions on Systems, Man and Cybernetics. Part C, Applications and Reviews, 36*(2), 203–207. doi:10.1109/TSMCC.2006.871151

Wang, Y., Wang, Y., Patel, S., & Patel, D. (2006). A layered reference model of the brain (LRMB). *IEEE Transactions on Systems, Man and Cybernetics. Part C, Applications and Reviews, 36*(2), 124–133. doi:10.1109/TSMCC.2006.871126

Wang, Y., Zadeh, L. A., & Yao, Y. (2009). On the system algebra foundations for granular computing. *International Journal of Software Science and Computational Intelligence, 1*, 1–17. doi:10.4018/jssci.2009040101

Wang, Y., Zhang, D., Latombe, J.-C., & Kinsner, W. (Eds.). (2008, August). *Proceedings of the 7th IEEE International Conference on Cognitive Informatics,* Stanford, CA. Washington, DC: IEEE Computer Society.

Wang, Y., Zhang, D., & Tsumoto, S. (2009). Cognitive informatics, cognitive computing, and their denotational mathematical foundations (1). *Fundamenta Informaticae, 90*(3), 1–7.

Wilson, R. A., & Keil, F. C. (Eds.). (2001). *The MIT encyclopedia of the cognitive sciences*. Cambridge, MA: MIT Press.

Zadeh, L. A. (1975). Fuzzy logic and approximate reasoning. *Syntheses, 30*, 407–428. doi:10.1007/BF00485052

Zadeh, L. A. (1999). From computing with numbers to computing with words – from manipulation of measurements to manipulation of perception. *IEEE Transactions on Circuits and Systems 1, 45*(1), 105-119.

Zhong, N. (2009, July). A unified study on human and web granular reasoning. In *Proceedings of the 8th International Conference on Cognitive Informatics*, Kowloon, Hong Kong (pp. 3-4). Washington, DC: IEEE Computer Society.

This work was previously published in the International Journal of Cognitive Informatics and Natural Intelligence, Volume 5, Issue 1, edited by Yingxu Wang, pp. 74-92, copyright 2011 by IGI Publishing (an imprint of IGI Global).

Chapter 2
Perspectives on the Field of Cognitive Informatics and its Future Development

Yingxu Wang
University of Calgary, Canada

Kenji Sugawara
Chiba Institute of Technology, Japan

Bernard Widrow
Stanford University, USA

Fuchun Sun
Tsinghua University, China

Bo Zhang
Tsinghua University, China

Jianhua Lu
Tsinghua University, China

Witold Kinsner
University of Manitoba, Canada

Thomas Weise
University of Science and Technology of China, China

Du Zhang
California State University, USA

ABSTRACT

The contemporary wonder of sciences and engineering has recently refocused on the beginning point of: how the brain processes internal and external information autonomously and cognitively rather than imperatively like conventional computers. Cognitive Informatics (CI) is a transdisciplinary enquiry of computer science, information sciences, cognitive science, and intelligence science that investigates the internal information processing mechanisms and processes of the brain and natural intelligence, as well as their engineering applications in cognitive computing. This paper reports a set of eight position statements presented in the plenary panel of IEEE ICCI'10 on Cognitive Informatics and Its Future Development contributed from invited panelists who are part of the world's renowned researchers and scholars in the field of cognitive informatics and cognitive computing.

DOI: 10.4018/978-1-4666-2476-4.ch002

1. INTRODUCTION

Cognitive informatics is a transdisciplinary enquiry of computer science, information science, cognitive science, and intelligence science, which investigates into the internal information processing mechanisms and processes of the brain and natural intelligence, as well as their engineering applications in cognitive computing (Wang, 2002a, 2003, 2006, 2007c, 2009c, 2009d; Wang & Kinsner, 2006; Wang & Wang, 2006; Wang & Chiew, 2010; Wang, Kinsner, & Zhang, 2009; Wang, Johnston, & Smith, 2002; Wang, Wang, Patel, & Patel, 2006; Wang, Zhang, Latombe, & Kinsner, 2008, Wang et al., 2009; Baciu et al., 2009; Chan et al., 2004; Kinsner et al., 2005; Patel et al., 2003; Yao et al., 2006; Zhang et al., 2007; Sun et al., 2010). Cognitive informatics is a cutting-edge and multidisciplinary research area that tackles the fundamental problems shared by computational intelligence, modern informatics, computer science, AI, cybernetics, cognitive science, neuropsychology, medical science, philosophy, formal linguistics, and life science (Wang, 2002a, 2003, 2007c, 2009c, 2010b, 2010d). The development and the cross fertilization among the aforementioned science and engineering disciplines have led to a whole range of extremely interesting new research areas known as Cognitive informatics, which investigates the internal information processing mechanisms and processes of the natural intelligence – human brains and minds – and their engineering applications in computational intelligence. Cognitive informatics studies the natural intelligence and internal information processing mechanisms of the brain, as well as processes involved in perception and cognition. Cognitive informatics forges links between a number of natural science and life science disciplines with informatics and computing science.

Definition 1: *Cognitive informatics* (CI) is a transdisciplinary enquiry of computer science, information science, cognitive science,

and intelligence science that investigates into the internal information processing mechanisms and processes of the brain and natural intelligence, as well as their engineering applications in cognitive computing.

The IEEE series of International Conferences on Cognitive Informatics (ICCI) has been established since 2002 (Wang, 2002a). The inaugural ICCI event in 2002 was held at University of Calgary, Canada (ICCI'02) (Wang, Johnston, & Smith, 2002), followed by the events in London, UK (ICCI'03) (Patel et al., 2003); Victoria, Canada (ICCI'04) (Chan et al., 2004); Irvine, USA (ICCI'05) (Kinsner et al., 2005); Beijing, China (ICCI'06) (Yao et al., 2006); Lake Tahoe, USA (ICCI'07) (Zhang et al., 2007); Stanford University, USA (ICCI'08) (Wang, Zhang, Latombe, & Kinsner, 2008); Hong Kong (ICCI'09) (Baciu et al., 2009); and Tsinghua University, Beijing (ICCI'10) (Sun et al., 2010). Since its inception, ICCI has been growing steadily in its size, scope, and depth. It attracts worldwide researchers from academia, government agencies, and industry practitioners. The conference series provides a main forum for the exchange and cross-fertilization of ideas in the new research field of CI toward revealing the cognitive mechanisms and processes of human information processing and the approaches to mimic them in cognitive computing.

The latest advances and engineering applications of CI have led to the emergence of *cognitive computing* (CC) and the development of *cognitive computers* that think, perceive, learn, and reason (Wang, 2006, 2009e, 2010a, 2010b; Wang et al., 2009; Wang, Kinsner, & Zhang, 2009). CI has also fundamentally contributed to autonomous agent systems (Wang, 2009a) and cognitive robots (Wang, 2010b). A wide range of applications of CI has been identified such as in the development of cognitive computers, cognitive robots, cognitive agent systems, cognitive search engines, cognitive learning systems, and artificial brains. The work in CI may also lead to a fundamental solution to computational linguistics, Computing with Natural Language (CNL), and Computing with Words (CWW) (Zedeh, 1975, 1999).

This paper is a summary of the position statements of panellists presented in the *Plenary Panel on Cognitive Informatics and Its Future Development* in IEEE ICCI 2010 at Tsinghua University held in July 2010 (Sun et al., 2010). It is noteworthy that the individual statements and opinions included in this paper may not necessarily be shared by all panellists.

2. COGNITIVE INFORMATICS (CI): THE SCIENCE OF ABSTRACT INTELLIGENCE AND COGNITIVE COMPUTING

CI establishes a systematical framework for brain and computational intelligence studies as well as their applications at hierarchical levels known as the functional level, the mathematical (logical) level, the cognitive level, and the neural (physiological) level. Corresponding to the four reductive levels as shown in Figure 1, the theoretical framework of CI (Wang, 2007c) encompasses

four main areas of basic and applied research on: (a) Abstract intelligence: fundamental theories of natural intelligence; (b) Denotational mathematics for modeling abstract intelligence; (c) Cognitive models of the brain; (d) Neural informatics; and e) Cognitive computing. These areas of CI are elaborated in the following subsections.

- **Abstract Intelligence (αI):** The studies on αI form a human enquiry of both natural and artificial intelligence at the reductive neural, cognitive, logical, and functional levels from the bottom up (Wang, 2009c). The paradigms of αI are such as natural, artificial, machinable, and computational intelligence. The studies in CI and αI lay a theoretical foundation toward revealing the basic mechanisms of different forms of intelligence (Wang, 2010a). As a result, cognitive computers may be developed, which are characterized as knowledge processors beyond those of data processors in conventional computing.

- **Cognitive Models of the Brain:** Fundamental theories developed in CI covers the Information-Matter-Energy

Figure 1. The architecture and theoretical framework of Cognitive Informatics (CI)

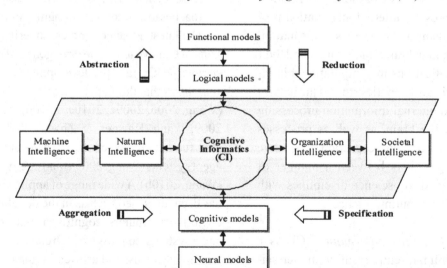

(IME) model (Wang, 2007a), the Layered Reference Model of the Brain (LRMB) (Wang, Wang, Patel, & Patel, 2006), the Object-Attribute-Relation (OAR) model of information/knowledge representation in the brain (Wang, 2007c), the cognitive informatics model of the brain (Wang, 2010b; Wang & Wang, 2006), and Natural Intelligence (NI) (Wang, 2007c). Recent studies on LRMB in cognitive informatics reveal an entire set of cognitive functions of the brain and their cognitive process models, which explain the functional mechanisms and cognitive processes of the natural intelligence with 43 cognitive processes at seven layers known as the sensation, memory, perception, action, meta-cognitive, meta-inference, and higher cognitive layers (Wang, Wang, Patel, & Patel, 2006).

- **Neural Informatics (NeI):** Neural informatics is an emerging interdisciplinary enquiry of the biological and physiological representation of information and knowledge in the brain at the neuron level and their abstract mathematical models (Wang, 2007c). NeI is an important branch of cognitive informatics, which reduces cognitive informatics theories and the studies on the internal information processing mechanisms of the brain onto the neuron and physiological level. In neural informatics, memory is recognized as the foundation and platform of any natural or machine intelligence based on the OAR model (Wang, 2007b; Wang & Wang, 2006) of information/knowledge representation. The theories of neural informatics explain a number of important questions in the study of natural intelligence. Enlightening findings in neural informatics are such as (a) Long-Tem Memory (LTM) establishment is a subconscious process; (b) The long-term memory is established during sleep-

ing; (c) The major mechanism for LTM establishment is by sleeping; (d) The general acquisition cycle of LTM is equal to or longer than 24 hours; (e) The mechanism of LTM establishment is to update the entire memory of information represented as an OAR model in the brain (Wang, 2007b); and (f) Eye movement and dreams play an important role in LTM creation.

- **Denotational Mathematics (DM):** *Denotational Mathematics* (DM) is a category of expressive mathematical structures that deals with high-level mathematical entities beyond numbers and sets, such as abstract objects, complex relations, perceptual information, abstract concepts, knowledge, intelligent behaviors, behavioral processes, and systems (Wang, 2008a). A number of DMs have been created and developed such as *concept algebra* (Wang, 2008b, 2010c), *system algebra* (Wang, 2008c; Wang, Zadeh, & Yao, 2009), *real-time process algebra* (RTPA) (Wang, 2002b, 2008d), *granular algebra* (Wang, 2010c), *visual semantic algebra* (VSA) (Wang, 2009e), and *formal causal inference methodologies* (Wang, in press). It is recognized that the maturity of a scientific discipline is characterized by the maturity of its mathematical (meta-methodological) means. DM provides a coherent set of contemporary mathematical means and explicit expressive power for CI, αI, NeI, CC, AI, and computational intelligence.

- **Applications of CI, αI, NeI, and DM:** The key applications of the fundamental theories and technologies of CI can be divided into two categories. The first category of applications uses informatics and computing techniques to investigate intelligence science, cognitive science, and knowledge science problems, such as abstract intelligence, memory, learning, and reasoning. The second category includes the areas that use cognitive informatics theories to investigate problems in informatics, computing, software engineering, knowledge engineering, and computational intelligence. CI focuses on the nature of information processing in the brain, such as information acquisition, representation, memory, retrieval, creation, and communication. Through the interdisciplinary approach and with the support of modern information and neuroscience technologies, mechanisms of the brain and the mind may be systematically explored (Wang & Chiew, 2010) within the framework of CI.

A key and exiting application of CI is its inspiration to and theoretical preparation for cognitive computers (cCs), which is an intelligent computer for knowledge processing as that of a conventional von Neumann computer for data processing. A cC is driven by a *cognitive CPU* with a cognitive learning engine and formal inference engine for intelligent operations on abstract concepts as the basic unit of human knowledge (Wang, 2009d, 2010c, 2010d). cCs are designed based on contemporary denotational mathematics (Wang, 2008a, 2009f), particularly concept algebra, as that of Boolean algebra for the conventional von Neumann architecture computers.

3. CI/CC AND COGNITIVE MEMORY

Cognitive computing is brain-like computing. Emulating mental processing is inherently difficult, since the overall working of the human brain is more or less unknown. Instead of trying to model the functioning of the entire brain, it is proposed that a portion be studied, an important portion, long-term memory. Without memory we can live, but we would not really exist as human beings.

Human memory is exceedingly complex. However, by observing our own behavior and that of fellow humans, electrical engineers can devise memory systems that exhibit human-like behavior. In my laboratory at Stanford University (Widrow & Aragon, 2010), we have constructed an elementary form of cognitive memory consisting of both software and hardware. We have used this memory to solve problems in the fields of pattern recognition, face recognition and control systems.

Taking inspiration from life experience, we have devised a new form of computer memory (Widrow & Aragon, 2010). Certain conjectures about human memory are keys to the central idea. We present the design of a practical and useful "cognitive" memory system. The new memory does not function like a computer memory where specific data is stored in specific numbered registers and retrieval is done by reading the contents of the specified memory register, or done by matching key words as with a document search. Incoming sensory data would be stored at the next available empty memory location, and indeed could be stored redundantly at several empty locations. The stored sensory data would neither have key words nor would it be located in known or specified memory locations. Retrieval would be initiated by a prompt signal from a current set of sensory inputs or patterns. The search would be done by a retrieval system that makes use of auto-associative artificial neural networks. A practical application of this cognitive memory system to human facial recognition has been implemented.

4. THE MULTIDISCIPLINARY RESEARCH FOR COGNITIVE COMPUTATION

Cognitive informatics should benefit from the multidisciplinary research among information science, cognitive science and brain science, etc. (Zhang, 2010). Taking computer vision as an example, human visual performances are much better than computer vision in many cases recently. Therefore, computer vision should learn something from the visual mechanism of human cognition in the brain.

As we known, the main approaches adopted in recent computer object recognition are statistics-based and data-driven. The objects (images) are represented in the data space composed by the computer's robustly detectable features. These features are generally less meaningful, i.e., so called low-level features, such as colors, textures, etc. Due to the big gap between semantics and low level features, image recognition, classification, and etc. are difficult to implement in computer vision. But human brain processes visual information in the conceptual space where the semantically meaningful features are extracted such as line-segments, boundaries, shapes, etc. Therefore, there is no semantic gap in human visual information processing. In order to endow computers with the human capacity, we need to learn from the human visual information processing mechanisms (Zhang, 2010).

In order to promote the multidisciplinary research, the Center for Neural and Cognitive Computation was established at Tsinghua University in 2009. It includes the fields of computational neuroscience, system neuroscience, information science, psychology, neural information, brain-computer interface, learning, and memory, etc.

5. EVOLUTION OF COGNITIVE DYNAMICAL SYSTEMS AND COGNITIVE INFORMATICS

Many developments of the last century focused on adaptation and adaptive systems. The focus in this century appears to be shifting towards cognition and cognitive dynamical systems with emergence. Although cognitive dynamical systems are always adaptive to various conditions in the environment where they operate, adaptive systems of the past have not been cognitive.

The evolving formulation of cognitive informatics (CI) (Wang, 2002a, 2003, 2007c; Wang & Kinsner, 2006; Wang, Kinsner, & Zhang, 2009; Wang, Zhang, & Tsumoto, 2009) has been an important step in bringing the diverse areas of science, engineering, and technology required to develop such cognitive systems. Current examples of various cognitive systems include autonomic computing, memetic computing, cognitive radio, cognitive radar, cognitive robots, cognitive networks, cognitive computers, cognitive cars, cognitive factories, as well as brain-machine interfaces for physically-impaired persons, and cognitive binaural hearing instruments. The phenomenal interest in this area may be due to the recognition that perfect solutions to large-scale scientific and engineering problems may not be feasible, and we should seek the best solution for the task at hand. The "best" means suboptimal and the most reliable (robust) solution, given not only limited resources (financial and environmental) but also incomplete knowledge of the problem and partial observability of the environment. Many exciting new theoretical, computational and technological accomplishments have been described at IEEE ICCI conferences and related journals.

The challenges in the evolving cognitive systems can be grouped into several categories: (a) theoretical, (b) technological, and (c) sociological. The first group of theoretical issues includes modeling, reformulation of information and entropy, multiscale measures and metrics, and management of uncertainty. Modeling of cognitive systems requires radically new approaches. Reductionism has dominated our scientific worldview for the last 350 years, since the times of Descartes, Galileo, Newton, and Laplace. In that approach, all reality can be understood in terms of particles (or strings) in motion. However, in this unfolding emergent universe with agency, meaning, values and purpose, we cannot prestate or predict all that will happen. Since cognitive systems rely on perceiving the world by agents, learning from it, remembering and developing the experience of self-awareness, feelings, intentions, and deciding how to control not only tasks but also communication with other agents, and to create new ideas, CI cannot only rely on the reductionist approach of describing nature. In fact, CI tries to expand the modeling in order to deal with the emergent universe where no laws of physics are violated, and yet ceaseless unforeseeable creativity arises and surrounds us all the time. This new approach requires many new ideas to be developed, including reformulation of the concept of cognitive information, entropy, and associated measures, as well as management of uncertainty, and new forms of cognitive computing.

As we have seen over the last decade, cognitive informatics is multidisciplinary (Wang, 2003, 2007c; Wang & Kinsner, 2006; Wang, Zhang, & Tsumoto, 2009), and requires cooperation between many subjects, including sciences (e.g., cognitive science, evolutionary computing, granular computing, computer science, game theory, crisp and fuzzy sets, mathematics, physics, chemistry, biology, psychology, humanities, and social sciences), as well as engineering and technology (computer, electrical, mechanical, information theory, control theory, intelligent signal processing, neural networks, learning machines, sensor networks, wireless communications, and computer networks). Many of the new algorithms replace the conventional concepts of second order statistics (covariance, L2 distances, and correlation functions) with scalars and functions based on information theoretic underpinnings (such as entropy, mutual information and correntropy) defined not only on a single scale, but also on multiple scales. Two recent special issues of the IEEE Proceedings are dedicated to cognitive systems with their practical perspectives (April 2009), and fundamental issues (May 2009).

6. COGNITIVE COMPUTING AND SYMBIOTIC COMPUTING

Cognitive Computing (CC) is an emerging paradigm of intelligent computing methodologies and systems based on cognitive informatics that implements computational intelligence by autonomous inferences and perception mimicking of the brain (Wang, 2009d; Wang et al., 2009). The Layered Reference Model of the Brain (LRMB) was proposed as a seven-layered model of the function layers of the brain for the fundamental cognitive mechanisms and processes of natural intelligence (Wang, Wang, Patel, & Patel, 2006).

Symbiotic Computing (SC) is a methodology based on cognitive computing, ubiquitous computing and agent-based computing to develop systems that have their significance of existence for serving humans, their goals of actions for obtaining the trust of them through their services, and their desire for growing up their capabilities to serve them, along with them through practice collaborating with them. The framework of SC is based on an agent framework that works on the Symbiotic-computing-based System Platform (SSP), which consists of partner tracking function, symbiotic zone control function, symbiotic zone sensing function and perception function corresponding to the low layers of the LRMB. The intelligent agents which work on the platform have functions of cognition corresponding to the upper layers of the LRMB and social knowledge to cooperate with humans and with other intelligent agents working on the platform and providing social norm, customs, conveniences and risks for humans.

The Symbiotic Zone (Symbiozone or SYZ) is a conceptual space surrounding a person with ubiquitous and wearable sensors and effectors in order for an intelligent agent called a Partner Agent (PA) to communicate with only person called a partner of the PA. The PA follows the partner by tracking him/her, finding sensors and effectors around him/her and making a network of the devices dynamically wherever possible. The partner can communicate with the PA when she/he wants and the PA gives advices if it considers that it should do it then for the partner.

The PA is an intelligent and complex software system based on a multi-agent system architecture, in which the following disciplines of behavior are implemented:

1. Significance of existence to serve its partner;
2. Goals of action to serve its partner;
3. Desire of co-growing with its partner.

In SYZ, a partner and a PA maintain close contact like the Licklider's view of Man-Computer symbiosis (Licklider, 1960) and promote mutual understanding to achieve the categories of (1) - (3). To do so in the SYZ, a PA has the following functions inside, (a) perceiving poses and actions of the partner, (b) perceiving surroundings of the partner, (c) acquiring partner's requirements, (d) understanding partner's intension, (e) serving the partner based on the intension and requirements, (f) advising the partner against risky actions based on assessment function of actions of the partner from social point of view.

In order to assess actions of both a partner and its PA from the social point of view, a Group Agent (GA) cooperates with PAs for a group or a community, which supports group dynamics (Grundin, 2002) using the Social Informatics (Wang, Carley, Zeng, & Mao, 2007). The social informatics stored in a GA is acquired by mining information from Web and is learned through actions that a PA and a partner have made in the group. This SC project aims at developing cognitive systems which work in an area ranging from network systems to physical places where ubiquitous devices are embedded, working like the cognitive machine (Kinsner, 2002).

7. ROBOTICS AND COGNITIVE INFORMATION PROCESSING

As an ideal platform for research on cognitive information processing, robots have been paid strong attention for many years and many successful progresses have been achieved (Sun et al., 2010; Wang, 2010b). In the National Laboratory of Information Science and Technology (TNLIST) at Tsinghua University, the only state laboratory in the field of information sciences at the national level in China, various types of robots have been developed, such as mobile robots (also named as intelligent vehicles), flexible-link manipulators, space robots, unmanned aerial vehicles, soccer robots, and so on. Based on these well-equipped platforms, researchers in this laboratory have developed comprehensive theory and approaches for robot sensing and control. Many novel approaches are deeply rooted at cognitive informatics, such as fuzzy control, neural network control, path planning using genetic algorithms and estimation distributed algorithm, visual serving using cognitive approach, object tracking based on machine learning.

Recently, a more collaborative project is being conducted in TNLIST, which is supported by the National Science Foundation of China. This project aims at controlling a mobile robot using brain-computer-interface (BCI). The BCI signal processing module is provided by the Medical School of Tsinghua University. Researchers in TNLIST will develop more rich local environmental information to aid human for teleoperation of the remote mobile robots. This project integrates many research fields including robots, computer vision, and brain sciences. We hope the advanced studies reported above provide more insights of TNLIST towards the future development of cognitive informatics and its engineering applications.

8. CI/CC AND EVOLUTIONARY COMPUTATION

Evolutionary Computation (EC) (Weise, 2009) comprises all Monte Carlo metaheuristics which iteratively refine sets (populations) of multiple candidate solutions. Most EC approaches are either Swarm Intelligence (SI) methods or Evolutionary Algorithms (EAs). SI is inspired by fact that natural systems of many independent, simple agents (such as ants or birds) are of tenable to find pieces of food or shortest distance routes very efficiently. EAs, on the other hand, copy the behavior of natural evolution and treat candidate solutions as individuals which compete and reproduce in a virtual environment defined by the user-provided objective function(s). Generation after generation, these individuals adapt to the environment and thus, tend to become suitable solutions for the problem at hand.

- **Past:** The roots of EC go back to the mid-1950s, where the biologist Barricelli (1954) began to apply computer-aided simulations in order to gain more insight into the natural evolution. Bremermann (1962) and Bledsoe (1961) were the first ones to use evolutionary approaches for solving optimization problems. In the early 1980s Genetic Programming emerged as the youngest member of the EA family (1980). The most common SI methods followed in the 1990s (Dorigo et al., 1996; Eberhart & Kennedy, 1995).
- **Present:** Evolutionary Computation now exists for almost 50 years. When taking a look on the current situation of this area, I get the impression that (1) countless algorithm variants and analyses have been published and EC became widely accepted in the research community. (2) Most of the evidence of the efficiency of EAs is based on experiments and empirical studies. Due to the many configuration parameters of EAs and the wide range of existing optimization problems, it is very hard to define meaningful boundaries for performance or required runtime. (3) A tendency towards hybridizing optimization techniques can be observed, resulting from this lack of knowledge about which algorithm is "good" for which problem. This trend began in the 1970s (Bosworth, 1972), lead to the development of Memetic Algorithms (Moscato, 1989), and now culminates in the emergence of portfolio methods (Peng et al., 2010), which choose the best methods from an algorithm portfolio during the actual process of solving a given problem. (4) Despite the available evidence for the high utility of EAs, practitioners who solve real-world optimization problems appear to often prefer traditional, exact methods. Large-scale problems, which these approaches cannot handle any more due to their computational complexity, are often approached manually instead of using meta-heuristics which could have provided much better solutions in shorter time (Weise et al., 2009). (5) The communication between researchers working on meta-heuristic optimization and those working on traditional, exact methods is low; both communities appear to be separated.
- **Future:** My humble opinion about the future development in the EC area is that (1) in the next ten to twenty years, metaheuristic optimization should undergo a slow transition from a research area to a service. Virtually every decision or design task in engineering and business is an optimization problem. Yet currently, only the fewest of them are recognized as such and even fewer are actually solved using a suitable technique. More joint projects between research and economy targeting real-world applications are necessary to improve the awareness and trust of practitioners in EC.

(2) EA research should thus focus on tasks which are interesting for practitioners, such as large-scale real-world problems (Weise et al., 2009), in order to become more attractive for them. (3) Up to date, in my opinion, no framework exists for analyzing EAs theoretically which provides results that are actually useful in practice. The development of a robust and simple analysis approach would be highly desirable since it would further increase the acceptance of EC. (4) A closer cooperation between the EC community and traditional/emerging areas, such as cognitive informatics (Wang, 2002a, 2007c) and cognitive computing (Wang, 2009d), should be pursued, since an exchange of ideas would be beneficial for both sides.

9. INCONSISTENCY, MEMORY, AND INCONSISTENCY-INDUCED LEARNING

The focus of this section of the position paper is on the interplay among inconsistency, long-term memory, and inconsistency-induced learning in a cognitive system. There are five conclusions that are summarized as follows.

Two important types of memories exist in a human memory model: short-term memory (STM) and long-term memory (LTM) (Wang, 2007b; Wang & Wang, 2006; Wang, Kinsner, & Zhang, 2009). STM is a working memory where reasoning takes place with activated beliefs. The reasoning in STM results in actions taken that will affect a human being's behavior. STM has limited storage capacity and duration. Activated beliefs in STM come from either sensory memory or LTM.

LTM, on the other hand, is where beliefs are retained for long-term purpose. In general, LTM does not have practical capacity limit and some information can be stored in LTM indefinitely (Cherniak, 1983). Beliefs in LTM do not have

direct impact on a person's behavior unless they are recalled or retrieved to STM. In traditional models, beliefs in LTM are referred to as "relatively inert" (Cherniak, 1983). Some recent constructive memory models regard beliefs in LTM not as completely inactive once retained, but as undergoing some sort of transformation after acquisition and before recall. According to (McGuire, 1960), beliefs in LTM are organized into various "compartments" in human thinking apparatus. Cognitions and beliefs that are not recalled or activated contemporaneously tend to be separated into different compartments (Cherniak, 1983).

LTM invariably contains inconsistent beliefs. There are several reasons for this. The vast number of beliefs a person possesses makes it a daunting and next to impossible task to keep track of inconsistent beliefs. The compartmentalized structure in LTM is such that logical relations and inconsistencies between beliefs in different compartments are far less likely to be identified than those of beliefs belonging to the same compartment. The following observation in (Shastri & Grannes, 1996) succinctly summarizes the phenomenon: "we often hold inconsistent beliefs in our long-term memory without being explicitly aware of such inconsistencies. But at the same time, we often recognize contradictions in our beliefs when we try to bring inconsistent knowledge to bear on a particular task."

The second sentence of the aforementioned quote from (Shastri & Grannes, 1996) describes the fact that inconsistent beliefs in LTM manifest their influence on a person's behavior through STM, because items from LTM cannot directly affect behavior, they have to be recalled or retrieved to STM to influence one's behavior. In general, inconsistency detection is intractable (Johnson-Laird et al., 2004). Detecting inconsistencies in STM hinges on several conditions: the presence of a triggering event from sensory memory that causes additional beliefs to be recalled from LTM; efficient recall, and organization of compartmentalized beliefs in LTM. Failure of detecting inconsistency at STM may have detrimental or costly consequence.

To facilitate inconsistency detection process in a cognitive system, we propose to augment the constructive memory model by tagging potential inconsistent beliefs in LTM. The tagging approach relies on (1) obtaining a fixpoint representation for each compartment of beliefs (Zhang, 2005, 2007, 2008); (2) fusing fixpoints for individual compartments into one that allows us to establish relevant inter-compartment logical relations and inconsistencies between or among beliefs of different compartments; (3) tagging a set of beliefs that are conflicting with each other according to the nature of the inconsistency involved; and (3) organizing tags with regard to different types of inconsistency (Zhang, 2009). We call such a transformed LTM a tagged LTM, denoted as tLTM.

Assume that two beliefs P_i and P_k in tLTM belong to either the same compartment or two different compartments and that P_i and P_k are conflicting with each other. When recalling beliefs from tLTM into STM for a particular task at hand, even though one of P_i and P_k, but not both contemporaneously, is in STM, the tag, say, P_i has would alert the detection mechanism at STM to ascertain if the reasoning and subsequent actions should proceed with tagged P_i in STM. This would help improve the inconsistency detection and subsequent handling process. Of course the upfront price we pay is the transformation of LTM into tLTM.

Finally, identified inconsistency at STM can serve as an impetus to belief revisions at LTM. As quoted in (Gotesky, 1968), Henri Poincare once said that contradiction is the prime stimulus for scientific research. So to conclude, we would like to end with the follow slogan: inconsistency in LTM is a terrible thing to waste.

10. CONCLUSION

Cognitive Informatics (CI) has been described as a transdisciplinary enquiry of computer science, information sciences, cognitive science, and intelligence science that investigates into the internal information processing mechanisms and processes of the brain and natural intelligence, as well as their engineering applications in cognitive computing. This paper summarizes the presentations of a set of eight position papers in the IEEE ICCI'10 *Panel on Cognitive Informatics and Its Future Development* contributed from invited panelists who are part of the world's preeminent researchers and scholars in the field of cognitive informatics and cognitive computing.

REFERENCES

Baciu, G., Yao, Y., Wang, Y., Zadeh, L. A., Chan, K., Kinsner, W., et al. (2009, June). Perspectives on cognitive informatics and cognitive computing: Summary of the panel. In *Proceedings of the 8th IEEE International Conference on Cognitive Informatics*, Kowloon, Hong Kong (pp. 9-27). Washington, DC: IEEE Computer Society.

Bledsoe, W. W. (1961). *Lethally dependent genes using instant selection* (Tech. Rep. No. PRI1). Palo Alto, CA: Panoramic Research Inc.

Bosworth, J. (1972). *Comparison of genetic algorithms with conjugate gradient methods* (Tech. Rep. No. TR00312-1-T). Ann Arbor, MI: University of Michigan.

Bremermann, H. J. (1962). Optimization through evolution and recombination. In Yovits, M. C., Jacobi, G. T., & Goldstein, G. D. (Eds.), *Self-organizing systems* (pp. 93–106). Washington, DC: Spartan Books.

Chan, C., Kinsner, W., Wang, Y., & Miller, D. M. (Eds.). (2004, August). *Proceedings of the 3rd IEEE International Conference on Cognitive Informatics*, Victoria, BC, Canada Washington, DC: IEEE Computer Society.

Cherniak, C. (1983). Rationality and the structure of human memory. *Synthese, 57*(2), 163–186. doi:10.1007/BF01064000

Dorigo, M., Maniezzo, V., & Colorni, A. (1996). The ant system: Optimization by a colony of cooperating agents. *IEEE Transactions on Systems, Man, and Cybernetics B, 26*(1), 29–41. doi:10.1109/3477.484436

Eberhart, R. C., & Kennedy, J. (1995). A new optimizer using particles warm theory. In *Proceedings of the 6th International Symposium on Micro Machine and Human* Science (pp. 39-43). Washington, DC: IEEE Computer Society.

Gotesky, R. (1968). The uses of inconsistency. *Philosophy and Phenomenological Research, 28*(4), 471–500. doi:10.2307/2105687

Grundin, J. (2002). Group dynamics and ubiquitous computing. *Communications of the ACM, 45*(12), 74–78.

Johnson-Laird, P. N., Legrenzi, P., & Girotto, V. (2004). Reasoning from inconsistency to consistency. *Psychological Review, 111*(3), 640–661. doi:10.1037/0033-295X.111.3.640

Kinsner, W. (2002). Towards cognitive machine: Multiscale measures and analysis. *International Journal of Cognitive Informatics and Natural Intelligence, 1*(1), 28–38. doi:10.4018/jcini.2007010102

Kinsner, W., Zhang, D., Wang, Y., & Tsai, J. (Eds.). (2005, August). *Proceedings of the 4th IEEE International Conference*, Irvine, CA. Washington, DC: IEEE Computer Society.

Licklider, J. C. R. (1960). Man-computer symbiosis. *IRE Transactions on Human Factors in Electronics, 1*, 4–11. doi:10.1109/THFE2.1960.4503259

McGuire, W. J. (1960). A syllogistic analysis of cognitive relationships. In Hovland, C., & Rosenberg, M. (Eds.), *Attitude organization and change*. New Haven, CT: Yale University Press.

Moscato, P. (1989). *On evolution, search, optimization, genetic algorithms and martial arts: Towards memetic algorithms* (Tech. Rep. No. C3P826). Pasadena, CA: California Institute of Technology.

Patel, D., Patel, S., & Wang, Y. (Eds.). (2003, August). *Proceedings of the 2nd IEEE International Conference*, London, UK. Washington, DC: IEEE Computer Society.

Peng, F., Tang, K., Chen, G., & Yao, X. (2010). Population-based algorithm portfolios for numerical optimization. *IEEE Transactions on Evolutionary Computation, 14*(5), 782–800. doi:10.1109/TEVC.2010.2040183

Shastri, L., & Grannes, D. J. (1996). A connectionist treatment of negation and inconsistency. In *Proceedings of the 18th Annual Conference of the Cognitive Science Society*, San Diego, CA (pp. 142-147).

Smith, S. F. (1980). *A learning system based on genetic adaptive algorithms*. Unpublished doctoral dissertation, University of Pittsburgh, PA.

Sun, F., Wang, Y., Lu, J., Zhang, B., Kinsner, W., & Zadeh, L. A. (Eds.). (2010, July). *Proceedings of the 9ᵗʰ IEEE International Conference on Cognitive Informatics*, Beijing, China. Washington, DC: IEEE Computer Society.

Wang, Y. (2002a, August). Keynote: On cognitive informatics. In *Proceedings of the 1ˢᵗ IEEE International Conference on Cognitive Informatics*, Calgary, AB, Canada (pp. 34-42). Washington, DC: IEEE Computer Society.

Wang, Y. (2002b). The Real-Time Process Algebra (RTPA). *Annals of Software Engineering, 14*, 235–274. doi:10.1023/A:1020561826073

Wang, Y. (2003). On cognitive informatics. *Brain and Mind: A Transdisciplinary Journal of Neuroscience and Neurophilosophy, 4*(3), 151-167.

Wang, Y. (2006, July). Keynote: Cognitive informatics - towards the future generation computers that think and feel. In *Proceedings of the 5ᵗʰ IEEE International Conference on Cognitive Informatics*, Beijing, China (pp. 3-7). Washington, DC: IEEE Computer Society.

Wang, Y. (2007a). *Software engineering foundations: A software science perspective (Vol. 2)*. Boca Raton, FL: Auerbach.

Wang, Y. (2007b). The OAR model of neural informatics for internal knowledge representation in the brain. *International Journal of Cognitive Informatics and Natural Intelligence, 1*(3), 64–75. doi:10.4018/jcini.2007070105

Wang, Y. (2007c). The theoretical framework of cognitive informatics. *International Journal of Cognitive Informatics and Natural Intelligence, 1*(1), 1–27. doi:10.4018/jcini.2007010101

Wang, Y. (2008a). On contemporary denotational mathematics for computational intelligence. *Transactions of Computational Science, 2*, 6–29. doi:10.1007/978-3-540-87563-5_2

Wang, Y. (2008b). On concept algebra: A denotational mathematical structure for knowledge and software modeling. *International Journal of Cognitive Informatics and Natural Intelligence, 2*(2), 1–19. doi:10.4018/jcini.2008040101

Wang, Y. (2008c). On system algebra: A denotational mathematical structure for abstract system modeling. *International Journal of Cognitive Informatics and Natural Intelligence, 2*(2), 20–42. doi:10.4018/jcini.2008040102

Wang, Y. (2008d). RTPA: A denotational mathematics for manipulating intelligent and computational behaviors. *International Journal of Cognitive Informatics and Natural Intelligence, 2*(2), 44–62. doi:10.4018/jcini.2008040103

Wang, Y. (2009a). A cognitive informatics reference model of autonomous agent systems (AAS). *International Journal of Cognitive Informatics and Natural Intelligence, 3*(1), 1–16. doi:10.4018/jcini.2009010101

Wang, Y. (2009b). Cognitive computing. *International Journal of Software Science and Computational Intelligence, 1*(3). doi:10.4018/jssci.2009070101

Wang, Y. (2009c). On abstract intelligence: Toward a unified theory of natural, artificial, machinable, and computational intelligence. *International Journal of Software Science and Computational Intelligence, 1*(1), 1–18. doi:10.4018/jssci.2009010101

Wang, Y. (2009d). On cognitive computing. *International Journal of Software Science and Computational Intelligence, 1*(3), 1–15. doi:10.4018/jssci.2009070101

Wang, Y. (2009e). On visual semantic algebra (VSA): A denotational mathematical structure for modeling and manipulating visual objects and patterns. *International Journal of Software Science and Computational Intelligence, 1*(4), 1–15. doi:10.4018/jssci.2009062501

Wang, Y. (2009f). Paradigms of denotational mathematics for cognitive informatics and cognitive computing. *Fundamenta Informaticae, 90*(3), 282–303.

Wang, Y. (2010a). A sociopsychological perspective on collective intelligence in metaheuristic computing. *International Journal of Applied Metaheuristic Computing, 1*(1), 110–128. doi:10.4018/jamc.2010102606

Wang, Y. (2010b). Cognitive robots: A reference model towards intelligent authentication. *IEEE Robotics and Automation, 17*(4), 54–62. doi:10.1109/MRA.2010.938842

Wang, Y. (2010c). On concept algebra for computing with words (CWW). *International Journal of Semantic Computing, 4*(3), 331–356. doi:10.1142/S1793351X10001061

Wang, Y. (2010d). On formal and cognitive semantics for semantic computing. *International Journal of Semantic Computing, 4*(2), 203–237. doi:10.1142/S1793351X10000833

Wang, Y. (in press). On cognitive models of causal inferences and causation networks. International Journal of Software Science and Computational Intelligence, 3*(1).*

Wang, Y., Baciu, G., Yao, Y., Kinsner, W., Chan, K., & Zhang, B. (2010). Perspectives on cognitive informatics and cognitive computing. *International Journal of Cognitive Informatics and Natural Intelligence, 4*(1), 1–29. doi:10.4018/jcini.2010010101

Wang, Y., Carley, K., Zeng, D., & Mao, W. (2007). Social computing: From social informatics to social intelligence. *IEEE Intelligent Systems, 22*(2), 79–83. doi:10.1109/MIS.2007.41

Wang, Y., & Chiew, V. (2010). On the cognitive process of human problem solving. *Cognitive Systems Research: An International Journal, 11*(1), 81–92. doi:10.1016/j.cogsys.2008.08.003

Wang, Y., Johnston, R., & Smith, M. (Eds.). (2002, August). *Proceedings of the 1st IEEE International Conference*, Calgary, AB, Canada. Washington, DC: IEEE Computer Society.

Wang, Y., & Kinsner, W. (2006). Recent advances in cognitive informatics. *IEEE Transactions on Systems, Man, and Cybernetics, 36*(2), 121–123. doi:10.1109/TSMCC.2006.871120

Wang, Y., Kinsner, W., Anderson, J. A., Zhang, D., Yao, Y., & Sheu, P. (2009). A doctrine of cognitive informatics. *Fundamenta Informaticae, 90*(3), 203–228.

Wang, Y., Kinsner, W., & Zhang, D. (2009). Contemporary cybernetics and its faces of cognitive informatics and computational intelligence. *IEEE Transactions on Systems, Man, and Cybernetics, 39*(4), 1–11.

Wang, Y., & Wang, Y. (2006). Cognitive informatics models of the brain. *IEEE Transactions on Systems, Man, and Cybernetics, 36*(2), 203–207. doi:10.1109/TSMCC.2006.871151

Wang, Y., Wang, Y., Patel, S., & Patel, D. (2006). A layered reference model of the brain (LRMB). *IEEE Transactions on Systems, Man, and Cybernetics, 36*(2), 124–133. doi:10.1109/TSMCC.2006.871126

Wang, Y., Zadeh, L. A., & Yao, Y. (2009). On the system algebra foundations for granular computing. *International Journal of Software Science and Computational Intelligence, 1*, 1–17. doi:10.4018/jssci.2009040101

Wang, Y., Zhang, D., Latombe, J.-C., & Kinsner, W. (Eds.). (2008, August). *Proceedings of the 7th IEEE International Conference on Cognitive Informatics.* Washington, DC: IEEE Computer Society.

Wang, Y., Zhang, D., & Tsumoto, S. (2009). Cognitive informatics, cognitive computing, and their denotational mathematical foundations (I). *Fundamenta Informaticae, 90*(3), 1–7.

Weise, T. (2009). *Global optimization algorithms – theory and application.* Retrieved from http://www.it-weise.de/

Weise, T., Podlich, A., Reinhard, K., Gorldt, C., & Geihs, K. (2009). Evolutionary freight transportation planning. In M. Giacobini, A. Brabazon, S. Cagnoni, G. Di Caro, A. Ekart, A. Esparcia et al. (Eds.), *Proceedings of the EvoWorkshops on Applications of Evolutionary Computing* (LNCS 5484, pp.768-777).

Widrow, B., & Aragon, J. C. (2010). Cognitive memory: Human like memory. *International Journal of Software Science and Computational Intelligence, 2*(4), 1–15. doi:10.4018/jssci.2010100101

Yao, Y. Y., Shi, Z., Wang, Y., & Kinsner, W. (Eds.). (2006, July). *Proceedings of the 5th IEEE International Conference*, Beijing, China. Washington, DC: IEEE Computer Society.

Zadeh, L. A. (1975). Fuzzy logic and approximate reasoning. *Syntheses, 30*, 407–428. doi:10.1007/BF00485052

Zadeh, L. A. (1999). From computing with numbers to computing with words – from manipulation of measurements to manipulation of perception. *IEEE Transactions on Circuits and Systems I, 45*(1), 105–119. doi:10.1109/81.739259

Zhang, B. (2010, July). Computer vision vs. human vision. In *Proceedings of the 9th IEEE International Conference on Cognitive Informatics*, Beijing, China (p. 3). Washington, DC: IEEE Computer Society.

Zhang, D. (2005). Fixpoint semantics for rule base anomalies. In *Proceedings of the 4th IEEE International Conference on Cognitive Informatics*, Irvine, CA, (pp.10-17). Washington, DC: IEEE Computer Society.

Zhang, D. (2007). Fixpoint semantics for rule base anomalies. *International Journal of Cognitive Informatics and Natural Intelligence, 1*(4), 14–25. doi:10.4018/jcini.2007100102

Zhang, D. (2008). Quantifying knowledge base inconsistency via fixpoint semantics. In M. Gavrilova, C. J. K. Tan, Y. Wang, Y. Yao, & G. Wang (Eds.), *Transactions on Computational Science II* (LNCS 5150, pp. 145-160).

Zhang, D. (2009). Taming inconsistency in value-based software development. In *Proceedings of the 21st International Conference on Software Engineering and Knowledge Engineering*, Boston, MA (pp.450-455).

Zhang, D., Wang, Y., & Kinsner, W. (Eds.). (2007, August). *Proceedings of the 6th IEEE International Conference on Cognitive Informatics.* Washington, DC: IEEE Computer Society.

This work was previously published in the International Journal of Cognitive Informatics and Natural Intelligence, Volume 5, Issue 1, edited by Yingxu Wang, pp. 1-17, copyright 2011 by IGI Publishing (an imprint of IGI Global).

Chapter 3
Role–Based Human–Computer Interactions

Haibin Zhu
Nipissing University, Canada

Ming Hou
Defence Research and Development Canada (DRDC) - Toronto, Canada

ABSTRACT

With increased understanding of cognitive informatics and the advance of computer technologies, it is becoming clear that human-computer interaction (HCI) is an interaction between two kinds of intelligences, i.e., natural intelligence and artificial intelligence. This paper attempts to clarify interaction-related terminologies through step-by-step definitions, and discusses the nature of HCI, arguing that shared models are the most important aspect of HCI. This paper also proposes that a role-based interaction can be taken as an appropriate shared model for HCI, i.e., Role-Based HCI.

1. INTRODUCTION

Interaction is an important element of social activity and is a key aspect of cognitive informatics. Good interaction skills may assist people who are striving for success in their careers. Conversely, individuals who lack strong interaction skills may suffer adverse effects throughout their lives.

HCI has been a topic of research for nearly half a century (Myers, 1988), and has been studied from various vantage points. For example, a large body of evolutionary interaction research currently exists, and the research includes empirical studies, sensors, input tools, and user models as pertaining to HCI. Notably, the major interaction style currently used is still WIMP (Windows, Icons, Menus, and Pointers). On the whole, however, the advances in these research areas are still disappointing. The growing number of research topics has given rise to special domain characteristics. For example, Human-Robot Interaction (HRI) has been proposed to concentrate on special interaction problems related with robotics (Yanco et al., 2004).

DOI: 10.4018/978-1-4666-2476-4.ch003

There is a definitive lack of research on the fundamental issues associated with interactions, which is a clear shortcoming in HCI research. Many investigators have built concrete tools without clarifying the fundamental concepts, theorems, and principles underlying them. This methodology may produce many evolutionary products and designs, but makes revolutionary improvements very difficult to obtain.

Cognitive informatics examines the fundamental principles of the natural and man-made world as relating to the mind or intelligence. This research clearly affects the development of computer technologies as computers have been designed from their inception to simulate human behaviour (Wang, 2007a, 2007b, 2009a). Thus, investigating the principles, models, and processes of interactions is one of the most important tasks for cognitive informatics.

This paper uses step-by-step definitions to clarify the fundamental issues underlying interactions from the viewpoint of cognitive informatics, and reveals important fundamental concepts for interactions. It proposes different categories of interaction, and argues that a successful interaction is based primarily on shared models. Finally, it argues that roles are a promising shared model for HCI.

The structure of this paper is as follows: Section 2 presents definitions related to intelligence; Section 3 classifies different types of interactions; Section 4 discusses the differences between human-human interactions (HHI) and HCI; Section 5 discusses shared models of interactions; Section 6 presents role-based HCI; Section 7 presents a case study that used role-based HCI methodology; Section 8 reviews related work; and Section 9 presents conclusions and directions for future work.

2. NATURAL INTELLIGENCE AND ARTIFICIAL INTELLIGENCE

Although the term "intelligence" has been in use for a long time, a widely-accepted definition for the term does not seem to exist. Many researchers agree that "intelligence" or "natural intelligence" are terms used to describe a property of a person's mind that encompasses many related abilities, such as the capacities to reason, plan, solve problems, think abstractly, comprehend ideas, use languages, and learn.

According to Dictionary.com (http://dictionary.reference.com/), intelligence is defined as:

1. A capacity for learning, reasoning, understanding, and similar forms of mental activity; aptitude in grasping truths, relationships, facts, meanings, etc.
2. A manifestation of a high mental capacity.
3. The faculty of understanding.
4. Knowledge of an event, circumstance, etc., received or imparted; news; information.
5. The gathering or distribution of information, esp. secret information.
6. An interchange of information.

Furthermore, Artificial Intelligence (AI) is defined as the intelligence of machines and it is a branch of computer science that aims to create intelligence (Boden, 1987; McCarthy, 2010). Russell and Norvig (2003) define AI as "the study and design of intelligent agents" where an intelligent agent is a system that perceives its environment and takes actions that maximize its chances of success. Intelligence may be defined in a gradual way.

Intelligence may be understood by defining the concepts step by step.

Definition 1: *Object.* Everything in the world is an object (Kay, 1993; Zhu & Zhou, 2003, 2006).

Example 1: People and computers are objects.

Definition 2: *Ability.* Ability is what an object can do.

Example 2: People can perceive and learn; computers can store and compute.

Definition 3: *Intelligence or Natural Intelligence (NI).* Intelligence refers to a whole set of all the abilities possessed by people (Lindblom & Ziemke, 2003; Wang, 2007b).

Example 3: Listening and speaking are elements of NI.

Definition 4: *Individual Intelligence (I^2) or Individual Natural Intelligence.* Individual intelligence is the intelligence presented by an individual person.

Example 4: If a person has good memorization skills and is able to learn quickly, we say that he or she has a high level of intelligence. The Intelligence Quotient (IQ) (Weinberg, 1989) is a widely accepted measure of I^2.

Definition 5: *Known Intelligence (KI).* Known Intelligence (KI) is the intelligence that is known by people.

It includes all of the abilities that people can represent using natural languages. Common intelligence is the KI possessed by most people.

The contents of commonly offered courses in schools are elements of KI—e.g., calculus, mechanics, and statistics.

Definition 6: *Abstract Intelligence (αI).* Abstract Intelligence (αI) is known intelligence that is expressed by formal mathematical tools (Wang, 2009a, 2009b). It can also be referred to as Expressed Intelligence.

Example 6: Newton's laws, Ohm's law and chemical reaction equations are elements of αI, as they are formally expressed.

Definition 7: *Artificial Intelligence (AI).* Artificial Intelligence (AI) is the intelligence possessed by a created object and created by a group of people according to αI, i.e., AI is the implemented αI.

Example 7: A computing algorithm to compute the first cosmic velocity implemented in a computer is such intelligence. An expert system that diagnoses motor problems also represents AI.

From the above definitions, the relationships among them are shown in Figure 1 and expressed by sets as:

1. $I^2 \subset NI$;
2. $KI \subset NI$;
3. $\alpha I \subset KI$;
4. $AI \subset \alpha I$;
5. $I^2 \neq KI$;
6. $I^2 \neq \alpha I$; and
7. $I^2 \neq AI$.

We know, for instance, that understanding is an ability that belongs to KI, but we cannot express it with abstract mathematical tools completely, and therefore we cannot implement it completely.

Real-time algebra (RTA) (Wang, 2007a, 2007b, 2009a, 2009b) may be used to express some intelligence items, but even with this method, intelligence cannot be completely implemented. Logic inference is one type of intelligence: it can

Figure 1. The relationships among different intelligences

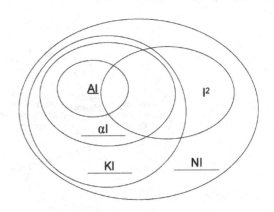

be expressed with first-order logic, and can be implemented as a logic system, i.e., PROLOG.

3. INTERACTION

Interaction is a reciprocal action, effect, or influence. Interaction includes communication which is: (1) the act or process of communicating; the fact of being communicated; (2) the imparting or interchange of thoughts, opinions, or information by speech, writing, or signs; and (3) something imparted, interchanged, or transmitted (Hollnagel, 2000).

The fundamental mechanism for interaction is communication. "Communication requires that information be: (1) encoded, (2) transmitted, (3) received, (4) decoded, and (5) understood" (Albus, 1991).

Based on Wang's General Intelligence Model (GIM) (Figure 2) (Wang, 2007a), it can be said that communication belongs to the first step of interaction. There is one information processing step from Stimuli Enquiries to the Short Buffer Memory and one information processing step to present behaviors with the information in the Behaviour Action Buffer Memory. Therefore, we have the following definitions.

Definition 8: *Communication.* Communication between objects is a process of exchanging information.

Definition 9: *Interaction.* Interaction is a process where one object affects another object's abilities through its own abilities.

For example, in the interaction between persons A and B, A hopes to convince B to buy A's product. In an interaction between a person A and a computer C, A compiles a file and stores the file into C's Disk (long term memory).

3.1 Characteristics of InteraCtion

An interaction includes communication, but communication does not have to be an interaction. This is because interaction involves more than just communication. Here, interaction is conducted through abilities. Therefore, objects may interact because objects have abilities.

An interaction should have the following characteristics:

- **Concision and Simplicity:** The two parties should provide simple interfaces in order to interact with one another. This is the rubric for understanding.

In the business world, purchasing involves a simple interaction, i.e., money is exchanged for

Figure 2. Wang's cognitive model

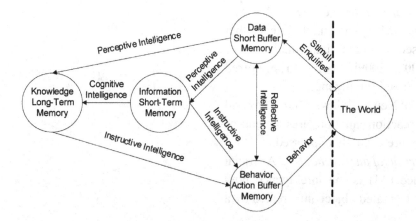

commodities. If every day purchases required one to sign a long contract, retail services would be impractical.

- **Understandability:** Each party should understand the other's scope, process, and goal.

If one side cannot understand the other, the interaction may not continue, even if communication has already been established. For example, language differences may cause a sale transaction to fail if the buyer and seller cannot understand one another.

- **Comfort:** The two sides should feel comfortable with one another.

If either side is uncomfortable, communication may not be established and, therefore, no interaction will occur. Research on human factors considers this principle when designing products such as, chairs, tables, stairs, etc (Hollnagel, 2000). That is, when designing these products, researchers keep in mind that users must be able to use the products safely, effectively, and easily. For example, a person sitting on a comfortable chair is taking part in a better interaction than a person who is sitting on an irregular stone.

3.2 Classification of Interactions

As gathered from the above definitions, interaction is a reciprocal activity between two capable objects. When a human is involved in an interaction, intelligence is required. Interactions involving humans are therefore different due to the special intelligences possessed by the participants in the interaction.

Definition 10: *Individual Intelligences Interaction (I^3)*. I^3 is a process where a person communicates with another person through his/her intelligence. It is also called Human-Human Interaction (HHI).

Example 10: A person may request of another person to explain why he or she was late.

Definition 11: *Artificial and Individual Intelligences Interaction (AI^3)*. AI^3 is a process where a person communicates with a man-made object by using his/her intelligences.

Example 11: A person may request a DVD player to play a video.

Definition 12: *Human-Machine Interaction (HMI)*. HMI is a process where a person interacts with a machine through his/her intelligence. HMI can be included in AI^3

Example 12: A person may require that a car needs to be moved.

Definition 13: *Human Computer Interaction (HCI)*. HCI is a process where a person interacts with a computer through his/her intelligence. HCI can be included in HMI.

Example 13: A person may request a computer to print a file.

Definition 14: *Human Robot Interaction (HRI)*. HRI is a process where a person interacts with a robot through his/her intelligence.

Example 14: A person may request a robot to collect data.

Definition 15: *Individual Interactions through Computers (I^2C)*. I^2C is a process by which a person interacts with another person through computers.

Example 15: A person may email a request to another person asking that a report be written.

It is worth noting that there is no Natural Intelligences Interaction (NI^2) because no single object in the world may possess the intelligence of all humans. To analyze I^3, understanding different intelligences serves as a starting point.

From Figure 3, interactions can be categorized as follows:

Figure 3. Different interactions

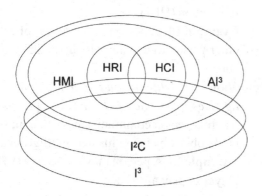

- I³ (also called HHI) occurs between humans where beliefs and trust are of significant importance as the context changes.

4. THE DIFFERENCES BETWEEN HCI AND AI³

From the definitions and classifications in Sections 2 and 3, HCI typically operates in a relative static or well-defined context based on assumptions, rules, and procedures. HCI differs from HHI (or I³) as the latter is dynamic and its context is sometimes unknown a priori. Although there are some overlaps between HCI and AI³, as AI is usually embedded in a computer, there are still some major differences between the two as highlighted in Table 1. Because of these differences, existing HCI research is likely still relevant but it is not sufficient for AI³ design. If designers are using current standards to design AI³, users may find these systems uncomfortable. A fresh look at AI³ may capture new principles or guidelines that do not appear in current standards. Thus, there is a strong need to develop guidelines to meet AI³ design challenges.

- HMI occurs between a human and a machine that is more than just a computer, operating in a fairly well-known environment.
- HCI occurs between a human and a computer typically in a sedentary environment.
- HRI occurs between a human and a robot in a dynamic environment.
- AI³ and I²C have overlaps with I³. They may improve I³ in some aspects by special AI mechanisms.

Table 1. Differences between HCI and AI³

Interaction Aspects	HCI	AI³
Predictability	Plans, actions, and system states are known within limits.	Plans and actions are not known a priori, and may produce unexpected system states.
Procedures	It is specific, systematic, and often associated with Standard Operating Procedures.	It is fuzzy and there may be many means to achieve the same end.
Belief	The human has beliefs (assumptions) about the machine and task. The machine design takes into consideration certain assumptions about the human and task.	The user has beliefs about the system and tasks. The system may have equivalent beliefs about the user and tasks.
Trust	Typically, trust is binary—the machine works or does not.	Trust must be built over time since there is a limited number of definite ways to judge whether the agent is performing well.
Levels of autonomy	Typically, there are two levels of autonomy—completely manual or completely automatic.	There are multiple levels of autonomy.
Awareness	The human knows how the system will process/display information.	The user does not necessarily know how the system will process/display information.
Context	The context is typically static or well-defined.	The context is dynamic and typically unknown a priori.

Figure 4. The different mental models between two persons (Credits to an Unknown Cartoonist)

Furthermore, AI[3] should be more human-like, i.e., similar to I[3], and it should simulate some forms of human behavior for effective interaction. These might include physiological attributes (i.e., eyes and other body parts), intellectual characteristics (e.g., capacity, recognition, learning, and decision), knowledge basis (knowing the environment, system, task, user, etc.), and psychological states (e.g., concentration, vigilance, fatigue, and patience).

Table 1 outlines the differences between HCI and AI[3] (Hou et al., 2011).

5. SHARED MODELS FOR INTERACTIONS

Interaction is not an easy task in many cases. In fact, I[3] (or HHI) is not always successful (Figure 4). People may fear interactions, as in society, misunderstandings can produce conflict, hate, and even war. As for AI[3], its major problem is still misunderstanding. Many humorous examples of the interactions between humans and computers (Hubpages, 2010; RinkWorks, 2010) show the different thinking models between human users and computers. Therefore, shared models

are required to provide a useful technique for supporting the properties that are required for a successful interaction.

Definition 16: *Shared Model.* A shared model is a model used by both parties of an interaction.

The shared model makes an interaction concise and simple. For example, even though mathematicians interact with one another when trying to solve a very complicated mathematical problem, their interactions are concise because they share a lot of mathematical models.

The shared model makes an interaction understandable. Again, from the example of mathematicians, a difficult problem is understandable between two parties because they share the same models.

The shared model makes an interaction comfortable. For example, musicians interact with one another comfortably through the use of musical symbols, i.e., a shared model. Mathematicians interact comfortably through the use of formal systems, i.e., a shared model.

A successful outcome may result from a model that is shared by both sides of an interaction

Figure 5. Norman's cognitive model (Norma, 1986)

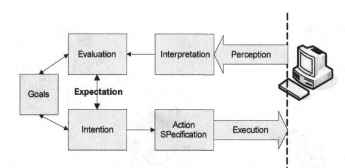

(Craik, 1943). For example, when one person asks another person for directions, it is easy to make this interaction successful if there is a map, i.e., the shared model. However, if there is no map at hand, the individuals involved lose the shared model and their interaction may be unsuccessful as shown in Figure 4.

Based on the cognitive model of Norman (1986) (Figure 5), there should be a general shared model in AI. Therefore the computer icon should be replaced by the cognitive process as shown in Figure 6. Based on the Generic Intelligence Model (GIM) of Wang (2007a) (Figure 2), there should be an interaction as shown in Figure 7.

Ideal interactions should have a symmetric sharing model, i.e., both parties share the same model and information flow is easily regulated (Figures 6 and 7). This is actually the interaction between αI and I^2, because a computer is built with αI and a human user possesses I^2.

In Figures 6 and 7, we hope to express an ideal way for two parties to interact, i.e., the two parties should have the same models not only for action specification and interpretation (Figure 6) or stimuli enquires and behavior (Figure 7) but also for the internal processing. For example, two normal persons with the same five senses (which are physical models for stimuli, behaviors, actions and interpretations) may fail in an interaction because they have different models for processing knowledge as based on each individual's unique prior training.

However, models may not reflect natural intelligence. As stated by Albus (1991), "much is unknown about intelligence, and much will remain beyond human comprehension for a very long time". Thus, the reality of interactions between AI and NI (or more exactly, AI^3) will be the case for a long time as shown in Figure 8. That is, computer-based systems do not have the same models as do human beings. The reasons for this are as follows:

1. Every individual is different.
2. Models in an individual mind may be different from those in a computer.
3. A model may not reflect the real NI, i.e., $\alpha I \subset NI$.
4. Even though a model is good for the NI, it may not be good for the I^2, i.e., $I^2 \subset NI$.
5. Even though a model is good for the NI, it may not have been completely implemented by people, i.e., $\alpha I \subset KI \subset NI$.
6. The two mentioned models from Norman and Wang are very abstract, and it is difficult for a computer to operate and implement these in a way that an individual can readily follow.

The normal situation for HCI is that individual models interact with a partial implementation of the man-made model. Therefore, it is still difficult to implement the interaction model shown in Figure 8 because the question marks express

Figure 6. Ideal interaction based on Norman's model

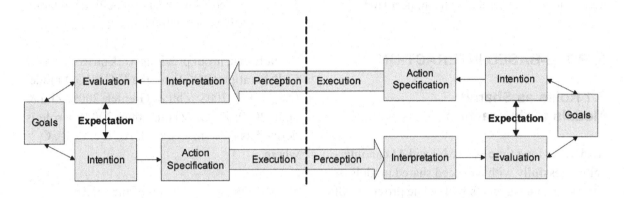

Figure 7. Ideal interaction based on Wang's model

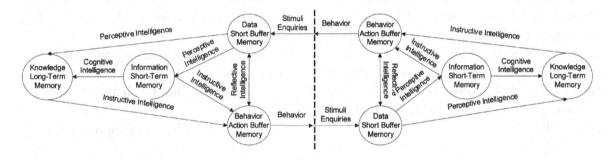

Figure 8. The asymmetric shared model

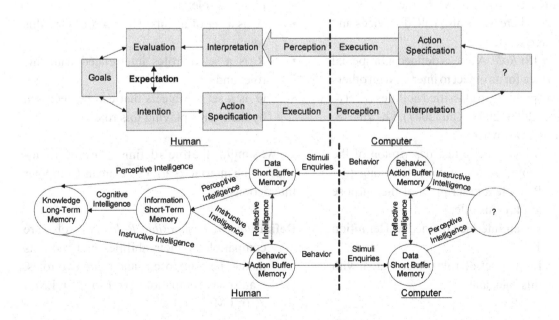

unknown processes. Thus, it is imperative that better models be developed to support HCI.

6. ROLE-BASED INTERACTION

6.1 Roles as Shared Models for Interaction

It will be a long time before NI and AI can interact successfully with balanced shared models. A challenge to researchers will be the provision of simple models that human users and computers can share.

Roles enable individuals to easily set interface preferences that are relevant to their jobs. Through roles, common interfaces for specialized information can be designed thus filtering out a great deal of irrelevant information. The use of roles provides both abstraction and analysis of complex system behaviors (Zhu, 2007, 2008; Zhu & Grenier, 2009; Zhu & Hou, 2009; Zhu & Zhou, 2006, 2008).

Definition 17: *Interface.* An interface is defined as a list of incoming messages (abilities or services) and outgoing messages (requests) of an object to interact with others, $I::=<M_{in}, M_{out}>$, where M_{in} is also called services and M_{out} requests.

Definition 18: *Role.* A role is defined as a special interface for an object to interact with others corresponding to classified abilities (Zhu & Zhou, 2006, 2008; Zhu, 2007), i.e., $r::=<S, P, I, R_x, O>$, where:

○   S is a concise text description of the subject to which this role belongs;

○   P is a text description of the purpose to play this role;

○   I is an interface defined in Definition 17;

○ R_x is a set of roles interrelated with this role; and

○   O a set of objects for role players to access in order to facilitate interactions through the interface I.

Note that, in our previous work on Role-Based Collaboration (RBC) and the E-CARGO model (Zhu, 2007, 2008; Zhu & Grenier, 2009; Zhu & Hou, 2009; Zhu & Zhou, 2006, 2008), a role is defined as $r::= <n, I, s, \alpha, \beta, A_c, A_p, A_o, R, O_r>$ where,

*   n is the identification of the role;
* $I::= < M_{in}, M_{out} >$ denotes a set of messages, where M_{in} expresses the incoming messages to the relevant agents, and M_{out} expresses a set of outgoing messages or message templates to roles, i.e., $M_{in}, M_{out} \subset M$;
*   s is the qualification (or called credit) requirement for an agent to play this role;
* α is the time limit for an agent to play this role;
* β is the space limit for an agent to play this role.
* A_c is a set of agents who are currently playing this role;
* A_p is a set of agents who are qualified to play this role;
* A_o is a set of agents who used to play this role;
* R_i is a set of roles interrelated with this role; and
* O_r is a set of objects that can be accessed by the agents playing this role.

We simplify the role's definition in order to emphasize the interface part, i.e., Human-Computer Interactions.

Definition 19: *Compatible Role.* Two roles are compatible if their abilities and requests match, i.e., suppose r_0 and r_1 are two roles, r_0 and r_1 are compatible if $(r_0.I.M_{out} \subseteq r_1.I.M_{in}) \wedge (r_1.I.M_{out} \subseteq r_0.I.M_{in})$.

For example, a student may pose a question to a professor and receive an answer. The professor may ask the student to hand in an assignment, and as such, he/she receives the submitted work. Therefore, the roles of student and professor are compatible.

In HCI, a person may request a computer to search for something and to display the result. During this process, the computer may request words input with the keyboard by the user. If the person cannot type the words on the keyboard, the interaction will be unsuccessful, i.e., the roles of the person and computer are not compatible.

Prior to an interaction, people must be aware of their roles and the roles played by their interaction participant(s). Therefore, interaction through sharing compatible roles helps set the stage for successful interaction between NI and AI.

Roles allow participants to be aware of the possibilities that are available through their interaction. For example, a person playing a reader role will not interact with another person who is playing a reporter role. A starting point for a possible interaction would be for the reporter to say "May I ask a question?" to the reader. With this question, the reporter can check if the reader would like to change his/her role to that of a respondent.

From Norman's model, it can be said that requests are the "execution" step. The ability is the process of perception, interpretation, evaluation, expectation, intention, and action specification (Figure 9).

From Wang's GIM model (Figure 2), it can be said that roles specify behavior through their requirements (Figure 9). They help implement perceptive, instructive, and reflective knowledge.

6.2 Scenarios of Role-Based Interaction

RBC is an emerging methodology that mainly uses roles as underlying mechanisms to facilitate an organizational structure, provide orderly system behavior, and consolidate system security for both

Figure 9. Roles as interfaces for interaction

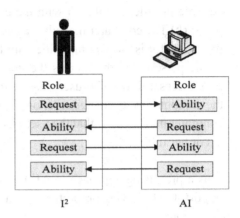

human and non-human entities that collaborate and coordinate their activities with or within systems. It is an approach that can be used to integrate the theory of roles into Computer-Supported Cooperative Work (CSCW) systems and other computer-based systems. It consists of a set of concepts, principles, mechanisms and methods. The properties of RBC are as follows (Zhu & Zhou, 2006): Clear role specification; Regulated role transition; Flexible role facilitation; and Flexible role negotiation. Therefore, RBC is a promising way to facilitate interaction.

Based on RBC (Zhu, 2007, 2008; Zhu & Grenier, 2009; Zhu & Hou, 2009; Zhu & Zhou, 2006, 2008), the scenario of role-based interaction (RBI) is as follows:

1. An individual user classifies his/her knowledge and ability into roles.
2. The computer holds different roles corresponding to the presumed user roles with its role engine (Zhu, 2008, 2010).
3. The user logs in by selecting his or her role.
4. The computer assumes a compatible role to interact with the human user.
5. If the user role is compatible with the role presented by the computer, the interaction continues.

6. If there are some incompatibilities between their roles: (a) The person may change his/her role in order to interact with the computer; (b) The computer may also adjust its role so that it is more compatible with that of the person, and this makes the interaction successful; (c) The user may instruct the computer to its change role; or (d) The computer may suggest that the user change his or her role.

RBI is a promising approach to AI[3] because RBI supports the following properties that facilitate interaction:

- **Abstraction:** Abstraction is an intelligence that allows details to be discarded while retaining major properties.

Roles provide an abstraction for an aspect (including behavior and states) of an object or an individual user, within a limited time slot, for example.

- **Classification:** Classification is an intelligence that groups objects with similar properties. Roles provide different aspects among different roles. Those people playing the same roles can be classified into one group.
- **Adaptation:** Adaptation (Hou et al., 2007) is an intelligence that allows an object to change itself in order to make interactions with others successful (Figure 10).

When a user wants to play the role of an instructor, the computer should adopt the role of student. If the computer plays the role of instructor, the interaction will fail. In contrast, if the user wishes to learn something from the computer, he/she actually plays a student role, and the computer must play the role of an instructor. When a human user opens a computer with a role of professor, what role should the computer play? There might

be three options: a student—the user teaches or inputs some knowledge to the computer; an assistant—the user searches for papers related to his/her research; or a professor—the user consults the computer and request suggestions pertaining to his/her research. These three optional roles must be considered for an effective interaction.

In RBC and role-based interactions, a role is formed directly from requirements. The composition relationship among roles is not taken as a key feature, i.e., one role may be composed of other roles. Role-Based Access Control (RBAC) is more concerned with this issue (Moyer & Ahamad, 2001). Role transfer problems have been investigated by the authors in their previous work. Interested readers may refer to (Zhu & Zhou, 2008, 2009, in press).

7. CASE STUDY: RESTRAIN MENTAL WORKLOAD WITH ROLES

Mental overload is not welcome in HCI. At the lowest level, people feel annoyed and frustrated. At the top level, mental overload is dangerous and critical. Sometimes, it may lead to loss of life, e.g., the mental overload of a pilot may compromise his/her command of the aircraft, and as a result, the aircraft may crash, resulting in casualties. That is why workload management is emphasized in in-vehicle information systems (IVIS) (Wiese & Lee, 2007). Currently, most research concerns *static workload* assessment, which comprises the assessment of workload prior to system application, and the design of the system by assuring that there is no overload. However, there are many situations (referred to as dynamic) where the assessment of workload cannot be predicted. Therefore, there are requirements for investigating how to restrain *dynamic workload* and how to avoid mental overload.

Generally, there are seven methods to deal with overload (Hancock & Chignell, 1988):

- **Omission:** To omit temporarily;
- **Reduced Precision:** To trade off precision in favour of speed and time;
- **Queuing:** To delay responses to some parts;
- **Filtering:** To neglect certain categories of information;
- **Cutting Categories:** To reduce the level of discrimination;
- **Decentralization:** To distribute the task; and
- **Escape:** To abandon the task.

RBC can be applied to restrain dynamic mental overload from all of the above directions by a specially designed role engine, a way to compute the mental workload, and suggested guidelines for interaction designs.

As shown in Figures 9, 10, and 11, a role-based interface can be designed as a set of roles including the current role and potential roles. Based on the principles of RBC, users' mental workload is significantly decreased because the messages are sent to roles but not users. The users do not need to think of the problems of workload balance and the qualifications of users who may accept the messages. All these jobs are completed by the role engine (Zhu, 2008) that is behind the interface. In

this case study, we assume that all the computers are networked and users are involved in I^2C.

Each role is composed of a list of messages that the user can send out, a list of received messages, and a group of objects that the user can access:

- **Incoming Messages:** Each should be accompanied with a weight to show how difficult it is to solve.
- **Outgoing Messages:** Each should be accompanied with a weight based on how difficult it is to prepare the argument objects.
- **Objects to be Accessed:** Each object should have a weight that expresses its level of complexity.

In the design of roles, regulations should be followed to avoid receiving too many messages, having the user send too many messages, and having too many objects to access. Through roles, messages sent to the potential roles will not interrupt the user's concentration on his or her current role, because these messages are buffered for future processing when the human user transfers his current role. Here, a role is a filter for incoming messages.

Figure 10. Role-Based AI³

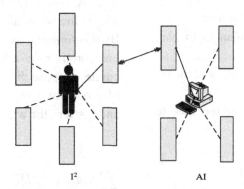

I^2 AI

Note: The rectangle bars denote roles, the solid lines current roles, and the dashed lines potential roles.

Figure 11. A user plays many roles by role transfers

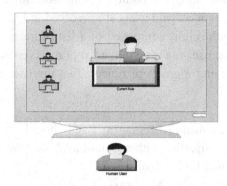

The current role (called a role profile in an interface) should be designed to occupy the whole screen of a monitor. The screen should be evenly distributed to include both messages and objects. To compute the load of a role, not only the direct control is considered but also, implicit messages and objects are kept in mind. For example, an icon for a complex object certainly expresses a heavier load than multiple icons for simple objects; a request message that needs deep thinking has a heavier load than a short request message that just needs a word to answer.

A potential role can be activated to become a current role under conditions that are checked by the role engine. Behind the role profiles, the role engine is designed to monitor the role play of the user from the following aspects:

- Limit the number of roles for the user to play.
- Check the qualification of the user.
- Evaluate the user's performances to play each role.
- Remind the user of the deadline and the effort required to play a role.
- Guarantee the least time for the user to play the current role.

7.1 Computation of the Mental Workload of a User

The overall workload is captured in the following power function (Hancock, 88):

$$w = \frac{1}{et^{s-1}} \tag{1}$$

where w equals the workload level, e is the effort expended by the individual operator, t is the effective time available for action, and s is the skill level of the performer under consideration.

Based on Formula (1), roles can be used to evaluate the effort e. Suppose the effort of a person is a total of 1. The more roles that he or she plays, the less effort he or she can put in each role. An increase in the number of roles may lead to information overload, because it will both increase the information to be considered and reduce the resources available to play the roles. The skills can be evaluated by the percentage of matching between the qualification of a user and the qualification requirement of a role (Zhu & Grenier, 2009), i.e., the q(a, r) in the following Formulas (4).

F may be used as the workload latch for a user (Zhu & Hou, 2009). It is the largest equivalent number of roles for a user to play. It is different for different human users to show that different users have different abilities. Simply for a system, $\forall\, a \in A \rightarrow |a.R_p|+1 \leq F$, where A is the set of users, a.R_p is the set of all the user a's potential roles. By this formula, we mean that user a may access one current role (a.r_c) at a time. To be more accurate, suppose that w(a, r) is the mental weight of user a for role r, n = $|a.R_p|$, $r_i \in R_p$ ($0 \leq i \leq n-1$) and r_n = a.r_c, the whole mental weight F of user a is:

$$\sum_{i=0}^{n} w(a, r_i) \times a.L[i] \,/\, a.\tau[i]) \tag{2}$$

where, $a.L[i] \rightarrow [0, 1]$ is a vector to show the time paid by user a for role r_i, $a.\tau[i] \rightarrow [0, 1]$ is a vector to show the time left for the user to play role r_i. Note: we use $[0, 1]$ to express the real number between 0 and 1 inclusively.

For a well-organized I²C system, there should be:

$$\forall\, a \in A \rightarrow \sum_{i=0}^{n} w(a, r_i) . \times a.L[i]\, /\, a.\tau[i] \leq F$$

(3)

Formula (3) tells that when a user is assigned to a role, the system should check if the user is overloaded to avoid an error-prone system.

Suppose that for role r, $nin = |M_{in}|$, $nout = |M_{out}|$, $nobj = |N_o|$, $M_{in} = \{i\_msg_0, i\_msg_1, ..., i\_msg_{nin-1}\}$, $M_{out} = \{o\_msg_0, o\_msg_1, ..., o\_msg_{nout-1}\}$, and $N_o = \{obj_0, obj_1, ..., obj_{nobj-1}\}$, $w(a, r)$ (i.e., the weight role r for user a) can be computed as follows:

$$w(a, r) =$$
$$k(a, r) \left(\sum_{i=0}^{nin-1} i\_msg_i \cdot w + \sum_{i=0}^{nout-1} o\_msg_i . w + \sum_{i=0}^{nobj-1} obj_i . w \right)$$
$$/\, q(a, r)$$

(4)

where,

$$k\,(a, r) = \begin{cases} 1 \\ \quad if | r.A_c | \leq r.l; \\ 0 \end{cases}$$

else.where $r.A_c$ denotes all the users currently playing role r, and r.l means the least number of users currently playing this role (Zhu & Hou, 2009; Zhu & Zhou, 2006).

- $q(a, r)$ is the qualification of user *a* for role *r (Zhu 09)*.

The followings help understand Formula (4):

- If a user is critical for a role, i.e., $k(a, r) = 1$, the role is definitely a heavy load for the user. If a user is not a critical user for a role, $k(a, r) = 0$, there is no mental weight for the user when concerning about that role.
- If a user is well-qualified for a role, the weight of this role for this user is low, i.e., it is an easy job for the user to play this role.
- The mental workload becomes larger when a user accesses more complex objects, send more difficult messages and accept more difficult messages.

Based on the above interface design, there are several steps that we can take in order to decrease the workload of a user:

- Limit the quantity of the messages that are outlined by a role.
- Limit the number of roles played by a user.
- Choose highly qualified users to play a role.
- Remind the users to discard some roles when a threshold is reached.
- Assist the users when scheduling roles to decrease the time pressures of roles.

7.2 Analysis of the Weight of Messages and Objects

From Section 7.1, the mental workload for a user is the sum of the roles' weights. A role's weight is computed based on the message weight and object weight. The question is: How can the weight of a message or an object be computed? This question must be answered in order to evaluate workload dynamically.

Based on standard problem solving and software development methods, decomposition is used to analyze the weight of a message or an object. Let sp be the span base of a tree hierarchy of decomposition. A message can be functionally decomposed by dividing it into sp sub-messages;

each sub-message can be further divided into sp sub-messages, ...; at last, each message can be processed directly. An object can be decomposed in the same way. The largest span number is set at 7 due to the fact that in many organizations, 5, 7, and 9 are the normal numbers for a board. Odd numbers are mainly used to make a decision based on majority. A chief of a board interacts mainly with other board members.

In decomposition, depth is also important. The levels (denoted as lv) of an object (or a message) are defined as: a level 0 object is a primitive object (message) that is not separable; level m object (message) is composed of sp level m-1 objects (message) ($lv \geq 0$, $1 \leq m \leq lv$).

As derived from the administrative style of our society, the largest number of levels lv of a message or an object is set at 3. For example, in a software company, a highly ranked administrator (e.g., a CEO) would not directly request an entry level programmer to complete a task, he or she may only interact with project managers (two levels lower than him or her). Therefore, three types of messages or objects are normally used, simple (lv=0), intermediate (lv=1) and complicated (lv =2).

The normal workload of a message or an object w should range between 1 (7^0) and 49 (7^2), $1 \leq w \leq 49$. If *w* is out of this range, the corresponding message or object is overloaded.

8. RELATED WORK

There is limited body of research that considers roles as a tool for HCI.

Albus (1991) systematically discusses his model, which integrates knowledge from research in both NI and AI. Many of his statements are still true today after 20 years. His model connects human factors with HCI including sensing, perception, value judgment, emotions, and machine learning. However, the importance of roles in NI, AI, and HCI is ignored.

Bales (2009) analyzes the process of interactions in small groups. He believes that there are 12 actions related with interactions including 3 positive, 6 neutral, and 3 negative categories. His research suggests that interaction processes are related to emotions and physical tasks. He also emphasizes that interaction occurs prior to communication.

Lebie et al. (1995) present a comparative description of interactions in computer-mediated communication (CMC) and face-to-face (FTF) groups. They conclude that the major difference between CMC and FTF is a pattern that strongly supports our statement pertaining to shared models.

Olson and Olson (1990) review cognitive models related with HCI. These researchers assert that there are still significant gaps in our understanding of the whole process of interacting with computers and that cognitive modeling is useful in initial design, evaluation, and training. They also believe that current cognitive models fail to capture the user's fatigue, individual differences, or mental workload (Hancock & Chignell, 1988; Zhu & Hou, 2009). This statement provides stimulation for our trial to use roles as a new shared model for HCI.

Ritter et al. (2000) state that cognitive models are computer programs that simulate the cognitive skills of humans. They show how models can interact with interfaces using an interaction mechanism that has been designed to apply to all interfaces generated within a User Interface Management System (UIMS). Their idea of providing a UIMS is similar to that of the proposed role-based interaction, i.e., roles as a tool to create interfaces.

As for interaction among agents, Cabri et al. (2003) propose a model for agent interactions based on the concept of a role. In their method, a role is modeled as a set of capabilities (actions) and an expected behavior (events). The translation from actions to events enables interactions between agents. This model achieves several advantages in terms of the separation of concerns, agent-oriented features, independence of communication mechanisms, and promotion of locality in interactions.

9. CONCLUSION

This paper argues that the nature of HCI is actually the interaction between NI and AI. Therefore, more research in this direction will be beneficial for cognitive informatics. Such research may provide a new approach that can be used in the challenging research field of HCI.

As based on the discussion in this paper, we can also state that the most meaningful research in HCI is to *understand how people interact* which is one major task of the research on cognitive informatics. The design of high-quality interactions will only be possible when we completely understand how people interact, in addition to understanding the advantages and the drawbacks of I³ or human-human interactions.

As an initial solution, from the discussed methodology, we propose that a shared model is the key to successful interactions, and furthermore, and we argue that roles are promising shared models for HCI.

Future work should focus on the abstraction of roles, interaction process models between roles, adaptation between role models and human users (i.e., user models), and the provision of an available role engine.

ACKNOWLEDGMENT

This research is supported in part by IBM Eclipse Innovation Grant and Defence Research and Development of Canada. Thanks also go to Mike Brewes of Nipissing University and Paul Bulas of DRDC – Toronto for their assistance in editing this article.

REFERENCES

Albus, J. S. (1991). Outline for a theory of intelligence. *IEEE Transactions on Systems, Man, and Cybernetics, 21*(3), 473–510. doi:10.1109/21.97471

Bales, R. F. (2009). Interaction process analysis. K. Krippendorff & M. A. Bock (Eds.), *The content analysis reader* (pp. 75-83). London, UK: Sage.

Boden, M. A. (1987). *Artificial intelligence and natural man* (2nd ed.). New York, NY: Basic Books.

Cabri, G., Leonardi, L., & Zambonelli, F. (2003, January). Implementing role-based interactions for internet agents. In *Proceedings of the Symposium on Applications and the Internet* Orlando, FL (pp. 380-387).

Craik, K. J. W. (1943). *The nature of explanation*. Cambridge, UK: Cambridge University Press.

Hancock, P. A., & Chignell, M. H. (1988). Mental workload dynamics in adaptive interface design. *IEEE Transactions on Systems, Man, and Cybernetics, 18*(4), 647–658. doi:10.1109/21.17382

Hollnagel, E. (2000). Modeling the orderliness of human actions. In Sarter, N. B., & Amalberti, R. (Eds.), *Cognitive engineering in the aviation domain* (pp. 65–98). Mahwah, NJ: Lawrence Erlbaum.

Hou, M., Kobierski, R. D., & Brown, M. (2007). Intelligent adaptive interfaces for the control of multiple UAVs. *Journal of Cognitive Engineering and Decision Making, 1*(3), 327–362. doi:10.1518/155534307X255654

Hou, M., Zhu, H., Zhou, M., & Arrabito, G. R. (2011). Optimizing operator-agent interaction in intelligent adaptive interface design: A conceptual framework. *IEEE Transactions on Systems, Man and Cybernetics. Part C, Applications and Reviews, 41*(2), 161–178. doi:10.1109/TSMCC.2010.2052041

Hubpages. (2010). *Funny stories about computers*. Retrieved from http://hubpages.com/hub/about-funny-stories-2

Kay, A. (1993). The early history of smalltalk. *ACM SIGPLAN Notice, 28,* 69–95. doi:10.1145/155360.155364

Lebie, L., Rhoads, J. A., & McGrath, J. H. (1995). Interaction process in computer-mediated and face-to-face groups. *Computer Supported Cooperative Work, 4*(2-3), 127–152. doi:10.1007/BF00749744

Lindblom, J., & Ziemke, T. (2003). Social situatedness of natural and artificial intelligence: Vygotsky and beyond. *Adaptive Behavior, 11*(2), 79–96. doi:10.1177/10597123030112002

McCarthy, J. (2010). *What is artificial intelligence?* Retrieved from http://www-formal.stanford.edu/jmc/whatisai/whatisai.html

Moyer, M. J., & Ahamad, M. (2001, April). Generalized role-based access control. In *Proceedings of the 21ˢᵗ IEEE International Conference on Distributed Computing Systems*, Mesa, AZ (pp. 391-398).

Myers, B. A. (1998). A brief history of human computer interaction technology. *Interactions (New York, N.Y.), 5*(2), 44–54. doi:10.1145/274430.274436

Norman, D. A. (1986). Cognitive engineering. In Norman, D. A., & Draper, S. W. (Eds.), *User centered system design: New perspectives on human-computer interaction* (pp. 32–65). Mahwah, NJ: Lawrence Erlbaum.

Olson, J. R., & Olson, G. M. (1990). The growth of cognitive modeling in human-computer interaction since GOMS. *Human-Computer Interaction, 5*(2), 221–265. doi:10.1207/s15327051hci0502&3_4

RinkWorks. (2010). *Computer stupidities*. Retrieved from http://www.rinkworks.com/stupid/

Ritter, F. E., Baxter, G. D., Jones, G., & Young, R. M. (2000). Supporting cognitive models as users. *ACM Transactions on Computer-Human Interaction, 7*(2), 141–173. doi:10.1145/353485.353486

Russell, S. J., & Norvig, P. (2003). *Artificial intelligence: A modern approach* (2nd ed.). Upper Saddle River, NJ: Prentice Hall.

Wang, Y. (2007a, August). Cognitive information foundations of nature and machine intelligence. In *Proceedings of the 6th IEEE International Conference on Cognitive Informatics*, Lake Tahoe, CA (pp. 3-12).

Wang, Y. (2007b). The theoretical framework of cognitive informatics. *International Journal of Cognitive Informatics and Natural Intelligence, 1*(1), 1–27. doi:10.4018/jcini.2007010101

Wang, Y. (2009a). A cognitive informatics reference model of autonomous agent systems (AAS). *International Journal of Cognitive Informatics and Natural Intelligence, 3*(1), 1–16. doi:10.4018/jcini.2009010101

Wang, Y. (2009b). On abstract intelligence: Toward a unifying theory of natural, artificial, machinable, and computational intelligence. *International Journal of Software Science and Computational Intelligence, 3*(1), 1–17. doi:10.4018/jssci.2009010101

Weinberg, R. A. (1989). Intelligence and IQ: Landmark issues and great debates. *The American Psychologist, 44*(2), 98–104. doi:10.1037/0003-066X.44.2.98

Wiese, E. E., & Lee, J. D. (2007). Attention grounding: A new approach to in-vehicle information system implementation. *Theoretical Issues in Ergonomics Science, 8*(3), 255–276. doi:10.1080/14639220601129269

Yanco, H. A., Drury, J. L., & Scholtz, J. (2004). Beyond usability evaluation: Analysis of human-robot interaction at a major robotics competition. *Human-Computer Interaction, 19*(1), 117–149. doi:10.1207/s15327051hci1901&2_6

Zhu, H. (2007). Role as dynamics of agents in multi-agent systems. *System and Informatics Science Notes*, *1*(2), 165–171.

Zhu, H. (2008, May). Fundamental issues in the design of a role engine. In *Proceedings of the International Symposium on Collaborative Technologies and Systems*, Irvine, CA (pp. 399-407).

Zhu, H. (2010). Role-based autonomic systems. *International Journal of Software Science and Computational Intelligence*, *2*(3), 32–51. doi:10.4018/jssci.2010070103

Zhu, H., & Grenier, M. (2009, July). Agent evaluation for role assignment. In *Proceedings of the IEEE 8th International Conference on Cognitive Informatics*, Hong Kong, China (pp. 405-411).

Zhu, H., & Hou, M. (2009, September). Restrain mental workload with roles in HCI. In *Proceedings of the IEEE Toronto International Conference – Science and Technology for Humanity*, Toronto, ON, Canada (pp. 387-392).

Zhu, H., & Zhou, M. C. (2003, November). Methodology first and language second: A way to teach object-oriented programming. In *Proceedings of the Educator's Symposium on Object-Oriented Programming, Systems, Languages and Applications*, Anaheim, CA (pp. 140-147).

Zhu, H., & Zhou, M. C. (2006). *Object-oriented programming with C++: A project-based approach*. Beijing, China: Tsinghua University Press.

Zhu, H., & Zhou, M. C. (2006). Role-based collaboration and its kernel mechanisms. *IEEE Transactions on Systems, Man and Cybernetics. Part C*, *36*(4), 578–589.

Zhu, H., & Zhou, M. C. (2008). Role mechanisms in information systems - a survey. *IEEE Transactions on Systems, Man and Cybernetics. Part C*, *38*(6), 377–396.

Zhu, H., & Zhou, M. C. (2008). Role transfer problems and their algorithms. *IEEE Transactions on Systems, Man, and Cybernetics. Part A, Systems and Humans*, *38*(6), 1442–1450. doi:10.1109/TSMCA.2008.2003965

Zhu, H., & Zhou, M. C. (2009). M–M role-transfer problems and their solutions. *IEEE Transactions on Systems, Man, and Cybernetics. Part A, Systems and Humans*, *39*(2), 448–459. doi:10.1109/TSMCA.2008.2009924

Zhu, H., & Zhou, M. C. (in press). Efficient role transfer. *IEEE Transactions on Systems, Man, and Cybernetics. Part A, Systems and Humans*.

This work was previously published in the International Journal of Cognitive Informatics and Natural Intelligence, Volume 5, Issue 2, edited by Yingxu Wang, pp. 37-57, copyright 2011 by IGI Publishing (an imprint of IGI Global).

Chapter 4
Main Retina Information Processing Pathways Modeling

Hui Wei
Fudan University, China

Qingsong Zuo
Fudan University, China

XuDong Guan
Fudan University, China

ABSTRACT

In many fields including digital image processing and artificial retina design, they always confront a balance issue among real-time, accuracy, computing load, power consumption, and other factors. It is difficult to achieve an optimal balance among these conflicting requirements. However, human retina can balance these conflicting requirements very well. It can efficiently and economically accomplish almost all the visual tasks. This paper presents a bio-inspired model of the retina, not only to simulate various types of retina cells but also to simulate complex structure of retina. The model covers main information processing pathways of retina so that it is much closer to the real retina. In this paper, the authors did some research on various characteristics of retina via large-scale statistical experiments, and further analyzed the relationship between retina's structure and functions. The model can be used in bionic chip design, physiological assumptions verification, image processing and computer vision.

INTRODUCTION

Understanding how biological visual systems cognize and represent the outside world is one of the important and intricate issues in many disciplines, such as physiology, neuroscience, cognitive informatics and artificial intelligence. As a kind of classical natural intelligences, the human visual system can easily and effortlessly handle many visual tasks that are very difficult to accomplish for contemporary computer vision and artificial cognitive model. Many investigations into the internal information processing mechanisms of visual system have shown that the visual natural

DOI: 10.4018/978-1-4666-2476-4.ch004

intelligence originates from the complex structure and parallel computation (Edelman, 1999; Grimson & Grimson, 1981). Therefore, modeling and simulating the human visual system becomes a good solution for some traditional tasks, such as cognition, object recognition and image representation (Palmeri & Gauthier, 2004; Riesenhuber & Poggio, 2000; Wang, 2009). In the human visual system, information is processed along the pathway from the retina through the lateral geniculate nucleus to the visual cortex. Not surprisingly, even in the early visual system, the retina, possessing intricate and delicate hierarchies, transforms and transmits visual information accurately and in real time.

In recent decades, computer vision has been widely applied to target tracking, object detection, position estimation, visual navigation, even missile guidance, and many other fields. However, many computer vision applications are still far from meeting all the requirements especially in real-time and accuracy balance. Meanwhile, artificial retina design confronts the similar balance issue that the chip must be capable of computing very quickly but still maintaining low power consumption and keeping enough accuracy in a limited space. For this issue, nature gives us a good solution. That's our visual system, which not only can meet real-time of information processing but also can ensure high precision of processing, not only can satisfy the different computing load requirements under various environments, but also can maintain low power consumption, not only can deal with a variety of visual tasks, but also can keep a relatively fixed structure. It might be an effective way to model retina's structure to try to solve the problem.

Actually, there have been a number of simulation models of retina structure and function. Since the retina is at the intersection of several subjects, different models may focus on different aspects and applications. These models can broadly be divided into three categories according to their fields and goal. The first category is for neuroscience. These retina models try to establish a model to analyze a number of physiological data to explain some physiological phenomena and verify some assumptions. These models must strictly consist with physiological data and fact; they even simulate ion channels and chemical regulation mechanism (Casti, Hayot, Xiao, & Kaplan, 2008; Dunn, Doan, Sampath, & Rieke, 2006; Niu & Yuan, 2007; Publio, Oliveira, & Roquea, 2006; Roska, Molnar, & Werblin, 2006; Schiller, Slocum, & Weiner, 2007; Shah & Levine, 1996a, 1996b). These models are often too detailed to show the whole retina information processing. The second category is for computer applications. The structure and characteristics of the retina are utilized in computer science fields such as edge detection (Niu & Yuan, 2008; Rong, Yi, & Qiang, 2004; Tang, Sang, & Zhang, 2005), image segmentation (Giorgio, Guido, & Francesco, 1997; Reddick, Glass, Cook, Elkin, & Deaton, 1997; Wu, McGinnity, Maguire, Valderrama-Gonzalez, & Dempster, 2010), motion detection and object tracking (Ishii, Deguchi, & Sasaki, 2004; Lee, Chae, Kim, Kim, & Cho, 2001; Risinger & Kaikhah, 2008), the depth calculation (Shimonomura, Kushima, & Yagi, 2008)and some other applications. Such models try to introduce the physiology idea to solve computer-related issues, not necessarily loyal to the physical structure and information processing of real retina. The third category is for the electronic engineering hardware design. The purpose of these retina models is to design artificial retina chips which can eventually replace real retina (Ahuja, Behrend, Kuroda, Humayun, & Weiland, 2008; Behrend, Ahuja, Humayun, Weiland, & Chowe, 2009; Caspi et al., 2009; Morillas et al., 2007; Wu, Yang, Basham, & Liu, 2008), but as mentioned, they face many problems such as power consumption, real-time, accuracy, compatibility, eyeball movement control, chip size, energy supply, and so on.

This paper presents a simulation retina model, not only simulating various types and distribution of retina cells, but also simulating the different connectivity and complex structure of retina, not only loyal to the physical structure but also covering main information processing pathways. Our results show that the simulation model can exhibit stable high accuracy and real time in processing of visual information just as the real retina does. In addition, experiments on retina performance show that a good balance point among those claims in actual applications can be acquired from the model. Based on this platform, several researches of different fields can be made. First, the model can guide the design of artificial retina chips. A series of quantitative and qualitative analysis can be made on conflicting requirements. Secondly, with the help of this model, we can study retina internal representation of the outside world and corresponding information pathways to process the internal signals. Through the optical-electronic conversion, the retina forms its internal representation of external world. So how is the effectiveness of such a representation? What kind of pathways are in charge of different representation signals? All these questions are very worth studying. Thirdly, the model can verify the theory of the neurobiology and physiology, such as verifying

information processing pathways or local neuron circuit assumption, and even can raise new questions to neurobiology and physiology to promote interdisciplinary development. There are a variety of synapses such as feed-forward synapse, feedback synapse, and horizontal connections synapse. Figure 1 is the overall structure of retina model which is abstracted from physiology and is used as a basis of our computing model.

The model illustrates the information flows in different layers and excitatory and inhibitory effect to different cells. Generally speaking, the retina model includes vertical and horizontal pathways. In the vertical pathway, the optical signal is sampled and converted to electronic signal via the photoreceptor layer, and then passed to the bipolar cell layer, and further transmitted to the ganglion cell layer, in which each layer contains a variety of subtypes dealing with different information. In the horizontal pathway, horizontal cells and amacrine cells can couple or decouple the horizontal connections among cells to regulate the size of the receptive field and strength of response. The horizontal cells and amacrine cells also have many subtypes. Take

Figure 1. Main information processing pathways model of retina

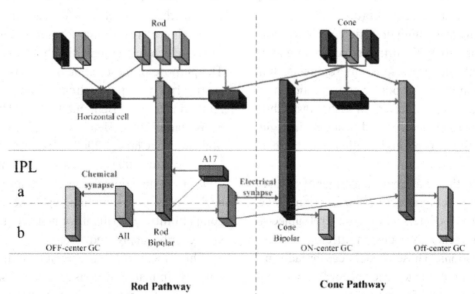

amacrine cells for example; it has been found that there are at least 22 different types. Many visual functions such as motion detection, texture recognition, and brightness adjustment are all relevant to amacrine cell. From function perspective, the retina can also be divided into another two categories: on-pathway and off-pathway which deal with brightness increase and decrease respectively. Based on this information processing diagram of retina, the paper presents a simulation retina model.

DESIGN AND IMPLEMENTATION OF THE RETINA MODEL

In this section, we will present the retina model including several types of cell. Generally, the retina includes photoreceptor cells, horizontal cells, bipolar cells, amacrine cells and ganglion cells, each of which includes a number of subtypes. The synaptic connections among cells constitute a very complex structure.

Simulation on Photoreceptor Layer

Photoreceptor layer samples external optical signals and converts the external information into electronic signals. One feature of the photoreceptor layer is the non-uniform distribution, namely high density in central area, low density in peripheral area. This distribution makes the center of the retina distinguish details, while in peripheral areas it gets contour information and keep awareness. But this distribution will not impact on the visual accuracy, because eyeball control muscles are very flexible and powerful, it make center of retina focus on an interesting thing in a very short period of time. This feature allows the retina to quickly access interesting information, while dramatically reduce information processing workload. This is one of examples of optimized balance between real-time and accuracy of retina.

Figure 2 is a flow chart of photoreceptor layer simulation. Actually, photoreceptor layer simulation is a complex process. Generally speaking, there are three steps. First, types and distribution of different photoreceptors are generated. Secondly, the corresponding wavelength is calculated according to the input RGB image information. Wavelength is the only factor to trigger photoreceptors to fire. Finally, calculating the strength of photoreceptors response based on the photoreceptor-wavelength sensitivity curve (for detailed information, please refer to Guan & Wei, 2009).

Simulation on Horizontal Cells and Bipolar Cells

Horizontal cell layer and bipolar cell layer are highly related to each other so they are put together to simulate. According to physiological studies, it is known that the famous center-surrounding antagonistic receptive field firstly appears in bipolar cells. Photoreceptors directly transmit the signals to bipolar to form "center" while horizontal cells collect photoreceptor signals in a large range and send the inhibitory feedback to bipolar to form "surrounding."

For the center-surrounding receptive field, Rodieck raised his famous DOG (Differences of Gaussian) model, as Figure 3. This model can simulate the center-surrounding receptive field very well. Based on two-dimensional DOG model, the discrete functions are designed. The central response of a bipolar cell is as Formula 1:

$$Center_{exc}(x_0, y_0) =$$
$$\sum_{i=1}^{n} P(x_i, y_i) * 1 / (\sqrt{2\pi} \cdot \sigma_{center}) \cdot e^{-\left((x_i - x_0)^2 + (y_i - y_0)^2\right)/2 \cdot \sigma_{center}^2}$$

$$(1)$$

The surrounding response of a bipolar cell is as Formula 2:

Figure 2. Flow chart of photoreceptors layer simulation

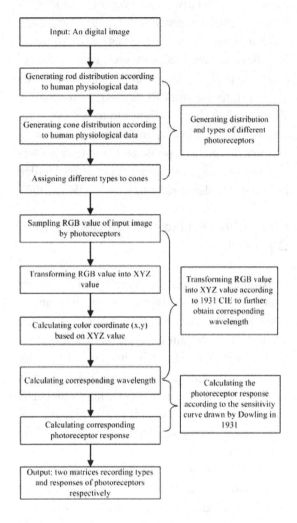

Physiological experiments show that when uniform light covers a certain circle range of a ganglion cell receptive field; the ganglion cell may not react or only have very weak response, because at this time, the excitatory input from photoreceptor and inhibitory input from horizontal are basically equal.

The integral of DOG function in continuous region can meet requirements that the central reaction equals to surrounding reaction. However, in the discrete case, although each layer cells are generated according to the real physiological data, it is difficult to ensure that central response always equals to the surrounding response because of random influence, so there should be some improvements on the formulas. A new weight assignment method is presented for each cell to make the central response and the surrounding response equal to each other.

The weight setting mechanism should meet the following requirements. First, the weight of cell should be relevant to total number of cells within the region concerned. The more the total amount of cells is, the smaller the weight of each cell is. Second, the weight of cell should be relevant to the location of cells. The closer the cell is from center, the greater the weight is. Third, in the same circumstances, the weight settings mechanism can always ensure that the central area equals to the surrounding area.

$Surround_{inh}(x_0, y_0) =$

$$\sum_{j=1}^{m} P(x_j, y_j) * 1 / (\sqrt{2p} * s_{surround}) * e^{-((x_j - x_0)^2 + (y_j - y_0)^2) / 2 * s_{surround}^2}$$

(2)

The final output of a bipolar cell is as Formula 3:

$$BipolarR(x0, y0) = Center_{exc}(x0, y0) \\ - Surround_{inh}(x0, y0)$$

(3)

Figure 3. Illustration of weight assignment

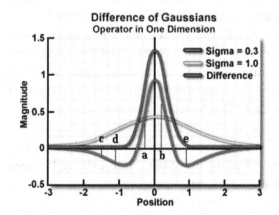

Take a simple example to illustrate the process of assigning weight. Blue curve in Figure 3 is cross-section diagram of a two-dimensional Gaussian function. Assuming there are two cells in central area a, b, and the corresponding function value is H_a, H_b, there are three cells in surrounding area c, d, e, and the corresponding function value is H_c, H_d, H_e. The weight of a in central area can be set $W_a = H_a / (H_a + H_b)$, and the weight of b is $W_b = H_b / (H_a + H_b)$. Similarly, the weight of c in surrounding areas is set as, $W_c = H_c / (H_c + H_d + H_e)$, the weight of d is $W_d = H_d / (H_c + H_d + H_e)$, the weight of e is $W_e = H_e / (H_c + H_d + H_e)$. When the uniform light exposures in this area, assuming that all cells have the same response strength R, then the center reaction $W_a * R + W_b * R$ is equal to surrounding reaction $W_d * R + W_e * R + W_f * R$. This setting method meets all requirements.

If there are n cells in central area of a bipolar cell's receptive field, the weight of i*th* cell is as Formula 4.

$$CWi = 1 / (\sqrt{2\pi} \cdot \sigma_{center}) \cdot e^{-\left((x_i - x_0)^2 + (y_i - y_0)^2\right)/2 \cdot \sigma_{center}^2} \Big/ \left(\sum_{i=1}^{n} 1 / (\sqrt{2\pi} \cdot \sigma_{center}) \cdot e^{-\left((x_i - x_0)^2 + (y_i - y_0)^2\right)/2 \cdot \sigma_{center}^2}\right)$$

(4)

Then the overall reaction of central area is as Formula 5:

$$Center_{exc}(x_0, y_0) = \sum_{i=1}^{n} P(x_i, y_i) * CWi$$

(5)

If there are m cells in surrounding area of a bipolar cell's receptive field, the weight of j*th* cell is as Formula 6:

$$SWj = 1 / (\sqrt{2\pi} \cdot \sigma_{surround}) \cdot e^{-\left((x_j - x_0)^2 + (y_j - y_0)^2\right)/2 \cdot \sigma_{surround}^2} \Big/ \left(\sum_{j=1}^{m} 1 / (\sqrt{2\pi} \cdot \sigma_{surround}) \cdot e^{-\left((x_j - x_0)^2 + (y_j - y_0)^2\right)/2 \cdot \sigma_{surround}^2}\right)$$

(6)

Then overall reaction of surrounding area is as Formula 7:

$$Surround_{inh}(x_0, y_0) = \sum_{j=1}^{m} P(x_j, y_j) * SWj$$

(7)

Now the final revised output of the bipolar cell is as Formula 8:

$$BipolarR(x0, y0) = Center_{exc}(x0, y0) - Surround_{inh}(x0, y0)$$

(8)

Simulation on Amacrine Cells

Amacrine cells play a significant role in visual signals integration and modulation. Dark vision, motion detection, non-classical receptive field formation, and time-domain information encoding are all relevant to amacrine cells. We selected AII amacrine to simulate because AII has largest number and is regarded as the most important types in all amacrine types.

There are three main input sources to AII, and we simulated all the three pathways. One of the inputs is from rod bipolar which accounts for 30% of the all input AII received. Another major input is from gap junctions between cone bipolar cells and amacrine cells. The gap junctions are electrical synaptic connections through which cone bipolar cells and AII cells can exchange information. This pathway accounts for about 20% of all input. Other 50% input comes from electrical synapses among AII connections. The more the AII cell connects to each other, the more ganglion cell receptive field will cover. We simulated the last pathway combined with ganglion cell layer like horizontal cells and bipolar cells. Here, we simulated first two pathways. The information AII received from off-rod bipolar cells is as Formula 9:

$$AamcrineR\ (x_0, y_0) =$$
$$\sum_{i=1}^{n}\left(RodBR\left(x_i, y_i\right) * 1 / \left(\sqrt{2\pi} \cdot \sigma_{center}\right) \cdot e^{-\left(\left(x_i - x_0\right)^2 + \left(y_i - y_0\right)^2\right)/2 \cdot \sigma_{center}^2}\right)$$

$$(9)$$

The information AII exchanged with on-cone bipolar cells

$$Temp = C * AamcrineR\ (x_0, y_0) \qquad (10)$$

$$AamcrineR\ (x_0, y_0) = (1 - C) * AamcrineR\ (x_0, y_0)$$
$$+ C * ConeBR(x_0, y_0)$$
$$(11)$$

$$ConeBR(x_0, y_0) = (1 - C) * ConeBR(x_0, y_0) + Temp$$
$$(12)$$

The C in the formula 12 represents exchange rate. According to physiological data, C could be set to 20%.

Simulation on Ganglion Cells

After a variety of processing in the horizontal level and vertical level, visual information ultimately is delivered to the ganglion cell layer. As the final layer in retina, ganglion cell is responsible for collecting all information and transmitting information to brain. When sampling optical signal, there are more than 100 million photoreceptor cells involved, but when transmitting information to brain, there are only 1 million ganglion output, in which the efficiency of visual information collection, representation effect, accuracy and real-time are very worthwhile to study. Certainly, the precondition of studying these characteristics is to establish a simulation retina model.

There are four types of ganglion cells considering on and off taxonomy and P and M taxonomy. P type and M type of ganglion cells are based on morphology. All M type ganglion cells are bigger and have larger receptive field than P type in the same position in retina. The specific input proportion and relationships among different cells are shown in Figure 4:

Figure 4. Input proportion and relationships among different cells

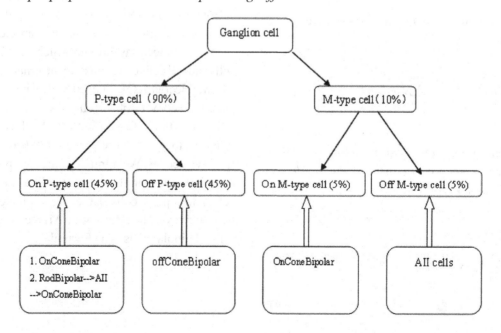

The on-center P type cells and on-center M type cells all receive On-Cone-Bipolar output, which means they may have same simulation function, but their receptive field radius is different. The receptive field radius is set according to physiological data. On-center P type cells and on-center M type cell's reaction function is as Formula 13:

$$GanlionR\ (x_0, y_0) =$$
$$\sum_{i=1}^{n} OnBipolarR(x_i, y_i) * 1 / (\sqrt{2\pi} \cdot \sigma_{center}) \cdot e^{-\left((x_i - x_0)^2 + (y_i - y_0)^2\right)/2 \cdot \sigma_{center}^2} \tag{13}$$

The off-center P type cells receive off-center bipolar cell input. Its response function is shown as Formula 14:

$$GanlionR\ (x_0, y_0) =$$
$$\sum_{i=1}^{n} OffBipolarR(x_i, y_i) * 1 / (\sqrt{2\pi} \cdot \sigma_{center}) \cdot e^{-\left((x_i - x_0)^2 + (y_i - y_0)^2\right)/2 \cdot \sigma_{center}^2} \tag{14}$$

The off-center M type cells receive AII input. Its response function is as Formula 15:

$$GanlionR\ (x_0, y_0) =$$
$$\sum_{i=1}^{n} A\operatorname{Re} sponse(x_i, y_i) * 1 / (\sqrt{2\pi} \cdot \sigma_{center}) \cdot e^{-\left((x_i - x_0)^2 + (y_i - y_0)^2\right)/2 \cdot \sigma_{center}^2} \tag{15}$$

This model realistically simulates ganglion cell layer response based on real physiological data.

EXPERIMENT DESIGN AND RESULTS ANALYSIS

This section will demonstrate experiment on receptive field size and distribution of different types of ganglion cells. As mentioned, ganglion cells can be divided into P type and M type. P type cells account for 90% of the whole ganglion cells. The highest density of P type is in the central area of retina, but P type proportion decreases to about 50% in surrounding area. As the distance from the fovea further and further, all ganglion cells receptive fields will become larger and larger. However, at the same position, P cell receptive field is always less than the M cell receptive field. Now we showed receptive field size and distribution of ganglion cells generated based on physiological data. In order to clearly demonstrate results, we used 1 / 200 density of real data to display. The result is shown in Figure 5.

In Figure 5, red point represents receptive field of P type cells and green point represents receptive field of M type cells. From the view of proportion of total number, P type cells take high proportion especially in the central area. Actually, there are no M cells in centre. From the view of the receptive field size, both P cells and M cells become larger with distance increase from the fovea, but at the same position, M cell receptive fields is always larger than P cell receptive fields.

Figure 5. Receptive field size and distribution of different types of ganglion cells

Perception Efficiency of Retina for Different Stimulus Size in Different Position of Retina

It is known that the photoreceptors are in non-uniform distribution. When an object can be perceived in center area of retina, it does not mean that the object can also be perceived in peripheral area. So what size of an object can be fully perceived by our brain in different position of retina? This question is important for us to know the effectiveness of visual system and is helpful for us to understand the world we live. So this experiment is designed to test what size of an object can be fully perceived in different position of retina.

Retina photoreceptors are divided into 10 rings according to distribution data. In each ring, following processes are carried out.

1. Retina is stimulated with the different size of object. Object size changes from small to large.
2. The same size stimulus is required to do many times to make the experiment can be analyzed statistically.

3. The activated ganglion cells are recorded. When a stimulus can be perceived, first of all it has to activate some photoreceptors. If the stimulus size is so small that it falls to blank areas between photoreceptors, it may be ignored. And further, even though it can activate photoreceptors, these photoreceptors' response must be strong enough to make ganglion cell response over a certain threshold to form an action potential.

During the object size changing process, if a certain size can always cause action potential, that means the size can be fully 100% perceived in this ring of retina. This size stimulation is called "threshold stimulation". If an object is smaller than this size, it cannot always cause action potential.

The experimental result is as shown in Figure 6. In Figure 6, the XYZ axes represent ring numbers, object size and perception percentage respectively.

If all threshold stimulation sizes were recorded, they can be drawn into a curve as Figure 7.

Physiology indicates that density of ganglion cells decreases with the increase of eccentricity. It seems that the sample and representation abil-

Figure 6. Illustration of perception efficiency of retina for different stimulus size in different position

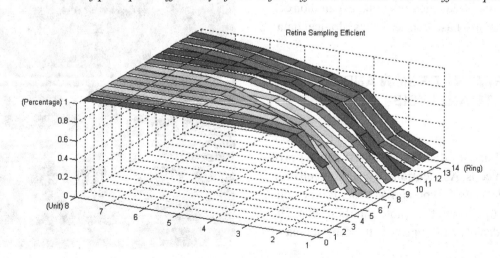

Figure 7. The threshold stimulation size in different position of retina

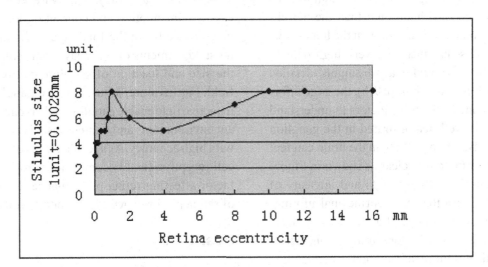

ity of retina may be weaker and weaker with increase of eccentricity. But it does not always accord with Figure 7, the reason is that we cannot just take ganglion cells density into account, but also need to consider the photoreceptor cells density. For example, in the central area, although ganglion cells density is very high but the final response depends on the collected signals within its receptive field, so how large the receptive field is and how many photoreceptors located within receptive field are the key to activate ganglion cells. That can explain why the threshold stimulation size in the 0.8mm distance dramatically becomes larger. That is because in this ring area the number of cone cells sharply declines, while the number of rod cells increases slowly, which makes the ring in a very low density of photoreceptors. From Figure 8, the most difficult places to capture the object are the peripheral area as well as 0.8mm away from the fovea. After conversion calculation based on reduced eye model, it indicates if an object size smaller than 0.1568mm appears in the front of your eye in 1m distance, you may ignore it just as it doesn't exist.

Ganglion Cells Representation of Objects

In ganglion cell layer, after horizontal and vertical connections, most of information has been classified into different pathways, so the final response output of ganglion cells may be quite different from original input image, but the important information such as boundary, brightness and complexity will still be shown in the ganglion cells output. Figure 8 is an input image, Figures 9(a), 9(b) are the corresponding ganglion cells output.

Figure 8. Input image: chess board

As seen from Figure 9(a), the central area responses were much stronger than the peripheral area, which is consistent with the human eye characteristics that there are very high density and resolution in central area to distinguish detailed and fine information. Figure 9(b), the edge information, as one of the most factors to understand the world, is well demonstrated in the ganglion cells layer. But it doesn't mean the brain can feel all edge information as clear as shown in Figure 9(b), because it is the strength and intensity of ganglion response that represent the final information to brain. In Figure 9(a), central area responses were obviously more strong and intensive, which indicates more information in central area is transmitted into higher cortex. Meanwhile distribution characteristic of ganglion cells mentioned above is also can be seen from the Figure 9(b). Further, the ganglion cells response can be utilized to study the meaning of neural info coding.

RETINA PERFORMANCE

According to the size of ganglion cell receptive field, this will analyze retina performance, such as real time, accuracy, hardware complexity, energy consumption and computing load. Since ganglion cells are the final layer cells in retina, most adjustments of other layers in retina affect the size and location of ganglion cell receptive field. Furthermore, adjustments of ganglion cell receptive field enable ganglion cell to adapt to variant environments and express visual information with high accuracy. As a result, radius of ganglion cell receptive field becomes the most important factor affecting retina performance. Parameters of retina performance are defined as followings.

Real Time

$$RT = \sum_{i=1}^{n} \left(\frac{layerT(i)*\alpha}{LayerA\,Conn(i)} + \frac{layerT(i)*(1-\alpha)}{LayerA\,Cell(i)} \right)$$

where the weight α ($0<\alpha<1$) is the ratio of the time of integrating information to the *layerT(i)*, which is the total time cost by the layer i ($1=<i<=n$). The number of activated connections of the layer i is shown as $LayerA\,Conn(i)$, meanwhile activated cells shown as $LayerA\,Cell(i)$. In natural retina, cells

Figure 9. Ganglion cell layer response; (a): side view of ganglion cell layer response; (b): vertical view of ganglion cell layer response

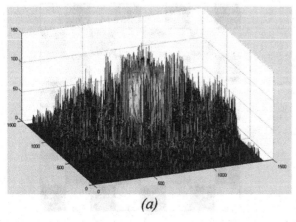
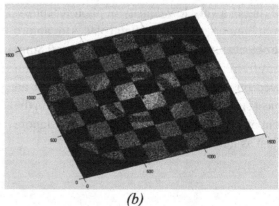

(a) (b)

and connections are working in parallel, so every cell and connection will consume time that equals $layerT(i) * a / LayerAConn(i)$ and $layerT(i) * (1-a) / LayerACell(i)$, spent in integrating information and transporting information, respectively. The total time is sum of cost time in all layers, so the Real Time (*RT*) is defined as above. Figure 10 shows the performance curves on size of ganglion cell receptive field; these curves are different from each other. This suggests that for different size of receptive field these parameters of retina performance are conflicting and we should select a good balance point among these parameters. Real time curve illustrates that Real Time increases when the size of receptive field increases. When the radius of receptive field r varies from 0.5 to 1.0, the real time varies slowly; in contrast, the real time varies quickly when r varies from 1.0 to 2.0.

Computing Load

$$CL = \sum_{i=1}^{n} (LayerACell(i) * R^2 + LayerAConn(i))$$

The Computing Load (CL) is relative to area of activated cell receptive field, where R is the mean radius of all activated cell receptive fields. This Computing Load measure is incorporated into activated cells and activated connections. Since quantity of information is closely related with the radius of receptive field, the bigger radius, the more computing load. As shown in Figure 10, Computing Load is an approximate linear function of the radius of receptive fields.

Complexity

$$Complexity = \sum_{i=1}^{n} (LayerTConn(i) + LayerTCell(i))$$

Figure 10. Performance curves on size of ganglion cell receptive field

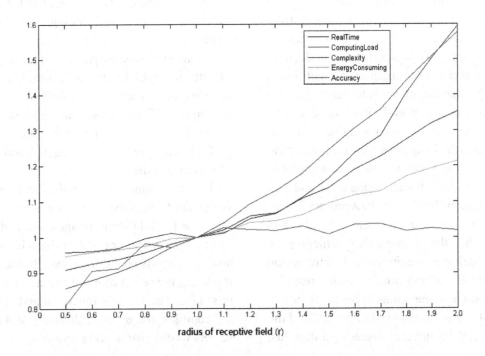

The Complexity is usually associated with total connections and total cells. The number of total connections in layer i is called as $LayerTConn(i)$, while total cells as $LayerTCell(i)$.

Energy Consumption

$$EC = \sum_{i=1}^{n} (LayerACell(i) * \beta + LayerAConn(i) * (1 - \beta))$$

Like Real Time, Energy Consumption of activated cell has different weight as EC of activated connection. We assign weight β as the weight of cell, meanwhile weight $1 - \beta$ functions as connection.

Accuracy

$$Accuracy = ActiveGanglionN$$

Perhaps the most important issue is what the trend of retinal accuracy is on the alteration of the receptive field. Visual information will be transmitted to the cortex, so the information processed by previous stages should maintain important and precise parts of original information. One interesting question is to ask how to measure accuracy of retina performance. Because output of retina is from the ganglion cell layer, we take the number of activated ganglion cell as Accuracy. In our natural visual system, cells represent visual information through cluster coding and group deciding. When the receptive field is very small, every cell maybe receive too little information to reach the activating level, and then lose information. This is why the Accuracy is very low when the radius of the receptive field is less than 1.0. When the receptive field is medium or big, every cell can acquire enough information to active itself and transmit processing result to later stages. Once the radius is bigger than 1.0, the quantity of information of every cell and the number of cells, which are covered by a stimulus,

are counterbalanced mutually. As a result of the counterbalance relationship, the accuracy remains constant relatively when the radius of the receptive field is bigger than 1.0. From the analysis above, we can conclude that the range of the radius of receptive field varying from 1.0 to 1.2 is a good balance point which the accuracy stays high while Real Time, Computing Load, Complexity and Energy Consuming stays relatively low.

CONCLUSION

In this paper, the retina model closely integrated physiology data with simulation model, not only simulating the hierarchical structure of retina, but also simulating the distribution and response of different type cells, which makes the model much closer to the real retina. The most interesting questions are what the applications of this model are and how far from it will be applied to human life. In fact, now there are many successful applications in medical fields. For example, Al-Atabany et al. applied their model and algorithms to help patients with retina degenerative disorders (Al-Atabany, Memon, Downes, & Degenaar, 2010). Researchers are cooperating with biologists and medical researchers on abnormal retina repairing, which will enable retinopathy patients to see the beautiful outside world. Furthermore, the model can be used in bionic chips design, physical verification assumptions, image processing, and many other fields. From the experiment results, the retina performance in our model can get a good balance of variety acquits.

However, some issues remain open to question. For example, the model just simulates the retinal response for 2-D static images, but it does not simulate the processing of 3-D visual information and moving visual information. Researches on modeling the retina of more features will be the next task. The retina's functions and structures are complex and diverse; there are still a lot of secrets waiting for us to be explored.

ACKNOWLEDGMENT

Supported by the 973 Program (Project No.2010CB327900) and Shanghai Science and Technology Development Funds (Project No.08511501703). The authors would like to thank the reviewers for their valuable comments and suggestions.

REFERENCES

Ahuja, A. K., Behrend, M. R., Kuroda, M., Humayun, M. S., & Weiland, J. D. (2008). An in vitro model of a retinal prosthesis. *IEEE Transactions on Bio-Medical Engineering, 55*(6), 1744–1753. doi:10.1109/TBME.2008.919126

Al-Atabany, W. I., Memon, M. A., Downes, S. M., & Degenaar, P. A. (2010). Designing and testing scene enhancement algorithms for patients with retina degenerative disorders. *Biomedical Engineering Online, 9,* 25. doi:10.1186/1475-925X-9-27

Behrend, M. R., Ahuja, A. K., Humayun, M. S., Weiland, J. D., & Chowe, R. H. (2009). Selective labeling of retinal ganglion cells with calcium indicators by retrograde loading in vitro. *Journal of Neuroscience Methods, 179*(2), 166–172. doi:10.1016/j.jneumeth.2009.01.019

Caspi, A., Dorn, J. D., McClure, K. H., Humayun, M. S., Greenberg, R. J., & McMahon, M. J. (2009). Feasibility study of a retinal prosthesis: Spatial vision with a 16-electrode implant. *Archives of Ophthalmology, 127*(4), 398–401. doi:10.1001/archophthalmol.2009.20

Casti, A., Hayot, F., Xiao, Y. P., & Kaplan, E. (2008). A simple model of retina-LGN transmission. *Journal of Computational Neuroscience, 24*(2), 235–252. doi:10.1007/s10827-007-0053-7

Dunn, F. A., Doan, T., Sampath, A. P., & Rieke, F. (2006). Controlling the gain of rod-mediated signals in the mammalian retina. *The Journal of Neuroscience, 26*(15), 3959–3970. doi:10.1523/JNEUROSCI.5148-05.2006

Edelman, S. (1999). *Representation and recognition in vision.* Cambridge, MA: MIT Press.

Giorgio, C., Guido, T., & Francesco, P. (1997). Neural networks for region detection. In *Proceedings of the 9th International Conference on Image Analysis and Processing* (Vol. 2).

Grimson, W. E. L., & Grimson, W. (1981). *From images to surfaces: A computational study of the human early visual system* (*Vol. 4*). Cambridge, MA: MIT Press.

Guan, X. D., & Wei, H. (2009). Realistic simulation on retina photoreceptor layer. In *Proceedings of the First International Joint Conference on Artificial Intelligence* (pp. 179-184).

Ishii, N., Deguchi, T., & Sasaki, H. (2004). Parallel processing for movement detection in neural networks with nonlinear functions. In Z. R. Yang, H. Yin, & R. M. Everson (Eds.), *Proceedings of the 5th International Conference on Intelligent Data Engineering and Automated Learning* (LNCS 3177, pp. 626-633).

Lee, J. W., Chae, S. P., Kim, M. N., Kim, S. Y., & Cho, J. H. (2001). A moving detectable retina model considering the mechanism of an amacrine cell for vision. In. *Proceedings of the IEEE International Symposium on Industrial Electronics, 1-3,* 106–109.

Morillas, C. A., Romero, S. F., Martinez, A., Pelayo, F. J., Ros, E., & Fernandez, E. (2007). A design framework to model retinas. *Bio Systems, 87*(2-3), 156–163. doi:10.1016/j.biosystems.2006.09.009

Niu, W. Q., & Yuan, J. Q. (2007). Recurrent network simulations of two types of non-concentric retinal ganglion cells. *Neurocomputing, 70*(13-15), 2576–2580. doi:10.1016/j.neucom.2007.01.008

Niu, W. Q., & Yuan, J. Q. (2008). A multi-subunit spatiotemporal model of local edge detector cells in the cat retina. *Neurocomputing, 72*(1-3), 302–312. doi:10.1016/j.neucom.2008.01.012

Palmeri, T. J., & Gauthier, I. (2004). Visual object understanding. *Nature Reviews. Neuroscience, 5*(4), 291–303. doi:10.1038/nrn1364

Publio, R., Oliveira, R. F., & Roquea, A. C. (2006). A realistic model of rod photoreceptor for use in a retina network model. *Neurocomputing, 69*(10-12), 1020–1024. doi:10.1016/j.neucom.2005.12.037

Reddick, W. E., Glass, J. O., Cook, E. N., Elkin, T. D., & Deaton, R. J. (1997). Automated segmentation and classification of multispectral magnetic resonance images of brain using artificial neural networks. *IEEE Transactions on Medical Imaging, 16*(6), 911–918. doi:10.1109/42.650887

Riesenhuber, M., & Poggio, T. (2000). Models of object recognition. *Nature Neuroscience, 3*, 1199–1204. doi:10.1038/81479

Risinger, L., & Kaikhah, K. (2008). Motion detection and object tracking with discrete leaky integrate-and-fire neurons. *Applied Intelligence, 29*(3), 248–262. doi:10.1007/s10489-007-0092-9

Rong, L., Yi, S., & Qiang, W. (2004, May 18-20). Edge detection based on early vision model incorporating improved directional median filtering. In *Proceedings of the 21st IEEE Conference on Instrumentation and Measurement Technology*.

Roska, B., Molnar, A., & Werblin, F. S. (2006). Parallel processing in retinal ganglion cells: How integration of space-time patterns of excitation and inhibition form the spiking output. *Journal of Neurophysiology, 95*(6), 3810–3822. doi:10.1152/jn.00113.2006

Schiller, P. H., Slocum, W. M., & Weiner, V. S. (2007). How the parallel channels of the retina contribute to depth processing. *The European Journal of Neuroscience, 26*(5), 1307–1321. doi:10.1111/j.1460-9568.2007.05740.x

Shah, S., & Levine, M. D. (1996a). Visual information processing in primate cone pathways. 1. A model. *IEEE Transactions on Systems, Man, and Cybernetics. Part B, Cybernetics, 26*(2), 259–274. doi:10.1109/3477.485837

Shah, S., & Levine, M. D. (1996b). Visual information processing in primate cone pathways. 2. Experiments. *IEEE Transactions on Systems, Man, and Cybernetics. Part B, Cybernetics, 26*(2), 275–289. doi:10.1109/3477.485878

Shimonomura, K., Kushima, T., & Yagi, T. (2008). Binocular robot vision emulating disparity computation in the primary visual cortex. *Neural Networks, 21*(2-3), 331–340. doi:10.1016/j.neunet.2007.12.033

Tang, Q. L., Sang, N., & Zhang, T. X. (2005). A neural network model for extraction of salient contours. In J. Wang, X.-F. Liao, & Z. Yi (Eds.), *Proceedings of the Second International Symposium on Advances in Neural Networks* (LNCS 3497, pp. 316-320).

Wang, Y. (2009). The cognitive informatics theory and mathematical models of visual information processing in the brain. *International Journal of Cognitive Informatics and Natural Intelligence, 3*(3), 1–11. doi:10.4018/jcini.2009070101

Wu, L. S., Yang, Z., Basham, E., & Liu, W. T. (2008). An efficient wireless power link for high voltage retinal implant. In *Proceedings of the IEEE Conference on Biomedical Circuits and Systems Conference - Intelligent Biomedical Systems* (pp. 101-104).

Wu, Q. X., McGinnity, T. M., Maguire, L., Valderrama-Gonzalez, G. D., & Dempster, P. (2010). Colour image segmentation based on a spiking neural network model inspired by the visual system. In D. S. Huang, Z. M. Zhao, V. Bevilacqua, & J. C. Figueroa (Eds.), *Proceedings of the International Conference on Advanced Intelligent Computing Theories and Applications* (LNCS 6215, pp. 49-57).

This work was previously published in the International Journal of Cognitive Informatics and Natural Intelligence, Volume 5, Issue 3, edited by Yingxu Wang, pp. 30-46, copyright 2011 by IGI Publishing (an imprint of IGI Global).

Chapter 5
Songs to Syntax:
Cognition, Combinatorial Computation, and the Origin of Language

Robert C. Berwick
Massachusetts Institute of Technology, USA

ABSTRACT

Language comprises a central component of a complex that is sometimes called "the human capacity." This complex seems to have crystallized fairly recently among a small group in East Africa of whom people are all descendants. Common descent has been important in the evolution of the brain, such that avian and mammalian brains may be largely homologous, particularly in the case of brain regions involved in auditory perception, vocalization and auditory memory. There has been convergent evolution of the capacity for auditory-vocal learning, and possibly for structuring of external vocalizations, such that apes lack the abilities that are shared between songbirds and humans. Language's recent evolutionary origin suggests that the computational machinery underlying syntax arose via the introduction of a single, simple, combinatorial operation. Further, the relation of a simple combinatorial syntax to the sensory-motor and thought systems reveals language to be asymmetric in design: while it precisely matches the representations required for inner mental thought, acting as the "glue" that binds together other internal cognitive and sensory modalities, at the same time it poses computational difficulties for externalization, that is, parsing and speech or signed production. Despite this mismatch, language syntax leads directly to the rich cognitive array that marks us as a symbolic species.

DOI: 10.4018/978-1-4666-2476-4.ch005

INTRODUCTION

It seems appropriate to address the full sweep of cognitive informatics and computing with an analysis of the origin and nature of that part of cognition that seems to be uniquely human, namely, language. There can be no doubt that language comprises a central component of what the co-founder of modern evolutionary theory, Alfred Russell Wallace, called "man's intellectual and moral nature," the human cognitive capacities for creative imagination, language and symbolism generally. In short, language makes us smart. In what follows, this article sketches how this remarkable ability might have arisen during the course of evolution and exactly how language boosts our cognitive capacity beyond that of all other animal species. To do this, it first outlines what we know about the evolution of modern humans. This will give us some important clues as to what marks out language as something uniquely human, leading naturally to a brief review of what it is that we humans have that other animals don't – what paleo-anthropologist Tattersall (1998) calls "flexibility instead of specificity in our behavior." After all, ants or bees can easily beat us at navigation, and it seems from recent studies that songbirds can do better than us at auditory production and perception. Yet we have surpassed them all in general intelligence.

Remarkably, as we shall see, it turns out that human language seems to arise from just a *single*, small evolutionary innovation, built on two already-available cognitive substrates, present separately in other animals, but brought together for the first time in modern humans. So human language is not just "more of the same," to use Tattersall's terms, but involves something entirely new, "how we integrate" cognitive competences that we share with other animals (Tattersall, 2010). In fact then, contrary to what is sometimes thought, human language is *not* complex – on the contrary, it is far simpler than anyone may have thought,

certainly simpler than what one reads about in standard linguistic textbooks. But it *is* novel. On reflection, this is not at all surprising, given the relatively short time scale involved in evolutionary terms – not millions of years but just 100-50 thousand years, according to current accounts. Complicated evolutionary change typically occurs over the time span of many thousands or millions of generations. Given this, we might anticipate that any evolutionary change leading to language would be relatively small, since it seems to have occurred within the time of a few hundred generations, and a hundred generations already takes us back to the founding of the Roman Republic. There simply was not enough time to evolve something as radically new and complex as, say, the wings of birds. As always, evolution by natural selection had to make do largely with what is at hand. Once unleashed, language served as a kind of *lingua franca*, a kind of "cognitive glue" that lets all our other cognitive faculties talk to each other, in a way that is not available to other animals. And applied to other human, digital cognitive domains, it leads to the number system, mathematics, and music. All this appears to be the result of a single evolutionary innovation. So how did this all happen?

Before beginning it is worthwhile to clear away two common misconceptions. First, this view does not entail that 'thought' is co-extensive with 'language.' Obviously they are not. Why? We all know that there can be language without thought, as is demonstrated to us every day by our rote conversations that take place seemingly without any reflection whatsoever – or failing that, the discourse of politicians. Conversely, there can be thought without language, as evidenced by the visual computation inherent in, for example, Feynman diagrams. Nonetheless, it is clear that language plays a large role in our mental lives. Second, I would also like to emphasize that the emergence of language cannot simply have been due to purely an expansion in brain size. While it is true that there has been a *general* increase in brain size throughout the primate lineage, as we shall see, language cannot be the result of brain size alone, since Neandertals were bigger brained than us.

Putting these two matters to one side then, let me turn to a brief review of the paleontological record regarding us and our immediate ancestors as it is currently understood, following Tattersall (2010). A picture of the 'family tree' of our recent homin ancestors, distinct species, with time stretching back to 5-7 million years ago reveals two crucial properties. First, like virtually any other tree of related species, all of the family *Hominidae* – it is very bushy, just as Darwin taught us. In all there may have in fact been at least over 100 distinct species in our immediate family tree. Many of these died out, after making their brief appearance on the evolutionary stage, just as Darwin suggested. Second, at any one time in the past there have typically been several, often many *Hominidae* species that co-existed, often for many hundreds of thousands of years – for instance, *Homo sapiens* (modern humans) and *Homo neandertalesis*. As Tattersall notes, what is unusual is that at their present time we have no other living relatives – there is just a single Homo species left alive in the world after millions of years of co-existence: us (Tattersall, 2010).

Perhaps most strikingly, the several-million year period before the appearance of clearly behaviorally modern Homo sapiens approximately 75 thousand years ago (kYA) is marked in general by a 'disconnect' (Tattersall, 2010) between the appearance of each new hominid species and new technologies as evidenced by differences in, e.g., stone tool making. That is, most often a new species appears on the scene (with a different body morphology and larger brain capacity, etc.), but *without* any concomitant innovative change in external behavior. For example, there is nearly a 1 million year 'gap' between the appearance of the first, Type 1 'scraping tools' about 2.5 million years ago and and the type 2 ovoid Acheulean tools,

appearing about 1.5 million years ago; crucially, these post-dated the appearance of the hominid that first made them, *Homo ergaster,* dating from about 1.9 million years ago. "As far as can be told, aside from the invention of the Acheulean in Africa (its spread beyond that continent occurred considerably later) the history of the genus Homo in the period between about two and one million years ago the Old World, but without radical physical or as far as we can tell cognitive innovation. It is not until 600 thousand years ago that we find, again first in Africa, a new kind of hominid with a significantly larger brain...This is *Homo heidelbergensis*" (Tattersall, 2008, p. 105).

Similarly, even though "anatomically recognizable" *Homo sapiens* appears about 200 – 150 thousand years ago in Africa, they come "bearing a technology that was basically indistinguishable from those of its contemporaries and immediate predecessors" (Tattersall, 2008), again a "disconnect" between anatomical and behavioral innovation. Remarkably, at one time in the Levant, at least 3 distinct species of *Homo* apparently lived side-by-side for a hundred thousand years – *Neandertals, heidelbergesis, and sapiens* – all at apparently the same level of tool-making.

All this changed starting about 75 thousand years ago, with the appearance of behaviorally modern humans in Europe, known informally as Cro-Magnons, who displaced Neandertals in Europe. Indeed, the contrast between the Neanderthals and Cro-Magnon provides a clear contrast of the differences between a cognitively competent species operating at the maximum level possible without something like language, the Neandertals, and a second species – us – who had already attained language and symbolic thought. A side-by-side look at the skeletons of *Neandertal* vs. *Homo sapiens* suffices to point out the remarkably different skull, thorax, and pelvis shapes; and in this regard, Neandertals more closely represents our common ancestor, highlighting the extensive, highly derived changes that took place yielding modern *Homo sapiens*.

Note, however, that if one considers simply brain capacity that in fact Neandertals had, if anything, a *larger* cranial capacity. They evidently hunted in groups, perhaps even more effectively than Cro-Magnons (consider the metabolic needs during the colder climate at that time); used animal skins and built shelters; and much more. But in contrast to the lack of evidence of any symbolic behavior in Neandertals, with the coming of modern Homo sapiens, there is a virtual explosion of artifacts that all say that this species was just us, including the first sculptures of remarkable aesthetic skill and beauty; sophisticated musical instruments; the first 'written' records on plaques and bones; and the astonishing images on the Chauvet caves in France that can be seen in Werner Herzog's film.

So what happened? What unleashed this astonishing creativity? Clearly, some other animals possess formidable cognitive skills. We saw that our pre-Homo sapiens ancestors made increasingly sophisticated tools, albeit at a glacial pace. It was once thought that tool-making distinguished us from other species, but this has long been proved false. Birds, such as the corvids (ravens, crows, and the like) all can make sophisticated tools and engage in what seems to be quite sophisticated causal reasoning. For example, the Western scrub jay can tie strings around a bit of rock to lower into a hole to catch ants. In Tokyo, Japan, carrion crows have been observed to get at the inside of walnuts by using automobile traffic to crack them: patiently waiting until the pedestrian "walk" signs turn green, then placing the walnuts where automobiles will crack them, stepping out of the way; then waiting until the "walk" signs turn green again to go and fetch their rewards. Finally, as Aristotle appreciated, songbirds are clearly superb at vocal production, perception, mimcry, and learning. In both birdsong and speech, auditory-vocal learning takes place during a sensitive period early in life, and there is a transitional phase of vocalization called 'babbling' in infants and 'subsong' in young songbirds. More recently, the parallels between speech and song have been extended to the neural

and genetic levels. As is now familiar, the juvenile males acquire their songs by listening to con-specific adult male 'tutors', apparently molding an initial 'babbling' template into a progressively more accurate form.

But as complex as birdsong is, it lacks two essential ingredients for human language. First, birdsongs are songs lacking *words*. Birdsong, as varied and as sophisticated as it might be, is not varied to convey distinct meanings, but rather maps directly to some attentive/hormonal state: it is a monoblock signal to mark territory (Me, me, me!) or sexual availability (Ready, ready, ready!). While people quite easily morph "Obama likes Palin" into "Palin likes Obama" to mean something radically different, no bird juggles its song components in any comparable way to result in distinct meanings. No words, no language. Second, as will be described in more detail below, birdsong is not hierarchically compositional in the way that human language is. Though birdsong is sometimes described as being decomposable into repetitive motifs, which are in turn broken down into syllables, this structure does not come close to the way in which human language is formed.

What about the great apes, our closest living relatives? They are good at many cognitive tasks, including cooperative behavior, causal reasoning, and the like. Other primates probably have conceptual structures are found in other primates: probably actor-action-goal schemata, categorization, possibly the singular-plural distinction, and others. These were presumably recruited for language, though the conceptual resources of humans that enter into language use appear to be far richer.

But again there is something missing. Much effort has been spent to 'teach' chimpanzees or gorillas 'language' either by using sign language or, in the case of pygmy chimpanzees, bonobos, by means by plastic tokens denoting actions or objects. These efforts all failed, and failed miserably. No other living non-human ape or dolphin has attained anything close to human language. Rather, as Prof. Laura-Anna Petitto notes, "while apes can string one or two 'words' [or signs] together in ways that seem patterned, they cannot construct patterned sequences of three, four, and beyond…After producing [a] matrix of two words they then– choosing from only the top five or so most frequently used words that they can produce (all primary food or contact words, such as *eat* or *tickle*) – randomly constructing a grocery list. There is no rhyme or reason to the list, only a word salad lacking internal organization" (Pettito, 2005, pp. 85-86).

So what do *we* have that the other animals don't? Here's the surprising answer in a single point: a pencil. While other animals make tools, there is apparently no other animal that makes a *combinatorial* tool. There is no other animal that stitches together separate 'bits' like an 'eraser' and a 'stick of lead' that then can be manipulated as if it were a new, single object, that can be labeled as such – a *pencil*.

What is the analog in the case of language? To answer that, we have to consider the properties of language. In essence, this comes down to Humboldt's famous aphorism that language makes 'infinite use of finite means'. The most elementary property of our shared language capacity is that it enables us to construct and interpret a discrete infinity of hierarchically structured expressions. The expressions are discrete (or 'digital') because there are 5 word sentences and 6 word sentences, but no 5½ word sentences; infinite because there is no longest sentence; and hierarchical because what our language capacity assembles are *structures*, not mere strings of sounds – what are called phrases. Language is therefore based on some generative procedure that takes elementary word-like elements from a mental store, and applies repeatedly to yield structured expressions, without bound. Operating unfettered, such a system can even arrive at astonishing language combinations such as this one: "Almost inconceivably, the gun into which she was now starting was clutched in the pale white hand of an enormous albino with long white hair."

This is this ability we immediately recognize as the hallmark of human language (even if it's not good language): the ability to produce a discrete infinity of possible meaningful 'signs' integrated with the human conceptual system, the algebraic closure of a recursive operator over our 'dictionary.' No other animal has this combinatorial promiscuity, an open-ended quality quite unlike the frozen 10-20 'word' vocalization repertoire that marks the maximum for any other animal species. Such combinatory promiscuity seemingly permeates all of human mental life, from our lexicon, to mathematics and music, as we shall see.

To account for the emergence of this new computational ability we have to account for its two key components. The first is the storehouse of words – commonly in the range of 30–50,000. The second comprises the computational properties of the language faculty. In turn, the computational properties of the language faculty that constructs internal mental representations may be subdivided further into two components, or *interfaces* with language-external (but organism-internal) systems: the system of thought, on the one hand, and also to the sensorimotor system, thus *externalizing* internal computations and thought. This is one way of reformulating the traditional conception that dates at least back to Aristotle, that language is sound with a meaning.

So what's the 'secret sauce' that lets us, but no other animal, grab any two individual 'words' and paste them together, assembling a new object that can itself be manipulated *as if* it were a single object? Whatever this procedure is, it can take two words, for example, *the* and *apples*, and glue them together into a single new object, here written as *the-apples,* This combinatory operation can in turn paste together a verb with this newly formed object, selecting the verb as 'most prominent' and yielding a verb-like chunk that forever after acts like a verb-like object and so on, yielding *ate the apples, John ate the apples, I know John ate the apples,* etc., the familiar open-ended creativity we associate with human language, an infinite number of (sound, meaning) pairs. If we assume, reasonably, that the human brain is finite, taking the computational theory of mind seriously, then all this must be produced by some *finite* number of rule or operators. But this logically entails that at least one of the operators or rules must apply to its own output, that is, the computational system must be *recursive*.

The simplest assumption is that this generative procedure emerged suddenly, in accordance with the archaeological evidence reviewed above. In that case we would expect the generative procedure to be very simple. Various kinds of generative procedures have been explored in the past 60 years. One approach familiar to linguists and computer scientists is context-free phrase structure grammar, developed in the 1950s. This fit very naturally into one of the several equivalent formulations of the mathematical theory of recursive procedures – Emil Post's rewriting systems – and it captured at least some basic properties of language, such as hierarchical structure and embedding. However, it was quickly recognized by the early 1960s that context-free phrase structure grammar is not only inadequate for natural language but is also quite a complex system with many arbitrary stipulations, and so unlikely to have emerged suddenly.

Over the years, research has found ways to reduce the complexities of such generative systems, and finally to eliminate them entirely in favor of the simplest possible mode of recursive application: an operation that takes two objects already constructed, say X and Y, and forms from them a new object that consists of the two objects unchanged, hence simply the set with X and Y as members, along with a *label* for the new object. Call this operation *cons*, after the familiar Lisp operation *constructor*. Provided with the conceptual atoms of words, the operation *cons* may be iterated without bound, yielding an infinity of hierarchically constructed expressions. If these can be interpreted by conceptual systems, the operation provides an internal "language of thought." Note that there is no room in this picture for any

precursors to true human language. To go from seven-word sentences to the infinity of human language requires the same recursive procedure as to go from zero to infinity. Further, there is no direct evidence for such "protolanguages."

Further, this tells us what the basic structure of human language is, akin to the spiral structure of DNA. But instead of DNA, the basic structure of language is an asymmetrical, hierarchical template. The right way to picture them is like this: as a pair of coat hangers stuck together, that are free to turn, in a mobile-like fashion, around the vertical axis. Thus, in the structure for *ate-the-apples*, the coat hanger unit corresponding to *the apples* is free to rotate around the higher coat hanger *ate*. So left to right order does not matter: indeed, in some languages, like German or Japanese, *the apples–ate* would be the correct order.

There are of course some constraints on *cons*. Not any two arbitrary objects can be combined: we cannot have *the the*, or *ate ate* for instance; this implies that one of the two objects X, Y glued together by *cons* has have what we might call and 'edge' feature, like the notch in a jigsaw puzzle piece, that matches up with the other object.

How do we know that these objects are hierarchical 'chunks' rather than, say, just flat strings? This is because one can show that the constraints and operations of human language respect hierarchical structure, not linear order. This is easy to demonstrate via simple examples. In the sentence, *Obama likes him*, 'him' cannot refer to Obama. However, if a chunk of hierarchical structure intervenes between 'him' and 'Obama', as in the example, *Obama thinks Palin likes him*, now 'him' can refer to Obama (but of course need not). It does not matter whether Obama is 'to the left' or 'to the right' of 'him'; what matters is the relationship between the two of *hierarchical* structure. But there is more that *cons* implies.

While in the previous cases of *cons*, the two items we combined, X and Y, were disjoint sets, suppose we have the case where Y is a subset of X (or vice-versa) where the set object Y is the structure corresponding to *the apples*, while the set object X corresponds to the structure associated with *John ate the apples*. In this case, Y is now a (proper) subset of X. In this case, *cons(X, Y)* yields the new set structure corresponding to, *the apples John ate the apples*. In effect, we have 'copied' the object of the verb *ate* to a position that is sometimes called the 'focus' of the sentence, to draw attention to it in the discourse. There is an additional principle at work that suppresses the pronunciation of the second copy of *the apples* when it is passed to the speech (or sign language) output machinery to get "flattened" onto a set of instructions to the speech apparatus in a left-to-right-fashion. So in fact what is pronounced, *the apples, John ate*, noting that internally the object of *ate* is in the proper place for interpretation. This is an important point: note that *the apples* must appear in *two* distinct places: one, the position for proper interpretation of *the apples* as the object of the verb (namely, directly after the verb); the second, the position for the proper interpretation of *the apples* as a 'focused' item for intonation (at the front of the sentence). The representation built by the generative apparatus is thus optimal in this regard: it yields exactly the right structure and no more.

More generally, the operation *cons* yields the familiar *displacement* property of language: the fact that we pronounce phrases in one position, but interpret them somewhere else. Thus in the sentence *guess what John is eating* we understand *what* to be the object of *eat* as even though it is pronounced somewhere else. This property has always seemed paradoxical, a kind of "imperfection" of language. It is by no means necessary in order to capture semantic facts. But it is found everywhere in human language. But it falls within *cons* automatically, as we have seen.

These observations generalize over a wide range of sentence types. The resulting representations are in the exact form needed for semantic interpretation: these *interior* mental representations yield a kind of 'logical form'. If you again

remember the lambda calculus, or better, programming languages built on *cons* like Scheme (or Lisp), then the representations are precisely those would we would expect to find if language takes *interior* representations, the interface to semantics, to be primary, so that these representations are 'easy' and transparent to process, involving no extra work for the semantic or interior representational apparatus. This asymmetry is pervasive. As a more complicated example, consider the question: *Which of his pictures did they persuade the museum that every painter likes best?* The answer to this question might be, 'his first one', crucially a *different* picture for each painter (Picasso, Manet, Rembrandt, ….). Now, this kind of answer is possible only if the human system of inference and interpretation constructs a representation that builds *two* instances of "his pictures", one that is logically present as the object of *likes* (and therefore hierarchically underneath), but is not pronounced, and one copy of "his pictures" that is the one you hear.

However, this dual representation, while making *semantics* easy does *not* yield representations that are equally transparent or easy to process for *external* processes like parsing or production. We do not say *guess what John is eating what*, but rather *guess what John is eating*. That is a universal property of displacement. The property follows from principles of computational efficiency. If we suppose serial motor activity to be computationally costly, a matter attested by the sheer quantity of motor cortex devoted to both motor control of the hands and for oro-facial articulatory gestures, then this follows. To externalize the internally generated expression *what did John eat what* it would be necessary to pronounce *what* twice, placing a heavier burden on computation, when we consider expressions of normal complexity and the actual nature of displacement *cons*. With all but one of the occurrences of *what* suppressed, the computational burden is greatly eased. The one occurrence that must be pronounced is the most prominent one, the last one created *cons*: otherwise there will be no indication that the operation has applied to yield the correct interpretation. It appears, then, that the language faculty recruits a general principle of computational efficiency for the process of externalization. The suppression of all but one of the occurrences of the displaced element is computationally efficient, but imposes a significant burden on interpretation, hence on communication. The person hearing the sentence has to discover the position of the gap where the displaced element is to be interpreted. That is a highly non-trivial problem in general, familiar from parsing programs. Sometimes the resulting sentence can be ambiguous, as in, *Who did you walk to the store,* and indeed, there are cases where externalization is impossible, even though the meaning is perfectly clear, as with the example, *Who is it you wonder about as to the reason why they quit school?*

There is, then, a conflict between computational efficiency and interpretive-communicative efficiency. Universally, languages seem to resolve the conflict in favor of computational efficiency and easier *internal* interpretation. That is, the system makes life easy for the *internal* system, rather than making life easy for the *external* system of parsing. These facts at once suggest that language evolved as an instrument of *internal thought*, with externalization by speech or sign a secondary process.

There are independent reasons for the conclusion that externalization is a secondary process. One is that externalization appears to be modality-independent, as has been learned from studies of sign language in recent years (Petitto, 2008). The structural properties of sign and spoken language are remarkably similar. Additionally, acquisition follows the same course in both, and neural localization seems to be similar as well. That tends to reinforce the conclusion that language is optimized for the system of thought, with the mode of externalization a secondary consideration.

The individual first endowed with *cons* would have had many advantages: capacities for complex thought, planning, interpretation, and so on. This capacity would presumably be partially transmitted to offspring, and because of the selective advantages it confers, it might come to dominate a small breeding group. What it implies is that the emergence of language in this sense could indeed have been a unique event, accounting for its species-specific character. Such 'founder effects' in population bottleneck situations like those often assumed about our ancestral population 50-75 thousand years ago are not uncommon.

Returning to the general theme of cognitive computing, it is important to see what this new combinatorial ability unleashed – that is, how language lit a bonfire under the rest of cognition. How? Recall that there is plenty of evidence for specialized cognitive "modules" in other animals – like bee navigation, or bat echolocation. But characteristically, these narrowly defined modules are unable to "talk" to one another – bats cannot press their echolocation abilities into the service of solving some *other* cognitive task. This is quite different from the "Swiss army knife" character of human cognition – the ability to cobble together novel mental representations of complex events, well beyond the power of any single 'module'.

It is natural to suggest, then, that language acts as a kind of 'cross-module' cognitive glue that links all *other* representations together. In other words, *cons* lets us hook together words into 'chunks' that can then act as single units, but we should recall what stands behind the words, namely, *concepts*. By enabling the construction of extremely complicated, novel conceptual objects and events, language enables the internal construction of representations *of* representations, cross-wiring other mental modules.

Evidence for this cross-modal coupling comes from recent brain imaging evidence regarding the interaction between the brain regions often cited to be active during syntactic processing, e.g., Broca's area (Brodmann's area 44/45), a "phylogenetically younger" part of the cortex (Friederici et al., 2011), and areas involved recognizing events semantically or in relevant motor processing. Two brain tracts as seem to form two streams: one, a top, *dorsal* tract or bundle of fibers, that is involved in coupling sound to its articulation; and two, a separate, bottom, ventral tract that connects syntax to the retrieval of stored representations of objects and actions. Note how this picture corresponds exactly to the two interfaces we mentioned earlier, as well as to similar dorsal/ventral processing streams found in the visual system. Further, it exhibits explicitly how visual representations are cross-wired by language. For example, Aziz-Zadeh et al. (2006) showed that when adults read about action sentences such as "eating a peach", the brain areas that lit up were not only the classical language ones but also the same areas activated when just viewing the action visually Both Aziz-Zadeh et al. (2006) and Pulvermüller and Fadiga (2010) have demonstrated that particular networks are activated for distinct body parts and actions, e.g., arm as opposed to legs, for throwing as opposed to kicking an object. One interpretation of activation patterns such as these is that it is language that 'binds' vision and action together.

A second line of evidence for such 'cross-modal' coupling enabled by language comes from the experimental work of Elizabeth Spelke and her colleagues at Harvard (Hermer-Vasquez, Spelke, & Katsnelson, 1999). In one typical experiment, they probed the influence of language on the ability to integrate distinct perceptual cues. Children and adults were placed in a room with four screens in each corner, in which a desired object (such as a toy, for the children), was placed behind a screen in one corner, while the subjects watched. Then the subjects were blindfolded and disoriented by turning them around, and afterwards had the job of finding the hidden toy. The experimenters wanted to see whether the subjects made use of other perceptual cues about the room: for example, if one wall of the room was painted blue, then that would serve as a landmark to reduce the search for the hidden

object, since a subject could simply remember whether the sought-for object was placed behind a blue-wall corner or not, reducing the search from four corners to two. If the perceptual cue is not used, for whatever reason, then performance changes, since any one of the four corners might be selected. The evidence that language is deeply involved with the perceptual integration of such cues is that if we 'overload' the language system when carrying out the search, by having the subjects perform a simultaneous language task, like reciting a poem, then performance degrades as if the blue wall were not even noticed. In this sense, it seems that language acts to bind

With *cons* then, humans have apparently been liberated to develop ever richer descriptions of the world, involving descriptions of actions, objects, and the like. But there is more than this, not limited to language. Applied repeatedly to the domain of a single element, *cons* acts like the successor function of Peano arithmetic: $cons(cons(cons(x))) = 3$. This immediately yields the number system of integers, with all its familiar properties. Intriguingly, as soon as children begin to acquire the rudiments of syntax, and *cons*, they apparently an open-ended quantificational ability with large numbers, unlike any other animals. While crows and chimpanzees can seemingly 'count' up to 5-7, beyond that, they deal with large numbers as though they were quantities of 'stuff' weighed on a scale – a more-or-less affair. But children by age 5, or as soon as they have acquired language syntax, seem to easily grasp that if there's a number 100, then there can be 101, as anyone who's ever played the game with a child, "I can find a bigger number than you can" might attest. This kind of ability lies beyond the reach of any other species we know.

Yet a third domain where the 'grouping' operations of *cons* again appears has to do with the domain of both rhythm and music. Consider the beat structure of a line of metrical poetry, for instance, *Tell me not in mournful numbers*. The pattern of strong and weak beats, and in fact that the strongest beat comes first on *tell*, then *not*, and then *mourn*, can be explained very simply by the same *cons* model as before. In this case, *cons* works its way left to right through the string, grouping together pairs of elements, here syllables, as before, just like *ate* and *the apples*, collecting them into a new group that is then supposed to be labeled as such. We then select one of the two units we have collected as the new label of the grouped hierarchical structure, obtaining a second level of representation. After one pass through the initial syllables, we make a second pass through this next level, as before, as shown, successively collecting groups of two, until we can do no more. Then the strongest 'beat' is just the stack with the greatest depth, and so on.

Note that there is a crucial difference here between full-fledged language and beat structure. The 'lexical items' here correspond to syllables, and are denoted simply as asterisks, because there are no elements in beat structure with features like 'verb' or 'noun.' There are just the 'marks' of the beats, corresponding to syllables. Thus, when two marks are collected together to form a new, higher level unit, there are no features to *label* the new unit, as we labeled *ate the apples* as a 'verb phrase'. That is, beat structure is what one gets if one applies the same *cons* operation as in the rest of language, but to a system *without* words. One and the same innovation leads to both language syntax and metrical structure. While there is no space to demonstrate it here, one can show that just a few different ways of 'passing through' the initial string of syllables – from right to left as opposed to right to left, perhaps alternating between levels, give rise to all the possible metrical patterns shown in all human languages (Fabb & Halle, 2008). We might even think of this 'beat structure' as the first glimmerings of language syntax – the platform for *cons* that existed before words were wired into language. If so, then a primitive ability like *cons* might actually be apparent in another species that exhibits metrical patterning to their vocal output – in particular, songbirds. And in fact, there is some suggestive evidence from ge-

nomic data, the famous FOXP2 gene, that this is the case: when this regulatory gene is disrupted in either songbirds or humans, then the ability to sequence either songs or speech into their proper linear arrangements, following the 'beats' of motor sequencing, is likewise disrupted (Haesler et al., 2007). We shall have to leave aside a detailed commentary on this point for now. What it suggests, though, is that the ability to carry out *cons* might have been available either many hundreds of millions of years ago, but that true language did not appear because there were no words to wire it to, or else that *cons* arose in songbirds by means of convergent evolution.

What we do know is that there is yet one more domain of human activity where *cons* surfaces, related to metrical structure, and it is this one: music. First, the beat structure of music works exactly like what we showed for metrical poetry: the same grouping-and-projection (without words). Second, extending this to a 'lexicon' that consists of melodic notes, leads to an analysis of 'tonal musical syntax' that looks a lot like language (with obvious differences again because melodic elements are not words). As an example, Katz and Pesetksy argue that language and tonal music are formally identical, differing only in the fundamental building blocks that they use. In the case of language, as we have seen, these 'atoms' are words (Katz & Pesetsky, 2011). In the case of music, Katz and Pesetsky maintain that these atoms are pitch-classes and chord qualities. Just as word-atoms may be combined into phrases by *cons*, Katz and Pesetsky demonstrate that the analog in music is so-called 'prolongation reduction' – "the hierarchical patterns of tension and relaxation in tonal harmony," which are also put together by the operation *cons*. So perhaps this was Mozart's deep secret: for whatever reason, the generative faculty for language that makes speaking for us so effortless, was somehow crosswired in Mozart's brain at a very early age so that literally, for him, making music was just as natural and easy as speaking is for us.

Let us just summarize briefly what seems to be the current best guess about the origin of language and human cognition. In some as yet unknown way, our ancestors developed human concepts, as opposed to what chimps, birds, and bees possess. At some time in the very recent past, perhaps about 75,000 years ago, an individual in a small group of hominids in East Africa underwent what was likely a small mutation that provided the operation *cons* – an operation that takes human concepts as computational atoms, and yields structured expressions that provide a rich language of thought. The innovation had obvious advantages, and took over the small group. At some later stage, this internal language of thought was connected to the sensory motor system, with its obvious consequences for communication, both for good and ill. In this way, human language provided the platform for its expansion into all areas of thought, yielding a good part of our "moral and intellectual nature," in Wallace's phrase (Wallace, 1871), extending far beyond language, to mathematics and music, indeed, much of what is distinctive about the human condition.

REFERENCES

Aziz-Zadeh, L., Wilson, S., Rizzolati, G., & Iacobani, M. (2006). Congruent embodied representations for visually presented actions and linguistic phrases describing actions. *Current Biology, 16,* 1818–1823. doi:10.1016/j.cub.2006.07.060

Fabb, N., & Halle, M. (2008). *Meter in poetry.* Cambridge, UK: Cambridge University Press. doi:10.1017/CBO9780511755040

Friederici, A., Bahlmann, J., Friedrich, R., & Makuuchi, M. (2011). The neural basis of recursion and complex syntactic hierarchy. *Biolinguistics, 5,* 87–104.

Haesler, S., Rochefort, C., Georgi, B., Licznerski, P., Osten, P., & Scharff, C. (2007). Incomplete and inaccurate vocal imitation after knockdown of FoxP2 in songbird basal ganglia nucleus Area X. *PLoS Biology, 5,* e321. doi:10.1371/journal.pbio.0050321

Hermer-Vasquez, L., Spelke, E. S., & Katsnelson, A. S. (1999). Sources of flexibility in human cognition: Dual-task studies of space and language. *Cognitive Psychology, 39,* 3–36. doi:10.1006/cogp.1998.0713

Katz, J., & Pesetsky, D. (2011). *The identity thesis for language and music.* Cambridge, MA: MIT. Retrieved from http://ling.auf.net/lingBuzz/000959

Petitto, L. (2005). How the brain begets language. In McGilvray, J. (Ed.), *The Cambridge Companion to Chomsky* (pp. 84–101). Cambridge, UK: Cambridge University Press. doi:10.1017/CCOL0521780136.005

Pulvermüller, F., & Fadiga, L. (2010). Active perception: sensorimotor circuits as a cortical basis for language. *Nature Reviews. Neuroscience, 11,* 351–360. doi:10.1038/nrn2811

Tattersall, I. (1998). *The origin of the human capacity (68th James Arthur Lecture on the Evolution of the Human Brain).* New York, NY: American Museum of Natural History.

Tattersall, I. (2008). An evolutionary framework for the acquisition of symbolic cognition by *Homo sapiens. Comparative Cognition Behavior Reviews, 3,* 99–114.

Tattersall, I. (2010). Human evolution and cognition. *Theory in Biosciences, 129,* 193–201. doi:10.1007/s12064-010-0093-9

Wallace, A. (1871). *Contributions to the theory of natural selection.* New York, NY: Macmillan.

This work was previously published in the International Journal of Cognitive Informatics and Natural Intelligence, Volume 5, Issue 4, edited by Yingxu Wang, pp. 22-32, copyright 2011 by IGI Publishing (an imprint of IGI Global).

Section 2
Cognitive Computing

Chapter 6
Cognitive Memory for Semantic Agents Architecture in Robotic Interaction

Sébastien Dourlens
Université de Versailles Saint Quentin, France

Amar Ramdane-Cherif
Université de Versailles Saint Quentin, France

ABSTRACT

Since 1960, AI researchers have worked on intelligent and reactive architectures capable of managing multiple events and acts in the environment. This issue is part of the Robotics domain. An extraction of meaning at different levels of abstraction and the decision process must be implemented in the robot brain to accomplish the multimodal interaction with humans in a human environment. This paper presents a semantic agents architecture giving the robot the ability to understand what is happening and thus provide more robust responses. Intelligence and knowledge about objects like behaviours in the environment are stored in two ontologies linked to an inference engine. To store and exchange information, an event knowledge representation language is used by semantic agents. This architecture brings other advantages: pervasive, cooperating, redundant, automatically adaptable, and interoperable. It is independent of platforms.

INTRODUCTION

This study has been done in the aim to solve multimodal interaction problems encountered in many applications like in robots or ambient intelligence where awareness, events storage, events relationships must highlight behavioral meaning in the environment. Lots of issues are researched such as how to manage a big amount of events, how to memorize them like human with all relationships between concepts, how to retrieve them regarding to a specific context. This work is unique because we try to answer these questions by using a knowledge representation

DOI: 10.4018/978-1-4666-2476-4.ch006

language, a multi agent architecture and a very specific memory structure. We also have tried to respect all current standards that we detail in the first section. Scientific literature refers to lots of articles in all different domains of Artificial Intelligence, Multi Agents Architectures, Ubiquitous network and Cognitive sciences. Our goal in this paper is not to be exhaustive and to compare these technologies but to build a multi disciplinary system using them. Related work presents these technologies and our choice. After defining the multimodal interaction problem, we present our architecture and its fusion and fission agents. Most important in this study is the cognitive memory which is the core of agents. As a proof of concept, we will show an application to human aid.

1. RELATED WORK

The study of intelligent systems have been done using multiple technologies in different scientific domains: Multi Agent systems (MAS), Ambient Intelligence (AmI), Architecture Modeling (AM), Artificial intelligence (AI), Formal logic (FL), Knowledge Representation Languages (KRL) from Description Logic (DL), Knowledge Management (KM), Information Retrieval (IR), Multi Agent Systems (MAS), Multimodal Interaction (MI), Human Robot Interaction (HRI) and Cognitive Informatics (CI). As Minsky said and Global AI researchers work now, intelligence or powerful reasoner will come from mixed technologies managing environment data fully or partially observable, deterministic or stochastic, episodic or sequential, dynamic or static, symbolic, possibilistic or probabilistic, in space and time. Our approach mixes all perceived events in a structured semantic architecture.

Lots of architectures have been designed in the aim to be embodied in a robot, in a house, in the city or to simply bring an intelligent software component into a system. We focus on intelligent architecture integrating semantic agents, semantic services as structural components, ontology

as knowledge base (Guarino, 1995), inference systems and KRL as communication protocol. Domain ontologies offer rich representations of machine-interpretable semantics ensuring interoperability and integration (Obrst, 2003) because of the non explicit information processed by existing architectures. The genericity of the components is a key of success of designing non dedicated applications opened to several domains and mixing different technologies.

MAS is a useful paradigm for distributed and ambient intelligence. It is an architectural choice permitting to model widely open, distributed and ubiquitous architecture (Macal & North, 2006; Allan, 2009). Agents are autonomous programming objects with the capacity to communicate. Intelligent robots often need awareness model and user fusion model, using KRL in a software agents' organization permit to these robots to reason (Blasch et al., 2006). This design view is very close to our ideal robotics platform. We will adapt this work to Human Robot Interaction.

Web services were designed by the W3C consortium in the goal to standardize software services in a distributed and interoperable way on the World Wide Web. They are connected using the Simple Object Access Protocol (SOAP) for Service Oriented Architecture (SOA) in XML format (W3C, 2007). To obtain semantic agents and services, a representation language component was required: ability to query, store or produce representative knowledge. Our objective is to really and simply represent and store the semantic events to be able to extract the meaning of a situation. Ontology Web language v2 (W3C, 2009) permits to describe relationships using classes, individuals, properties in a hierarchical structure called domain ontology. It appears to be a useful knowledge base and may be coupled to a reasoner. Reasoners are inference systems based on description logic like KAON2 (n. d.), Pellet (Clark & Parsia, 2004), JESS (2008), Fact++ (2007), and so on to realize the matching operation. SRWL (W3C, 2004) is a language that combines OWL-DL with RuleML (2004) to add rules into the ontology.

Recent languages appear to try to solve this issue, new Extensible Multimodal Annotation (EMMA) (Johnston et al., 2009). It is a standard language proposal from the W3C consortium to represent multimodal fusion and multimodal fission in a very procedural way. (Johnston, 2009) also presents an example of multimodal applications on iphone mobile as a proof of concepts. MurML (Kranstedt et al., 2002), MMIL (Landragin et al., 2004), MutliML (Giuliani & Knoll, 2008) are pure XML and are based on a natural language processing (NLP) parser made by a combinatory categorical grammar (Steedman & Baldrige, in press) at an intermediate level of recognition of gestures and speech utterances. It is an interesting approach to mix a grammar and semantic knowledge but it is more adapted for speech recognition. We wanted a solution human language grammar independent. (Kwak et al., 2006) also propose a robot mark-up language approach based on standard XML technology. Other languages more suitable, powerful and simple as KRL like KIF (Knowledge Interchange Format, n.d.) or Narrative Knowledge Representation Language (NKRL) (Zarri, 2009) exist. NKRL is a language very close to the frames and slots idea proposed by (Minsky, 1965) and refined by (Quillian, 1968). Frame is the choice we made to resolve our interpretation problem. The language we use in this article is very close to frames and NKRL but our ontologies and the inference system are different in terms of development, the result will be more adapted to robotic architectures. And more than that use of narrative languages and to respect the formal and mathematical background behind OWL (completeness calculus and consistency checking), we introduced in the memory of our agent some similar meta-concepts like properties, relationships and individuals types. One more important difference is the *n*-ary relationships implied by the use of frames and slots. This gives us the ability to import or export any OWL knowledge bases into the agent memory. Moreover, the transition from frame to XML is very easy. To conclude

this paragraph about languages, we will add here that the natural language is not so important to extract the meaning but more the structuring of the knowledge base and the relationships between models of action, models of situation linked to concepts in the representation of facts. For physical agents managing our material services (sensors and controllers), we also refer to the Foundation for Intelligent Physically Agent (FIPA, n. d.) standards from IEEE Computer Society. Our Event Knowledge Representation Language (EKRL) will be also used for the communication between agents and services even if they may use public methods of web services.

2. HUMAN ROROT MULTIMODAL INTERACTION

In the application of interaction, our architecture is composed of intelligent agents able to manage several multimodal services. Services are directly connected to the environment controlling sensors and actuators (Figure 1). They are inside the robot or they manage inputs and outputs in ubiquitous network. Thereby, semantic agents integrated into the architecture can use semantic services embedded in the robot, in the house, in the city or anywhere on Internet. These intelligent Agents contain a memory. They are able to reason and communicate with each other's. Human makes part of the environment and can interact with the robot by input modalities like gestures, body movements, voice or touch of a screen. Robot will be able to imitate, to dialog or to assist the human in his tasks, choosing the good modalities to answer and acting in the human environment with security.

It is obvious that our intelligent robot has to manage multimodal inputs and outputs related to different contexts. Robot software needs to interact with a number of sensors and actuators in real time facing uncertain, noise and complex tasks to realize in different situations. Architecture must

Figure 1. Multi-agents multimodal interaction architecture

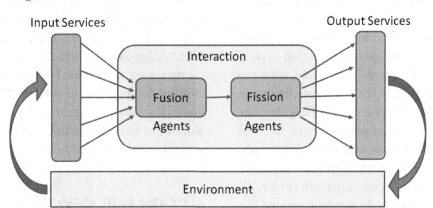

be well-conceived to reduce this complexity, events combination and knowledge must be well organized. Meaning of the situation must be quickly extracted to take a reactive decision. This meaning is also very important to obtain a correct understanding of what is happening as well as ontological storage of the events. Meaning of the situation and situation refinement require elaboration of a description of the current relationships among entities and events in the environment.

Human Robot Interaction involves acquisition and awareness context, interpretation context and execution context. These are the three main parts of the multimodal interaction. They define scenarios of interaction to realize the precondition part called *multimodal fusion*, and the post condition part called *multimodal fission* of the cognitive process of our assistant robot. Multimodal Interaction is well presented in (Landragin, 2007) but to resume his view: Multimodal fusion starts from low level integration (signal information) to high level storage of the meaning (semantic event information) by composing and correlating previous data coming from multiple sources in the case of a human machine interactions. Therefore, information fusion refers to particular mathematical functions, algorithms, methods and procedures for data combination. Multimodal fission is the

process to physically act or show any reactions to the inputs or the current situation. According to decision rules taking following the fusion process, the fission will split semantic results of the decision into single actions to be sent to the actuators. Multimodal fusion is a central question to solve and provide effective and advanced human-computer interaction by using complementary or redundant modalities. Multimodal fusion helps to provide more informative, exact, complete, reliable interpretation. The cross-modal dependency between modalities allows reciprocal disambiguation and improves recognition in the interpretation of the scene or the state of the world at a high semantic level. Multimodal fission will define the best modalities and actions to do in the environment depending to the current context and evaluation of events resulting of the fusion step. Essential requirements for multimodal fusion and fission engines are the synchronization of modalities by sending events, cognitive algorithms, contexts representation considering all concepts and actions, even internal constraints to respect. In this document, we bring functional and exchangeable components to fulfill these above requirements. The objective of our agents is to realize the fusion and fission processes using all useful knowledge present in the environment.

3. ARCHITECTURE COMPONENTS

In our architecture, all components of the architecture are agents: fusion agents, fission agents, input service agents (hardware sensors, software webservices) and output service agents (hardware actuators, software webservice). They interact with the environment in two manners: the network between all of them and the hardware parts or webservices. But we have modeled semantic agents and semantic services differently (Figure 2).

As shown on Figure 2, semantic service has code and memory (properties and methods in generic programming object model), the communication module and the hardware controller module. The hardware controller enables the service to receive information from a sensor or to drive an actuator. The communication module contains the network card and its semantic functionalities to write and send the events in EKRL or receive it. Input services role is to send any information perceived from the environment using hardware sensors. Output services role is to call a webservices or to make orders to be executed by actuators. Services have no cognitive part but enough exchangeable knowledge and required code to realize the process they are designed to.

Semantic Agent contains its knowledge base, its inference engine and its communication module (Figure 3). Semantic Agents are more complex because they are cognitive and are able to process the matching operations. They possess their own abilities and program to achieve their tasks and goals. They are intelligent agents with cognitive abilities to answer queries. Scenarios or execution schemes are stored in their knowledge base.

4. COGNITIVE MEMORY FOR SEMANTIC AGENTS

4.1. Knowledge Base

Semantic agents and semantic services communicate with events. These events are written in EKRL using concepts (formal T-BOX of concepts) and instances of concepts (formal A-BOX of concepts). They are directly included in the agent memory. This memory is a knowledge base with an embedded inference system. Knowledge base a domain ontology called *Concepts Ontology* and a second ontology called *Models Ontology*. The second one embeds templates of events (formal T-BOX of models) under the form of predicates and instances of events called *facts* (formal A-BOX).

Figure 2. Services & agents components

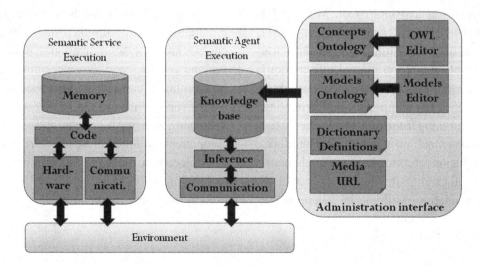

Figure 3. Storage and querying the agent memory

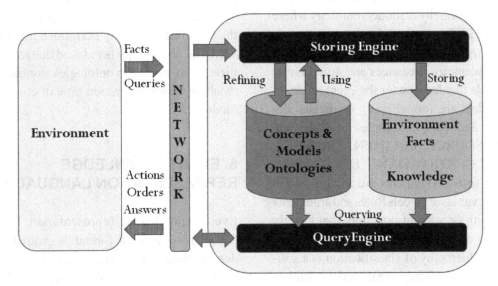

Concepts, Events models, Query models and instances are stored in ontologies. Instances are facts, happened scenario and context knowledge. All inserted facts coming from the network fully link, in a rational way, concepts, models and instances between us storing hierarchically the knowledge and intelligence of what's happening (Figure 4). In addition, concept ontology is fully OWL2 compatible. OWL editor can be Protégé, Swoop or any others OWL2 compliant editors. Thus OWL files can be imported. Concepts ontology in the knowledge base can be exported. The models editor has been developed by us to insert and modify models. It is our agent memory editor (Figure 2). Models are filled with concepts and instances of concepts to give the facts.

4.2. Ontologies

In Figure 5, we find that agent memory is composed of a Meta ontology, a Concepts ontology and a Models ontology: *Meta ontology* is only used to build the 2 next ontologies. It contains meta-concepts like types of node, types of relationship, types of data, possible roles, possible modifiers and properties. *Concepts ontology* is a standard

multi-domains ontology where classes of concepts are organized hierarchically and contains instances and properties. Concepts are entities (like humans, agents, robots, objects, parts and other living organisms), domains (arts, legal, education, science, and so on), types of activities (processes, domains, evaluations, relations) and situations (contexts,

Figure 4. Knowledge base content

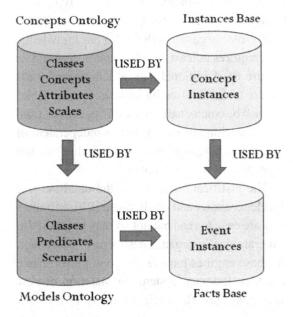

causes, states, scales). OWL compliant. *Models ontology* is a hierarchy of models ontology where a model is a EKRL frame and a slot corresponds to a Role-Argument pair. Models are of events, query and scenarios. Instances are stored under these models like facts under the event models.

All models are formal *n-ary* predicates described in Section 5. Examples of Roles are OBJECT, SOURCE, RECIPIENT, SUBJECT, PATIENT, LOCATION, DATE1, DATE2, MODIFIER and so on. Arguments are a composition of concepts, values or models Roles and arguments of a fact with or without operators that exactly match roles and arguments of the parent event model. This hierarchy of classification is a simplification very much important for the inference engine speed.

4.3. Database

All 3 Ontologies are stored in a database under a graph where unique meta-concepts, concepts and models references are nodes, and relationships are links between two nodes. To design our concepts and models ontologies, each node has a metatype among metaconcept, concept, concept instance, event model, query model, event instance, relationship, modifier type, role. Frames of models ontology are special sublinks called roles-arguments (RA). So our knowledge base requires at least 3 tables: *nodes, link* and *ra* (Figure 6). Additional natural languages words (*labels* table) and media sources (*media* table) can also be connected to nodes. They are not used in the matching process which is independent of the natural language used. Instead, they are not dependent of node reference.

The justification to use of the database is a developer choice and also because OWL or frames files are easy to read but not so easy to modify in a hierarchically organized structure. Moreover, database engines have really good performance and their indexing system for insertion of new data makes search or selection very fast. In our

code, one frame query is equivalent to one SQL query sent to the database (select in *ra* table) so the matching is directly performed in a very fast way. The complexity is reduced due to our organization of storage in ontologies storage into the database tables at creation time of concepts and models.

5. EVENT KNOWLEDGE REPRESENTATION LANGUAGE

Event Knowledge Representation Language (EKRL) is a semantic formal language L that can describe events in a narrative way. Formal system is composed of the formal language based on variable arity relations in logic of predicates (event frames). It permits to realize semantic inference in order to extract the meaning of the situation. As seen in previous section, ontologies are useful and powerful structures to store the events and extract this meaning. Inference system may use models to match the instances of the ontologies. In EKRL, frames are predicates with slots that represent pieces of information. A slot is represented by a role associated to an argument. A predicate P is a semantic *n-ary* relationship between Roles and Arguments and represents a simple event *SE* or a composed event *CE*; it is denoted by the following formula:

$$P((R_1 A_1) ... (R_n A_n)) \tag{1}$$

where R is a list of roles and A is a list of arguments. Roles R_i are the possible roles (dedicated variables storing arguments) in the event and A_j are the possible values or instances of each role in the stored fact. Figure 7 shows the EKRL syntax to write an event model into the Model ontology only.

The list of all roles is defined in Meta ontology. They have been presented in the section 4.2. Models of events are models of predicates and instances of predicates specific to a situation. For

Figure 5. Ontologies in the agent memory

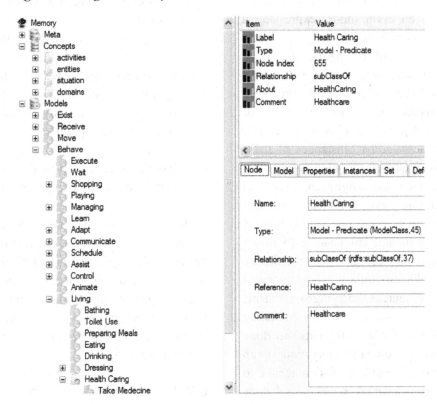

Figure 6. Database tables

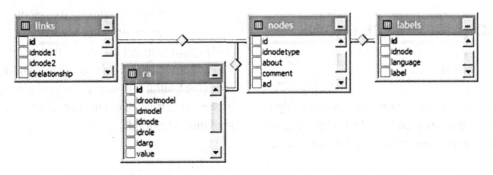

Figure 7. Event model description

Name: <RootPredicate>:<PredicateName>
Natural language description: '<Predicate Description>'
 <Role1> : <Arguments1>
 <Role2> : <Arguments2>
 <Role3> : <Arguments3>

example, *Move* is an informative term, one of root predicates. It is an event model expressing a movement of anything. Examples of sub predicates of *Move* class could be *Move:DoAStep, Move:Walk, Move:Run, Move:Avoid* and *Move:Stop*. The pattern of Move event is preserved but some role is changed to be more accurate, for example, *Move:Stop* is *Move:Move* with *Speed=0* or with *Modifier=negative*. Instances of these classes are perceived facts stored under them.

We consider agents exchange events messages using EKRL and store events in their knowledge base. It implies that no conversion software is necessary to use natural language or to find a matching event (root predicate or sub predicate). We are not searching for similarities between event coming from other sources and trying to match this event or its predicate name with ours. Matching operations presented further in this paper are done to find facts corresponding to a query model or an event model. For example, for fusion agents, to find if several facts have taken place, so deduce a meaning of a situation and compose an event of higher level of abstraction.

6. MEMORY INFERENCE ENGINE

6.1. Objective

Memory of an agent is composed of a knowledge base and its inference engine. Inference engine has the role to process memory information and is used to:

1. Store events using events models already in memory;
2. Query the memory (using query models) to find direct answers (direct matching) or to find indirect answers (matching needs operations execution) using operations on concepts as arguments of a role and operation on events (other predicates) as arguments of a role;
3. Execute scenarios (send/broadcast several events to one or several service agents).

We include some events models corresponding to actions or scenarios in the memory to make this memory to recognize facts of this nature and send meaning of higher level. It is more a modeling and agent programming memory than a self adapting memory. We wanted a well organized symbolic memory that can fulfill robotic interaction requirements.

6.2. StoreFact Algorithm

This function permits to store a fact into memory under an event model and can be called as is: *[Result]← StoreFact(Fact)*. We denote *[data]* a vector or an array of data (Algorithm 1).

If result is true, fact is stored under its model. Else, event is ignored. If no corresponding event model is found, it means agent is not designed to process this type of events reducing its memory-load and workload.

Algorithm 1. StoreFact

```
[ParseError,RootPredicate, Predicate, [Roles], [Arguments]]←Split(Fact)
If ParseError Then Return False
EventPredicateID ←Matching(RootPredicate, Predicate)
IF EventPredicateID >0 Then
       StoreRA(EventPredicateID, [Roles], [Arguments])
       Return True
Else
       Return False
End If
```

6.3. QueryEvents Function Algorithm

This function (Algorithm 2) queries memory and is called as is: *[Result, [MatchEvents]]←QueryEvents(QueryModel or QueryModelID)*

If result is true, events found are sent to all other agents. Otherwise, no event is sent.

6.4. Matching Algorithm

This function (Algorithm 3) does matching operations between predicates and roles and is called as is: *[EventsPredicateID]←Matching(RootPredicate, Predicate, [Roles], [Arguments])*

StoreRA() and ReadRA() are respectively SQL Insert and Select operations in the *ra* table filtered by ID arguments given to these two functions. Modifier is a role to modify the sense of an event.

Depending on modifiers, matching result can vary. If sense of the fact is negative, event's sense is inverted so matching must take into account of this too.

6.5. MatchArguments Algorithm

This function (Algorithm 4) does matching operations between two arguments of a role and is called as is: *[MatchingResult]←MatchArguments(QueryArgID,EventArgID) where Result is boolean*

ReadConcept(QueryArgID) is SQL Select operation in the *nodes* table where type of node is concept classes or instances, and where these nodes are under the given node using the subsumption relationship of *links* table. The SQL request gives all nodes of subtree sorted.

Algorithm 2. QueryEvents

```
If (QueryModelID) Then QueryModel←Get(QueryModelID)
[ParseError, RootPredicate, Predicate, Roles, Arguments]←Split(QueryModel)
If ParseError Then Return [False, EmptySet]
[EventsPredicateID]←Matching(RootPredicate, Predicate, [Roles],[Arguments])
IF count(EventsPredicateID)>0 Then
        [MatchEvents]←GetTextEvents([EventsPredicateID])
        Return [True, [MatchEvents]]
Else
        Return [False, EmptySet]
End If
```

Algorithm 3. Events Matching

```
RootPredicateID← GetNodeID(RootPredicate)
PredicateID←GetNodeID (Predicate)
[RolesID] ← GetNodeID ([Roles])
[ArgsID] ← GetNodeID ([Arguments])
[EventsID]←ReadRA(RootPredicateID, PredicateID,[RolesID])
For Each EventID in [EventsPredicatesID]
    [[EventRolesID],[EventArgsID]] ←ReadRA(EventID)
    For Each ArgID in [ArgsID]
        EventArgID←EventsArgsID[rank(ArgID)]
        If not MatchArguments(ArgID,EventArgID) Then Next EventID
    Next ArgID
    If ApplyModifier(EventRolesID) != ApplyModifier(RoleID) Then Next EventID
    [EventsPredicateID]← [EventsPredicateID EventID]
Next EventID
Return [EventsPredicateID]
```

Algorithm 4. Arguments Matching

```
If QueryArgID=EventArgID Then Return True
// Check if argument in the same class (subtree search)
[ArgsID]=ReadConcept(QueryArgID)
For all ArgId in ArgsID
      If ArgId=EventArgID then Return True
Next
Return False
```

7. FUSION AGENT

A fusion agent (FA) is a semantic agent which has the role to extract a specific meaning. The matching operation of the inference engine will extract the required meaning in the dialog context. Once information are extracted from the knowledge base, the fusion agent will store a composed event into and send it to others agents in the network. Each fusion agent is specialized to a task or a domain to compose new events or to act in the environment using linked semantic services. Composition of events is realized by matching operations. All fusion agents process consists to execute a loop of these five following steps:

1. Store new events;
2. Take a model of events & Fill roles with known or wanted arguments;
3. Query the knowledge base;
4. Get list of matching events;
5. Send composed event(s) if exist.

The following example represents a model (Figure 8) and one of its instance (Figure 9) of a composed event describing "a human saying and doing Hello". In this example, one instance of our fusion agents is in charge to merge events that happen in a same period of time. It uses a model that let it read that two events "Gesture Arm Up" and "Speech Hello" sent at the index time *11:34* by the two services "GestureDectectionVideo1" and "VocalRecognition1". The first service is in charge of the detection of gestures produced by human when the human is near by the camera1 sensor. The second service uses one or several mikes to recognize a speech sentence. These services have sent their basic event in parallel to any agents able to store them into their memory. These two services are embedded in the robot or are parts of the house, no problem until they are connected to the robot agents. At a scheduled interval of time and after new events took place in its memory, our agent can compose an instance of the Behave:SayHello model. Interpretation of events happening in environment is very simple and fast. The matching operation following a query will give a true description of the event. We may compose several fusion agents (Figure 10) to build complex events that will be sent on the network and stored by the memory of each other agents (past facts world representation) in order to be used to think, to act or to adapt the architecture itself.

Figure 8. "Behave: SayHello" model

```
Behave: SayHello
      SUBJECT: COORD(GestureArmUp, SpeechHello)
      SENDER: COORD(GestureSensors, VocalSensors)
      DATE:date time
      LOCATION: location
```

Figure 9. "Behave: SayHello" fact

```
Behave: SayHello
      SUBJECT: COORD(Gesture Arm Up, Speech Hello)
      SENDER:COORD(GestureDetectionVideo1, VocalRecognition1)
      DATE: 09/09/2009 11:34
      LOCATION: sidewalk
```

Figure 10. Composition of fusion agents

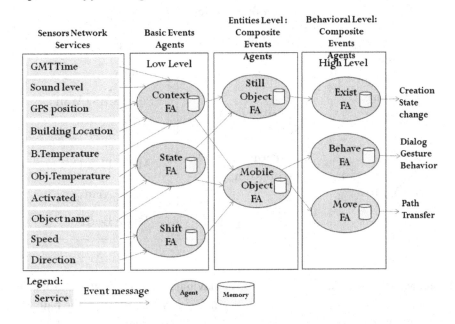

8. FISSION AGENT

A fission agent is a semantic agent which has the role to manage actuators services. Fission agent acts exactly like the fusion agent except that in addition they will produce events sent to services to be executed directly or at a specific time. So fission agent model is the same that fusion agent model presented in the previous section. Only the meaning will be of different types because events will be orders or plans called "execution events" and sent to actuators. Figure 11 presents an example of composition of fission agents and actuators services realized by three fission agents (settings agent, communication agent and move agent) and a planning agent.

Agents are represented as disks, services as boxes and arrows represent the composition (events communications that are taken into account by the agents). Past facts are the previous events sent on the network and so available in the memories of agents. On this figure, we notice two types of services: the software services containing useful methods and the hardware services driving actuators making part of effectors network. The choice of software services and hardware services also depends on the conceived application.

Figure 11. Composition of fission agents

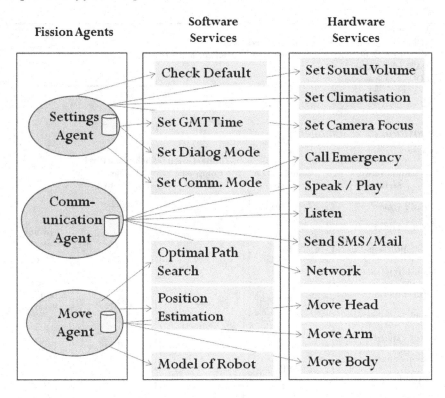

9. HUMAN AID APPLICATION

As a proof of concept, we developed an application to assist the elderly or disabled person at home. The idea is that the little humanoid robot Nao (Figure 12) may help blind people to detect objects and warn the user using all visual sensors in a domotic house. Visual sensors can also detect human gestures and human in trouble positions. Nao is a powerful robot with lots of sensors and capabilities like objects recognition, vocal synthesis, wifi connexion, and so on. In our application, Nao, as a good companion, can start and stop a vacuum cleaner, start and stop any controllable devices like TV, Wifi coffee machine, washing machine, alarm clock. It can make Internet search, read texts, read and send vocal emails, phone to first aid services, to any friends using its internal video camera. For house safety, a second wifi robot Spykee (Figure 12) is in charge to monitor the entrance door and recognize a visitor. Nao

navigates into the house and listens any strange noise as a water leak in the bathroom.

All these scenarios are interesting to highlight ubiquitous discovery, environment awareness, natural interaction and execution context. In this paper, we will present a simple and illustrative scenario where a man called Jim moves at home and Nao helps him to avoid obstacles. Nao is connected to the domotic network and may discover all available agents and services (smart sensors) with the two KRL messages "Exist:Available Services" and "Exist:Available Agents". It is useful to check all available modalities to interact with Jim but it is not necessary to know which sensors are sending messages because each time something is detected by a sensor, an event is sent to all other agents (Figures 13 and 14). Nao focuses on video camera sensors sending events about Jim and the location of Jim, and understand Jim will hurt the table. With his reasoner, Nao chooses the best modality to prevent Jim (Figure 15).

Figure 12. Nao and Skypee Robots

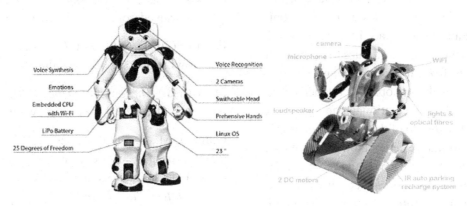

Figure 13. "Move: Walk" event

```
Move:Walk
    SUBJECT: Jim (instance of human being)
    POSITION: Between Sofa and TV (x=25,y=10)
    DIRECTION: Left (d=45°)
    SPEED: Slow (s=1m/s)
    SENDER: GestureDetectionVideo1
    DATE: 09/01/2010 10:30:10
    LOCATION: living room (r=3)
```

Figure 14. "Exist:Existing Objects" event

```
Exist:Existing Objects
    SUBJECT: Table (instance of furnitures)
    POSITION: Between Sofa and TV (x=27,y=12)
    SENDER: GestureDetectionVideo2
    DATE: 09/01/2010 10:30:11
    LOCATION: living room (r=3)
```

Figure 15. "Behave:Speak" event

```
Behave:Speak
    SUBJECT: Obstacle Warning
    CONTENT: "Be careful, you may hurt the table !"
    SENDER: NaoRobot1 (instance of robots)
    DATE: 09/01/2010 10:30:11
    LOCATION: living room (r=3)
```

Nao should also send orders to several modalities depending on the rules. Jim knows about the obstacle and thanks Nao. Even without this acknowledgement, Nao continues to be aware to the all events from the environment and well reacts. As a result, we observe all behavioral scenarios are fully operational and simple to implement. The design of the architecture is reliable. The performance depends on the robots and sensors hardware. The discovery and communication between agents and services are effective. This ambient and robotic architecture is suitable for multimodal interaction. Memory of agent is fast and really permits to understand and reason about the environment.

10. CONCLUSION

In this work, capabilities of this human-robot interaction architecture have been extended with an additional ontology memory. This layer describes the semantic conceptual models of the objects in the environment. It permits the integration mechanisms based on deductive rules and derivation new information from existing information or adapt the architecture. For future work, we prepare more complex applications and a model checking proof with Kripke models of the reliability of the memory system.

REFERENCES

Allan, R. (2009). *Survey of agent based modeling and simulation tools* (Tech. Rep. No. TR-2010-007). Warrington, UK: STFC Daresbury Laboratory.

Blasch, E., Kadar, I., Salerno, J., Kokar, M. M., Das, S., & Powell, G. M. (2006). Issues and challenges in situation assessment (level 2 fusion). *Journal of Advances in Information Fusion, 1*(2).

Clark & Parsia. (2004). *Pellet: OWL 2 reasoner for Java.* Retrieved from http://clarkparsia.com/pellet

FaCT++ (2007). *OWL: FaCT++.* Retrieved from http://owl.man.ac.uk/factplusplus/

FIPA. (n. d.). *The foundation for intelligent physical agents.* Retrieved from http://www.fipa.org/

Giuliani, M., & Knoll, A. (2008). MultiML - a general purpose representation language for multimodal human utterances. In *Proceedings of the 10th International Conference on Multimodal Interfaces* (pp. 165-172).

Guarino, N. (1995). Formal ontology, conceptual analysis and knowledge representation. *Human-Computer Studies, 43*(5-6), 625–640. doi:10.1006/ijhc.1995.1066

JESS. (2008). *The rule engine for the Java platform.* Retrieved from http://www.jessrules.com/

Johnston, M. (2009). Building multimodal applications with EMMA. In *Proceedings of the International Conference on Multimodal Interfaces and Workshop on Machine Learning for Multimodal Interfaces*, Cambridge, MA (pp. 47-54).

Johnston, M., Baggia, P., Burnett, D. C., Carter, J., Dahl, D. A., MacCobb, G., et al. (2009). *EMMA: Extensible multimodal annotation markup language.* Retrieved from http://www.w3.org/TR/emma/

KAON2. (n. d.). *Ontology management for the semantic web.* Retrieved from http://kaon2.semanticweb.org/

Knowledge Interchange Format. (n. d.). *The KIF specification.* Retrieved from http://www.ksl.stanford.edu/knowledge-sharing/kif/

Kranstedt, A., Kopp, S., & Wachsmuth, I. (2002). *Murml: A multimodal utterance representation markup language for conversational agents.* Paper presented at the First International Autonomous Agents and Multiagent Systems Workshop on Embodied Conversational Agents - Let's Specify and Evaluate Them, Bologna, Italy.

Kwak, J.-Y., Yoon, J. Y., & Shinn, R. H. (2006). An intelligent robot architecture based on robot mark-up languages. In *Proceedings of the IEEE International Conference on Engineering of Intelligent Systems* (pp. 1-6).

Landragin, F., Denis, A., Ricci, A., & Romary, L. (2004). Multimodal meaning representation for generic dialogue systems architectures. In *Proceedings of the Fourth International Conference on Language Resources and Evaluation* (pp. 521-524).

Macal, C. M., & North, M. J. (2006). Tutorial on agent-based modeling and simulation part 2: How to model with agents. In *Proceedings of the 38th Winter Simulation Conference* (pp. 73-83).

Minsky, M. (1965). Matter, mind and models. In *Proceedings of the International Federation for Information Processing Congress*, New York, NY (pp. 45-49).

Obrst, L. (2003). Ontologies for semantically interoperable systems. In *Proceedings of the Twelfth International Conference on Information and Knowledge Management*, New Orleans, LA (pp. 366-369).

Quillian, R. (1968). Semantic memory. In Minsky, M. (Ed.), *Semantic Information Processing*. Cambridge, MA: MIT Press.

Rule, M. L. (2004). *The rule markup initiative.* Retrieved from http://ruleml.org/

Steedman, M., & Baldridge, J. (in press). Combinatory categorial grammar. In Borsley, R., & Borjars, K. (Eds.), *Non-transformational syntax: Formal and explicit models of grammar.* New York, NY: John Wiley & Sons.

W3C. (2004). *SWRL: A semantic web rule language combining OWL and RuleML.* Retrieved from http://www.w3.org/Submission/SWRL/

W3C. (2007). *SOAP: Version 1.2 part 1: Messaging framework.* Retrieved from http://www.w3.org/TR/soap12-part1/

W3C. (2009). *OWL: Web ontology language.* Retrieved from http://www.w3.org/TR/owl-features/

Zarri, G. (2009). *Representation and processing of complex events.* Paper presented at the Association for the Advancement of Artificial Intelligence Spring Symposium, Stanford, CA.

This work was previously published in the International Journal of Cognitive Informatics and Natural Intelligence, Volume 5, Issue 1, edited by Yingxu Wang, pp. 43-58, copyright 2011 by IGI Publishing (an imprint of IGI Global).

Chapter 7

Interactive Feature Visualization and Detection for 3D Face Classification

Jason McLaughlin
Indiana University-Purdue University, USA

Shiaofen Fang
Indiana University-Purdue University, USA

Sandra W. Jacobson
Wayne State University, USA, & University of Cape Town, South Africa

H. Eugene Hoyme
Sanford School of Medicine, USA

Luther Robinson
State University of New York, USA

Tatiana Foroud
Indiana University, USA

ABSTRACT

A new visual approach to the surface shape analysis and classification of 3D facial images is presented. It allows the users to visually explore the natural patterns and geometric features of 3D facial scans to provide decision-making information for face classification which can be used for the diagnosis of diseases that exhibit facial characteristics. Using surface feature analysis under a digital geometry analysis framework, the method employs an interactive feature visualization technique that allows interactive definition, modification and exploration of facial features to provide the best discriminatory power for a given classification problem. OpenGL based surface shading and interactive lighting are employed to generate visual maps of discriminatory features to visually represent the salient differences between labeled classes. This technique will be applied to a medical diagnosis application for Fetal Alcohol Syndrome (FAS) which is known to exhibit certain facial patterns.

INTRODUCTION

In recent years, interactive visualization and 3D computer graphics techniques have started to play significant roles in data mining and data analysis applications. The feedback and involvement of human intuition and knowledge through graphical interfaces and interactive visualization can often lead to faster and more reliable data mining and analysis outcomes. This visual analytics approach is particularly effective for the analysis of visual/spatial data from 3D sensors, as spatial data have

DOI: 10.4018/978-1-4666-2476-4.ch007

natural visual representations. On the other hand, the rapid advances of sensory technologies have led to the enormous proliferation of multimedia data and have created a great challenge and an urgent need for new and more effective data mining and analysis techniques for data from multimedia sources. Many applications, such as medical diagnosis and biometrics, can benefit greatly from more reliable and efficient 3D data analysis methods and techniques.

This paper presents a new visual approach to the analysis of polygon mesh surfaces of human faces for a medical diagnosis application: Fetal Alcohol Syndrome (FAS). FAS is a neurological disorder resulting from prenatal exposure to alcohol. It is the most common nonhereditary cause of mental retardation and is often associated with growth deficiencies and developmental abnormalities of the central nervous system, and a pattern of various facial anomalies (Moore, Ward, Jamison, Morris, Bader, & Hall, 2002). It is estimated that the prevalence of FAS in the general population is likely to be between 0.5 and 2.0 per 1,000 births (May & Gossage, 2001). While there are several FAS diagnostic criteria used in the medical community (Jones & Smith, 1973; Moore, Ward, Jamison, Morris, Bader, & Hall, 2002; Jacobson, 2008), identification of the distinctive pattern of facial features anomalies is a key and necessary component for the diagnosis and the only one at this time that can be readily detected and determined and potentially automated.

In this study, 3D facial images are collected using an eye-safe laser scanner. Multiple scans are made and then stitched together to form a dense and irregular polygon mesh surface. The analysis of polygon mesh surfaces is a new research approach in medical imaging, as existing 3D medical imaging technologies focus on volumetric images, such as CT and MRI scans. There is, however, a tremendous need in medical applications for the image analysis of surface scans that capture detailed surface geometry and texture information.

For some neurological disorders, facial feature anomalies are highly characteristic and unique. 3D data analysis techniques that capture the correlation between face and disease can potentially provide effective diagnosis tools for medical research and clinical studies, especially in the pre-screening process and in early diagnosis of children. It has also been shown in Jain and Hoffman (2005) that the traditional 2D facial analysis approach is not effective for this type of diagnosis problems, since the features affecting the facial dysmorphology are highly three-dimensional and cannot be easily captured by 2D photos.

3D facial scans of both children with FAS and normal controls that were not exposed prenatally to alcohol were collected and processed. The analysis can be carried out using supervised classification techniques. A critical component in this type of data analysis is the identification of salient features that are most discriminatory in separating FAS faces and controls. Traditionally, the initial features are defined based on some pre-determined feature models that usually represent a small fraction of all available features. When sufficient prior knowledge is available, this feature model can be established as an "educated guess." But in most cases, the initial features are determined based on highly subjective assumptions, which can potentially lead to inaccurate or incomplete analysis results. To further add to the difficulties, a very large initial feature set can make the analysis even worse due to the infamous "curse of dimensionality." Since there are infinite numbers of potential features, a guided and intelligent mechanism needs to be installed into the feature search process. One way to provide an effective feature search is to utilize human knowledge and intuition by applying interactive visualization techniques to allow the users to make decisions based on visual feedbacks. Human vision combined with intuition and knowledge is often superior to machine vision in identifying color and geometric patterns. This approach can also provide an exploratory environment for physicians and medical researchers to build, test, and evaluate hypotheses for both theoretical and clinical studies.

In the following, we will first discuss the related work associated with surface image analysis and feature visualization. In Section 2, 3D facial data processing techniques including alignment and a morphing based face correspondence technique will be described. The interactive feature visualization technique for region-based feature selection and exploration will be presented in Section 3. In Section 4, some application background and experimental results will be shown. Final remarks and future work will be given in Section 5.

1. RELATED WORK

The processing of polygon mesh data is sometimes called Digital Geometry Processing (DGP). Much of the promise of DGP lies on a systematic treatment of signal processing styled operations with geometry mesh data (Schroder, 2001; Taubin, 2000), such as smoothing and compression. There is relatively little work in discriminatory feature detection and analysis using polygon mesh surface data.

There are several previous works addressing the problem of feature identification and extraction in 3D images, Gordon (1991) proposed a 3D face recognition method using curvature calculation based on range image data obtained from a rotating laser scanner. Tanaka et al. (1998) extended the concept of free-form curved surface in 3D shape recognition problem to 3D face recognition application. Beumier and Acheroy (2000) used central and lateral facial profiles to generate curvature values as feature vector for face authentication. Range image is a special case of polygon mesh data. Some of the range image processing operations can be extended to irregular mesh data. For example, segmentation of range images and extraction of 3D features have been intensely

studied (Besl & Jain, 1988; Zhao, Zhao, & Chen, 1996). These segmentations and identification of object surfaces are then used for object recognition (Arman & Aggarwal, 1993; Besl & Jain, 1985; Jain, 1988). Object recognition strategies based on surface properties, such as surface area, Gaussian curvatures, and mean curvatures, have also been used (Besl & Jain, 1988).

Machine learning and pattern recognition techniques (Vapnik, 1998) are commonly applied for various analysis tasks such as classification in the feature space. The first step in any classification process is to choose candidate discriminatory features and to evaluate them for their usefulness. Typical techniques include using Principal Component Analysis (PCA) to extract eigenfaces (Turk & Pentland, 1991), and using Independent Component Analysis (ICA) (Barlett & Sejnowski, 1997) to project training images onto a lower dimensional subspace for feature selection.

There have been many feature visualization techniques developed in the visualization community (Post, van Walsum, Post, & Silver, 1995; Silver & Wang, 1997; Weiler et al., 2005). Most focus on visualizing the results of some feature extraction and tracking algorithms. Visualization is not directly involved in and does not contribute to the feature detection/discovery process. The use of tightly coupled visual human-computer interaction for difficult machine learning problems has been a relatively recent area of research (Zhao, Chen, & Yao, 2008). Cognitive Informatics provides a potential framework for better understanding human-computer interaction as applied to machine learning applications (Wang, 2002, 2007).

2. FACE ALIGNMENT AND CORRESPONDENCE

Facial datasets are collected by various groups under different conditions. In order to properly compare 3D facial datasets, all face scans need to be precisely aligned in a common coordinate system. This can be done by using a template face (a standard face dataset), to which each new dataset will be aligned. The problem of 3D alignment has been studied extensively. The most popular and effective solution is the Iterative Closest Point (ICP) algorithm (Besl & McKay, 1992; Lu, Colbry, & Jain, 2004), which computes the optimal transformation by iteratively finding a local minimum of a mean-square distance metric. In our ICP algorithm, the cost function is defined as the sum of the least square distances from the vertices of one face dataset to the other face surface. This alignment induces a correspondence between points from the template to a new face.

A space encoding method is designed to allow fast minimum distance computation by localizing the search space. We use a simple uniform space subdivision in the volume space of the template dataset to encode the vertices of the template dataset first. A bounding box of the template face is subdivided into a 3D regular grid. The vertices of template face as well as the face to be compared with will be registered with the corresponding grids to facilitate fast search and cost function computation during iteration.

This distance function allows us to define a mapping between a vertex on the template face to a vertex on the surface of another face dataset that is aligned with the template, hence establish a one-to-one correspondence between vertices of all faces. Unfortunately, directly mapping vertices based on a minimum distance criterion between aligned faces can lead to misalignments of local features, as the shape features (e.g. sizes and ratios) of different faces can be different.

To avoid local feature misalignment, we apply an additional morphing process to generate an intermediate face between two aligned faces. A set of landmark points is first manually selected on both the template face and the aligned face to capture prominent and easy to define facial features. The template face is then morphed into the aligned face by interpolating their corresponding landmark points. The morphing is implemented using the Hardy's scattered data interpolation function (Hardy 1971), which computes the displacement of each point on the face by interpolating the displacements at the landmarks points:

$$H(P) = \sum_i h_i (d_i^2(P) + r_i^2)^\alpha$$

where $\{d_i\}$ are the distances from P to the landmarks, $\{r_i\}$ are the stiffness radii, $\alpha > 0$ is an exponent parameter, and $\{h_i\}$ are the Hardy's coefficients that can be solved using the interpolation conditions.

This morphed face serves as an intermediate step in building a feature-preserving correspondence between points across different faces. A correspondence function is a mapping from a vertex on the template face to a point on another face that was aligned with the template. Let $\{P_i\}$ be the vertices of the template dataset, and $\{Q_i\}$ be the vertices of an aligned face dataset. The correspondence function of a point P_i is defined as the vertex on $\{Q_i\}$ with the minimal distance to the morphed point of P_i; i.e.

$$f(P_i) = \min dist(H(P_i), \{Q_i\})$$

This correspondence function allows us to define certain geometric features, such as landmarks, regions and boundary lines, on the template dataset first, and then automatically map these features, to another face dataset. Our tests show that the correspondences are highly accurate with key features correctly preserved.

3. INTERACTIVE FEATURE VISUALIZATION

The goal of our analysis task is to identify a set of salient features that can effectively distinguish FAS and non-FAS faces and, therefore, can be used as part of the diagnostic criteria. This is traditionally done by first selecting a set of potential features, and then using machine learning techniques with labelled (FAS or non-FAS) data to determine the salient features through a system training process. Since there are potentially an infinite number of possible features, the initial feature selection process is often very subjective. Without an effective feature space search and iteration strategy, the feature selection result is unlikely to be optimal. A second problem of machine learning based feature selection methods is that machine learning often requires a large number of samples available in a training set and a separate set of samples of about half of the size of the training set for validation and testing. In most human subject related research, it is very difficult, expensive, and time-consuming to collect a large set of data samples.

Feature visualization and interaction techniques can be effective in overcoming these problems, as human vision is often able to recognize visual patterns more quickly and accurately than machine vision. Interactive feature visualization can also provide a feature exploration environment for the more sophisticated users (e.g., physicians and medical researchers) to explore and evaluate various hypotheses and alternatives that may not be available in a fully automatic analysis process.

3.1 Features and Regions

Our main goal is to detect facial regions that are significant in discriminating FAS and non-FAS faces. (There are other facial features including distances and shape ratios, which have been studied in other literatures, such as (Moore, Ward, Jamison, Morris, Bader, & Hall, 2002)). We define a feature as a pair, (r, v), where r is a region on a face surface, and v is a quantity (measurement, statistics, etc.) computed within region r that represents some property of the region. We also call v a feature value. For a given region r, k types of feature values can be computed. If n regions

are identified, the feature vector will then consists of the set of all pairs:

$$(r_i, v_j),\ i = 1 \cdots n, j = 1 \cdots k\,.$$

A region is usually defined on the template face by a sequence of points on the boundary of the region, and is then mapped to each sample face by the correspondence function. Feature values computed in each region often represent some geometric properties of the mathematical surface in that region. For a polygon mesh surface, discrete differential geometry algorithms may be applied to compute the surface properties (Meyer, Desbrun, Schroder, & Barr, 2002; Schroder, 2001). Typical feature values include:

- The average values of surface curvatures, such as Gaussian and Mean curvatures.
- The area and aspect ratio of a surface region.
- Flatness measurement, which can be computed by fitting a planar surface to the region using a least square mesh fitting.

3.2 The Feature Visualization Algorithm

The visualization process serves two purposes:

- It provides an interactive environment for the users to explore and identify important regions that may relate to the classification problem. This could lead to the discovery of new hypotheses for further research and studies (e.g., relations between brain damages and facial feature changes).
- The regions identified can be used as part of the initial features (with k feature values computed in each region) for feature analysis and automated diagnosis.

The regions that are most significant are the ones that exhibit the largest differences between datasets in the FAS group and datasets in the non-FAS group. While these differences for a feature value may be computed for a given region, there is no easy way to know in advance which regions to use and how many to use. Furthermore, such computations are often expensive, hence not realistic to view and compare the features interactively for exploration purpose.

The main idea of our approach is to interactively render shaded images of corresponding faces under various lighting conditions, and compare the shading results to identify patterns of color and intensity differences. OpenGL shading functions are employed to support interactive explorations with dynamically changing lighting conditions. The shading differences are carefully measured and visualized to provide visual cues that may reveal salient regions and features for a given classification problem.

The visualization algorithm is designed to visually represent and display the cumulative shape differences of two groups of facial surfaces. An average face is first constructed and rendered for each group, and the color differences of the two average faces, normalized by the average differences within each group, are then displayed to illustrate patches of large differences. The average face in a group is defined on the base mesh of a template face with normal vectors that are averaged over the normals of all corresponding points in the dataset.

Let the faces in the FAS group be $\{P_i^k\}$ $(i = 1 \cdots n_p; k = 1 \cdots m_p)$ and the faces in the control (non-FAS) group be $\{Q_i^k\}$ $(i = 1 \cdots n_q; k = 1 \cdots m_q)$, where index i represents the vertices of each face, and the index k represents the different faces in each group. The algorithm works as follows:

- **Constructing Average Faces:** To construct the average face of the FAS group $\{\overline{F}_i\}$, the template face is used as the common rendering model, and the normal vector of each vertex (on the template face) is computed as the average of the unit normal vectors of all corresponding points on faces in the FAS group. The average face for the non-FAS group, $\{\overline{C}_i\}$, is computed similarly.

- **Computing a Difference Face:** The two average faces are rendered by OpenGL, using a common set of lighting condition, materials properties, and camera setting. The difference face, $\{D_i\}$, is constructed by computing the positive intensity differences between corresponding vertices of the two average faces, normalized by their sum of their intensity values, i.e.

$$I(Di) = \frac{|\, I(\overline{F}_i) - I(\overline{C}_i)\,|}{I(\overline{F}_i) + I(\overline{C}_i)}$$

On each image of the difference face, the regions that exhibit higher intensity values represent the ones that are more likely to be discriminatory in separating FAS and control (non-FAS) faces. An example of the normalized difference face is shown in Figure 1(a) under a sample lighting condition. When rendering each difference face, the same base mesh from the template face is used. Texture data that was captured by the cameras is not displayed for several reasons: additional color information could reduce the clarity of colors rendered by our algorithm; the 2D data is not always perfectly matched with the underlying mesh and may give the user and incorrect view of where certain physical features may be located; there are also privacy concerns when dealing with data that could identify subjects. It is much less likely that a mesh with texture data removed could be used to easily identify someone.

The above algorithm visualizes the shape differences using only the normal information of the surfaces. Other surface features such as curvatures can also be embedded in this visualization process. This is particularly important when the differences of two classes are more subtle and involving higher order surface properties. In order to visualize the differences in a feature value, such as Gaussian curvature, the visualization algorithm can be modified by normal vector perturbation (similar to the Bump Mapping approach in computer graphics) in the direction of normal difference vector.

Let the normal vectors of the average face of the FAS group be $\{\vec{N}_i\}$, the normal vectors of the average face of the non-FAS group be $\{\vec{M}_i\}$, the curvature values on the average face $\{F_i\}$ be $\{u_i\}$, and the curvature values on the average face $\{C_i\}$ be $\{v_i\}$. At each vertex of the template face, the normal vector, $\{\vec{V}_i\}$, is perturbed by a displacement vector, $\vec{V}_i = \vec{V}_i + \vec{D}_i$

$$\vec{D}_i = \frac{t_i(\vec{N}_i - \vec{M}_i)}{|\vec{N}_i - \vec{M}_i|}$$

where t_i is the normalized curvature difference:

$$t_i = \frac{(u_i - v_i)}{|u_i| - |v_i|}$$

Both the original template face and the perturbed template face are rendered under the same lighting condition, and the normalized differences of their corresponding intensity values are then computed and displayed to show the visual effect of the perturbation. Figure 1(b) shows a sample image of this perturbed difference face.

A key aspect of the algorithm is that different lighting conditions affect the calculation of the average difference face. Figure 2(a), (b), and (c), shows examples of the average faces for the FAS group and the non-FAS group, along with their difference faces under different lighting conditions. Figure 2(d) and 2(e) show the perturbed template faces and the original template face, and their difference faces under different lighting conditions.

Figure 1. Sample difference faces

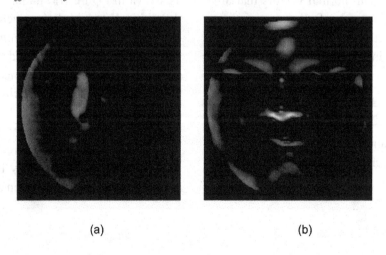

(a)   (b)

Figure 2. (a, b, and c) Average faces of FAS and non-FAS groups, rendered with different lighting conditions, and their difference faces. (d and e) The perturbed and the original template faces, rendered with different lighting conditions, and their difference faces.

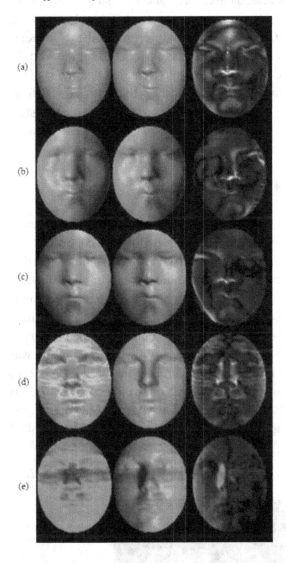

3.3 Interactive Feature Selection and Exploration

Using the above feature visualization algorithm, users can interactively manipulate and adjust the visualization parameters to search for facial regions that consistently exhibit patterns of high intensities in the *normalized difference faces*.

Since some features can only be revealed when viewed at a certain angle, or with a certain lighting condition, it is imperative to allow the users to dynamically change the rendering parameters in a systematic and efficient way.

Our user interaction strategy adopts part of the *design galleries* idea (Marks et al., 1997), which computes an optimal set of rendering parameters for a set of rendering outputs that are optimally dispersed. In order to effectively navigate the rendering parameters, we use an iterative parameter selection approach. In this approach, a user only makes change to one type of parameters (e.g. light position) at a time, and the system will automatically select the rest of the rendering parameters to generate an optimally dispersed *normalized difference faces*. The iteration works as follows:

1. Select a certain type of parameter vector to manipulate.
2. For each value of the selected parameter vector, apply a dispersion algorithm (Marks et al., 1997) to automatically select an optimal set of values for the rest of the parameters and generate a set of optimally dispersed *normalized difference face* images.
3. Manually define regions on one or more of these images if desired.
4. If desired, the user can select one of the images to continue the iteration with narrower parameter ranges.
5. Select another type of parameter vector, go back to (b).

In the dispersion algorithm, a distance function is defined between each pair of *difference images*. Since all difference faces are defined on a common template face, the distance function can be easily computed as the sum of the color differences between corresponding vertices. As in (Marks et al., 1997), an iterative process can be applied to find a set of parameters such that the smallest distance between two different images is the maximum.

It should be noted that, though we are using design galleries to guide our search, we are not necessarily concerned with finding a minimal number of optimal lighting conditions for feature detection. Once the initial average faces are computed, difference faces can be calculated under new lighting conditions within two seconds, and additional iterations of the feature selection process are not a significant overhead to the overall analysis. It would be more desirable to perform additional iterations without selecting features in step c, than to miss features with possible discriminatory power.

In Figure 3, we show an example of feature selection on two difference faces created with different light positions. The user sees that the black and dark blue regions are unlikely to contain discriminatory features. The user also uses knowledge of the problem domain to ignore regions that are very unlikely to be indicators of FAS, such as the nose, even though it may be highlighted. The regions selected represent areas that are both brightly colored and consistent with the user's knowledge of FAS diagnosis

4. EXPERIMENTAL RESULTS

Our experiment used datasets collected by an NIH-NIAAA consortium: The Collaborative Initiative on Fetal Alcohol Spectrum Disorders (CIFASD). Facial datasets are collected using two cameras: Early images were captured with a Minolta Vivid 910 laser scanner, and more recent data is collected with a 3dMD facial camera. Using the Minolta Camera, three scans (front, left and right) of the face were made, and then stitched together to form a large polygon mesh surface. The 3dMD system uses stereo cameras, and is able to generate a 3D mesh surface from a single scan.

4.1 Minolta Scanner

The subject datasets used in testing this particular technique were collected from a cohort of Cape Coloured (CC; mixed ancestry) children participating in a longitudinal study on the effects of heavy prenatal alcohol exposure in Cape Town, South Africa (Jacobson et al., 2008). FAS diagnosis was determined during a clinic held at the 5-year follow-up visit in which each child was independently examined for growth and FAS anomalies using a standard protocol by two expert FAS dysmorphologists, who subsequently reached agreement regarding the diagnosis. A total of 67

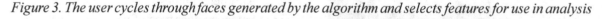

Figure 3. The user cycles through faces generated by the algorithm and selects features for use in analysis

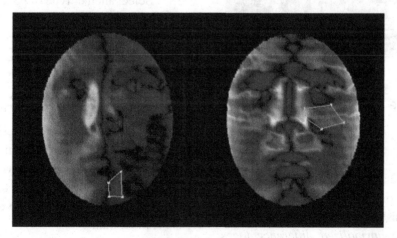

datasets were used, among which 36 were FAS and 31 were controls (non-FAS, non-alcohol exposed). Approximately 50% of each group (FAS and non-FAS) are females. The age ranges are also very similar between the two groups (the mean ages are 5.09 and 4.48 years old, respectively, in the FAS and non-FAS groups).

Through the interactive feature selection process from section 4, we selected 18 regions. Curvature values were used when calculating perturbed difference faces, and difference faces calculated without perturbation were also used. Both face types were explored under a variety of lighting conditions. Once the regions were selected, several feature values were then defined and computed within each selected facial region to represent the geometric properties of the region. These include *Area, Aspect Ratio, Flatness*, and *Curvatures*.

1. **Area:** It measures the surface area of the region. Because a region on a face is defined by a mapping from the corresponding region on the template face through an intermediate morphed face, the differences in areas of various regions reflect the shape differences (primarily in size) of the faces in a given facial region.
2. **Aspect Ratio:** It measures the ratio of the width to the height of a region. Similar to "area", differences in *aspect ratio* reflect shape differences in terms of width compared to height for a certain common region.
3. **Flatness:** It measures how flat the surface is within the region. It can be computed by fitting a planar surface to the region using a least square minimization.
4. **Curvatures:** Curvatures measure the local curving effect of the surface at each point. The average curvatures for all points in a

region provide some information about the shape variations of a region. For polygon mesh surfaces, curvatures are computed using discrete differential geometry operators (Meyer, Desbrun, Schroder, & Barr, 2002). Two typical types of curvatures were computed: *Gaussian Curvature* $\kappa_G = \kappa_1 \times \kappa_2$ and *Mean Curvature* $K_H = (K_1 + K_2)/2$, where κ_1 and κ_2 are *Principal Curvatures* representing the maximum and minimum curvatures in the 2 principal directions.

Once we extracted the 5 feature types in each of the selected regions, the feature analysis was carried out using the Radial Basis Function Networks (RBFN) classifier. The RBFN used was part of the Waikato Environment for Knowledge Analysis (Weka), an open-source collection of machine learning and data-mining tools. We applied a Leave-One-Out cross validation approach. In this approach, the features were first selected using all *n* datasets in each group. Then each individual dataset was used as the test dataset one at a time, and the remaining *n-1* datasets became the training set. This process was repeated *n* times by rotating the test dataset, and the average performance was measured. The classification results are shown in Table 1. The overall classification rate is 88.1%.

In a previous study, a fully automated feature selection technique was used on this dataset (Fang et al., 2008). The overall classification rate for the automated technique was slightly higher, at

Table 1. Classification results

Clinical Diagnosis	Classification (CC)	
	FAS (+)	FAS (-)
FAS	33	3
Controls	5	26
Class. Rate	88.1%	

90.1%, using a test set validation, rather than Leave-One-Out. The initial feature vectors for that analysis contained several thousand features defined on a mix of individual vertices and regular, rectangular regions. The feature selection process itself did not use available domain knowledge and the process of reducing the feature space for analysis was completely automated. In both cases, our region-based features analysis methods gave much better results than early attempts to apply PCA and LDA learning methods for FAS diagnosis (Huang, Jain, Fang, & Riley, 2005).

4.2 3dMD Camera

Using the 3dMD camera, 96 datasets from the Cape Town group were analyzed. From this group, 22 were FAS, and 74 were controls. Using the interactive feature selection technique, we extracted five feature types from 21 facial regions as described (Figure 4).

Results using the RBFN Classifier and a Leave-One-Out cross validation are shown in Table 2: The overall classification rate was 92.7%

5. CONCLUSION

3D surface image analysis has a great potential to impact applications in medicine, engineering, forensics and many other fields. The complexity of the 3D surface data, however, makes the feature selection process difficult. This paper shows that feature visualization provides a very effective feature space searching mechanism and can lead to more accurate and intuitive 3D features that cannot be easily extracted in traditional data analysis methods. Taking advantages of modern graphics hardware, with careful design of data alignment and rendering, the visualization process can be highly interactive, and therefore allow the users to dynamically explore the feature space and

Figure 4. Example faces generated by the algorithm using an updated color scheme for clarity

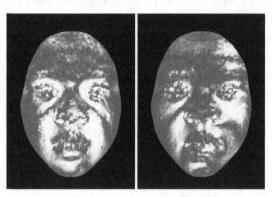

Table 2. Classification results using the RBFN classifier and a leave-one-out cross validation

Clinical Diagnosis	Classification (CC)	
	FAS (+)	FAS (-)
FAS	18	4
Controls	3	71
Class. Rate	92.7083%	

reveal useful and interesting features for more effective and accurate data analysis. This approach can also be a viable alternative for analysis tasks that do not have sufficient number of datasets to effectively apply traditional machine learning and pattern recognition based methods.

In the future, we would like to continue developing the visual feature exploration environment and provide better graphical interfaces and user interaction tools. We would also like to extend the feature types to include non-regional features, such as curve length and volumes.

ACKNOWLEDGMENT

The FAS diagnosis application is supported in part by NIH-NIAAA grants U24AA014809 and U01AA014790; two supplements to R01-

AA09524; NIH Office of Research on Minority Health; the Foundation for Alcohol Related Research, Cape Town, South Africa; and the Joseph Young, Sr., Fund from the State of Michigan. We would like to thank the members of CIFASD for their support in data collection. We also want to thank Christopher Molteno, Joseph Jacobson, and the University of Cape Town Child Development Research Laboratory staff for their help in recruiting and collecting the FAS dysmorphology clinic data. This project is also partly supported by the Open Project Program of the State Key Lab of CAD&CG, Zhejiang University, China.

REFERENCES

Arman, F., & Aggarwal, J. K. (1993). Model-based object recognition in dense-range images—a review. *ACM Computing Surveys, 25*(1), 5–43. doi:10.1145/151254.151255

Barlett, M., & Sejnowski, T. (1997, May 17). Independent components of face images: A representation for face recognition. In *Proceedings of the 4th Annual Joint Symposium on Neural Computation*, Pasadena, CA.

Besl, P. J., & Jain, A. (1988). Segmentation through variable-order surface fitting. *IEEE Transactions on Pattern Analysis and Machine Intelligence, 10*(2), 167–192. doi:10.1109/34.3881

Besl, P. J., & Jain, R. C. (1985). Three-dimensional object recognition. *ACM Computing Surveys, 17*(1), 75–145. doi:10.1145/4078.4081

Besl, P. J., & McKay, N. (1992). A method for registration of 3D shapes. *IEEE Transactions on Pattern Analysis and Machine Intelligence, 14*(2), 239–256. doi:10.1109/34.121791

Beumier, C., & Acheroy, M. (2000). Automatic 3D face authentication. *Image and Vision Computing, 18*(4), 315–321. doi:10.1016/S0262-8856(99)00052-9

Fang, S., McLaughlin, J., Fang, J., Huang, J., Foroud, T., & Autti-Rämö, I. (2008). Automated diagnosis of fetal alcohol syndrome using 3D facial image analysis. *Orthodontics & Craniofacial Research, 11*, 162–171. doi:10.1111/j.1601-6343.2008.00425.x

Gordon, G. (1991). Face recognition based on depth maps and surface curvature. In *Proceedings of the SPIE Conference on Geometric Method in Computer Vision, 1570*, 234–247.

Hardy, R. L. (1971). Multiquadric equations of topography and other irregular surfaces. *Journal of Geophysical Research, 76*, 1905–1915. doi:10.1029/JB076i008p01905

Huang, J., Jain, A., Fang, S., & Riley, E. P. (2005). Using facial images to diagnose fetal alcohol syndrome (FAS). In *Proceedings of the IEEE International Conference on Information Technology: Coding and Computing* (pp. 66-71).

Jacobson, S. W., Stanton, M. E., Molteno, C. D., Burden, M. J., Fuller, D. S., & Hoyme, H. E. (2008). Impaired eyeblink conditioning in children with fetal alcohol syndrome. *Alcoholism, Clinical and Experimental Research, 32*, 365–372. doi:10.1111/j.1530-0277.2007.00585.x

Jain, A. K., & Hoffman, R. (1988). Evidence based recognition of 3D objects. *IEEE Transactions on Pattern Analysis and Machine Intelligence, 10*(6), 783–802. doi:10.1109/34.9102

Jones, K. L., & Smith, D. W. (1973). Recognition of the fetal alcohol syndrome in early infancy. *Lancet, 302*(7386), 999–1001. doi:10.1016/S0140-6736(73)91092-1

Lu, X., Colbry, D., & Jain, A. (2004, July). Matching 2.5D scans for face recognition. In *Proceedings of the International Conference on Biometric Authentication*, Hong Kong, China (pp. 30-36).

Marks, J., Andalman, B., Beardsley, P. A., Freeman, W., Gibson, S., Hodgins, J., et al. (1997). Design galleries: A general approach to setting parameters for computer graphics and animation. In *Proceedings of the 24th Annual Conference on Computer Graphics and Interactive Techniques* (pp. 389-400).

May, P. A., & Gossage, J. P. (2001). Estimating the prevalence of fetal alcohol syndrome. A summary. *Alcohol Research & Health, 25*(3), 159–167.

Meyer, M., Desbrun, M., Schroder, P., & Barr, A. (2002). *Discrete differential geometry operators for triangulated 2-manifolds.* Paper presented at the Visual Mathematics Workshop, Berlin, Germany.

Moore, E. S., Ward, R. E., Jamison, P. L., Morris, C. A., Bader, P. I., & Hall, B. D. (2002). New perspectives on the face in fetal alcohol syndrome: What anthropometry tells us. *American Journal of Medical Genetics, 109*(4), 249–260. doi:10.1002/ajmg.10197

Post, F. J., van Walsum, T., Post, F. H., & Silver, D. (1995). Iconic techniques for feature visualization. *IEEE Visualization,* 288-295.

Schroder, P., & Sweldens, W. (2001). Digital geometry processing. In *Proceedings of the Sixth Annual Symposium on Frontiers of Engineering* (pp. 41-44).

Silver, D., & Wang, X. (1997). Tracking and visualizing turbulent 3D features. *IEEE Transactions on Visualization and Computer Graphics, 3*(2), 129–141. doi:10.1109/2945.597796

Tanaka, H. T., Ikeda, M., & Chiaki, H. (1998). Curvature-base face surface recognition using spherical correlation – principal directions for curved object recognition. In *Proceedings of the International Conference on Automatic Face Gesture Recognition* (pp. 372-377).

Taubin, G. (2000). *Geometric signal processing on polygonal meshes.* Paper presented at the Eurographics Workshop State of the Art Report (STAR), Interlaken, Switzerland.

Turk, M., & Pentland, A. (1991). Eigenfaces for recognition. *Journal of Cognitive Neuroscience, 3*(1), 71–86. doi:10.1162/jocn.1991.3.1.71

Vapnik, V. N. (1998). *The statistical learning theory.* New York, NY: Springer.

Wang, Y. (2002). On cognitive informatics. In *Proceedings of the International Conference on Cognitive Informatics* (pp. 34-42).

Wang, Y. (2007). The theoretical framework of cognitive informatics. *International Journal of Cognitive Informatics and Natural Intelligence, 1*(1), 1–27. doi:10.4018/jcini.2007010101

Weiler, M., Botchen, R., Huang, J., Jang, Y., Stegmaier, S., & Gaither, K. P. (2005). Hardware-assisted feature analysis and visualization of procedurally encoded milti-level volumetric data. *IEEE Computer Graphics and Applications, 25*(5), 72–81. doi:10.1109/MCG.2005.106

Zhao, C., Zhao, D., & Chen, Y. (1996). Simplified Gaussian and mean curvatures to range image segmentation. In *Proceedings of the International Conference on Pattern Recognition* (pp. 427-431).

Zhao, Y., Chen, Y., & Yao, Y. (2008). User-centered interactive data mining. *International Journal of Cognitive Informatics and Natural Intelligence, 2*(1), 58–72. doi:10.4018/jcini.2008010105

This work was previously published in the International Journal of Cognitive Informatics and Natural Intelligence, Volume 5, Issue 2, edited by Yingxu Wang, pp. 1-16, copyright 2011 by IGI Publishing (an imprint of IGI Global).

Chapter 8

A Computational Simulation of the Cognitive Process of Children Knowledge Acquisition and Memory Development

Jeff Bancroft
University of Calgary, Canada

Yingxu Wang
University of Calgary, Canada

ABSTRACT

The cognitive mechanisms of knowledge representation, memory establishment, and learning are fundamental issues in understanding the brain. A basic approach to studying these mental processes is to observe and simulate how knowledge is memorized by little children. This paper presents a simulation tool for knowledge acquisition and memory development for young children of two to five years old. The cognitive mechanisms of memory, the mathematical model of concepts and knowledge, and the fundamental elements of internal knowledge representation are explored. The cognitive processes of children's memory and knowledge development are described based on concept algebra and the object-attribute-relation (OAR) model. The design of the simulation tool for children's knowledge acquisition and memory development is presented with the graphical representor of memory and the dynamic concept network of knowledge. Applications of the simulation tool are described by case studies on children's knowledge acquisition about family members, relatives, and transportation. This work is a part of the development of cognitive computers that mimic human knowledge processing and autonomous learning.

DOI: 10.4018/978-1-4666-2476-4.ch008

1. INTRODUCTION

With the ever increasing storage capacity of new electronic devices, some individuals may suggest that we no longer need to learn how the brain stores information because we will soon be able to have all the information we need in personal computers. While the storage space on a computer tends to double every eighteen to twenty four months, this assumption is completely false. Researches in neural science, biopsychology, and cognitive informatics have discovered that the average human brain possesses approximately 10^{11} neurons and each neuron has an average of 10^3 synaptic connections (Pinel, 1997; Gabrieli, 1998; Sternberg, 1998; Matlin, 1998; Wang, 2009b; Wang & Wang, 2006). The observation on the generally unchanging number of neurons over the life span of an adult leads researchers to believe that information in the brain is stored as relationships between neurons via the creation of synaptic connections (Gabrieli, 1998; Wilson & Keil, 2001; Wang & Wang, 2006). Based on these factors, Wang and his colleagues find that the maximum capacity of human memory, i.e., the possible number of synaptic connections among neurons in the brain, is up to $10^{8,432}$ bits based on a rigorous mathematical model (Wang et al., 2003).

The current size of a desktop computer with dual terabyte drives holds close to 10^{12} bits of information. When we compare how miniscule the amount of information the desktop computer can hold and that of the human brain, we quickly realize how impressive the brain is. We must also consider the accessibility of that information, how quickly we can sort though the information to recall specific knowledge, and how humans are still much better at understanding patterns than a computer. When all of these observations are taken into account, we not only see that the idea of the computer being better than the brain as ridiculous, but it also demands that the brain be studied so that the current computers may be improved and the future generation of cognitive computers may be developed.

All around the globe and throughout history, many people such as Plato, Socrates, and Chuang Tsui have wondered about the cognitive ability of the human mind (Tsui, 400BC; Wang, 2003). This desire for answers led to the development of fields of study such as philosophy, psychology, life science, and knowledge engineering. A new field of enquiry, cognitive informatics, was initiated by Wang and his colleagues in 2002, which establishes a trans-disciplinary study on cognitive science, computer science, information science, cybernetics, and life science (Wang, 2002, 2003, 2007b; Wang et al., 2009). Cognitive informatics investigates natural intelligence, i.e., how the brain acquires, processes, interprets, expresses, and utilizes information, its applications in cognitive computing, and the denotational mathematical means for both natural and computational intelligence (Wang, 2003).

It is recognized that studies about human knowledge acquisition, memory development, and internal knowledge representation can be enriched by observations on mechanisms of young children learning and memory development. Findings in this approach may improve the understanding about human memory and knowledge representation. Based on this study, a computational simulation of the cognitive process of children knowledge acquisition and memory development is designed and implemented. The project reported in this paper uses the theories of the formal concept model of memory to model the growing understanding of a small child. A child of eighteen months should have a vocabulary of three to twenty words and be able to comprehend approximately fifty words. A child of twenty-four months should have a vocabulary of approximately two hundred words (Grizzle & Simms, 2005). This project helps to demonstrate the growth and complexity of the growing amount of information a child knows in order to empirically simulate the relational memory theories (Baddeley, 1990; Squire et al., 1993; Wang, 2009b; Hu et al., 2010) and the mathematical model of formal concepts as the basic unit of human knowledge (Wang, 2008a, 2009a, 2009b).

This paper presents a simulation tool for knowledge acquisition and memory development particularly for young children between 2 to 5 years old. Related work on the physiological and logical models of memory is reviewed in Section 2. The cognitive processes of children memory and knowledge development are described based on concept algebra in Section 3, where the cognitive mechanisms of memory, the mathematical model of concepts and knowledge, as well as the fundamental elements of internal knowledge representation are explained. The design of the tool for children knowledge acquisition and memory development simulation is presented in Section 4 with a graphical representor of memory and the dynamic concept network of knowledge. The implementation of the simulation tool is described in Section 5 where a case study of children knowledge acquisition about transportation is presented.

2. BACKGROUND AND RELATED WORK

According to the cognitive models of memory (Baddeley, 1990; Pinel, 1997; Gabrieli, 1998; Sternberg, 1998; Matlin, 1998; Wang, 2009b; Wang & Wang, 2006), there are four types of memory known as the sensory buffer memory (SBM), short-term memory (STM), long-term memory (LTM), and sensory buffer memory

(SBM). STM contains all the temporal information during mental operation and reasoning. LTM is where all the permanent information and knowledge are stored. This paper focuses on the study on the mechanisms and simulations of LTM, particularly that of children, at the neural, logical, and mathematical levels.

The first level of memory modeling is at the *neural* layer that explains how knowledge is physiologically represented and memorized in the brain. The neural model is a web-like structure created by the neurons and their synaptic connections as shown in Figure 1, where the micro and macro views of memory are illustrated. It is noteworthy that, although it is well understood that the nervous play an important role in the establishment of knowledge in LTM, it is not explained in neuroscience and cognitive science that: (a) Why do adults use fewer neurons (300×10^{11}) to represent more knowledge in memory than that of children (at the peak of 1000×10^{11}) in 8 month-old (Pinel, 1997; Sternberg, 1998)? and (b) Why may increased knowledge acquisition in an adult's brain result in no change of the number of neurons in it? The observation in neurophysiology that the number of neurons is kept stable during life-long growth of knowledge in adult brains is an indirect evidence for supporting the relational cognitive model of information representation in human memory.

Figure 1. The physiological model of memory for knowledge representation

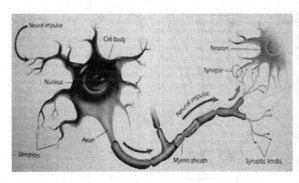

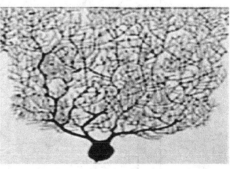

(a) The micro model of memory   (b) The macro model of memory

Based on the relational metaphor that knowledge is represented and memorized by the synaptic connections of neurons, the Object-Attribute-Relation (OAR) model for internal knowledge representation was developed by Wang in 2007. The OAR model (Wang, 2007c) creates a *logic* level of explanation of human memory where information is stored as a finite set of objects, a finite set of attributes, and the relational connection. The OAR model is illustrated in Figure 2 where O is a set of objects, A is a set of attributes, and r is a set of relations such as r(O, A), r(O, O), and r(A, A). The OAR model provides an explanation for internal knowledge representation by creating new relations rather than requiring more neurons. It also provides a rigorous estimation of the maximum capacity of human memory (Wang et al., 2003) as discussed earlier in the introduction section.

The highest level of human memory models beyond the physiological and logical levels is the *mathematical* level. This level can be described by the formal concept model developed by Wang in 2008. This new model adds concepts to the OAR model and dictates that a physiological and logical object can be modeled as a universal mathematical entity known as the formal concept (Wang, 2008a, 2010b). Detailed description of the concept model and concept algebra will be provided in Section 3.2. Related denotational mathematical studies may be referred to (Wang, 2007a, 2008a, 2008b).

Although there are many computational simulations on processes of neural network aggregations, there is no model explaining how knowledge is stored within human memory, particularly LTM, in the brain. Currently, the closest project out there is WordNet (Miller et al., 1990; Miller, 1995), which is a linguistic network of words founded upon how they are connected because of their semantic relations in the English language. If we are able to better understand how information is stored and retrieved from the human brain, it will help us to create better ways to store and use the wealth of information available online in the web and the Internet.

3. DESCRIBING CHILDREN MEMORY DEVELOPMENT BY CONCEPT ALGEBRA

Investigations into the cognitive models of information and knowledge representation in the brain is perceived to be one of the fundamental research areas that help to unveil the mechanisms

Figure 2. OAR: The logical model of memory and internal knowledge representation

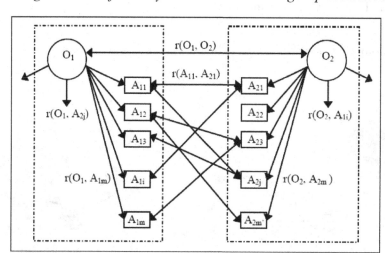

of the brain (Wang, 2003, 2009a, 2010a; Wang et al., 2006, 2009, 2010; Tian et al., 2009, in press). The OAR model (Wang, 2007c) describes human memory, particularly LTM, by using the *relational* metaphor, rather than the traditional *container* metaphor that used to be adopted in psychology, computing, and information science. The OAR model shows that human memory and knowledge are represented by relations, i.e., connections of synapses between neurons, rather than by the neurons themselves as the traditional container metaphor suggested. The OAR model reveals the biological and physiological foundations of human concept and knowledge formation. On the basis of OAR, the logical and mathematical models of concepts and knowledge may be rigorously derived in the following subsections.

3.1 The Cognitive Mechanisms of Memory

The cognitive process of memorization encompasses encoding (knowledge representation), retention (store in LTM), retrieve (knowledge reallocation and LTM search), and decoding (knowledge reformation) as shown in Figure 3 (Wang, 2009b). The sign of a successful memory process in cognitive informatics is that the same information can be correctly recalled or retrieved. Therefore, memorization may need to be repeated for a number of cycles before it is completed.

The memorization process is a closed-loop between STM and LTM, where it may be divided into the establishment and reconstruction phases. The *establishment phase* of memorization is a memory creation process that represents a certain information in the form of a sub-OAR in STM via encoding, and then creates relations with the entire OAR in LTM via retention. The *reconstruction phase* of memorization is a retrieval process that searches the entire OAR in LTM via content patterns or keywords, and then reconfigures the information in STM via decoding.

The tremendous difference of memory magnitudes between human beings and computers demonstrates the efficiency of information representation, storage, and processing in human brains. Computers store data in a direct and unconsumed manner; while the brain stores information by relational neural clusters. The former can be accessed directly by explicit addresses and can be sorted; while the latter may only be retrieved by content-sensitive search and matching among neuron clusters where spatial connections and configurations themselves represent information.

3.2 The Mathematical Model of Concepts and Knowledge

A concept is a cognitive unit to identify and/ or model a real-world concrete entity and a perceived-world abstract subject. A concept can

Figure 3. The cognitive process of memorization

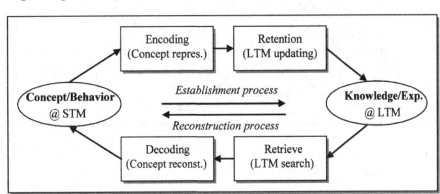

be identified by its intension and extension (Smith & Medin, 1981; Murphy, 1993; Codin et al., 1995; Ganter & Wille, 1999; Medin & Shoben, 1988; Wang, 2008a). The *intension* of a concept is the attributes or properties that a concept connotes, while the *extension* of a concept is the members or instances that the concept denotes. Based on the formal structure of concepts and their relations, meanings of real-world concrete entities may be represented and semantics of abstract subjects may be embodied.

A *formal concept C* is modeled as a 5-tuple (Wang, 2008a), i.e.:

$$C \triangleq (O, A, R^c, R^i, R^o) \qquad (1)$$

where O is a nonempty set of objects of the concept, $O = \{o_1, o_2, ..., o_m\}$, A is a nonempty set of attributes, $A = \{a_1, a_2, ..., a_n\}$, $R^c \subseteq O \times A$ is a set

of internal relations, $R^i \subseteq A' \times A$ is a set of input relations with external concepts. For convenience, R^i may be simply denoted as $R^i = \Re \times C$ where K denotes existing knowledge, and $R^o = C \times \Re$ is a set of output relations.

The most important properties of the formal concept model, as defined in Equation 1, are the capture of a set of essential attributes as its intension; the classification of a set of instantiation objects as its extension; and the adaptive capability to autonomously interrelate itself to other concepts in existing knowledge. According to Equation 1, the general schema of concepts can be modeled as shown in Figure 4.

On the basis of the formal concept model and the OAR model, human *knowledge* is a hierarchical concept network interconnected by a set of concept associations R, i.e.:

Figure 4. The hierarchical relations of concepts and their internal structures

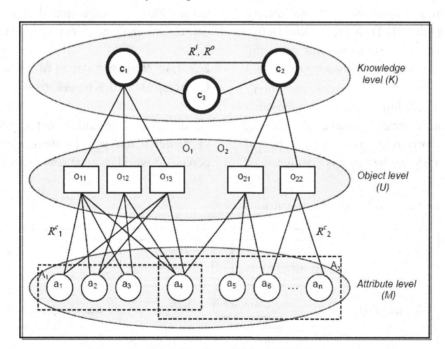

$$\Re \triangleq OAR = \Re : \overset{n}{\underset{i=1}{X}} C_i \rightarrow \overset{n}{\underset{i=1}{X}} C_i \qquad (2)$$

where R has been formally defined in concept algebra (Wang, 2008a).

Concept algebra provides a denotational mathematical means for algebraic manipulations of structured concepts (Wang, 2008a). Concept algebra can be used to model, specify, and manipulate knowledge systems and models of domain ontologies in cognitive informatics, cognitive computing, knowledge engineering, and computational intelligence (Wang, 2006, 2009c, 2010a, 2011).

3.3 The Fundamental Elements of Internal Knowledge Representation

To build a model of how a child's brain may represent and store information, the fundamental blocks of memory must be built. These memory blocks include the classes of concept, object, attribute, and relation. Each instance of a class needs a name and identification number. To help make sure there are no duplicate instances of a class, except relations, the use of a text string class is used. All instance names used in the program would be contained in the class of strings. Every time a new instance is added, a search of the text strings would indicate if the name has been used already. When a relation is to be added, a search through the text strings indicates if the relation of both instances has already exists in the memory model.

To increase the speed of searching through the information stored within the memory model, pointers were used within the fundamental classes. Within each instance of a class, a set of arrays containing pointers that indicates all other instances that are related to it. Every time a new relation connecting node A to node B, both A and B are updated with pointers containing the new relationship. Although this system may increase the time needed to add new information to the model, it removes the need to search through all the relations each time we desire to see how a node is connected within the model.

Within the concept model of memory, the intension of a concept includes all attributes contained within all the objects belonging to the concept. An attribute must belong to at least one object associated to the concept to be considered an attribute of that concept. An example of this can be seen in Figure 5. In Figure 5, one can observe that both objects "Fay" and "Janenne" belong to the concept of "a_mother". The attributes of Fay are that she is a "girl", she is "nice", and she is "fun". Janenne has the attributes of being a "girl" and being "nice". Figure 5 shows that the concept "a_mother" has two related objects, "Fay" and "Janenne", that both are related to the attributes of "girl" and "nice" for the data set as current knowledge. As the data set grows, the graph must continue to reflect the current set of data. With the additional information in Figure 5, that "Lola" is also "a_mother", the concept graph of "a_mother" changes to include "Lola", as seen in Figure 6.

A concept may inherit another concept according to concept algebra. In this case, the newly derived concept will also inherit the same set of attributes belonging to the original concept. As seen in Figure 6, one must be "a_mother" to be "a_grandma". Figure 6 shows that the concept "a_grandma" contains all the attributes contained in the concept "a_mother", plus the additional attribute "fun".

The total set of attributes belonging to a concept is the union of the subsets of attributes associated to all objects of the target concept. Figure 6 indicates the subconcept inherited from the concept "a_mother" as well as the subset of object obtained from the data set in Figure 5.

Figure 5. Concept graph of knowledge about relatives in memory

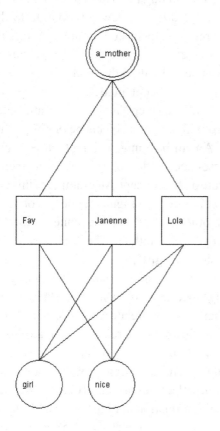

4. DESIGN OF THE TOOL FOR CHILDREN MEMORY DEVELOPMENT SIMULATIONS

The knowledge acquisition and memory development simulation tool encompasses two components: the *graphical representor* of cognitive memory and the *manipulator* of dynamic concept networks for knowledge acquisition and memory development. The design of the simulation tool is focused on young children of 2 to 5 year-old in order to highlight the fundamental mechanisms of the brain in early development. This work forms a part of the entire project on cognitive computers, cognitive computing, and computational intelligence (Wang, 2006, 2009c, 2010a, 2011).

4.1 The Graphical Representor of Cognitive Memory

The easiest way to understand a set of complex information is via graphical representations of the relations among given concepts, objects, and attributes. According to the generic OAR model, the concept graph for memory modeling is a special graph where its nodes represent concepts with double circles, objects with squires, or attributes circles as shown in Figure 7. In the concept graph of Figure 7, the concept instances are represented at the top, object instances are shown in the middle, and attribute instances are indicated at the bottom. Each of these nodes is labeled by the name of the instance. The order from left to right indicates the sequence as to when each node was added to the memory model.

Figure 6. Concept graph developed based on Figure 5

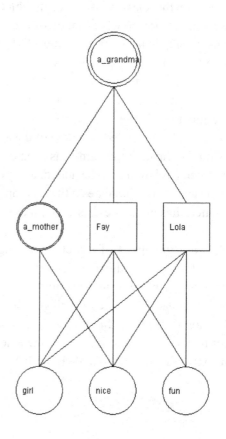

Figure 7. Knowledge about mother and grandma in the concept graph

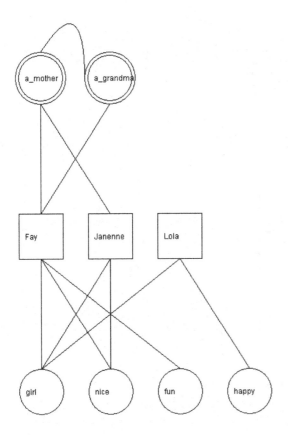

According to concept algebra, a specific concept can be elicited or tailored from a complex concept graph with all directly related objects and attributes pertaining to the derived concept. For example, as shown in Figure 8, the simulation tool can automatically tailor a given concept, a_mother, based on the memory model as shown in Figure 7. In the concept graph of Figure 8, only shared attributes among all objects are displayed.

4.2 The Acquisition and Development of Knowledge as a Dynamic Concept Network

To mimic how new knowledge is stored within the mind of a small child, the memory models are built as a *dynamic concept network* (DCN).

Figure 8. An automatically tailored concept based on Figure 7

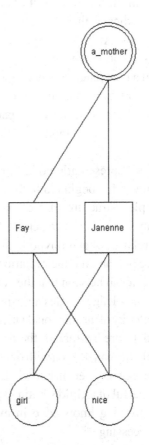

The instances of relations within the model are shown by links (edges) between the nodes. As new information is continually added to the model, the graphs will continue to become more complex, thus the names of each relation is not shown in the graph. Most links indicating a relation are a straight line between two nodes. However, when a relation connects two concepts, a curve is used to represent the relation from the top of the concept to the side of the other concept. This indicates that the first concept specifies, or is a subconcept of, the second concept. An example of the subconcept-concept relation is the link between the concepts a_mother and a_grandma as shown in Figure 7, which denotes that to be a grandmother one must first be a mother, thus a_mother is a subconcept of a_grandma.

The ability of the memory model to grow as new information is added is one of the major distinctions of DCN from neural networks, where in the latter the basic model does not change once the nodes are set in place. With the memory model of DCN, users can learn new knowledge by typing it into the prompt window of the simulation tool. Then, the tool checks for duplicate instance names of concepts, objects, or attributes in order to prevent that an entity is created in the memory model for more than once.

The simulation tool receives input of new relations from users in the format of a three-word pattern as follows:

$$\text{<NodeID> <Relation> <Attribute>} \qquad (3)$$

In Equation 3, the first word in the sentence indicates a node of a concept, object, or attribute; the second word denotes an instance of attribute of the node. A simple example is "Lola is nice". This statement indicates that: NodeID = "Lola", Relation = "is", and Attribute = "nice". The tool creates two given instance nodes, "Lola" and "nice", and add a new relation between them. An example of the creation of a new concept node and new relations can be seen by comparing the differences in the memory models of Figure 7 and Figure 9.

The graphical representation of all the newly memorized knowledge begins to help us understand the complex structure of memory that a small child may have. The data set used in this example is the relatives that the hypothetical child knows. As more persons and their relations to the child are added, it can be seen that the continued acquisition of knowledge creates an exponential growth of complexity of the relational DCN. When Figures 10 and 11 are compared, the drastic increase of complexity can be seen with only a small addition to the existing set of knowledge. This shows how powerful of a child's brain to process exponentially growing knowledge in everyday information processing.

Figure 9. Additional concept and relations added based on Figure 7

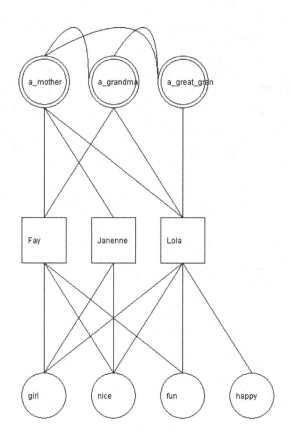

5. IMPLEMENTATION OF THE CHILDREN MEMORY SIMULATION TOOL

The implementation of the knowledge and memory simulation tool is illustrated in Figure 12. The tool supports a number of queries about the learning results represented in the internal DCN model, which includes: (a) What are all the concepts in current memory? (b) What are all the relations in memory? (c) What is a certain <Concept> identified by its attributes and objects? (d) What are the attributes of a certain <Concept>? e) What are the objects of a certain <Concept>? and (f) What are all the relationships involving a given <Concept>?

The tool can be applied to simulate a wide range of children knowledge acquisition and

Figure 10. An incremental knowledge model

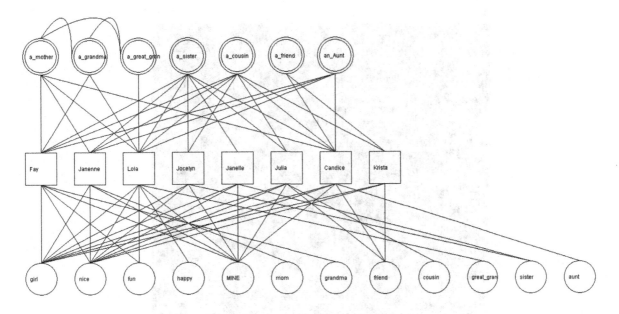

Figure 11. A complex knowledge model with additional objects and relations

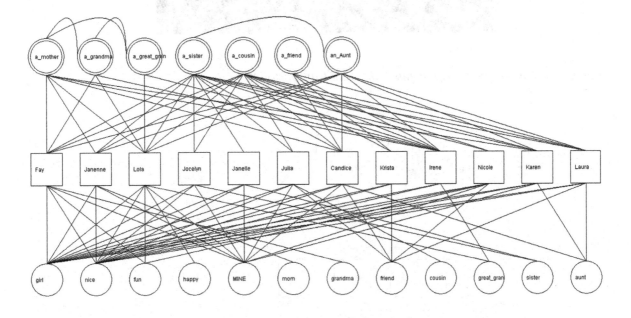

memory development. For instance, a case study on concept development on transportation is shown in Figure 13. Figure 13 simulates and illustrates how children learn and understand information pertaining to vehicles and modes of transportation.

The concept graph as shown in Figure 13 contains five concepts, i.e., transportation, a_vehicle, a_car, a_truck, and a_horse. The first four concepts can be automatically tailored by the tool as shown in Figures 14, 15, 16, and 17, respectively. We can see from Figures 13 and 14 that

Figure 12. Screenshot of the knowledge and memory simulation tool

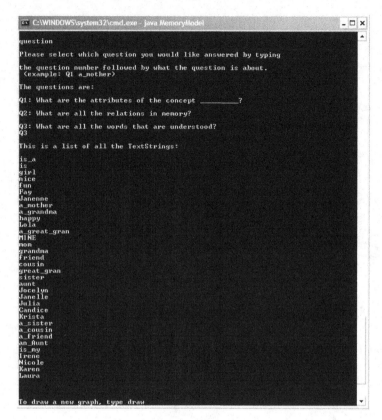

Figure 13. A case of knowledge acquisition about transportation

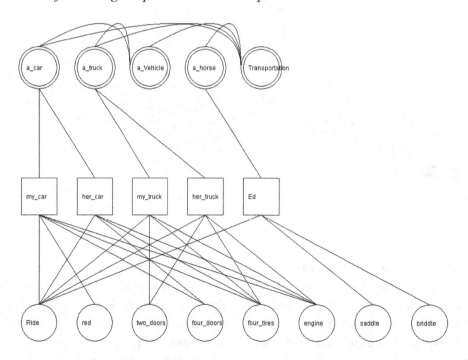

the concept "a_car" has four attributes. From Figures 12 and 14, we can see that the "a_truck" concept also has four attributes. Figure 15 show that the concept "a_vehicle" as a superconcept has only three attributes that belong to both "a_car" and "a_truck". Therefore, the concept "a_vehicle" is more generic or abstract in the knowledge hierarchy according to concept algebra.

The last concept graph on "Transportation" illustrates a high-level superconcept as shown in Figure 17. Figure 17 indicates that there are many modes of transportation by cars, trucks, and bikes, but there are still others such as by a horse. The only common attribute of transportation is that it gives a person a ride, which is shared by all modes of transportation. This very idea shows that a totally ordered *series* of decreasing intensions in a serial concept aggregation is reversely proportional to a totally ordered series of increasing

Figure 14. The concept graph of "a_car"

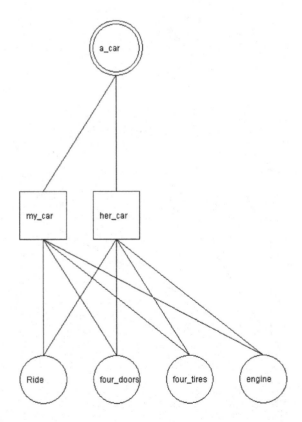

Figure 15. The concept graph of "a_truck"

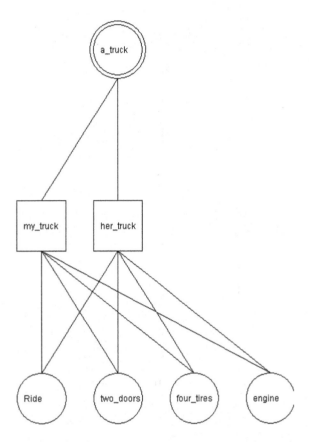

extensions (Wang, 2008a). In other words, the higher the level of a concept in the concept network hierarchy, the fewer the attributes the concept may contain.

The project on the development of the knowledge acquisition and memory development simulation tool will be further extended in order to increase its processing capacity, to provide greater choice of graphs, and to add more intelligent support to users. To increase the processing capacity of the simulation tool, an enhanced implementation of memory nodes and DCN needs to use a linked list to model internal knowledge. Although a child of twenty-four months has an approximate vocabulary of 200 words, a child of thirty-six months should have an approximate vocabulary

Figure 16. The concept graph of "a_vehicle"

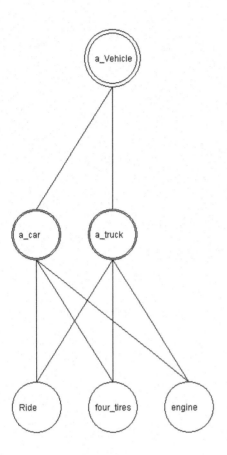

of 900 words (Grizzle & Simms, 2005). As it has been shown that a small concept network with thirty-four words becomes complex because of the large number of relations generated, the number of relations for such a large vocabulary will be immense. In addition, the physical DCN model in computer storage as a permanent knowledgebase and the logical DCN model in memory as an efficient run-time knowledge representation will be fully implemented.

6. CONCLUSION

The work reported in this paper has designed to explore the cognitive process of memory particularly little kids as well as AI entities such as cogni-

tive robots, software agents, and computational intelligent systems. The cognitive mechanisms of memory, the mathematical model of concepts and knowledge, and the fundamental elements of internal knowledge representation have been systematically studied. This paper has presented a simulation tool for knowledge acquisition and memory development particularly for young children between 2 to 5 years old. The cognitive processes of children memory and knowledge development have been explained based on concept algebra and the object-attribute-relation (OAR) model. This work has revealed the fundamental mechanisms of memory, knowledge representation, and the result of learning. It has also revealed that how quickly the complexity increases in knowledge representation and memory establish-

Figure 17. The concept graph of "Transportation"

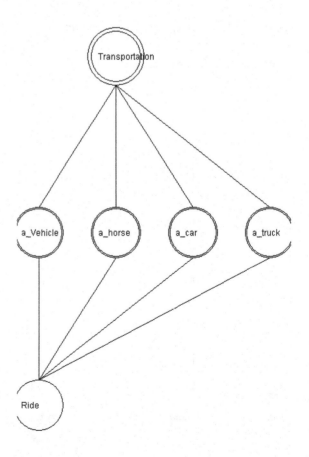

ment as an NP problem. This work is a part of the development towards cognitive computers that mimic human knowledge processing and autonomous learning. The knowledge acquisition and memory development simulation tool supports the dynamic simulation and explicit visualization of simple knowledge acquisition and memory development in children's brain and cognitive robots' knowledgebase.

REFERENCES

Baddeley, A. (1990). *Human memory: Theory and practice*. Needham Heights, MA: Allyn and Bacon.

Codin, R., Missaoui, R., & Alaoui, H. (1995). Incremental concept formation algorithms based on Galois (concept) lattices. *Computational Intelligence*, *11*(2), 246–267. doi:10.1111/j.1467-8640.1995. tb00031.x

Gabrieli, J. D. E. (1998). Cognitive neuroscience of human memory. *Annual Review of Psychology*, *49*, 87–115. doi:10.1146/annurev.psych.49.1.87

Ganter, B., & Wille, R. (1999). *Formal concept analysis*. Berlin, Germany: Springer-Verlag.

Grizzle, K. L., & Simms, M. D. (2005). (in Review). Early language development and language learning disabilities. *Pediatrics*, *26*, 274–283.

Hu, K., Wang, Y., & Tian, Y. (2010). A web knowledge discovery engine based on concept algebra. *International Journal of Cognitive Informatics and Natural Intelligence*, *4*(1), 80–97. doi:10.4018/jcini.2010010105

Matlin, M. W. (1998). *Cognition* (4th ed.). Orlando, FL: Harcourt Brace College Publishers.

Medin, D. L., & Shoben, E. J. (1988). Context and structure in conceptual combination. *Cognitive Psychology*, *20*, 158–190. doi:10.1016/0010-0285(88)90018-7

Miller, G. A. (1995). WordNet: A lexical database for English. *Communications of the ACM*, *38*(11), 39–41. doi:10.1145/219717.219748

Miller, G. A., Beckwith, R., Fellbaum, C. D., Gross, D., & Miller, K. (1990). WordNet: An online lexical database. *International Journal of Lexicograph*, *3*(4), 235–244. doi:10.1093/ijl/3.4.235

Murphy, G. L. (1993). Theories and concept formation. In Mechelen, I. V. (Ed.), *Categories and concepts, theoretical views and inductive data analysis* (pp. 173–200). New York, NY: Academic Press.

Pinel, J. P. J. (1997). *Biopsychology* (3rd ed.). Needham Heights, MA: Allyn and Bacon.

Smith, E. E., & Medin, D. L. (1981). *Categories and concepts*. Cambridge, MA: Harvard University Press.

Squire, L. R., Knowlton, B., & Musen, G. (1993). The structure and organization of memory. *Annual Review of Psychology*, *44*, 453–459. doi:10.1146/annurev.ps.44.020193.002321

Sternberg, R. J. (1998). *In search of the human mind* (2nd ed.). Orlando, FL: Harcourt Brace & Co.

Tian, Y., Wang, Y., Gavrilova, M. L., & Ruhe, G. (in press). A formal knowledge representation system for the cognitive learning engine. In *Proceedings of the 10th IEEE International Conferences on Cognitive Informatics and Cognitive Computing*. Washington, DC: IEEE Computer Society.

Tian, Y., Wang, Y., & Hu, K. (2009). A knowledge representation tool for autonomous machine learning based on concept algebra. *Transactions of Computational Science, 5*, 143–160. doi:10.1007/978-3-642-02097-1_8

Tsui, C. (400BC). Autumn water (Ch. 17). In *Outer chapters.*

Wang, Y. (2002, August). Keynote: On cognitive informatics. In *Proceedings of the 1st IEEE International Conference on Cognitive Informatics*, Calgary, AB, Canada (pp. 34-42). Washington, DC: IEEE Computer Society.

Wang, Y. (2003). On cognitive informatics. *Brain and Mind: A Transdisciplinary Journal of Neuroscience and Neurophilisophy, 4*(3), 151-167.

Wang, Y. (2006, July). Keynote: Cognitive informatics - Towards the future generation computers that think and feel. In *Proceedings of the 5th IEEE International Conference on Cognitive Informatics*, Beijing, China (pp. 3-7). Washington, DC: IEEE Computer Society.

Wang, Y. (2007a). *Software engineering foundations: A software science perspective (Vol. 2)*. Boca Raton, FL: CRC Press.

Wang, Y. (2007b). The theoretical framework of cognitive informatics. *International Journal of Cognitive Informatics and Natural Intelligence, 1*(1), 1–27. doi:10.4018/jcini.2007010101

Wang, Y. (2007c). The OAR model of neural informatics for internal knowledge representation in the brain. *International Journal of Cognitive Informatics and Natural Intelligence, 1*(3), 64–75. doi:10.4018/jcini.2007070105

Wang, Y. (2008a). On concept algebra: A denotational mathematical structure for knowledge and software modeling. *International Journal of Cognitive Informatics and Natural Intelligence, 2*(2), 1–19. doi:10.4018/jcini.2008040101

Wang, Y. (2008b). On contemporary denotational mathematics for computational intelligence. *Transactions of Computational Science, 2*, 6–29. doi:10.1007/978-3-540-87563-5_2

Wang, Y. (2009a). Toward a formal knowledge system theory and its cognitive informatics foundations. *Transactions of Computational Science, 5*, 1–19. doi:10.1007/978-3-642-02097-1_1

Wang, Y. (2009b). Formal description of the cognitive process of memorization. *Transactions of Computational Science, 5*, 81–98. doi:10.1007/978-3-642-02097-1_5

Wang, Y. (2009c). On cognitive computing. *International Journal of Software Science and Computational Intelligence, 1*(3), 1–15. doi:10.4018/jssci.2009070101

Wang, Y. (2010a). Cognitive robots: A reference model towards intelligent authentication. *IEEE Robotics and Automation, 17*(4), 54–62. doi:10.1109/MRA.2010.938842

Wang, Y. (2010b). On concept algebra for computing with words (CWW). *International Journal of Semantic Computing, 4*(3), 331–356. doi:10.1142/S1793351X10001061

Wang, Y. (2011). On cognitive models of causal inferences and causation networks. *International Journal of Software Science and Computational Intelligence, 3*(1), 50–60.

Wang, Y., Kinsner, W., Anderson, J. A., Zhang, D., Yao, Y., & Sheu, P. (2009). A doctrine of cognitive informatics. *Fundamenta Informatica, 90*(3), 203–228.

Wang, Y., Kinsner, W., & Zhang, D. (2009). Contemporary cybernetics and its faces of cognitive informatics and computational intelligence. *IEEE Transactions on System, Man, and Cybernetics (B), 39*(4), 823–833. doi:10.1109/TSMCB.2009.2013721

Wang, Y., Liu, D., & Wang, Y. (2003). Discovering the capacity of human memory. *Brain and Mind: A Transdisciplinary Journal of Neuroscience and Neurophilosophy, 4*(2), 189-198.

Wang, Y., & Wang, Y. (2006). On cognitive informatics models of the brain. *IEEE Transactions on Systems, Man and Cybernetics. Part C, Applications and Reviews, 36*(2), 203–207. doi:10.1109/TSMCC.2006.871151

Wang, Y., Wang, Y., Patel, S., & Patel, D. (2006). A layered reference model of the brain (LRMB). *IEEE Transactions on Systems, Man and Cybernetics. Part C, Applications and Reviews, 36*(2), 124–133. doi:10.1109/TSMCC.2006.871126

Wilson, R. A., & Keil, F. C. (2001). *The MIT encyclopedia of the cognitive sciences*. Cambridge, MA: MIT Press.

This work was previously published in the International Journal of Cognitive Informatics and Natural Intelligence, Volume 5, Issue 2, edited by Yingxu Wang, pp. 17-36, copyright 2011 by IGI Publishing (an imprint of IGI Global).

Chapter 9
A Novel Emotion Recognition Method Based on Ensemble Learning and Rough Set Theory

Yong Yang
Chonggqing University of Posts and Telecommunications, China

Guoyin Wang
Chonggqing University of Posts and Telecommunications, China

ABSTRACT

Emotion recognition is a very hot topic, which is related with computer science, psychology, artificial intelligence, etc. It is always performed on facial or audio information with classical method such as ANN, fuzzy set, SVM, HMM, etc. Ensemble learning theory is a novelty in machine learning and ensemble method is proved an effective pattern recognition method. In this paper, a novel ensemble learning method is proposed, which is based on selective ensemble feature selection and rough set theory. This method can meet the tradeoff between accuracy and diversity of base classifiers. Moreover, the proposed method is taken as an emotion recognition method and proved to be effective according to the simulation experiments.

INTRODUCTION

In recent years, cognitive informatics (CI) emerges as a profound interdisciplinary research area that consists of modern informatics, computation, software engineering, artificial intelligence (AI), neural psychology and cognitive science. It studies the internal information processing mechanisms and natural intelligence of the brain (Wang, 2009, 2007a; Wang & Kinsner, 2006).

Affective computing is also an interdisciplinary research area, which related with computer science, psychology, artificial intelligence, etc. It is proposed by Picard, which handles with recognition, expressing, modeling, communicating and responding to emotion (Ahn & Picard, 2006; Picard, 2003). As for the relationship of emotion and cognition, at first, most cognitive psychologists ignore the issue of the effects of emotion on cognition by trying to ensure that all their participants are in a relatively neutral emotional state. Nowadays, it is widely accepted that affective information would selectively influences cognitive procedure, such as attention, learning, and memory. On the other hand, there are numer-

DOI: 10.4018/978-1-4666-2476-4.ch009

ous studies showing that emotional experience is influenced by cognitive appraisal. Although the relationship between cognitive appraisals and specific emotional experience may sometimes be weak because any given emotion can be produced by various combinations of appraisals, the relationship is widely accepted and is researched by different researchers (Wang, 2007b; Picard, 2003). In a sense, the research on emotion and cognition will be helpful for each other.

In the research works on affective computing, emotion recognition is one of the most fundamental and important modules. Usually, emotion recognition is studied by the methods of artificial neural network (ANN), fuzzy set (FS), support vector machine (SVM), hidden Markov model (HMM), rough set (RS), and the recognition rate often arrives at 64% to 98% (Picard, 1997, 2003; Picard, Vyzas, & Healey, 2001).

Until now, research on emotion recognition is mainly according to the basic type of emotion, such as happiness, sadness, surprise, anger, disgust, fear and neutral. Some applications focus on the particular emotion states, for example, sleepy is focused in a driver monitor system. Research on complicated and mixed emotion recognition, such as bittersweet, and slight emotion are very difficult, since they can't be taken as a simple classification problems, and they are hardly solved based on traditional classification methods. As for the complicated and mixed emotion, we prefer to model how it can be mixed by basic emotion states and how it can be changed over time, but it is beyond the scope of this paper.

It is a long way to achieve a computer act as a human in emotion recognition since there are many problems unsolved in psychology and cognitive theories, for example, how does emotion come into being, what is the essence of emotion, and what is the feature of emotion. Among these problems, it is an open question that which features are important and essential for emotion, and which features are crucial for emotion recognition.

Since it is hardly to get the inner features without any interfere on human, and at the same time, guarantee human be in a natural states. Research on emotion recognition is always taken based on facial or speech features. In this paper, emotion recognition is researched based on facial features according to seven classical types, i.e., happiness, sadness, surprise, anger, disgust, fear and neutral.

Ensemble learning has been a hot research topic in machine learning since 1990s' (Ditterrich, 1997). Ensemble methods are based on learning algorithms that construct a set of base classifiers and then classify new objects by integrating the prediction of the base classifiers. An ensemble system is often much more accurate than each base classifiers. Ditterrich proved the effectiveness of ensemble methods from viewpoint of statistic, computation and representation (Ditterrich, 2001). As a popular machine learning method, ensemble methods are often used in pattern recognition, network security, medical diagnosis, etc. (Freund, 1995; Breiman, 1996; Tsymbal, Pechenizkiy, & Cunningham, 2005).

A necessary and sufficient condition for an ensemble system to be more accurate than any of its base classifiers is that the base classifiers are accurate and diverse. An accurate classifier is one that has an error rate less than random guessing on new instance. Two classifiers are diversity if they make different errors on unseen objects (Ditterrich, 2001). Besides accuracy and diversity, another important issue for creating an effective ensemble system is the choice of the function for combining the predictions of the base classifiers. There are many techniques for the integration of an ensemble system, such as majority voting, weighted voting, reliability-based weighted voting, etc (Tsymbal, Pechenizkiy, & Cunningham, 2005).

There are so many methods proposed for ensemble. The most popular way for ensemble is to get different subset of the original dataset by resampling the training data set many times. Bagging (Breiman, 1996), boosting (Freund,

1995) and cross- validation are all such ensemble methods. These methods work well especially for unstable learning algorithms, such as decision trees, neural network. Some other methods are also studied, such as manipulating the output targets (Diettench, 1995), injecting randomness into classifiers (Ditterrich, 2001). Besides these methods, there is another effective approach for ensemble, and it is called ensemble feature selection (Opitz, 1999). Ensemble feature selection (EFS) is also a classical ensemble method. It is a method focusing on the diversity of features for ensemble.

There are two methods for generating base classifiers and integrating the predictions of base classifications. One is called direct strategy; the other is called over produce and choose strategy. The direct strategy aims to generate an ensemble of base classifiers directly in the training period. Usually, some heuristic rules are used in direct strategy, and it will make the method more efficient. The over produce and choose strategy is also called selective ensemble, which creates a lot of base classifiers at first, and then select a subset of the most suitable base classifiers and generate the final prediction. Although the direct strategy is more efficient, but maybe some good candidate classifiers would be missed, therefore, the effectiveness of an ensemble system would be affected. In this paper, the selective ensemble strategy is considered.

Rough set (RS) is a valid mathematical theory for dealing with imprecise, uncertain, and vague information. It has been applied successfully in such fields as machine learning, data mining, pattern recognition, intelligent data analyzing and control algorithm acquiring, etc. (Pawlak, 1984a, 1984b; Wang, 2001, 2003; Skowron & Pal, 2003; Zhong, Dong, & Ohsuga, 2001), since it was developed by Professor Z. Pawlak in 1980s (Pawlak, 1984a,1984b). The most advantage of rough set is its great ability for attribute reduction (feature selection). In this paper, owing to the great ability of feature selection based on rough

set theory, a novel selective ensemble feature selection method is proposed.

The rest of this paper is organized as follows. At first, some related works are reviewed. Secondly, based on the basic concepts and methods of rough set theory and diversity of ensemble learning, an algorithm for selective ensemble feature selection is proposed. Thirdly, simulation experiments are taken. Finally, conclusion and future works are discussed.

Related Works

In this section, related works included ensemble feature selection, selective ensemble, and some ensemble methods based on rough set theory are introduced.

Ensemble Feature Selection

By changing the feature subsets used to generate the base classifiers, ensemble feature selection aims to produce base classifiers which tend to err in different objects and promote diversity among these base classifiers. Traditional feature selection algorithms have the goal of finding the best feature subset which is relevant to both the learning task and the learning algorithm. Ensemble feature selection algorithm has an additional goal of finding feature subsets that will maximize disagreement among the base classifiers.

There are many EFS methods proposed by different researchers. T. K. Ho showed that simple random selection of feature subsets may be an effective technique for ensemble feature selection in Ho (1998). This method is usually taken as a base of ensemble feature selection strategies. Salcedo and Whitley improved Ho's approach by improving the quality of ensemble members. They used a genetic algorithm (GA) to produce the ensemble members and they showed that this almost always improve random ensemble methods (Salcedo & Whitley, 1999).

Selective Ensemble

Selective ensemble is an effective strategy for ensemble, but it is not always suitable for all the ensemble methods. It is effective when base classifiers are parallelized generated, like bagging. Zhou et al. proposed a selective ensemble method based on neural network, in which the genetic algorithm was used for selecting the suitable subset of base classifiers (Zhou, Wu, & Tang, 2002). Giacinto and Roli (2001) proposed another ensemble method based on neural network, in which all the candidates were clustering, and suitable base classifiers were chosen from each cluster. In this method, a selective strategy for clustering and choosing is also used.

Ensemble Methods Based on Rough Set Theory

Since different feature subsets can be found based on rough set theory, rough set can be combined with ensemble methods. Hu (2001) proposed an ensemble method base on rough set theory and database operations. It is a method of EFS based on direct strategy. Firstly, a set of reducts was computed directly, which included all the indispensable attributes required for the decision categories. Next, a novel reduction induction algorithm was used to compute the maximal generalized rules for each reduct table and a set of classifiers was formed based on the corresponding reducts. In the method, each reduct was a minimum subset of attributes, and had the same classification ability as the entire attributes, each classifier constructed from the corresponding reduct had a minimal set of classification rules, and was as accurate as possible and at the same time as diverse as possible from the other classifiers, the simulation experiments showed that the number of classifiers used to improve the accuracy was much less than that of other methods. There are some other research works similar to Hu's work. For example, Wroblewski (2001) proposed an ensemble method based on approximate reducts. Mei et al. (2005) proposed an ensemble method based on rough set theory and SVM.

Q. D. Wang, et al proposed a different ensemble method based on rough set. It is a method of manipulating the training set (Wang, Wang, & Huang, 2004). In this method, an approximate reduct is presented firstly which can reflect the changes of the weight distribution on the training set. Then, based on the reduct, a new ensemble learning algorithm similar to Adaboost was designed. The algorithm maintained the weight distribution on the training set. During each iteration of the algorithm, this distribution was adjusted, an approximate reduct of feature set was generated, and a new weak classifier was constructed with the approximate reduct. The ensemble classifier was weighted voting on all weak classifiers. Experiments on UCI datasets demonstrated the efficiency of this algorithm.

It is acknowledged that the most advantage of rough set is its great ability of attribute reduction (feature selection). The ability of feature selection of rough set should be promoted when rough set theory combined with ensemble learning theory. Unfortunately, rough set theory was just used for generating a base classifier in Wang's research work, more useful feature suitable for ensemble did not consider continually. The research work proposed by Hu promoted the ability feature selection, but it was based on direct strategy, maybe it would miss some useful base classifiers for ensemble in the process of directly producing base classifiers. In this method, an ensemble feature selection method is proposed which can embody the ability of feature selection of rough set, and all the possible candidate base classifiers are considered.

Selective Ensemble Feature Selection Based on Rough Set Theory

In this section, based on the related works, a novel selective ensemble feature selection based on rough set theory is proposed. At first, basic concept of rough set theory and diversity in ensemble learning are surveyed. And then, the proposed method is introduced.

Basic Concept of Rough Set Theory

Some basic concepts of rough set are introduced here for the convenience of following discussion.

Defintion 1: A decision information system is a formal representation of a data set to be analyzed. It is defined as a pair $S=(U, R, V, f)$, where U is a finite set of objects and $R = C \cup D$ is a finite set of attributes, C is the condition attribute set and $D=\{d\}$ is the decision attribute set. With every attribute $a \in R$, set of its values V_a is associated. Each attribute a determines a function $f_a : U \to V_a$.

Definition 2: For a subset of attributes $B \subset A$, the indiscernibility relation is defined as $Ind(B) = \{(x, y) \in U \times U : a(x) = a(y), \forall a \in B\}$

Definition 3: The lower approximation $B\_(X)$ and upper approximation $B^-(X)$ of a set of objects $X \subseteq U$ with reference to a set of attributes $B \subseteq A$ are defined in terms of the classes in the indiscernibility relation as follows:

$$B\_(X) = \cup\{E \in U \ / \ Ind(B) \mid E \subseteq X\},$$
$$B^-(X) = \cup\{E \in U \ / \ Ind(B) \mid E \cap X \neq \Phi\}$$

They are called the B_lower and B^- upper approximation of X respectively.

Definition 4: $POS_P(Q) = \cup_{x \in U / Ind(Q)} P\_(X)$ is the P positive region of Q, where P and Q are both attribute sets of an information system.

Definition 5: A reduction of P of an information system is a set of attributes $S \subseteq P$ such that all attributes $a \in P - S$ are dispensable, all attributes $a \in S$ are indispensable and $POS_S(Q) = POS_P(Q)$. We use the term $RED_Q(P)$ to denote the family of all reductions of P. $CORE_Q(P) = \cap RED_Q(p)$ is called the Q-core of the attribute set P.

Definition 6: The discernibility matrix $M_D C = \{c_{ij}\}_{n*n}$ of an information system S is defined as:

$$c_{ij} = \begin{cases} \{a \in C : a(x_i) \neq a(x_j)\}, & D(x_i) \neq D(x_j) \\ 0, & D(x_i) = D(x_j) \end{cases} \quad i = 1, 2, ..., n$$

where $D(x_i)$ is the attribute value of the decision attribute.

Based on the dicernibility matrix, all possible reducts can be generated. An attribute reduction algorithm based on dicernibility and logical operation is proposed (Wang, 2001). See Algorithm 1.

Any attributes combination of C_0 as well as a conjunctive term of P' can be an attribute reduct of the original information system. All possible reducts of the original information system can be generated. Assume $|U|= n$, $|C|=m$, the complexity of algorithm 1 is O(mn^2) (Wang, 2001).

Diversity in Ensemble Method

Theoretically speaking, if the based classifiers are more diversity between each other, an ensemble system will be more accurate than its base classifiers. Therefore, there are a number of ways to measure the diversity of ensemble methods. The first type is called pairwise diversity measures, which are able to measure the diversity in predictions of a pair of classifiers, and the total ensemble diversity is the average of all the classifier pairs of the ensemble. There are some pairwise diversity measures, for example, plain disagreement, the fail/non-fail disagreement measure, the Q statistic, the correlation coefficient, the kappa statistic and double fault measure (Kuncheva, 2003; Tsymbal, Pechenizkiy, & Cunningham, 2005; Brown, Wyatt, Harris, & Yao, 2005). The second one is called non-pairwise diversity measures, which measure the diversity in predictions of the whole ensemble only. The entropy measure, the measure of difficult, coincident failure diversity, and the generalized diversity belongs to the non-pirwise diversity measures (Kuncheva, 2003; Tsymbal, Pechenizkiy, & Cunningham, 2005; Brown, Wyatt, Harris, & Yao, 2005).

As discussed, there are many diversity measure and integration methods. What is the relationship between a diversity measure and an integration method? Is it efficient when we choose a diversity measure and an integration method randomly for an ensemble system? The relationship between 10 diversity measure and 5 integration methods are discussed in Shipp and Kuncheva (2002). The authors found little correlation between the integration methods and diversity measure. In fact, most of them showed independent relationship. Only double fault measure and the measure of difficult showed some correlations greater than 0.3. The measure of difficult showed stronger correlation with the integration methods than the double fault measure, unfortunately it is more computationally expensive. Therefore, double fault measure and integration method of majority are used in this proposed ensemble method since they showed higher correlation and they are both computationally simple.

The double fault measure (DF) was proposed by Giacinto and Roli (2001). It is the ratio between the number of observations on which two classifiers are both incorrect. It is defined as follows.

$$Div_{i,j} = \frac{N^{00}}{N^{11} + N^{10} + N^{01} + N^{00}} \quad (1)$$

where N^{ab} is the number of instances in the data set, classified correctly (a=1) or incorrectly (a=0) by the classifier i, and correctly (b=1) or incorrectly (b=0) by the classifier j. The denominator in (1) is equal to the total number of instances N.

Algorithm 1. Attribute reduction algorithm based on dicernibility matrix and logical operation.

```
Inputs: An information system S with its discernibility matrix M_DC.
Output: Reducts of S.
Step 1: Find all core attributes in the discernibility matrix M_DC, that is, Redut=C_0.
Step 2: Find the set (T) of elements (C_ij's) of M_DC that is nonempty and does not contain
any core attribute.
T={C_ij:C_ij∩ C_0=Φ∧ C_ij ≠Φ}
Step 3: A logic function is generated by treating each attribute as a Boolean variable
and then forming Boolean conjunctions of disjunctions of components belonging to each
element (C_ij) of T. That is,
P=∧_ij{∨_k{a_k}:a_k∈C_ij∧C_ij∈T}
Step 4: Express the logic function P in a simplified form (P') of a disjunction of mini-
mal conjunctive expressions by applying the distributivity and absorption laws of Bool-
ean algebra.
Step 5: Select suitable reducts for the problem.
```

Selective Ensemble Feature Selection Method Based on Rough Set Theory (SEFSBRS)

Based on rough set theory and the diversity measure of the double fault measure, a novel selective ensemble feature selection method based on rough set theory is proposed and introduced in this section.

In the first step, multiple candidate base classifiers can be gotten based on rough set theory. Firstly, all possible reducts are generated on the training set based on algorithm 1. That is to say, all the possible feature subsets can be found and each subset is equal to the original feature set. Secondly, all candidate base classifiers can be trained based on the different feature subsets. According to this method, multiple classifiers can be generated based on rough set theory, and each classifier can be seen equal with the classifier trained on the original feature set. On the other hand, multiple classifiers are coming from different feature subset, therefore, the diversity of the base classifiers can be guaranteed. It is the merit of rough set theory for the proposed method that it can result multiple accurate base classifiers, meanwhile, it can guarantee the diversity of different base classifiers.

In the second step, the most diverse base classifier can be selected. Firstly, diversity between two classifiers can be measured on validation set according to Equation 1. Secondly, each classifier can be seen as a sample, and all the classifiers can be clustered based on the diversity measurement results. Finally, a pair of base classifiers, which are most diversity among two clusters, are chosen from each two clusters.

In the third step, the selected base classifiers are integrated. Majority voting is taken as the integration method for ensemble, and the final prediction can be taken on the testing set.

The detailed algorithm is introduced in Algorithm 2.

The complexity of Algorithm 2 is mainly depending on step 1.1, step 2.1 and step 2.3. Assume $|U|= n$, $|C|=m$, the number of classifiers is p, the number of cluster is q. the complexity of step 1.1 is $O(mn^2)$, the complexity of step 2.1 is $O(p^2)$, the complexity of step 2.3 is $O(q^2)$. Since $p,q <<n$, the complexity of algorithm 2 is $O(mn^2)$. Since the complexity of algorithm 2 is mainly depend on algorithm 1, if some other fast algorithm is used, it can guarantee to find all the reducts and accelerate the speed of algorithm 2.

Experiment Results and Analysis

In this section, the proposed method is used as an emotion recognition method. Some experiments are carried out. Since there are few open facial emotional dataset included all the races, sex, ages, three facial emotional datasets are intendedly selected and used in the experiments. The first dataset comes from the Cohn-Kanade AU-Coded Facial Expression (CKACFE) database (Kanade, Cohn, & Tian, 2000) and the dataset is a representation of western people in some extent. The second one is the Japanese female facial expression (JAFFE) database (Lyons, Akamatsu, Kamachi, & Gyoba, 1998) and it is a representation of eastern women in some extent. The third one named CQUPTE (Yang, 2008) and it is collected from 6 graduate students in Chongqing University of Posts and Communications in China, in which three are three female and three male. Details of the datasets are listed in Table 1. Some samples in the three datasets are shown in Figure 1.

In this section, three comparative experiments are taken. At first, the three datasets are divided into a training set, a validation set and a test set according to the ratio of 3:2:2 for all the experiments. 5-fold cross validation is carried out for each experiment. For all the three experiments, SVMs are trained as base classifiers.

In the first experiment, the proposed method (SEFSBRS) is used for emotion recognition.

Algorithm 2. Selective ensemble feature selection based on rough set theory (SEFSBRS).

```
Input: Decision tables consisting of the training set, validation set and testing set.
Output: Final ensemble prediction.
Step 1.1:  Generate all the reducts on the training set based on Algorithm1.
Step 1.2:  Construct all the classifiers based on the reducts.
Step 2.1:
For each classifier Do
Calculate div(i,j) of each two classifiers on the validation set according to equation
(1).
End for
Step 2.2:  Based on all the calculated div(i,j), all the classifiers are clustered.
Step 2.3:
For each two clusters Do
Select a pair of classifiers which are most diversity among all pairwise classifiers of
the two clusters.
End for
Step 3: Generate the final prediction of the ensemble system based on majority voting of
the selected classifiers on the testing set.
```

In the second experiment, a single classify is used for emotion recognition. Firstly, a classical reduction algorithm named CEBARKNC is used as a feature selection method, and some important features are selected. Secondly, a SVM classifier is trained based on the selected feature subset.

In the third experiment, an ensemble method based on all the classifiers is used for emotion recognition, and this method is called ensemble all in the following. Firstly, multiple classifiers can be trained based on all reducts according to Algorithm 1. Secondly, all the classifiers can be integrated based on majority voting.

The correct recognition rates of each method for the three datasets are shown in Table 2.

Firstly, when we compared the first experiment with the second one, we can found that the proposed method can use more classifiers and get more correct recognition rate of nearly 10%. From the viewpoint of recognition system, ensemble method is proved an effective method since it can get better recognition results. Secondly, when we compared the first experiment with the third one,

we can found that SEFSBRS can use nearly half classifiers and get higher correct recognition ratio, therefore, we can draw a conclusion that the proposed selective ensemble strategy based on rough set reducts is effective for emotion recognition. Accordingly, the proposed method SEFSBRS is proved effective especially for the real time emotion recognition system, since the method can use less system resources and get better recognition results. In a word, the proposed method SEFSBRS is an effective emotion recognition method.

From the experiment results, we can found that the proposed system can be taken as a suitable emotion classification method. The effectiveness of the proposed method is also proved on some UCI datasets (Yang, Wang, & He, 2007). That is to say, the proposed method can be taken as a good method for classification. Although seven classical emotion states are discussed in this paper, the proposed method can be used for other emotion classification in some real applications.

Figure 1. Facial emotion samples

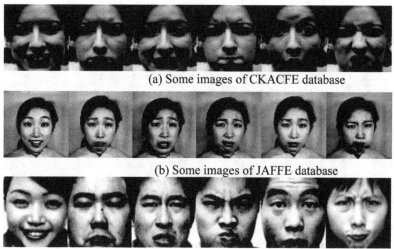

(a) Some images of CKACFE database

(b) Some images of JAFFE database

(c) Some images of CQUPTE database JAFFE database

Table 1. Three facial emotional datasets

Dataset Name	Samples	People	Emotion Classes
CKACFE	405	97	Happiness, Sadness, Surprise, Anger, Disgust, Fear, Neutral
JAFFE	213	10	Happiness, Sadness, Surprise, Anger, Disgust, Fear, Neutral
CQUPTE	652	6	Happiness, Sadness, Surprise, Anger, Disgust, Fear, Neutral

Table 2. Experiment results

Dataset	SEFSBRS		CEBARKNC		Ensemble All	
	Classifier Number	Correct Recognition Ratio	Classifier Number	Correct Recognition Ratio	Classifier Number	Correct Recognition Ratio
CKACFE	8.75	83.674	1	73.07	17.75	81.35
JAFFE	5.88	70.16	1	63.17	11.50	68.88
CQUPTE	5.38	89.566	1	78.83	10.75	90.63
Average	6.67	81.13	1	71.69	13.33	80.29

CONCLUSION AND FUTURE WORK

In this paper, a novel ensemble learning method is proposed, which is based on selective ensemble feature selection and rough set theory. Moreover, the proposed method is used for emotion recognition and proved to be effective for emotion recognition based on the simulation experiments. In the future work, the ensemble strategy based on rough set theory will be researched continually, the proposed method will be used in a real emotion recognition system and non-basic emotion recognition, such as sleepy, would be researched based on the proposed method.

ACKNOWLEDGMENT

The paper is supported by Natural Science Foundation of China under Grant No. 60773113, Natural Science Foundation Grant of Chongqing under Grant No. 2008BA2041, No. 2008BA2017 and No. 2007BB2445, Science & Technology Research Program of Chongqing Education Commission under grant No. KJ110522, Chongqing Key Lab of Computer Network and Communication Technology Foundation under Grant No. CY-CNCL-2009-02, Chongqing University science foundation under Grant No. A2009-26, No. JK-Y-2010002.

REFERENCES

Breiman, L. (1996). Bagging predictors. *Machine Learning*, *24*(2), 123–140. doi:10.1007/BF00058655

Breiman, L. (2001). Random forests. *Machine Learning*, *45*(1), 5–32. doi:10.1023/A:1010933404324

Brown, G., Wyatt, J., Harris, R., & Yao, X. (2005). Diversity creation methods: A survey and categorisation. *Journal of Information Fusion*, *6*(1), 1–28.

Dietterich, T. G. (1997). Machine learning research: four current direction. *Artificial Intelligence Magazine*, *18*(4), 97–136.

Dietterich, T. G. (1998). An experimental comparison of three methods for constructing ensembles of decision trees: Bagging, boosting, and randomization. *Machine Learning*, *40*(2), 139–157. doi:10.1023/A:1007607513941

Dietterich, T. G. (2001). Ensemble methods in machine learning. In J. Kittler & F. Roli (Eds.), *Proceedings of the First International Workshop on Multiple Classifier Systems* (LNCS 1857, pp. 1-15).

Dietterich, T. G., & Bakiri, G. (1995). Solving multi-class learning problem via error-correcting output codes. *Journal of Artificial Intelligence Research*, *2*, 263–286.

Freund, Y. (1995). Boosting a weak algorithm by majority. *Information and Computation*, *121*(2), 256–285. doi:10.1006/inco.1995.1136

Giacinto, G., & Roli, F. (2001). Design of effective neural network ensembles for image classification purposes. *Image and Vision Computing*, *19*(9), 699–707. doi:10.1016/S0262-8856(01)00045-2

Hansen, L. K., & Salamon, P. (1990). Neural network ensembles. *IEEE Transactions on Pattern Analysis and Machine Intelligence*, *12*(10), 993–1001. doi:10.1109/34.58871

Ho, T. K. (1998). The random subspace method for constructing decision forests. *IEEE Transactions on Pattern Analysis and Machine Intelligence*, *20*(8), 832–844. doi:10.1109/34.709601

Hu, X. H. (2001). Using rough sets theory and database operations to construct a good ensemble of classifiers for data mining applications. In *Proceedings of the IEEE International Conference on Data Mining* (pp. 233-240).

Kanade, T., Cohn, J. F., & Tian, Y. L. (2000). *The Cohn-Kanade AU-coded facial expression database*. Retrieved from http://vasc.ri.cmu.edu/idb/html/face/ facial_expression/index.html

Kuncheva, L. (2003). That elusive diversity in classifier ensembles. In F. J. Perales, A. J. C. Campilho, N. P. de la Blanca, & A. Sanfeliu (Eds.), *Proceedings of the First Iberian Conference on Pattern Recognition and Image Analysis* (LNCS 2652, pp. 1126-1138).

Kuncheva, L. I., & Whitaker, C. J. (2003). Measures of diversity in classifier ensembles and their relationship with the ensemble accuracy. *Machine Learning*, *51*(2), 181–207. doi:10.1023/A:1022859003006

Lyons, M., Akamatsu, S., Kamachi, M., & Gyoba, J. (1998). *The Japanese female facial expression (JAFFE) database*. Retrieved from http://www.kasrl.org/jaffe.html

Mei, S. Y., Liu, Y., Wu, G. F., & Zhang, B. F. (2005). Rough reducts based SVM ensemble. In *Proceedings of the IEEE International Conference on Granular Computing* (pp. 571-574).

Opitz, D. (1999). Feature selection for ensembles. In *Proceedings of the 16th National Conference on Artificial Intelligence* (pp. 379-384).

Pawlak, Z. (1984a). On rough sets. *Bulletin of the EATCS, 24*, 94–108.

Pawlak, Z. (1984b). Rough classification. *International Journal of Man-Machine Studies, 20*(5), 469–483. doi:10.1016/S0020-7373(84)80022-X

Picard, R. W. (1997). *Affective computing*. Cambridge, MA: MIT Press.

Picard, R. W. (2003). Affective computing: Challenges. *International Journal of Human-Computer Studies, 59*(1), 55–64. doi:10.1016/S1071-5819(03)00052-1

Picard, R. W., Vyzas, E., & Healey, J. (2001). Toward machine emotional intelligence: Analysis of affective physiological state. *IEEE Transactions on Pattern Analysis and Machine Intelligence, 23*(10), 1175–1191. doi:10.1109/34.954607

Salcedo, C. G., & Whitley, D. (1999). Feature selection mechanisms for ensemble creation: A genetic search perspective. In Freitas, A. A. (Ed.), *Data mining with evolutionary algorithms: Research directions*. Menlo Park, CA: AAAI Press.

Shipp, C. A., & Kuncheva, L. I. (2002). Relationships between combination methods and measures of diversity in combining classifiers. *Information Fusion, 3*(2), 135–148. doi:10.1016/S1566-2535(02)00051-9

Skowron, A., & Pal, S. K. (2003). Rough sets, pattern recognition, and data mining. *Pattern Recognition Letters, 24*(6), 829–933.

Tsymbal, A., Pechenizkiy, M., & Cunningham, P. (2005). Diversity in search strategies for ensemble feature selection. *Information Fusion, 6*(1), 83–98. doi:10.1016/j.inffus.2004.04.003

Wang, G. Y. (2001). *Rough set theory and knowledge acquisition*. Shaanxi, China: Xi'an Jiaotong University Press.

Wang, G. Y. (2003). Rough reduction in algebra view and information view. *International Journal of Intelligent Systems, 18*(6), 679–688. doi:10.1002/int.10109

Wang, Q. D., Wang, X. J., & Huang, H. (2004). Rough set based feature ensemble learning. In *Proceedings of the 5th World Congress on Intelligent Control and Automation* (pp. 1890-1894).

Wang, Y. (2007a). The theoretical framework of cognitive informatics. *International Journal of Cognitive Informatics and Natural Intelligence, 1*(1), 1–27. doi:10.4018/jcini.2007010101

Wang, Y. (2007b). On the cognitive processes of perception with emotions, motivations, and attitudes. *International Journal of Cognitive Informatics and Natural Intelligence, 1*(4), 1–13. doi:10.4018/jcini.2007100101

Wang, Y. (2009). On cognitive computing. *International Journal of Software Science and Computational Intelligence, 1*(3), 1–15. doi:10.4018/jssci.2009070101

Wang, Y., & Kinsner, W. (2006). Recent advances in cognitive informatics. [C]. *IEEE Transactions on Systems, Man, and Cybernetics, 36*(2), 121–123. doi:10.1109/TSMCC.2006.871120

Wroblewski, J. (2001). Ensemble of classifiers based on approximate reducts. *Fundamenta Informaticae, 47*(3), 351–360.

Yang, Y. (2008). *Chongqing University of Posts and Telecommunications Emotional database (CQUPTE)*. Retrieved from http://cs.cqupt.edu.cn/users/ 904/docs/9317-1.rar

Yang, Y., Wang, G. Y., & He, K. (2007). An approach for selective ensemble feature selection based on rough set theory. In *Proceedings of the Second International Conference on Rough Sets and Knowledge Technology* (pp. 518-525).

Zhong, N., Dong, J. Z., & Ohsuga, S. (2001). Using rough sets with heuristics for feature selection. *Journal of Intelligent Information Systems, 16*(3), 199–214. doi:10.1023/A:1011219601502

Zhou, Z. H., Wu, J. X., & Tang, W. (2002). Ensembling neural networks: Many could be better than all. *Artificial Intelligence, 137*(1-2), 239–263. doi:10.1016/S0004-3702(02)00190-X

This work was previously published in the International Journal of Cognitive Informatics and Natural Intelligence, Volume 5, Issue 3, edited by Yingxu Wang, pp. 61-72, copyright 2011 by IGI Publishing (an imprint of IGI Global).

Chapter 10
Cognitive Informatics and Cognitive Computing in Year 10 and Beyond

Yingxu Wang
University of Calgary, Canada

Witold Kinsner
University of Manitoba, Canada

Robert C. Berwick
Massachusetts Institute of Technology, USA

George Baciu
Hong Kong Polytechnic University, Hong Kong

Simon Haykin
McMaster University, Canada

Du Zhang
California State University, Sacramento, USA

Witold Pedrycz
University of Alberta, Canada

Virendrakumar C. Bhavsar
University of New Brunswick, Canada

Marina Gavrilova
University of Calgary, Canada

ABSTRACT

Cognitive Informatics (CI) is a transdisciplinary enquiry of computer science, information sciences, cognitive science, and intelligence science that investigates into the internal information processing mechanisms and processes of the brain and natural intelligence, as well as their engineering applications in cognitive computing. The latest advances in CI leads to the establishment of cognitive computing theories and methodologies, as well as the development of Cognitive Computers (CogC) that perceive, infer, and learn. This paper reports a set of nine position statements presented in the plenary panel of IEEE ICCI*CC'11 on Cognitive Informatics in Year 10 and Beyond contributed from invited panelists who are part of the world's renowned researchers and scholars in the field of cognitive informatics and cognitive computing.

DOI: 10.4018/978-1-4666-2476-4.ch010

1. INTRODUCTION

The theories of informatics and their perceptions on the object of information have evolved from the classic information theory, modern informatics, to cognitive informatics in the last six decades. The *classic information theories* (Shannon & Weaver, 1949; Bell, 1953; Goldman, 1953), particularly Shannon's information theory (Shannon, 1948), are the first-generation informatics, which study signals and channel behaviors based on statistics and probability theory. The *modern informatics* studies information as properties or attributes of the natural world that can be distinctly elicited, generally abstracted, quantitatively represented, and mentally processed (Wang, 2002a, 2003a, 2003b). The first- and second-generation informatics put emphases on external information processing, which are yet to be extended to observe the fundamental fact that human brains are the original sources and final destinations of information. Any information must be cognized by human beings before it is understood, comprehended, and consumed.

The aforementioned observations have led to the establishment of the third-generation informatics, *cognitive informatics* (CI), a term coined by Wang in a keynote in 2002 (Wang, 2002a). CI is defined as the science of cognitive information that investigates into the internal information processing mechanisms and processes of the brain and natural intelligence, and their engineering applications via an interdisciplinary approach. It is recognized in CI that *information* is the third essence of the natural world supplementing to matter and energy. *Informatics* is the science of information that studies the nature of information, its processing, and ways of transformation between information, matter and energy.

The IEEE series of *International Conferences on Cognitive Informatics and Cognitive Computing* (ICCI*CC) has been established since 2002

(Wang, 2002a; Wang et al., 2002). The inaugural ICCI event in 2002 was held at University of Calgary, Canada (ICCI'02) (Wang et al., 2002), followed by the events in London, UK (ICCI'03) (Patel et al., 2003); Victoria, Canada (ICCI'04) (Chan et al., 2004); Irvine, USA (ICCI'05) (Kinsner et al., 2005); Beijing, China (ICCI'06) (Yao et al., 2006); Lake Tahoe, USA (ICCI'07) (Zhang et al., 2007); Stanford University, USA (ICCI'08) (Wang et al., 2008); Hong Kong (ICCI'09) (Baciu et al., 2009); Tsinghua University, Beijing (ICCI'10) (Sun et al., 2010); and Banff, Canada (ICCI*CC'11) (Wang et al., 2011). Since its inception, the ICCI*CC series has been growing steadily in its size, scope, and depth. It attracts worldwide researchers from academia, government agencies, and industry practitioners. The conference series provides a main forum for the exchange and cross-fertilization of ideas in the new research field of CI toward revealing the cognitive mechanisms and processes of human information processing and the approaches to mimic them in cognitive computing.

A series of fundamental breakthroughs have been recognized and a wide range of applications has been developed in cognitive informatics and cognitive computing in the last decade. The representative paradigms and technologies developed in cognitive informatics are such as cognitive computing, cognitive computers, abstract intelligence, formal knowledge representation, cognitive learning engines, denotational mathematics for cognitive system modeling, and applicants in cognitive systems.

This paper is a summary of the position statements of panellists presented in the *Plenary Panel on Cognitive Informatics in Year 10 and Beyond* in IEEE ICCI*CC 2011 held in Banff, Alberta, Canada during August 18-20, 2011 (Wang et al., 2011). It is noteworthy that the individual statements and opinions included in this paper may not necessarily be shared by all panellists.

2. THE FRAMEWORK OF COGNITIVE INFORMATICS AND COGNITIVE COMPUTING

The framework of cognitive informatics (Wang, 2003a, 2007b) and cognitive computing (Wang, 2006, 2009b, 2010a; Wang, Zhang, & Kinsner, 2010) can be described by the following theories, mathematical means, cognitive models, computational intelligence technologies, and applications.

2.1 Fundamental Theories of Cognitive Informatics

Cognitive Informatics (CI) is a transdisciplinary enquiry of computer science, information science, cognitive science, and intelligence science that investigates into the internal information processing mechanisms and processes of the brain and natural intelligence, as well as their engineering applications in cognitive computing (Wang, 2002a, 2003a, 2006, 2007b, 2007d, 2009a, 2009b; Wang & Kinsner, 2006; Wang & Wang, 2006; Wang, Zhang, & Kinsner, 2010; Wang, Kinsner, & Zhang, 2009; Wang, Kinsner et al., 2009).

CI is a cutting-edge and multidisciplinary research area that tackles the fundamental problems shared by computational intelligence, modern informatics, computer science, AI, cybernetics, cognitive science, neuropsychology, medical science, philosophy, formal linguistics, and life science (Wang, 2002a, 2003a, 2007b). The development and the cross fertilization among the aforementioned science and engineering disciplines have led to a whole range of extremely interesting new research fields known as CI, which investigates the internal information processing mechanisms and processes of the natural intelligence – human brains and minds – and their engineering applications in computational intelligence. CI is a new discipline that studies the natural intelligence and internal information processing mechanisms of the brain, as well as processes involved in perception and cognition. CI forges links between a number of natural science and life science disciplines with informatics and computing science.

Fundamental theories developed in CI covers the Information-Matter-Energy-Intelligence (IME-I) model (Wang, 2007a), the Layered Reference Model of the Brain (LRMB) (Wang et al., 2006), the Object-Attribute-Relation (OAR) model of internal information representation in the brain (Wang, 2007c), the cognitive informatics model of the brain (Wang & Wang, 2006), natural intelligence (Wang, 2007b), abstract intelligence (Wang, 2009a), neuroinformatics (Wang, 2007b), denotational mathematics (Wang, 2002b, 2007a, 2008a, 2008b, 2008c, 2008d, 2009c, 2009d, 2009e, 2010b, 2011, in press; Wang, Zadeh, & Yao, 2009), and cognitive systems (Berwick, 2011; Haykin, 2011; Kinsner, 2011; Pedrycz, 2011; Wang, 2011). Recent studies on LRMB in cognitive informatics reveal an entire set of cognitive functions of the brain and their cognitive process models, which explain the functional mechanisms and cognitive processes of the natural intelligence with 43 cognitive processes at seven layers known as the sensation, memory, perception, action, metacognitive, meta-inference, and higher cognitive layers (Wang et al., 2006).

2.2 Cognitive Computing for Cognitive Computers

Computing systems and technologies can be classified into the categories of *imperative, autonomic,* and *cognitive* computing from the bottom up. The imperative computers are a passive system based on stored-program controlled behaviors for data processing (Wang, 2009b). The autonomic computers are goal-driven and self-decision-driven machines that do not rely on instructive and procedural information (Pescovitz, 2002; Wang, 2007d). Cognitive computers are more intelligent computers beyond the imperative and autonomic computers, which embody major natural intelligence behaviors of the brain such as thinking, inference, and learning.

Cognitive Computing (CC) is a novel paradigm of intelligent computing methodologies and systems based on CI that implements computational intelligence by autonomous inferences and perceptions mimicking the mechanisms of the brain (Wang, 2006, 2009b, 2009c, 2010a; Wang, Tian, & Hu, 2011). CC is emerged and developed based on the multidisciplinary research in CI (Wang, 2002a, 2003, 2007b; Wang, Zhang, & Kinsner, 2010; Wang, Kinsner et al., 2009).

The latest advances in CI and CC, as well as denotational mathematics, enable a systematic solution for the future generation of intelligent computers known as *cognitive computers* (CogCs) that think, perceive, learn, and reason (Wang, 2006, 2009b, 2009c, 2010a; Wang, Zhang, & Kinsner, 2010; Wang, Widrow et al., 2011). A CogC is an intelligent computer for knowledge processing as that of a conventional von Neumann computer for data processing. CogCs are designed to embody *machinable intelligence* such as computational inferences, causal analyses, knowledge manipulation, machine learning, and autonomous problem solving.

Recent studies in cognitive computing reveal that the computing power in computational intelligence can be classified at four levels: *data, information, knowledge,* and *intelligence* from the bottom up. Traditional von Neumann computers are designed for imperative data and information processing by stored-program-controlled mechanisms. However, the increasing demand for advanced computing technologies for knowledge and intelligence processing in the high-tech industry and everyday lives require novel cognitive computers for providing autonomous computing power for various cognitive systems mimicking the natural intelligence of the brain.

2.3 Abstract Intelligence (αI)

The studies on abstract intelligence (αI) form a human enquiry of both natural and artificial intelligence at reductive levels of the neural, cognitive,

functional, and logical layers from the bottom up (Wang, 2009a). αI is the general mathematical form of intelligence as a natural mechanism that transfers information into behaviors and knowledge.

The *Information-Matter-Energy-Intelligence* (IME-I) model (Wang, 2003a, 2007c) states that the natural world (*NW*) which forms the context of human and machine intelligence is a dual: one aspect of it is the *physical* world (*PW*), and the other is the *abstract* world (*AW*), where *intelligence* (αI) plays a central role in the transformation between information (*I*), *matter* (*M*), and *energy* (*E*). In the IME-I model as shown in Figure 1, αI plays an irreplaceable role in the transformation between information, matter, and energy, as well as different forms of internal information and knowledge. Typical paradigms of αI are natural intelligence, artificial intelligence, machinable intelligence, and computational intelligence, as well as their hybrid forms. The studies in CI and αI lay a theoretical foundation toward revealing the basic mechanisms of different forms of intelligence. As a result, cognitive computers may be developed, which are characterized as knowledge processors beyond those of data processors in conventional computing.

Figure 1. The IME-I model and roles of abstract intelligence

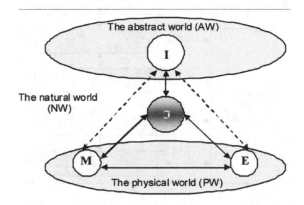

2.4 The Layered Reference Model of the Brain (LRMB)

The Layered Reference Model of the Brain (Wang et al., 2006) is developed to explain the fundamental cognitive mechanisms and processes of natural intelligence. Because a variety of life functions and cognitive processes have been identified in CI, psychology, cognitive science, brain science, and neurophilosophy, there is a need to organize all the recurrent cognitive processes in an integrated and coherent framework. The LRMB model encompasses 43 cognitive processes at seven layers known as the *sensation, memory, perception, action, metacognitive, metainference,* and *higher cognitive layers* from the bottom-up as shown in Figure 2.

LRMB explains the functional mechanisms and cognitive processes of the natural and artificial brains with the interactive processes at the seven layers (Wang et al., 2006). LRMB elicits the core and highly repetitive recurrent cognitive processes from a huge variety of life functions,

which may shed light on the study of the fundamental mechanisms and interactions of complicated mental processes as well as of cognitive systems, particularly the relationships and interactions between the inherited and the acquired life functions at the subconscious and conscious layers.

2.5 Denotational Mathematics (DM)

The needs for complex and long-series of causal inferences in cognitive computing, αI, computational intelligence, software engineering, and knowledge engineering have led to new forms of mathematics collectively known as denotational mathematics (Wang, 2002b, 2007a, 2008a, 2008b, 2008c, 2008d, 2009c, 2009d, 2009e, 2010b, 2011, in press; Wang, Zadeh, & Yao, 2009). *Denotational Mathematics (DM)* is a category of expressive mathematical structures that deals with high-level mathematical entities beyond numbers and sets, such as abstract objects, complex relations, perceptual information, abstract concepts, knowledge, intelligent behaviors, behavioral processes, and systems (Wang, 2008a, 2009c, 2010b).

It is recognized that the maturity of any scientific discipline is characterized by the maturity of its mathematical (meta-methodological) means, because the nature of mathematics is a generic meta-methodological science (Wang, 2008a). In recognizing mathematics as the *metamethodology* of all sciences and engineering disciplines, a set of DMs have been created and applied in CI, αI, AI, CC, CogC, soft computing, computational intelligence, and computational linguistics. Typical paradigms of DM are such as *concept algebra* (Wang, 2008b; Wang, Widrow et al., 2011), *system algebra* (Wang, 2008c; Wang, Zadeh, & Yao, 2009), *real-time process algebra* (Wang, 2002b, 2007a, 2008d), *granular algebra* (Wang, 2009e), *visual semantic algebra* (Wang, 2009d), and *inference algebra* (Wang, 2011, in press). DM provides a coherent set of contemporary mathematical means and explicit expressive power for cognitive informatics, cognitive computing, artificial intelligence and computational intelligence.

Figure 2. The layered reference model of the brain

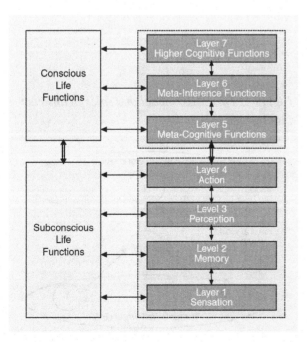

2.6 Formal Knowledge Representation and Cognitive Learning Systems

An internal knowledge representing theory known as the *Object-Attribute-Relation* (OAR) model is proposed by Wang in 2007, which reveals the logical foundation of concepts and their attributes based on physiological and biological observations. The *OAR* model explains the mechanism of long-term memory (LTM) of the brain. It can be described as a triple (*O, A, R*), where *O* is a finite set of objects identified by unique symbolic names; *A* is a finite set of attributes for characterizing the objects; and *R* is a set of relations between an object and other objects or their attributes.

The OAR model explains the logic structure and configurations of knowledge based on physiological observations (Wang, 2007b). According to the OAR model, the mechanism and result of learning are the updating of the entire OAR in LTM, which can be formally modeled by a compositional operation (â) between the existing OAR and the newly created sub-*OAR* (sOAR), i.e.: *OAR' = OAR* â *sOAR* (Wang, 2008b; Wang, Tian, & Hu, 2011).

A Cognitive Learning Engine (CLE) (Tian et al., 2011), known as the "CPU" of CogCs, is under developing in the Cognitive Informatics and Cognitive Computing Lab on the basis of concept algebra (Wang, 2008b), which implements the basic and advanced cognitive computational operations of concepts and knowledge for CogCs. The work in this area may also lead to a fundamental solution to computational linguistics, computing with natural language (CNL), and computing with words (CWW) (Zadeh, 1965, 1975, 1999, 2008; Wang, 2010a, 2010c, 2010d).

Because CI and CC provide a common and general platform for the next generation of cognitive computing, a wide range of applications of CI, αI, CC, CogC, and DM are expected toward the implementation of highly intelligent machinable thought such as formal inference, symbolic reasoning, problem solving, decision making, cognitive knowledge representation, semantic searching, and autonomous learning. Some expected innovations that will be enabled by CI and CC are as follows, *inter alia*: (a) A *reasoning machine* for complex and long-series of inferences, problem solving, and decision making beyond traditional logic and if-then-rule based technologies; (b) An *autonomous learning system* for cognitive knowledge acquisition and processing; c) A novel *search engine* for providing comprehendable and formulated knowledge via the Internet; (d) A *cognitive medical diagnosis system* supporting evidence-based medical care and clinical practices; (e) A *cognitive computing node* for the next generation of the intelligent Internet; and (f) A *cognitive processor* for cognitive robots (Wang, 2010e) and cognitive agents (Wang, 2009f).

3. PSYCHOLOGICALLY REALISTIC COGNITIVE COMPUTING BEYOND 2011

Language's recent evolutionary origin suggests that the computational machinery underlying syntax arose via the introduction of a single, simple, combinatorial operation. Further, the relation of a simple combinatorial syntax to the sensory-motor and thought systems reveals language to be asymmetric in design: while it precisely matches the representations required for inner mental thought, acting as the "glue" that binds together other internal cognitive and sensory modalities, at the same time it poses computational difficulties for externalization, that is, parsing and speech or signed production. Despite this mismatch, language syntax leads directly to the rich cognitive array that marks us as a symbolic species, including mathematics, music, and much more (Berwick, 2011).

Engineers have long appreciated the wisdom of the approach known as "KISS" – short for "Keep it Simple Stupid." But what about cognitive computing? Recent years have seen the rise of ever-more sophisticated and computationally intensive statistical models drawn from the analysis of biostatistics and the social sciences, now extended to the domain of human cognition. In particular, these models have recently been applied to human language acquisition, with the claim that they overcome previously insurmountable obstacles. However, there are two problems with these methods. First, they require computational resources well beyond the known bounds available to children. Second, one can show that far simpler models suffice to solve the same learning problems. In the domain of language acquisition at least, the KISS approach still prevails.

4. NEW VISION FOR THE WORLD OF WIRELESS COMMUNICATIONS ENABLED WITH COGNITION

During the past 10 years or so, much has been written on the application of human cognition in a variety of diverse fields. This new multidisciplinary subject is called Cognitive Systems. From an engineering perspective, Cognitive Systems may be categorized into three broadly defined classes: a) Cognitive dynamic systems (Haykin, 2011); b) Cognitive informatics (Wang, 2002a, 2003a, 2006, 2007b, 2007d, 2009a, 2009b; Wang & Kinsner, 2006; Wang & Wang, 2006; Wang, Zhang, & Kinsner, 2010; Wang, Kinsner, & Zhang, 2009; Wang, Kinsner et al., 2009); c) Cognitive computing (Wang, 2006, 2009b, 2009c, 2010a; Wang, Zhang, & Kinsner, 2010; Wang, Widrow et al., 2011; Modha et al., 2011);

In this section, I will discuss the four principles of cognition inspired by the human brain: perception-action cycle, memory, attention, and intelligence. In applying cognition to wireless communications, the current status of traditional

cognitive radio is: (1) spectrum sensing: the multitaper method, (2) transmit-power control: Nash equilibrium based on iterative water-filling, and (3) dynamic spectrum management: brain-inspired allocation of underutilized subbands of the radio spectrum in a multi-user network (Haykin, 2011). In my view, beyond traditional cognitive radio, new vision for the world of wireless communications includes: (1) principle of cognition, (2) Femtocells: improved indoor reception, higher data rate, lower power consumption, and benefits to network providers. In developing cognitive Femtocell networks, the requirements are scalability, stability and heterogeneous coexistence community, and the tools and solutions are self-organized dynamic spectrum management, transmit-power control, spectrum identification and synchronization.

5. GRANULAR COMPUTING AND COGNITIVE INFORMATICS

As lucidly emphasized in Wang, Kinsner et al. (2009), Cognitive informatics is a transdisciplinary enquiry of cognitive and information sciences that brings together the mechanisms of information processing and processes of the brain and natural intelligence along with their engineering applications. Some interesting linkages between Cognitive Informatics and cyberntics are drawn in Wang, Kinsner, and Zhang (2009). An effective human – system interaction is one of the essential facets that becomes visible here especially in the context of applications. Information granules along with their numerous ways of formalization give rise to the discipline of Granular Computing (Bargiela & Pedrycz, 2003, 2008, 2009). In a nutshell, Granular Computing delivers a cohesive framework supporting a formation of information granules (as well as their ensuing hierarchical structures) and facilitating their processing. We elaborate on important facets of Granular Computing, which are also essential to

Cognitive Informatics. This concerns a hierarchy of information granules, which contributes to a formation of a suitable cognitive perspective. Human centricity of Granular Computing is supported by a variety of formal ways in which information granules are represented, say fuzzy sets or rough sets. Tradeoffs between precision (and associated processing overhead) and interpretability of constructs of Granular Computing are formed by invoking a suitable level within the hierarchy of information granules.

6. DEALING WITH EMERGENT COGNITIVE SYSTEMS

Many developments of the last century focused on modeling of adaptation and adaptive systems. The focus in this century appears to have been shifting towards cognition and cognitive dynamical systems with emergence. Although cognitive dynamical systems are always adaptive to various conditions in the environment where they operate, adaptive systems of the past have not been cognitive.

The evolving formulation of cognitive informatics (CI) (Kinsner, 2007a, 2007b, 2009; Wang, 2002a, 2003a, 2007b; Wang, Widrow et al., 2011) has been an important step in bringing the diverse areas of science, engineering, and technology required to develop such cognitive computing (CogC) and cognitive systems. Cognitive computing focuses on the development a coherent, unified, universal, and system-based models inspired by the mind's nonlinear and evolutionary capabilities (Wang, 2009b; Wang, Zhang, & Kinsner, 2010; Modha et al., 2011). The intent of CogC is that by combining neuroscience, intelligent signal processing, supercomputing, nanotechnology and other non-standard technological developments, some insight into the brain's core algorithms might be possible.

Current examples of various cognitive systems include autonomic computing, memetic computing, cognitive radio, cognitive radar, cognitive robots, cognitive networks, cognitive computers, cognitive cars, cognitive factories, as well as brain-machine interfaces for physically-impaired persons, and cognitive binaural hearing instruments. The increasing interest in this area may be due to the recognition that perfect solutions to large-scale scientific and engineering problems may not be feasible, and we should seek the best solution for the task at hand. The "best" means suboptimal and the most reliable (robust) solution, given not only limited resources (financial and environmental) but also incomplete knowledge of the problem and partial observability of the environment. Many exciting new theoretical, computational and technological accomplishments have been described in recent literature.

The challenges in the evolving cognitive systems can be grouped into several categories: (a) theoretical, (b) technological, and (c) sociological. The first group of theoretical issues include modelling, reformulation of information and entropy, multi-scale measures and metrics, and management of uncertainty. Modelling of cognitive systems requires radically new approaches. *Reductionism* has dominated our scientific worldview for the last 350 years, since the times of Descartes, Galileo, Newton, and Laplace. In that approach, all reality can be understood in terms of particles (or strings) in motion. However, in this nonlinear (Enns, 2010) unfolding emergent universe with agency, meaning, values and purpose, we cannot predict all that will happen. Since cognitive systems rely on perceiving the world by agents, learning from it, remembering and developing the experience of self-awareness, feelings, intentions, and deciding how to control not only tasks but also communication with other agents, and to create new ideas, CI cannot rely on the reductionist approach of describing nature. In fact, CI tries to expand the modeling in order to deal with the emergent universe where no laws of physics are violated, and yet ceaseless unforeseeable creativity arises and surrounds us all the time. This new approach requires many new ideas to be developed, including reformulation of the concept of cognitive information, entropy, and associated measures, as well as management of uncertainty, and new forms of cognitive computing.

As we have seen over the last decade, cognitive informatics is multidisciplinary, and requires cooperation between many subjects, including sciences (e.g., cognitive science, computer science, evolutionary computing, granular computing, multi-criteria decision making, multi-objective evolutionary optimization, game theory, crisp and fuzzy sets, mathematics, physics, chemistry, biology, psychology, humanities, and social sciences), as well as engineering and technology (computer, electrical, mechanical, information theory, control theory, intelligent signal processing, neural networks, learning machines, sensor networks, wireless communications, and computer networks). Many of the new algorithms replace the conventional concepts of second order statistics (covariance, L2 distances, and correlation functions) with scalars and functions based on information theoretic underpinnings (such as entropy, mutual information and correntropy) defined not only on a single scale, but also on multiple scales. A serious challenge is the modeling and measuring of complexity of complex dynamical system (Kinsner, 2010). The continuing progress in the field may lead to some useful solutions to pressing problems.

7. COGNITIVE TEXTURE: A UNIFIED MULTI-SENSORY FEEDBACK FRAMEWORK

The human brain has an uncanny capacity to assimilate, integrate and fuse the multiple modalities from our visual, auditory, tactile, taste and olfactory sensors. The multi-sensory feedback modalities could be attributed to the perception of the change of environmental conditions. Observers often associate these changes to textures. Often, one can relate taste to texture, background sound to texture, as well as visual and tactile feedback as textures. The fusion of multi-sensory feedback seems to take the form of multi-dimensional textures. For example, tactile feedback is often correlated to visual perception of a surface. The two modalities can enforce each other into a richer form of multi-dimensional texture. We refer to these multi-sensory textures as CogTex, or Cognitive Texture. In its simplest form, CogTex relates any two or more channels of sensory feedback into a multi-dimensional cognitive texture. Multi-dimensional cognitive textures, specifically, visual, auditory and tactile, have recently become increasingly important in user interfaces through multi-touch visual displays. Here we will look at a few applications that are currently on a fast growing curve.

The world is now officially in a multi-touch, multitasking, multi-streaming mode. The number of *Apps* on the iPhone/iPad and Android systems is literally on an exponential growing curve. Software development for both iPhone OS and Android are in the hundreds. Developers are porting all the conceivable tools to tablets and smart phones. The latest entry into the frenzy, Nokia X7, possibly signals the capitulation of traditional PC/Laptop operating environments as the world converges onto the fastest growing computational platforms of all times as shown in Figure 3.

One of the latest contenders to the iPad2 generation, the Samsung's Galaxy pad 10.1 (or Galaxy 2) has made its debut at electronic shows and it is ready for launch. What is really interesting about the race is that despite the similar form factors and expected performance measurements, the tablet enthusiasts are looking for attributes that have not been considered in traditional computational technology before, especially portable ones. These are touch, feel, color and sound. More specifically, the devices are scrutinized in much more detail at a level of cognitive texture or CogTex for short.

Figure 3. Native multi-touch system for weaving design of virtual fabric

The manifestation for human computer interaction has now taken another dimension in the tactile feel, image manipulation and the accompanying embedded sound feedback (Brooks et al., 2008; Cai & Baciu, 2011; Chen & Pappas, 2005; He & Pappas, 2010; Pappas et al., 2009). That is, all our sensory preceptors (minus one: olfactory), tactile, visual, auditory, are directly engaged in information processing (Rolls, 2005, 2008a, 2008b). However, these are currently very difficult modes to measure. For example, in video comparisons between iPad2 and Galaxy 10.1, the comments refer to the fact that the iPad2 "feels smoother" and the colors are "softer." The touch response is "less jerky" and "more fluid."

7.1 COGTEX: Cognitive Texture Modeling

Both these devices are state of the art multitouch tablet machines, but hardly anyone is paying attention to the IO throughput (other than video streaming), or the CPU, or GPU performance. The category of interest is now "texture" and not just "image texture" but also tactile, haptic (force feedback), and auditory. We have now transcended into a new era of cognitive texture information

processing. Some of the not so new problems that we are fast converging to are taking a new flavor. Among these we can identify the following:

1. Cognitive processing under multimodal sensory feedback;
2. Balancing the load between feedback response (visual, tactile, auditory) and contextual processing;
3. Predictive cognitive behavior in the new multi-touch, multi-sensory environments.

An interesting question that arises from the above is under what conditions a common ground could be established for studying the cognitive interactions between these (at least) three modes and the brain. The answer could be found in the realm of "textures", or more appropriately "cognitive textures" or CogTex.

Based on preliminary studies (Bargiela & Pedrycz, 2009; Liang et al., 2010; Zhang et al., 2010; Zheng et al., 2010) that we are undergoing with digitizing textile materials, the iTextile project in the GAMA Lab at the Hong Kong Polytechnic University, Figure 4, we find that the CogTex platform could potentially unify the multimodal sensory feedback for cognitive processes. Supplemented by a theoretical behavior model such as cognitive algebra (Wang, 2003b, 2003c, 2008b) it can lead to a unifying theory of multi-modal cognitive sensory information processing.

8. BOUNDED RATIONALITY AND INCONSISTENCY IN COGNITIVE COMPUTING SYSTEMS

Cognitive computing systems are systems of bounded rationality. Toward the goal of building bounded rational cognitive computing systems, a whole host of human cognitive skills and abilities should be brought to bear. In this section, we call

Figure 4. Dual touch tablets on exponential growth

(a) Samsung Galaxy Pad   (b) Motorola XOOM

for attention on the interplay between inconsistency and bounded rationality, and emphasize on incorporating into the cognitive computing systems the human cognitive capability in handling inconsistency.

The theory of *bounded rationality*, developed by Simon (1982), underpins human intelligent behaviors. When engaged in problem solving, human beings' decision-making process is confined by the following constraints: the knowledge or information they possess, the cognitive limitations they have, and the time limit within which a decision needs to be made (Russell & Norvig, 2010). As a result, the human decision-making process in practice really consists in a search through a finite number of options, the fewer the better. People tend to identify with sub-goals rather than with global aims, exploit pre-existing structures or regularities in the environment, apply approximate or heuristic approaches to problems, and be content with good enough solutions.

In a nutshell, practical decision-making process is not a perfect rational process of finding an optimal solution (a solution that maximizes the expected utility) given the information available from the environment. Agents of bounded rationality exhibit *satisfying*, rather than optimizing, behavior: (1) seeking satisfactory solutions rather than optimal ones; (2) adopting simplified choices; (3) deliberating only long enough; and (4) relying on heuristic approaches rather than rigid rule of optimization.

To a large extent, building cognitive computing systems amounts to developing systems that possess bounded rationality. Toward the goal of bounded rational cognitive computing systems, a whole host of human cognitive skills and abilities should be brought to bear. In this section, we call for attention on the interplay between inconsistency and bounded rationality, and emphasize on incorporating into the cognitive computing systems the human cognitive capability in handling inconsistency.

Inconsistency is an important phenomenon that exists ubiquitously in human behaviors and in various aspects of real life (Gotesky, 1968). Inconsistent phenomena manifest themselves in data, information, knowledge, meta-knowledge, and expertise. Inconsistent or conflicting assumptions, beliefs, evidences, or options can serve as important heuristics in the decision-making process of a bounded rational agent (Zhang, 2011; Zhang & Lu, 2011). Using the consistency-rationality dichotomy, we can categorize heuristics as show in Table 1.

Table 1. Heuristics.

	Consistent	Inconsistent
Rational	RAC	RINC
Irrational	IRRAC	IRRINC

Of particular interest is the category of *RINC* (inconsistent but rational heuristics), which can be of more powerful tools in the decision-making process.

9. COGNITIVE INFORMATICS TOWARDS 2031

Cognitive Informatics (CI) has been based on many disciplines, for example computer science, cognitive science and information sciences. CI has both scientific and engineering goals: (a) to understand and explain "intelligent" behavior living organisms, and (b) to develop systems that have some form of "intelligence".

CI consists of a multitude of techniques, theories, systems and applications. Some of the theories use numeric or sub-symbolic representations and often use vector-space models. Although they have been applied in a number of domains, they have certain inherent limitations due to their vector-space representations. On the other hand, other theories operate in the symbolic domain and these have been also applied in many applications. However, the emergence of symbols is not "natural" in these theories when we consider inputs from the real world through various sensors. Thus, it is clear that there is a lack of integrative and unified theory of cognition and intelligence that is applicable across various aspects and levels of cognition and intelligence.

Ideally, a unified theory of CI should be developed that will be able to explain various sensory perceptions, reasoning, intuition and all other intelligent processes. The technologies and systems based on such a theory would be applicable to construct systems applicable to all these domains. These systems would also exhibit some fundamental features of evolution in nature. Further, we believe that since biological systems embody a marked degree of parallel and distributed functioning, the CI architectures and/or their implementations should also embody some forms of intrinsic parallel and distributed computing. We hope that by 2031 we will have such a unified theory.

10. COMPUTATIONAL INTELLIGENCE IN BIOMETRIC

The area of computational intelligence in general and cognitive informatics in particular has experienced tremendous growth in the past decade. Research on neural networks, evolutionary computing, fuzzy logic, intelligent design and decision-making has influenced, in turn, growth in numerous application areas, such as pattern recognition, image processing, and biometric authentication. Biometric research specifically is one of the most dynamic areas which benefitted from such developments, which recently has displayed a gamut of broader links to other fields of sciences. Among those are visualization, robotics, multi-dimensional data analysis, computational geometry, computer graphics, e-learning, data fusion and data synthesis. The topic of this talk is reviewing state-of-the-art in multi-modal data fusion and neural networks and its recent connections to advanced biometric research. Application examples in this area demonstrate high potentials of this research symbiosis.

11. CONCLUSION

This paper has summarized nine position statements presented in the plenary panel of IEEE ICCI*CC'11 on *Cognitive Informatics in Year 10 and Beyond* contributed by invited panelists who are part of the world's renowned researchers and scholars in the field of cognitive informatics and cognitive computing. Cognitive Informatics (CI) has been described as a transdisciplinary enquiry of computer science, information sciences, cognitive science, and intelligence science that investigates into the internal information processing mechanisms and processes of the brain and natural intelligence, as well as their engineering applications in cognitive computing. Cognitive Computing (CC) has been recognized as an emerging paradigm of intelligent computing methodologies and systems that implements computational intelligence by autonomous inferences and perceptions mimicking the mechanisms of the brain. It has been elaborated that the theoretical foundations underpinning cognitive computing and cognitive computers (CogC) are cognitive informatics – the science of cognitive and intelligent information and knowledge processing.

A series of fundamental breakthroughs have been recognized and a wide range of applications has been developed in cognitive informatics and cognitive computing in the last decade. The representative paradigms and technologies developed in cognitive informatics and cognitive computing have been recognized as prototypes of cognitive computers, cognitive systems, cognitive knowledge bases, cognitive robots, cognitive learning engines, and autonomous inference systems and frameworks.

REFERENCES

Baciu, G., Yao, Y., Wang, Y., Zadeh, L. A., Chan, K., & Kinsner, W. (Eds.). (2009, June). *Proceedings of the 8th IEEE International Conference on Cognitive Informatics*. Washington, DC: IEEE Computer Society.

Bargiela, A., & Pedrycz, W. (2003). *Granular computing: An introduction*. Dordrecht, The Netherlands: Kluwer Academic.

Bargiela, A., & Pedrycz, W. (2008). Toward a theory of granular computing for human-centered information processing. *IEEE Transactions on Fuzzy Systems, 16*(2), 320–330. doi:10.1109/TFUZZ.2007.905912

Bargiela, A., & Pedrycz, W. (Eds.). (2009). *Human-centric information processing through granular modelling*. Heidelberg, Germany: Springer-Verlag. doi:10.1007/978-3-540-92916-1

Bell, D. A. (1953). *Information theory*. London, UK: Pitman.

Berwick, R. C. (2011, August). Keynote: Songs to syntax: Cognition, combinatorial computation, and the origin of language. In *Proceedings of the 10th IEEE International Conference on Cognitive Informatics and Cognitive Computing*, Banff, AB, Canada (p. 1). Washington, DC: IEEE Computer Society.

Brooks, A. C., Zhao, X., & Pappas, T. N. (2008). Structural similarity quality metrics in a coding context: Exploring the space of realistic distortions. *IEEE Transactions on Image Processing*, 1261–1273. doi:10.1109/TIP.2008.926161

Cai, Y., & Baciu, G. (2011, June). Detection of repetitive patterns in near regular texture images. In *Proceedings of the 10th IEEE Conference on Perception and Visual Signal Analysis*, Ithaca, NY (pp. 60-65). Washington, DC: IEEE Computer Society.

Chan, C., Kinsner, W., Wang, Y., & Miller, D. M. (Eds.). (2004, August). *Proceedings of the 3rd IEEE International Conference on Cognitive Informatics*. Washington, DC: IEEE Computer Society.

Chen, J., Pappas, T. N., Mojsilovic, A., & Rogowitz, B. E. (2002). Adaptive image segmentation based on color and texture. In *Proceedings of the IEEE International Conference on Information Processing* (Vol. 3, pp. 777-780). Washington, DC: IEEE Computer Society.

Chen, J., Pappas, T. N., Mojsilovic, A., & Rogowitz, B. E. (2005). Adaptive perceptual color-texture image segmentation. *IEEE Transactions on Image Processing*, 1524–1536. doi:10.1109/TIP.2005.852204

Enns, R. H. (2010). *It's a nonlinear world*. New York, NY: Springer.

Goldman, S. (1953). *Information theory*. Upper Saddle River, NJ: Prentice Hall.

Gotesky, R. (1968). The uses of inconsistency. *Philosophy and Phenomenological Research, 28*(4), 471–500. doi:10.2307/2105687

Haykin, S. (2011, August). Keynote: Cognitive dynamic systems: An integrative field that will be a hallmark of the 21st century. In *Proceedings of the 10th IEEE International Conference on Cognitive Informatics and Cognitive Computing*, Banff, AB, Canada (p. 2). Washington, DC: IEEE Computer Society.

He, L., & Pappas, T. N. (2010, September). An adaptive clustering and chrominance-based merging approach for image segmentation and abstraction. In *Proceedings of the IEEE 17th International Conference on Image Processing*, Hong Kong (pp. 241-244). Washington, DC: IEEE Computer Society.

Kinsner, W. (2007a). A unified approach to fractal dimensions. *International Journal of Cognitive Informatics and Natural Intelligence, 1*(4), 26–46. doi:10.4018/jcini.2007100103

Kinsner, W. (2007b). Towards cognitive machines: Multiscale measures and analysis. *International Journal of Cognitive Informatics and Natural Intelligence, 1*(1), 28–38. doi:10.4018/jcini.2007010102

Kinsner, W. (2009). Challenges in the design of adaptive, intelligent and cognitive systems. *International Journal of Software Science and Computational Intelligence, 1*(3), 16–35. doi:10.4018/jssci.2009070102

Kinsner, W. (2010). System complexity and its measures: How complex is complex. In Wang, Y., Zhang, D., & Kinsner, W. (Eds.), *Advances in cognitive informatics and cognitive computing* (*Vol. 323*, pp. 265–295). Berlin, Germany: Springer-Verlag. doi:10.1007/978-3-642-16083-7_14

Kinsner, W. (2011, August). Keynote: It's time for multiscale analysis and synthesis in cognitive systems. In *Proceedings of the 10th IEEE International Conference on Cognitive Informatics and Cognitive Computing*, Banff, AB, Canada (pp. 7-10). Washington, DC: IEEE Computer Society.

Kinsner, W., Zhang, D., Wang, Y., & Tsai, J. (Eds.). (2005, August). *Proceedings of the 4th IEEE International Conference on Cognitive Informatics*. Washington, DC: IEEE Computer Society.

Liang, S., Chan, E., Baciu, G., & Li, R. (2010, July). Cognitive garment design interface using user behavior tree model. In *Proceedings of the 9th IEEE International Conference on Cognitive Informatics*, Beijing, China (pp. 496-500). Washington, DC: IEEE Computer Society.

Modha, D. S., Ananthanarayanan, R., Esser, S. K., Ndirango, A., Sherbondy, A. J., & Singh, R. (2011). Cognitive computing. *Communications of the ACM, 54*(8), 62–71. doi:10.1145/1978542.1978559

Pappas, T. N., Tartter, V. C., Seward, A. G., Genzer, B., Gourgey, K., & Kretzshmar, I. (2009). Perceptual dimensions for a dynamic tactile display, human vision and electronic imaging. *Proceedings of the Society for Photo-Instrumentation Engineers, 7240*.

Patel, D., Patel, S., & Wang, Y. (Eds.). (2003, August). *Proceedings of the 2nd IEEE International Conference on Cognitive Informatics*. Washington, DC: IEEE Computer Society.

Pedrycz, W. (2011, August). Keynote: Human centricity and perception-based perspective of architectures of granular computing. In *Proceedings of the 10th IEEE International Conference on Cognitive Informatics and Cognitive Computing*, Banff, AB, Canada (p. 3). Washington, DC: IEEE Computer Society.

Pescovitz, D. (2002). Autonomic computing: Helping computers help themselves. *IEEE Spectrum, 39*(9), 49–53. doi:10.1109/MSPEC.2002.1030968

Rolls, E. T. (2005). *Emotion explained*. Oxford, UK: Oxford University Press. doi:10.1093/acprof:oso/9780198570035.001.0001

Rolls, E. T. (2008a). *Memory, attention, and decision-making: A unifying computational neuroscience approach*. Oxford, UK: Oxford University Press.

Rolls, E. T. (2008b). The affective and cognitive processing of touch, oral texture, and temperature. *Journal of Neuroscience and Biobehavioral Reviews, 34*, 237–245. doi:10.1016/j.neubiorev.2008.03.010

Russell, S., & Norvig, P. (2010). *Artificial intelligence: A modern approach*. Upper Saddle River, NJ: Prentice Hall.

Shannon, C. E. (1948). A mathematical theory of communication. *The Bell System Technical Journal, 27*, 379–423, 623–656.

Shannon, C. E., & Weaver, W. (1949). *The mathematical theory of communication*. Urbana, IL: Illinois University Press.

Simon, H. A. (1982). *Models of bounded rationality*. Cambridge, MA: MIT Press.

Sun, F., Wang, Y., Lu, J., Zhang, B., Kinsner, W., & Zadeh, L. A. (Eds.). (2010, July). *Proceedings of the 9th IEEE International Conference on Cognitive Informatics*, Beijing, China. Washington, DC: IEEE Computer Society.

Tian, Y., Wang, Y., Gavrilova, M., & Rehe, G. (2011). A formal knowledge representation system for the cognitive learning engine. *International Journal of Software Science and Computational Intelligence, 3*(4).

Wang, Y. (2002a, August). Keynote: On cognitive informatics. In *Proceedings of the 1st IEEE International Conference on Cognitive Informatics*, Calgary, AB, Canada (pp. 34-42). Washington, DC: IEEE Computer Society.

Wang, Y. (2002b). The Real-Time Process Algebra (RTPA). *Annals of Software Engineering, 14*, 235–274. doi:10.1023/A:1020561826073

Wang, Y. (2003a). On cognitive informatics. *Brain and Mind: A Transdisciplinary Journal of Neuroscience and Neurophilosophy, 4*(2), 151-167.

Wang, Y. (2003b). Cognitive informatics: A new transdisciplinary research filed. *Brain and Mind, 4*(2), 115–127. doi:10.1023/A:1025419826662

Wang, Y. (2003c). Using process algebra to describe human and software behavior. *Brain and Mind, 4*(2), 199–213. doi:10.1023/A:1025457612549

Wang, Y. (2006, July). Keynote: Cognitive informatics - Towards the future generation computers that think and feel. In *Proceedings of the 5th IEEE International Conference on Cognitive Informatics*, Beijing, China (pp. 3-7). Washington, DC: IEEE Computer Society.

Wang, Y. (2007a). *Software engineering foundations: A software science perspective* (*Vol. 2*). Boca Raton, FL: Auerbach.

Wang, Y. (2007b). The theoretical framework of cognitive informatics. *International Journal of Cognitive Informatics and Natural Intelligence, 1*(1), 1–27. doi:10.4018/jcini.2007010101

Wang, Y. (2007c). The OAR model of neural informatics for internal knowledge representation in the brain. *International Journal of Cognitive Informatics and Natural Intelligence, 1*(3), 64–75. doi:10.4018/jcini.2007070105

Wang, Y. (2007d). Towards theoretical foundations of autonomic computing. *International Journal of Cognitive Informatics and Natural Intelligence, 1*(3), 1–16. doi:10.4018/jcini.2007070101

Wang, Y. (2008a). On contemporary denotational mathematics for computational intelligence. *Transactions of Computational Science, 2*, 6–29. doi:10.1007/978-3-540-87563-5_2

Wang, Y. (2008b). On concept algebra: A denotational mathematical structure for knowledge and software modeling. *International Journal of Cognitive Informatics and Natural Intelligence, 2*(2), 1–19. doi:10.4018/jcini.2008040101

Wang, Y. (2008c). On system algebra: A denotational mathematical structure for abstract system modeling. *International Journal of Cognitive Informatics and Natural Intelligence, 2*(2), 20–42. doi:10.4018/jcini.2008040102

Wang, Y. (2008d). RTPA: A denotational mathematics for manipulating intelligent and computational behaviors. *International Journal of Cognitive Informatics and Natural Intelligence, 2*(2), 44–62. doi:10.4018/jcini.2008040103

Wang, Y. (2009a). On abstract intelligence: Toward a unified theory of natural, artificial, machinable, and computational intelligence. *International Journal of Software Science and Computational Intelligence, 1*(1), 1–18. doi:10.4018/jssci.2009010101

Wang, Y. (2009b). On cognitive computing. *International Journal of Software Science and Computational Intelligence, 1*(3), 1–15. doi:10.4018/jssci.2009070101

Wang, Y. (2009c). Paradigms of denotational mathematics for cognitive informatics and cognitive computing. *Fundamenta Informaticae, 90*(3), 282–303.

Wang, Y. (2009d). On Visual Semantic Algebra (VSA): A denotational mathematical structure for modeling and manipulating visual objects and patterns. *International Journal of Software Science and Computational Intelligence, 1*(4), 1–15. doi:10.4018/jssci.2009062501

Wang, Y. (2009e, June). Granular algebra for modeling granular systems and granular computing. In *Proceedings of the 8th IEEE International Conference on Cognitive Informatics*, Hong Kong (pp. 145-154). Washington, DC: IEEE Computer Society.

Wang, Y. (2009f). A cognitive informatics reference model of Autonomous Agent Systems (AAS). *International Journal of Cognitive Informatics and Natural Intelligence, 3*(1), 1–16. doi:10.4018/jcini.2009010101

Wang, Y. (2010a, July). Keynote: Cognitive computing and World Wide Wisdom (WWW+). In *Proceedings of the 9th IEEE International Conference on Cognitive Informatics*, Beijing, China. Washington, DC: IEEE Computer Society.

Wang, Y. (2010b, August). Keynote: Cognitive informatics and denotational mathematics means for brain informatics. In *Proceedings of the 1st International Conference on Brain Informatics*, Toronto, ON, Canada.

Wang, Y. (2010c). On concept algebra for Computing with Words (CWW). *International Journal of Semantic Computing, 4*(3), 331–356. doi:10.1142/S1793351X10001061

Wang, Y. (2010d). On formal and cognitive semantics for semantic computing. *International Journal of Semantic Computing, 4*(2), 203–237. doi:10.1142/S1793351X10000833

Wang, Y. (2010e). Cognitive robots: A reference model towards intelligent authentication. *IEEE Robotics and Automation, 17*(4), 54–62. doi:10.1109/MRA.2010.938842

Wang, Y. (2011). Inference Algebra (IA): A denotational mathematics for cognitive computing and machine reasoning (I). *International Journal of Cognitive Informatics and Natural Intelligence, 5*(4).

Wang, Y. (in press). Inference Algebra (IA): A denotational mathematics for cognitive computing and machine reasoning (II). *International Journal of Cognitive Informatics and Natural Intelligence, 6*(1).

Wang, Y., Baciu, G., Yao, Y., Kinsner, W., Chan, K., & Zhang, B. (2010). Perspectives on cognitive informatics and cognitive computing. *International Journal of Cognitive Informatics and Natural Intelligence, 4*(1), 1–29. doi:10.4018/jcini.2010010101

Wang, Y., Celikyilmaz, A., Kinsner, W., Pedrycz, W., Leung, H., & Zadeh, L. A. (Eds.). (2011, August). *Proceedings of the 10th IEEE International Conference on Cognitive Informatics and Cognitive Computing*, Banff, AB, Canada. Washington, DC: IEEE Computer Society.

Wang, Y., Johnston, R., & Smith, M. (Eds.). (2002, August). *Proceedings of the 1st IEEE International Conference on Cognitive Informatics*, Calgary, AB, Canada. Washington, DC: IEEE Computer Society.

Wang, Y., & Kinsner, W. (2006). Recent advances in cognitive informatics. *IEEE Transactions on Systems, Man and Cybernetics. Part C, Applications and Reviews, 36*(2), 121–123. doi:10.1109/TSMCC.2006.871120

Wang, Y., Kinsner, W., Anderson, J. A., Zhang, D., Yao, Y., & Sheu, P. (2009). A doctrine of cognitive informatics. *Fundamenta Informaticae, 90*(3), 203–228.

Wang, Y., Kinsner, W., & Zhang, D. (2009). Contemporary cybernetics and its faces of cognitive informatics and computational intelligence. *IEEE Transactions on System, Man, and Cybernetics. Part B, 39*(4), 823–833.

Wang, Y., Tian, Y., & Hu, K. (2011, August). The operational semantics of concept algebra for cognitive computing and machine learning. *Proceedings of the 10th IEEE International Conference on Cognitive Informatics and Cognitive Computing*, Banff, AB, Canada. Washington, DC: IEEE Computer Society.

Wang, Y., & Wang, Y. (2006). Cognitive informatics models of the brain. *IEEE Transactions on Systems, Man and Cybernetics. Part C, Applications and Reviews, 36*(2), 203–207. doi:10.1109/TSMCC.2006.871151

Wang, Y., Wang, Y., Patel, S., & Patel, D. (2006). A Layered Reference Model of the Brain (LRMB). *IEEE Transactions on Systems, Man, and Cybernetics. Part C, 36*(2), 124–133.

Wang, Y., Widrow, B., Zhang, B., Kinsner, W., Sugawara, K., & Sun, F. (2011). Perspectives on the field of cognitive informatics and its future development. *International Journal of Cognitive Informatics and Natural Intelligence, 5*(1), 1–17. doi:10.4018/jcini.2011010101

Wang, Y., Zadeh, L. A., & Yao, Y. (2009). On the system algebra foundations for granular computing. *International Journal of Software Science and Computational Intelligence, 1*(1), 64–86. doi:10.4018/jssci.2009010105

Wang, Y., Zhang, D., & Kinsner, W. (Eds.). (2010). *Advances in cognitive informatics and cognitive computing*. Berlin, Germany: Springer-Verlag.

Wang, Y., Zhang, D., Latombe, J.-C., & Kinsner, W. (Eds.). (2008, August). *Proceedings of the 7th IEEE International Conference on Cognitive Informatics*, Stanford, CA. Washington, DC: IEEE Computer Society.

Yao, Y. Y., Shi, Z., Wang, Y., & Kinsner, W. (Eds.). (2006, July). *Proceedings of the 5th IEEE International Conference on Cognitive Informatics*, Beijing, China. Washington, DC: IEEE Computer Society.

Zadeh, L. A. (1965). Fuzzy sets and systems. In Fox, J. (Ed.), *Systems theory* (pp. 29–37). Brooklyn, NY: Polytechnic Press.

Zadeh, L. A. (1975). Fuzzy logic and approximate reasoning. *Syntheses, 30*, 407–428. doi:10.1007/BF00485052

Zadeh, L. A. (1999). From computing with numbers to computing with words – from manipulation of measurements to manipulation of perception. *IEEE Transactions on Circuits and Systems I, 45*(1), 105–119. doi:10.1109/81.739259

Zadeh, L. A. (2008, August). Toward human level machine intelligence – Is it achievable? In *Proceedings of the 7th IEEE International Conference on Cognitive Informatics*, Stanford, CA (p. 1). Washington, DC: IEEE Computer Society.

Zhang, D. (2011). Inconsistency-induced heuristics for problem solving. In *Proceedings of the 23rd International Conference on Software Engineering and Knowledge Engineering*, Miami, FL (pp. 137-142).

Zhang, D., & Lu, M. (2011, August). Inconsistency-induced learning: A step toward perpetual learners. *Proceedings of the 10th IEEE International Conference on Cognitive Informatics*, Banff, AB, Canada. Washington, DC: IEEE Computer Society.

Zhang, D., Wang, Y., & Kinsner, W. (Eds.). (2007, August). *Proceedings of the 6th IEEE International Conference on Cognitive Informatics*, Lake Tahoe, CA. Washington, DC: IEEE Computer Society.

Zhang, J., Baciu, G., Liang, S., & Liang, C. (2010, November). A creative try: Composing weaving patterns by playing on a multi-input device. In *Proceedings of the 17th ACM Symposium on Virtual Reality Software and Technology* (pp. 127-130).

Zheng, D., Baciu, G., & Hu, J. (2010). Weave pattern accurate indexing and classification using entropy-based computing. *International Journal of Cognitive Informatics and Natural Intelligence, 4*(4), 76–92. doi:10.4018/jcini.2010100106

This work was previously published in the International Journal of Cognitive Informatics and Natural Intelligence, Volume 5, Issue 4, edited by Yingxu Wang, pp. 1-21, copyright 2011 by IGI Publishing (an imprint of IGI Global).

Section 3
Denotational Mathematics

Chapter 11
Inference Algebra (IA):
A Denotational Mathematics for Cognitive Computing and Machine Reasoning (I)

Yingxu Wang
University of Calgary, Canada

ABSTRACT

Inference as the basic mechanism of thought is one of the gifted abilities of human beings. It is recognized that a coherent theory and mathematical means are needed for dealing with formal causal inferences. This paper presents a novel denotational mathematical means for formal inferences known as Inference Algebra (IA). IA is structured as a set of algebraic operators on a set of formal causations. The taxonomy and framework of formal causal inferences of IA are explored in three categories: (a) Logical inferences on Boolean, fuzzy, and general logic causations; (b) Analytic inferences on general functional, correlative, linear regression, and nonlinear regression causations; and (c) Hybrid inferences on qualification and quantification causations. IA introduces a calculus of discrete causal differential and formal models of causations; based on them nine algebraic inference operators of IA are created for manipulating the formal causations. IA is one of the basic studies towards the next generation of intelligent computers known as cognitive computers. A wide range of applications of IA are identified and demonstrated in cognitive informatics and computational intelligence towards novel theories and technologies for machine-enabled inferences and reasoning.

DOI: 10.4018/978-1-4666-2476-4.ch011

1. INTRODUCTION

Inference is a reasoning process that derives a causal conclusion from given premises. Formal inferences are usually symbolic and mathematical logic based, in which a causation is proven true by empirical observations, logical truths, mathematical equivalence, and/or statistical norms. Conventional logical inferences may be classified into the categories of deductive, inductive, abductive, and analogical inferences (Zadeh, 1965, 1975, 1999, 2004, 2008; Schoning, 1989; Sperschneider & Antoniou, 1991; Hurley, 1997; Tomassi, 1999; Wilson & Clark, 1988; Wang, 2007b, 2008a, 2011a; Wang et al., 2006), as well as *qualification* and *quantification* (Zadeh, 1999, 2004; Wang, 2007b, 2009c).

Studies on mechanisms and laws of inferences can be traced back to the very beginning of human civilization, which formed part of the foundations of various disciplines such as philosophy, logic, mathematics, cognitive science, artificial intelligence, computational intelligence, abstract intelligence, knowledge science, computational linguistics, and psychology (Zadeh, 1965, 1975, 2008; Mellor, 1995; Ross, 1995; Bender, 1996; Leahey, 1997; Wang, 2007c). Aristotle (1989) established *syllogism* that formalized inferences as logical arguments on propositions in which a conclusion is deductively inferred from two premises. Syllogism was treated as the fundamental methodology for inferences by Bertrand Russell in *The Principles of Mathematics* (Russell, 1903). Causality is a universal phenomenon of both the natural and abstract worlds because any rational state, event, action, or behavior has a cause. Further, any sequence of states, events, actions, or behaviors may be identified as a series of causal relations. In his classic work, *Principia: Mathematical Principles of Natural Philosophy* (Newton, 1687), Isaac Newton described a set of rules for inferences about nature known as the *experimental philosophy of causality* as follows:

- "**Rule 1:** We are to admit no more causes of natural things than such as are both true and sufficient to explain their appearances."
- "**Rule 2:** Therefore to the same natural effects we must, as far as possible, assign the same causes."
- "**Rule 3:** The qualities of bodies, which admit neither intension nor remission of degrees, and which are found to belong to all bodies within the reach of our experiments, are to be esteemed the universal qualities of all bodies whatsoever."
- "**Rule 4:** In experimental philosophy we are to look upon propositions collected by general induction from phenomena as accurately or very nearly true, notwithstanding any contrary hypotheses that may be imagined, till such time as other phenomena occur, by which they may either be made more accurate, or liable to exceptions."

In *A System of Logic*, John S. Mill identified six forms of *causal connections* between events in philosophy known as the methods of agreement, inverse agreement, double agreement, difference, residues, and concomitant variation (Mill, 1847). George Boole in *The Laws of Thought* studied the mathematical and logical laws of human thinking mechanisms and processes, where he perceived inference as operations of the mind based on logical and probability laws (Boole, 1854). Lotfi A. Zadeh created the fuzzy set theory and fuzzy logic since 1960s (Zadeh, 1965, 1975, 1999, 2004, 2008), which become one of the most applied theory for building fuzzy reasoning models in modern sciences and engineering. Fuzzy inferences are a powerful denotational mathematical means for rigorously dealing with degrees of matters, uncertainties, and vague semantics of linguistic variables, as well as for precisely reasoning the semantics of fuzzy causations.

Although there are various inference schemes and methods developed in a wide range of disciplines and applications, the framework of formal inferences can be described in five categories known as the *relational, rule-based, logical, fuzzy logical,* and *causal inferences.* With a higher expressive power, causal inferences are a set of advanced inference methodologies building upon the other fundamental layers, which is one of the central capabilities of human brains which plays a crucial role in thinking, perception, reasoning, and problem solving (Zadeh, 1975; BISC, 2010; Wang, 2009c, 2011a; Wang, Zadeh, & Yao, 2009). The coherent framework of formal inferences reveals how human reasoning may be formalized and how machines may rigorously mimic the human inference mechanisms. The central focus of formal inferences is to reveal causations implied in a thread of thought beyond the semantics of a natural language expression.

Definition 1: Let $\mathfrak{S}$ be a finite nonempty set of states or facts, $\mathfrak{R}$ be a finite set of relations between a pair of states. The *discourse of causality* $\mathfrak{U}$ is a 2-tuple, i.e.:

$$\mathfrak{U} \triangleq (\mathfrak{S}, \mathfrak{R}) \tag{1}$$

where $\mathfrak{R}$ is a Cartesian product, $\mathfrak{R} = \mathfrak{S} \times \mathfrak{S}$.

On the basis of the causal discourse $\mathfrak{U}$, a *causation* is a relation of a logical consequence between a sole or multiple causes and a single or multiple effects. A causation is usually a pair of (*Cause, Effect*). The causal relations may be 1-1, 1-n, n-1, and n-m, where n and m are integers greater than 1 that represent multiple relations. The *cause* (C) in a causation on $\mathfrak{U}$ is a premise state such as an event, phenomenon, action, behavior, or existence. Related to a cause, a *reason* is a premise of an argument in support of a belief or causation. However, the *effect* (E) in a causation on $\mathfrak{U}$ is a consequent or conclusive state such an event, phenomenon, action, behavior, or existence.

Definition 2: A *formal causation* ξ, $\xi \in \mathfrak{R}$, on $\mathfrak{U}$ is a relation that maps a set of causes C onto a set of effects E, i.e.:

$$\xi \triangleq f_{\xi} : C \rightarrow E, \ \xi \in \mathfrak{R}, C \subseteq \mathfrak{S}, E \subseteq \mathfrak{S} \tag{2}$$

Based on the causal discourse $\mathfrak{U}$ and the formal causation ξ, an inference is a cognitive process that deduces a conclusion, particularly a causation, based on evidences and reasoning.

Definition 3: An *inference* κ on $\mathfrak{U}$ is a reasoning process that deduces a causation ξ between a set of causes $C \subseteq \mathfrak{S}$ and a set of effects $E \subseteq \mathfrak{S}$ denoted by $\vdash$ (reads *turnstile*), i.e.:

$$\begin{aligned} \kappa &\triangleq C \vdash E \\ &= C \vdash f_{\xi}(C) \end{aligned} \tag{3}$$

Reasoning is a synonym of inference as an action to think, understand, and form judgments logically. Conventional method for inference and reasoning are such as relational, logical, functional, rule-based, neural networks, concept algebra, Bayesian networks, and causal networks.

Lemma 1: The *forms of inferences* on $\mathfrak{U}$ are determined by the types of causes C and effects E in Boolean variables or real numbers, which result in the *Boolean* (κ_b), *real* (κ_r), *qualification* (κ_{ql}), and *quantification* (κ_{qt}) inferences as follows:

$$\kappa_b \triangleq C\mathbf{BL} \vdash E\mathbf{BL} \tag{4}$$

$$\kappa_r \triangleq C\mathbf{R} \vdash E\mathbf{R} \tag{5}$$

$$\kappa_{ql} \triangleq C\mathbf{R} \vdash E\mathbf{BL} \tag{6}$$

$$\kappa_{qt} \triangleq C\mathbf{BL} \vdash E\mathbf{R} \tag{7}$$

where **BL** and **R** are the Boolean or real type suffixes, respectively.

Theorem 1: The *generality of causality* states that every natural and mental effect has a cause.

Proof: According to Definition 1, because nothing exists as absolutely independent on the discourse of causality $(\mathfrak{U})$, the formal causation $\xi \in \mathfrak{R}$ is universal.

This paper presents a theory of formal inferences and the framework of causal inferences in a denotational mathematical structure known as *Inference Algebra* (IA). In the remainder of this paper, the taxonomy and framework of formal inferences are explored in Section 2. IA is created as a set of algebraic operators on a set of formal causations in Section 3, where a set of general formal causations are rigorously modeled. A set of nine algebraic inference operators of IA, such as the Boolean, fuzzy, logical functional, analytic functional, correlative, linear-regression, nonlinear-regression, qualification, and quantification inferences, are formally described in Sections 4 and 5 on corresponding models of the formal causations. IA elicits and formalizes the common and empirical reasoning processes of humans in a rigorous form, which enable artificial and computational intelligence systems to mimic and implement similar inference abilities of the brain by cognitive computing. Applications of IA are demonstrated in cognitive computing and computational intelligence in Section 6 towards novel theories and technologies for machine-enabled inference and reasoning. Due to its excessive length and complexity, this paper presents the first part of IA on the structure of formal inference, the framework of IA, and the mathematical models of formal causations in Sections 1, 2, and 3. It will be followed by another paper that presents the second part of IA, Sections 4 through 7, on the inference operators of IA as well as their extensions and applications (Wang, 2012).

2. THE FRAMEWORK OF FORMAL INFERENCES AND RELATED WORK

Formal inferences are a deducing process that reasons a rational causation between a pair of cause and effect by proven empirical observations, theoretical rules, and/or statistical regulations (Bender, 1996; Wang, 2011a). A various inference schemes and methods have been developed in philosophy, logic, cognitive science, artificial intelligence, computational intelligence, computational linguistics, and psychology as reviewed in Section 1 (Zadeh, 1965, 2008; Sperschneider & Antoniou, 1991; Tomassi, 1999; Wilson & Clark, 1988; Wang, 2007b, 2011a; Wang et al., 2006). The framework of formal inferences can be modeled as a hierarchical structure as shown in Figure 1 with five categories of methodologies such as the relational, rule-based, logical, fuzzy, and causal inferences from the bottom up. Specific technologies in each category of the framework are summarized in Table 1, which will be further described in the following subsections.

2.1 Relational Inferences

Relation is one of the most important concepts in inference and reasoning because the result of inference is a relation between a pair of cause and effect. Relations also play an important role in explaining human internal knowledge representation and the natural intelligence. The simplest form of inference is relational inference that can be classified in the categories of binary, *n*-nary, reflexive, chain, and loop inferences (Wang, 2011a).

Definition 4: A *binary inference* κ_b deduces a one-to-one (1-1) relation between a pair of events or states as the cause $c\mathbf{BL}$ and effect $e\mathbf{BL}$, i.e.:

$$\kappa_b \triangleq c\mathbf{BL} \vdash e\mathbf{BL} \tag{8}$$

Figure 1. The hierarchical framework of formal inferences

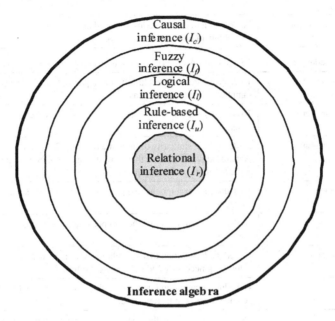

Table 1. The framework of formal inferences

No.	Category	Method	Description
1	Relational inference	Binary	To deduce a 1-to-1 relation between a pair of cause and effect.
		N-nary	To deduce an *n*-to-1 causal relation.
		Reflexive	To deduce a mutual causal relation between two states.
		Chain	To deduce a series of causal relations.
		Loop	To deduce a closed causal chain where any state is both a cause of its succeeding state and an effect of its preceding state.
2	Rule-based inference	Conditional rule	To deduce an appropriate action based on a Boolean condition.
		Numerical rule	To deduce an appropriate action based on a numerical expression.
		Event-driven rule	To deduce an appropriate action based on a predefined event dispatch scheme.
		Time-driven rule	To deduce an appropriate action based on a prescheduled time dispatch scheme.
3	Logical inference	Argument	To deduce a rational conclusion based on the conjunction of a set of known premises via syllogism.
		Deduction	To deduce a specific conclusion based on general premises.
		Induction	To deduce a general conclusion of a sequence of recurring patterns by three specific samples.
		Abduction	To deduce the most likely cause(s) and reason(s) of an observed effect.
		Analogy	To deduces the similarity of the same relation holds between different domains or systems, and/or examines that if two things agree in certain respects then they probably agree in others.

continued on following page

Table 1. Continued

4	Fuzzy logical inference	Fuzzy argument	To deduce a rational conclusion based on the conjunction of a set of known premises via syllogism on fuzzy expressions.
		Fuzzy deduction	To deduce a specific conclusion based on general premises on fuzzy expressions.
		Fuzzy induction	To deduce a general conclusion of a sequence of recurring patterns by three specific samples on fuzzy expressions.
		Fuzzy abduction	To deduce the most likely cause(s) and reason(s) of an observed effect on fuzzy expressions.
		Fuzzy analogy	To deduce the similarity of the same relation holds between different domains or systems, and/or examines that if two things agree in certain respects then they probably agree in others on fuzzy expressions.
5	Causal inference	Agreement	To deduce a causation that the presence of a cause necessarily results in the presence of an effect.
		Inverse agreement	To deduce a causation that the absence of a cause sufficiently results in the absence of an effect.
		Double agreement	To deduce a causation that a combination of agreement and inverse agreement for identifying necessary and sufficient conditions.
		Difference	To deduce a causation that the only difference of the status of a cause among a set of potential ones results in the effect.
		Residues	To deduce a causation that the effect(s) of an element of a complex cause can be elicited by the residual causation when all other elements are removed.
		Concomitant variation	To deduce a causation that the degrees of effects are typically proportional to those of their causes when the causation is defined on $\mathbb{R}$.
6	Inference algebra	Boolean inference	To deduce a $\mathbf{BL} \times \mathbf{BL}$ causation by causal differential.
		Fuzzy logical inference	To deduce an $\mathbf{FZ} \times \mathbf{BL}$ causation by causal differential.
		General logical inference	To deduce a general $\mathbf{BL} \times \mathbf{BL}$ logical functional causation by causal differential.
		General analytic inference	To deduce a general $\mathbf{R} \times \mathbf{R}$ analytic functional causation by causal differential.
		Correlative inference	To deduce a specific $r(c\mathbf{R}, e\,\mathbf{R})$ correlative causation by causal differential.
		Linear regression inference	To deduce a specific $e\mathbf{R} = \alpha_0 \mathbf{R} + \alpha_1 c\mathbf{R}$ linear regression causation by causal differential.
		Nonlinear regression inference	To deduce a specific $e\mathbf{R} = \alpha_n c_i{}^n \mathbf{R} + \alpha_{n-1} c_i{}^{n-1} \mathbf{R} + \ldots + \alpha_1 c_i \mathbf{R} + \alpha_0$ nonlinear regression causation by causal differential.
		Qualification inference	To deduce a specific $\mathbf{R} \times \mathbf{BL}$ qualification causation by causal differential.
		Quantification inference	To deduce a specific $\mathbf{BL} \times \mathbf{R}$ quantification causation by causal differential.

where $\vdash$ denotes a causation that $c\mathbf{BL}$ yields $e\mathbf{BL}$, and $\mathbf{BL}$ is the Boolean type suffix.

The binary inference as given in Definition 4 can be extended to complex ones such as *n*-nary and chain inferences.

Definition 5: An *n-nary inference* κ_n is a composite form of binary inference where an effect $e\mathbf{BL}$ is deduced from multiple causes $c_1\mathbf{BL}, c_2\mathbf{BL}, \ldots, c_n\mathbf{BL}$, i.e.:

$$\kappa_n \triangleq (c_1\mathbf{BL} \wedge c_2\mathbf{BL} \wedge \ldots \wedge c_n\mathbf{BL}) \vdash e\mathbf{BL}$$
$$= (\bigwedge_{i=1}^{n} c_i\mathbf{BL}) \vdash e\mathbf{BL} \tag{9}$$

Definition 6: A *reflexive inference* κ_f is a bidirectional binary inference where two states $p\mathbf{BL}$ and $q\mathbf{BL}$ are mutually cause and effect with each other, i.e.:

$$\kappa_f \triangleq (p\mathbf{BL} \vdash q\mathbf{BL}) \wedge (q\mathbf{BL} \vdash p\mathbf{BL}) \tag{10}$$

Definition 7: A *chain inference* κ_c is a series of n binary inference triggered by a cause $p\mathbf{BL}$ where each effect $q_i\mathbf{BL}$ is the cause of a succeeding effect $q_{i+1}\mathbf{BL}$, $1 \le i \le n$, i.e.:

$$\kappa_c \triangleq p\mathbf{BL} \vdash q_1\mathbf{BL} \vdash q_2\mathbf{BL} \vdash \ldots \vdash q_n\mathbf{BL}$$
$$= p\mathbf{BL} \underset{i=1}{\overset{n}{\vdash}} q_i\mathbf{BL} \tag{11}$$

Definition 8: A *loop inference* κ_l is a closed chain inference where a state $q_i\mathbf{BL}$ is both the cause of $q_{i+1}\mathbf{BL}$ and the effect of $q_{i-1}\mathbf{BL}$, $1 \le i \le n$, i.e.:

$$\kappa_l \triangleq Q_0\mathbf{BL} \vdash Q_1\mathbf{BL} \vdash Q_2\mathbf{BL} \vdash \ldots \vdash Q_n\mathbf{BL} \vdash Q_0\mathbf{BL}$$
$$= Q_0\mathbf{BL} \underset{i=1}{\overset{n}{\vdash}} Q_i\mathbf{BL} \vdash Q_0\mathbf{BL} \tag{12}$$

2.2 Rule-Based Inferences

A *rule* is a predefined production that determines a target action according to the given condition (Bender, 1996; Pearl, 2009; Wang, 2007c). Rule-based inferences are conventional and relatively simple form of inferences in which the effect is deduced by evaluating a specific condition. The difficulty in rule-based inference is that not all rules and conditions are always known in inferences, because there are complex, uncertain, unpredictable and indeterministic conditions and causations. The following subsections describe rule-based inferences with the conditional, iterative, event-driven, and time-driven rules.

2.2.1 Conditional-Rule-Based Inferences

Definition 9: A *conditional-rule-based inference* κ_{cr} is a selective inference denoted by ◆ that deduces an appropriate action $e_i\mathbf{PC}$ based on the Boolean expression $exp\mathbf{BL}$, i.e.:

$$\kappa_{cr} \triangleq (\ \ ◆\ exp\mathbf{BL} = \mathbf{T} \vdash e_0\mathbf{PC}$$
$$| ◆ \sim \qquad \vdash e_1\mathbf{PC} \tag{13}$$
$$)$$

where '$\sim$' means '$exp\mathbf{BL} = \mathbf{F}$,' or more general, 'otherwise;' and $\mathbf{BL}$ and $\mathbf{PC}$ denote the type suffixes of Boolean and process type suffixes, respectively.

When the *else* branch is optional, Equation 13 is equivalent to the following where $\varnothing$ denotes doing nothing but exit:

$$\kappa_{cr} \triangleq (\ \ ◆\ exp\mathbf{BL} = \mathbf{T} \vdash e\mathbf{PC}$$
$$| ◆ \sim \qquad \vdash \varnothing \tag{14}$$
$$)$$

2.2.2 Numerical-Rule-Based Inferences

Definition 10: A *numerical-rule-based inference* κ_{nr} is a multiple selective inference that deduces an appropriate action $e_i\mathbf{PC}$ based on the numerical expression $exp\mathbf{N}$, i.e.:

$$\kappa_{nr} \triangleq (\ ◆exp\mathbf{N} = \ \ 0: \ \vdash e_0\mathbf{PC}$$
$$| \ 1: \ \vdash e_1\mathbf{PC}$$
$$| \ \ldots$$
$$| \ n\text{-}1: \vdash e_{n-1}\mathbf{PC} \tag{15}$$
$$| \sim: \ \vdash \varnothing$$
$$)$$

The numerical-rule-based inference is equivalent to a set of embedded conditional-rule-based inferences as defined in Equation 13. Both conditional and numerical rule-based inferences can be perceived as a *production* that determines a target action according to the given condition (McDermid, 1991; Wang, 2007c).

2.2.3 Event-Driven-Rule-Based Inferences

Definition 11: An *event-driven-rule-based inference* κ_{er} is a reflexive inference that deduces an appropriate action $e_i\textbf{PC}$ based on a predefined event dispatch scheme $c_i\textbf{BL}$, i.e.:

$$\kappa_{er} \triangleq \mathop{R}_{i=1}^{n} c_i\textbf{BL} \vdash e_i\textbf{PC} \tag{16}$$

where $\mathop{R}_{i=1}^{n}$ is the big-R notation of Real-Time Process Algebra (RTPA) that denotes an iterative operation as a loop (Wang, 2002, 2007c, 2008b).

2.2.4 Time-Driven-Rule-Based Inferences

Definition 12: A *time-driven-rule-based inference* κ_{tr} is a timed inference that deduces an appropriate action $e_i\textbf{PC}$ based on a prescheduled time dispatch scheme $c_i\textbf{TM}$, i.e.:

$$\kappa_{tr} \triangleq \mathop{R}_{i=1}^{n} c_i\textbf{TM} \vdash e_i\textbf{PC} \tag{17}$$

where **TM** is the type suffix of time, **TM = hh:mm:ss:ms**, in RTPA.

2.3 Logical Inferences

Logical inferences are most classical approach in formal reasoning, which is suitable to inference on logical propositions. However, it is noteworthy that not all inferences may be modeled as logical inferences, and not all causes, effects, and their relations are logical variables or relations.

Definition 13: A *logical inference* is a form of formal inferences that deduces a logical causation between a pair of cause and effect modeled by logical variables or expressions.

The taxonomy of logic inferences can be classified as causal argument, deductive inference, inductive inference, abductive inference, and analogical inference (Hurley, 1997; Ross, 1995; Schoning, 1989; Tomassi, 1999; Wang, 2007a, 2007c). *Argumentative* inference is a form of logical inferences that deduces a rational conclusion based on the conjunction of a set of known premises via syllogism. *Deductive* inference is a form of logical inferences where a specific conclusion necessarily follows from a set of general premises. *Inductive* inference is a form of logical inferences where a general conclusion is drawn from a set of specific premises based on a set of samples in reasoning or experimental evidences. *Abductive* inference is a form of logical inferences that deduces the best explanation or most likely reason of an observation or event. *Analogy* inference is a form of logical inferences, which deduces similar relations hold between different domains or systems, and/or examines that if two things agree in certain respects then they probably agree in others.

A summary of logical inferences is provided in Table 2. Further descriptions, examples, and case studies may refer to Wang (2007a, 2011a).

2.4 Fuzzy Logical Inferences

Although logic inferences may be carried out on the basis of abstraction and symbolic reasoning with sets and Boolean logic, more inference mechanisms and rules such as those of intuitive, empirical, heuristic, and perceptive inferences, are

Table 2. Formal models of logical inferences

No.	Operator	Description	Definition
1	⊢	Argument inference	$\mathbf{ABL} \triangleq P_1\mathbf{BL} \land P_2\mathbf{BL} \land \dots \land P_n\mathbf{BL} \vdash Q\mathbf{BL}$
2	⇓	Deductive inference	$\mathfrak{D}\mathbf{BL} \triangleq \forall x \in \mathbf{X}, p(x) \Downarrow \exists a \in \mathbf{X}, p(a)$
3	⇑	Inductive inference	$\mathfrak{I}\mathbf{BL} \triangleq ((\exists a \in \mathbf{X}, p(a)) \land (\exists k, succ(k) \in \mathbf{X}, (p(k) \Rightarrow p(succ(k))))) \Uparrow \forall x \in \mathbf{X}, p(x)$
4	↲	Abductive inference	$\mathfrak{B}\mathbf{BL} \triangleq (\forall x \in \mathbf{X}, p(x) \Rightarrow q(x)) \lrcorner (\exists a \in \mathbf{X}, q(a) \Rightarrow p(a))$
5	~	Analogy inference	$\mathfrak{N}\mathbf{BL} \triangleq \exists a \in \mathbf{X}, p(a) \sim \exists b \in \mathbf{X}, p(b)$

Table 3. Formal models of fuzzy logical inferences

No.	Operator	Description	Definition
1	$\hat{\vdash}$	Fuzzy argument	$\mathbf{AFZ} \triangleq (P_1\mathbf{FZ} \land P_2\mathbf{FZ} \land \dots \land Pn\mathbf{FZ}) \hat{\vdash} Q\mathbf{FZ}$
2	$\hat{\Downarrow}$	Fuzzy deductive	$\mathfrak{D}\mathbf{FZ} \triangleq \forall x \in \tilde{S}(x, \mu(x)), p(x) \hat{\Downarrow} \exists a \in \tilde{S}(a, \mu(a)), p(a)$
3	$\hat{\Uparrow}$	Fuzzy inductive	$\mathfrak{I}\mathbf{FZ} \triangleq$ $\exists a \in \tilde{S}(a, \mu(a)), p(a) \land \exists k, succ(k) \in \tilde{S}((k, \mu(k)), (succ(k), \mu(succ(k))), p(k) \Rightarrow p(succ(k))$ $\hat{\Uparrow} \forall x \in \tilde{S}(x, \mu(x)), p(x)$
4	$\hat{\lrcorner}$	Fuzzy abductive	$\mathfrak{B}\mathbf{FZ} \triangleq \forall x \in \tilde{S}(x, \mu(x)), p(x) \Rightarrow q(x) \hat{\lrcorner} \exists a \in \tilde{S}(a, \mu(a)), q(a) \Rightarrow p(a)$
5	$\hat{\sim}$	Fuzzy analogy	$\mathfrak{N}\mathbf{FZ} \triangleq$ $\forall x \in \tilde{S}(x, \mu(x)), p(x) \land \exists a \in \tilde{S}(a, \mu(a)), p(a) \hat{\sim} \exists b \in \tilde{S}(b, \mu(b)) \land a \neq b, p(b)$

fuzzy and uncertain, which are yet to be studied by fuzzy inferences on the basis of fuzzy logic (Zadeh, 1965, 1975, 2008; Wang, 2009c). Lotfi A. Zadeh extends the scope of inferences and traditional mathematical means to fuzzy sets (Zadeh, 1965) and fuzzy logic (Zadeh, 1975, 2008). The fuzzy set theory is an extended set theory for dealing with uncertainty and imprecision in reasoning, qualification, and quantification, particularly when where involves vague linguistic variables.

Definition 14. A *fuzzy* set $\tilde{S}$ on the universal discourse U is an extended set in which each element is a pair with a member x and its degree of membership determined by a membership function $\mu(x)$ in the domain (0, 1], i.e.:

$$\tilde{S} \triangleq \{(x, \mu(x)) \mid 0 < \mu(x) \leq 1\} \quad (18)$$

where $\mu(x) \triangleq f : x \to \mathbb{R}', 0 < \mathbb{R}' \leq 1$, and f($x$) is case dependent.

It is noteworthy in Definition 14 that the domain of $\mu(x)$ is specified in (0, 1] rather than [0, 1], which allows a fuzzy set to avoid the following paradoxes:

There is infinitive number of members x' of $\tilde{S}$ on $\mathfrak{U}$ where its degree of membership is negative (zero), i.e.:

$$\exists x' \in \tilde{S}(x, \mu(x)) \Rightarrow \mu(x') = 0 \land |\tilde{S}| = \infty \quad (19)$$

Table 4. Mill's methods for identifying causal connections

No.	Method	Description	Potential course					Effect $e\mathbf{BL}$	Causation $C'\mathbf{BL} \vdash e\mathbf{BL}$
			$c_1\mathbf{BL}$	$c_2\mathbf{BL}$	$c_3\mathbf{BL}$	$c_4\mathbf{BL}$	$c_5\mathbf{BL}$		
1	Agreement	The presence of a cause necessarily results in the presence of an effect.	F	T	F	F	T	T	$\{c_2\mathbf{BL}\} \vdash e\mathbf{BL}$
			T	T	T	F	F	T	
2	Inverse agreement	The absence of a cause sufficiently results in the absence of an effect.	F	T	F	F	T	F	$\{c_3\mathbf{BL}\} \vdash e\mathbf{BL}$
			T	F	F	T	F	F	
3	Double agreement	A combination of Methods 1 and 2 for identifying necessary and sufficient conditions.	F	F	F	T	F	T	$\{c_4\mathbf{BL}\} \vdash e\mathbf{BL}$
			T	F	T	F	T	F	
4	Difference	The only difference of the status of a cause among a set of potential ones results in the effect.	T	F	T	F	T	T	$\{c_5\mathbf{BL}\} \vdash e\mathbf{BL}$
			T	F	T	F	F	F	
5	Residues	The effect(s) of an element of a complex cause can be elicited by the residual causation when all other elements are removed.	F	X	X	X	X	F	$\{c_1\mathbf{BL}\} \vdash e\mathbf{BL}$
			T	X	X	X	X	T	
6	Concomitant variation	The degrees of effects are typically proportional to those of their causes when the causation is defined on $\mathbb{R}$.	$c_1\mathbf{R}\uparrow$	-	-	-	-	$e\mathbf{R}\uparrow \vee e\mathbf{R}\downarrow$	$\{c_1\mathbf{R}\} \vdash e\mathbf{R}$
			$c_1\mathbf{R}\downarrow$	-	-	-	-	$e\mathbf{R}\downarrow \vee e\mathbf{R}\uparrow$	

Any arbitrary nonmember x' of $\tilde{S}$ on $\mathfrak{U}$ is a member of $\tilde{S}$ with the degree of membership as zero, i.e.:

$$\forall x' \notin \tilde{S}(x, \mu(x)) \wedge \mu(x') = 0 \Rightarrow x' \in \tilde{S}(x, \mu(x)) \tag{20}$$

However, for convenience and compatibility, the domain of $\mu(x)$ will be denoted as [0, 1] when there is no confusion in discussions.

On the basis of fuzzy logic, traditional logical inferences can be extended to fuzzy inferences where the premises of reasoning become fuzzy expressions rather than Boolean ones in traditional logical inferences.

Definition 15: A *fuzzy inference* is an extended form of formal inferences that deduces a fuzzy causation between a pair of cause and effect modeled as fuzzy expressions.

Fuzzy inferences are powerful denotational mathematical means for rigorously dealing with degrees of matters, uncertainties, and vague semantics of linguistic entities, as well as for precisely reasoning the semantics of fuzzy causations. Typical fuzzy inferences are the *fuzzy argument, deductive, inductive, abductive,* and *fuzzy analogical* inferences (Zedeh, 2008; Wang, 2009c). A summary of fuzzy logical inferences is provided in Table 3.

2.5 Causal Inferences

A classical study on causal inferences was presented by John S. Mill in philosophy and logic (Mill, 1847). Mill developed six approaches to deal with *causal connections* between events known as the methods of *agreement, inverse agreement, double agreement, difference, residues,* and *concomitant variation* as formalized in Table 4. In Table 4, the logical/real variables T, F, −, and X represent true (presence), false (absence), no change, and

Table 5. Properties of causal conditions

No.	Necessary condition	Sufficient condition
1	A cause is not a necessary condition for an effect if it occurs without the cause to be true, i.e.: $$e_i \mathbf{BL} = \mathbf{T} \wedge c_i \mathbf{BL} = \mathbf{F} \Rightarrow c_i \mathbf{BL} \notin P^n$$	A course is not a sufficient condition for an effect if the cause occurs without the effect to be true, i.e.: $$c_i \mathbf{BL} = \mathbf{T} \wedge e_i \mathbf{BL} = \mathbf{F} \Rightarrow c_i \mathbf{BL} \notin P^s$$
2	The absence of a cause is necessary for the absence of an effect, iff the cause is necessary for the effect, i.e.: $$c_i \mathbf{BL} \in P^n \Leftrightarrow (c_i \mathbf{BL} = \mathbf{F} \Rightarrow e_i \mathbf{BL} = \mathbf{F})$$	The absence of a cause is sufficient to the absence of an effect, iff the effect is sufficient to the cause, i.e.: $$e_i \mathbf{BL} \in P^s \Leftrightarrow (c_i \mathbf{BL} = \mathbf{F} \Leftrightarrow e_i \mathbf{BL} = \mathbf{F})$$
3	A necessary condition is a member of sufficient conditions, i.e.: $P^n \subseteq P^s$.	A sufficient condition includes all necessary conditions, i.e.: $P^s \supseteq P^n$.
4	When an effect is true, all the necessary conditions are true, i.e.: i.e.: $e_i \mathbf{BL} = \mathbf{T} \Rightarrow \forall P^n = \mathbf{T}$.	When an effect is true, at least one sufficient condition is true, i.e.: $e_i \mathbf{BL} = \mathbf{T} \Rightarrow \exists P^s = \mathbf{T}$.

eliminated (ignore), respectively. Almost all causal arguments are in Boolean type except Method 6 that is in real type. Examples of Mill's methods may refer to (Hurley, 1997).

Various studies on causations, causality, and causal reasoning have been reported (Lewis, 1973; Beebee et al., 2009; Turner, 2010; Wang, 2011a). In causal inference, a logical condition may be classified as either necessary or sufficient based on the following definitions.

Definition 16: A *necessary condition P^n* of a causal inference is a valid condition that represents one of the cause $c_i \mathbf{BL}$ for an effect $e_i \mathbf{BL}$ where if $e_i \mathbf{BL}$ is true, then $c_i \mathbf{BL}$ must be true, i.e.:

$$P^n \triangleq \forall (c_i \mathbf{BL} \vdash e_i \mathbf{BL}), e_i \mathbf{BL} = \mathbf{T} \Rightarrow c_i \mathbf{BL} = \mathbf{T} \tag{21}$$

Definition 17: A *sufficient condition P^e* of a causal inference is a valid condition that represents one of the causes $c_i \mathbf{BL}$ for an effect $e_i \mathbf{BL}$ where if $c_i \mathbf{BL}$ is true, then $e_i \mathbf{BL}$ must be true, i.e.:

$$P^s \triangleq \forall (c_i \mathbf{BL} \vdash e_i \mathbf{BL}), c_i \mathbf{BL} = \mathbf{T} \Rightarrow e_i \mathbf{BL} = \mathbf{T} \tag{22}$$

Definition 18. A *necessary and sufficient condition P^{ne}* of a causal inference is a composite condition that includes the least sufficient set of necessary conditions, i.e.:

$$P^{ns} \triangleq \forall (c_i \mathbf{BL} \vdash e_i \mathbf{BL}), c_i \mathbf{BL} \Leftrightarrow e_i \mathbf{BL} \tag{23}$$

A set of properties of causal conditions is identified as summarized in Table 5, which can be applied in causal inferences. It is noteworthy that the first five causation identification methods proposed by Mill can be generally covered by a more general Boolean n-to-1 causal inference method, which will be discussed in Section 4.1.

Corollary 1: When an effect is true, at least one sufficient condition is true, or all the necessary conditions are true.

For a binary causation, the cause is both necessary and sufficient, because it is the only condition in the causation.

3. INFERENCE ALGEBRA AND THE MODEL OF FORMAL CAUSATIONS

In the preceding section the framework of formal inferences was explored. It has been shown that the formal inference methodologies were involved from the simple relational, rule-based, logical, and fuzzy logical inferences to causal inferences. However, conventional causal inference technologies were application specific, which only cover the qualitative causal inferences on Boolean (as well as fuzzy) expressions. This problem indicates that a formal treatment of generic causal inferences is yet to be studied.

Table 6 contrasts the coverage of Mill's methods for causal arguments and inference algebra. The limitations of Mill's methods for causal arguments are described in the left column. The enhancements to these limitations by inference algebra are presented in the right column of Table 6.

This section introduces the mathematical models of formal causations and formalizes causal inferences by inference algebra. The architecture of inference algebra is a 2-tuple encompassing a finite set of formal causations and a finite set of inference operators. The former are the entity and operand of inference algebra, which are modeled in this section; while the later are the operators of inference algebra that will be described in Section 4.

3.1 The Framework of Inference Algebra

Definition 19: Let C be a finite nonempty set of *causes*, and E be a finite nonempty set of *effects*. The *discourse of causality*, U, as given in Definition 1 can be extended as follows:

$$\mathfrak{U} \triangleq (\mathfrak{S}, \mathfrak{R})$$
$$= (C, \mathrm{E}, \mathfrak{R}), C \subseteq \mathfrak{S}, E \subseteq \mathfrak{S} \qquad (24)$$

where $\mathfrak{R}$ is a Cartesian product, $\mathfrak{R} = \mathfrak{S} \times \mathfrak{S} = C \times E$.

On the basis of the causal discourse $\mathfrak{U}$ and the general properties of the five categories of inference methodologies as summarized in Table 1, inference algebra can be created as a universal form of algebraic operations on the formal models of causations.

Table 6. Comparative analysis of mill's causal argument and inference algebra

No.	Causal argument	Inference algebra
1	Informal and oriented to human inference.	Adapt a formal and rigorous way for causal inferences in order to enable machines to mimic the advanced human inference ability.
2	Only Boolean causal mechanism is considered. Methodology is focused on the effect of the presence or absence of potential causes.	Two new categories of the analytic and hybrid causal inferences are added beyond logical inferences. In addition, a more general logical inference known as logical-function-based inference is developed.
3	No method is provided for guiding the finding of a complete hypothetic causation set.	Support the derivation of the hypothetic causation set by searching the entire causal discourse.
4	When n-1 causations are processed, only conjunction of n causes is considered rather than other forms of relations such as *or* relations and inverse causations where the absence of a cause is the reason of an effect.	Cover all forms of multi-causal relations and the inverse causations.
5	Only 1-1 and partially n-1 causations are considered. The 1-m and n-m causations could not be dealt with.	Cover all four categories of causal inferences, particularly n-1 and n-m causal inferences.

Definition 20: *Inference algebra, IA,* is a denotational mathematical structure of a 2-tuple, on the causal discourse $\mathfrak{U}$, with a set of formal causations Ξ and a set of finite inference operators K on Ξ, i.e.:

$$IA \triangleq (\Xi, K)$$
$$= (\Xi, \{\kappa_b, \kappa_{fz}, \kappa_{lf}, \kappa_f, \kappa_r, \kappa_{lr}, \kappa_{nr}, \kappa_{ql}, \kappa_{qt}\})$$
(25)

where the nine inference operators in K denote the *Boolean, fuzzy, general logical, general functional, correlative, linear-regression, nonlinear-regression, qualification,* and *quantification* inferences.

The structure of IA can be perceived as a set of algebraic operators on a set of formal causations as the fundamental operands Ξ. The formal causations of IA are modeled in the following subsections; while the nine algebraic operators K of IA will be elaborated in Section 4.

3.2 The Mathematical Model of Formal Causations

According to Definition 2, the *formal causation* Ξ, $\Xi \subseteq \mathfrak{R}$, on $\mathfrak{U}$ is a relation that maps a set of causes C onto a set of effects E, i.e.:

$$\xi \triangleq f_\xi : C \to E, \ \xi \in \Xi \subseteq \mathfrak{R}, C \subseteq \mathfrak{S}, E \subseteq \mathfrak{S}$$
(26)

Definition 21: The *categories of formal causations* Ξ of IA on $\mathfrak{U}$ can be classified into four forms of Cartesian products between *Boolean-Boolean* (ξ_{BB}), *real-real* (ξ_{RR}), *Boolean-real* (ξ_{BR}), and *real-Boolean* (ξ_{RB}) causations, i.e.:

$$\Xi \triangleq \{\xi_{BB}(C\mathbf{BL} \times E\mathbf{BL}), \xi_{RR}(C\mathbf{R} \times E\mathbf{R}), \xi_{RB}(C\mathbf{R} \times E\mathbf{BL}), \xi_{BR}(C\mathbf{BL} \times E\mathbf{R})\}$$
(27)

The properties of different types of formal causations in Ξ are different. The Boolean-to-Boolean causation ξ_{BB} provides a formal model for conventional logical inferences. The real-to-real causation ξ_{RR} provides a formal model for more general analytic and quantitative inferences. The real-to-Boolean ξ_{RB} and Boolean-to-real ξ_{BR} causations provide formal models for hybrid inferences in complex contexts. It is noteworthy that the conventional inference method may have only covered qualitative causal inferences on Boolean (as well as fuzzy) expressions. Therefore, both of their reasoning power and coverage are quite limited.

Theorem 2: The principle of *invariant causality* states that, in an invariant layout of an *n-m* causal discourse $\mathfrak{U}$, similar effects arise from similar causes; similar causes result in similar effects.

Proof: According to Definition 19, a property of the causal discourse $\mathfrak{U}$ is that the causal relations $\mathfrak{R}$ on $\mathfrak{U}$ are invariant. Hence, all known and unknown causations, $\xi \in \Xi \subseteq \mathfrak{R}$, are invariant.

Corollary 2: Different effects arise from different causes or different combinations of them; different causes may result in the same effect.

3.3 The Discrete Causal Differential

In order to rigorously describe the IA operators, two new mathematical operations known as the qualitative causal differential and the quantitative causal differential are introduced (Wang, 2011a). Both discrete causal differentials will serve as the core calculi of all inference operators in IA.

Definition 22: A *qualitative causal differential* $\dfrac{\partial C\mathbf{BL}}{\partial c_i}$ on a set of n potential *Boolean* causes $C\mathbf{BL} = \{c_1\mathbf{BL}, c_2\mathbf{BL}, ..., c_n\mathbf{BL}\}$ with respect to its individual elements c_iBL,

$1 \leq i \leq n$, is a *partial discrete differential* that elicits a subset of effective causes $C'\mathbf{BL} = \{c_i\mathbf{BL} \mid 1 \leq i \leq n' \leq n\} \subseteq C\mathbf{BL}$, where the change of $c_i\mathbf{BL} \in C'\mathbf{BL}$ results in the target *effect* $e\mathbf{BL}$, i.e.:

$$C'\mathbf{BL} \triangleq \mathop{R}_{i=1}^{n} (\frac{\partial}{\partial c_i} C\mathbf{BL} \to e\mathbf{BL})$$

$$= \mathop{R}_{i=1}^{n} (\frac{\partial}{\partial c_i} \{c_1\mathbf{BL}, c_2\mathbf{BL}, ..., c_n\mathbf{BL}\} \to e\mathbf{BL})$$

$$= \{\mathop{R}_{i=1}^{n} c_i\mathbf{BL} \mid \frac{\partial}{\partial c_i} C\mathbf{BL} \to e\mathbf{BL}\}$$

(28)

where $\mathop{R}_{i=1}^{n}$ is the big-*R* notation of RTPA (Wang, 2002, 2008b) that denotes a series of repetitive individual evaluation of discrete partial causal differentials.

The qualitative causal differential is designed to individually identify the effective causes of a potential causation with a set of Boolean variables or expressions. In Equation 28 valid causations can be derived via observing change of $e\mathbf{BL}$ by changing the value of a $c_i\mathbf{BL}$, which is denoted by $\frac{\partial}{\partial c_i} C\mathbf{BL} \to e\mathbf{BL}$. The result of the qualitative causal differential is the elicitation of a set of effective and valid causes $C'\mathbf{BL}$ from the potential causation set $C\mathbf{BL}$.

Definition 23: A *quantitative causal differential* $\frac{\partial C\mathbf{R}}{\partial c_i}$ on a set of *n* potential *real* causes $C\mathbf{R} = \{c_1\mathbf{R}, c_2\mathbf{R}, ..., c_n\mathbf{R}\}$ with respect to its individual elements $c_i\mathbf{R}$, $1 \leq i \leq n$, is a *partial discrete differential* that evaluates the extend of effect of each valid causes $C'\mathbf{R} = \{c_i\mathbf{R} \mid 1 \leq i \leq n' \leq n\} \subseteq C\mathbf{R}$, where the change of $c_i\mathbf{R} \in C'\mathbf{R}$ results in the target *effect* $e\mathbf{R}$, $e\mathbf{R} = f(C\mathbf{R}) = f(c_1\mathbf{R}, c_2\mathbf{R}, ..., c_n\mathbf{R})$, i.e.:

$$C'\mathbf{R} \triangleq \mathop{R}_{i=1}^{n} (\frac{\partial}{\partial c_i} C\mathbf{R} \to e\mathbf{R})$$

$$= \mathop{R}_{i=1}^{n} (\frac{\partial}{\partial c_i} \{c_1\mathbf{R}, c_2\mathbf{R}, ..., c_n\mathbf{R}\} \to e\mathbf{R})$$

$$= \{\mathop{R}_{i=1}^{n} c_i\mathbf{R} \mid \frac{\partial}{\partial c_i} f(C\mathbf{R}) = \frac{\partial}{\partial c_i} e\mathbf{R} \neq 0\}$$

(29)

The quantitative causal differential is designed to evaluate the extent of effect of each cause as a real variable or expression in a given set of potential causations. In Equation (29) valid causations can be derived by testing if the effect $e\mathbf{R}$ is sensitive to $c_i\mathbf{R}$, which is denoted by $\frac{\partial}{\partial c_i} e\mathbf{R} \neq 0$ when the function $e\mathbf{R} = f(C\mathbf{R})$ is known. The result of the quantitative causal differential is the quantitative evaluation of a set of effective and valid causes $C'\mathbf{BL}$ from the potential causation set $C\mathbf{BL}$.

According to Definitions 22 and 23, the process of formal inferences can be conducted from a set of *hypothetic* causations to a set of *valid* causations, and then, to a set of *concrete* causations. This formal inference process can be illustrated by Examples 1, 2, and 3.

Example 1: Let $e\mathbf{BL}$ be an effect of wet surface of road, and $C\mathbf{BL}$ be the hypothetic causation set, $C\mathbf{BL} = \{$rained$\mathbf{BL}$, sprinkled$\mathbf{BL}$, flooded$\mathbf{BL}$, high_temperature$\mathbf{BL}\}$. A qualitative causal differential on $C\mathbf{BL}$ with respect to the individual potential causes $c_i\mathbf{BL} \in C\mathbf{BL}$, $C'\mathbf{BL}$, can be inferred according to Definition 22 to determine the set of valid causations by testing the change of an individual causation in $C\mathbf{BL}$ in Box 1, Equation 30.

The inference result by the quantitative causal deferential indicates that the valid causation set for the target effect as deduced from the hypothetic causation set is:

$\{$rained$\mathbf{BL}$, *sprinkled*$\mathbf{BL}$, *flooded*$\mathbf{BL}\} \vdash wet\_road\mathbf{BL}$

Box 1. Equation 30

$$
\begin{aligned}
C'\mathbf{BL} &= \mathop{R}_{i=1}^{n}(\frac{\partial}{\partial c_i}C\mathbf{BL} \to e\mathbf{BL}) \\
&= \mathop{R}_{i=1}^{4}(\frac{\partial}{\partial c_i}\{rained\mathbf{BL},\ sprinkled\mathbf{BL},\ flooded\mathbf{BL},\ high\_temperature\mathbf{BL}\} \to wet\_road\mathbf{BL}) \\
&= \{\mathop{R}_{i=1}^{4}c_i\mathbf{BL}\ |\ [\frac{\partial C\mathbf{BL}}{\partial rained\mathbf{BL}} \to wet\_road\mathbf{BL}),(\frac{\partial C\mathbf{BL}}{\partial sprinkled\mathbf{BL}} \to wet\_road\mathbf{BL}), \\
&\qquad (\frac{\partial C\mathbf{BL}}{\partial flooded\mathbf{BL}} \to wet\_road\mathbf{BL}),(\frac{\partial C\mathbf{BL}}{\partial high\_temperature\mathbf{BL}} \to wet\_road\mathbf{BL})]\} \\
&= \{\mathop{R}_{i=1}^{4}c_i\mathbf{BL}\ |\ [\mathbf{T,T,T,F}]\} \\
&= \{rained\mathbf{BL},\ sprinkled\mathbf{BL},\ flooded\mathbf{BL}, \varnothing\} \\
&= \{rained\mathbf{BL},\ sprinkled\mathbf{BL},\ flooded\mathbf{BL}\}
\end{aligned}
\tag{30}
$$

Box 2. Equation 31

$$
\begin{aligned}
C''\mathbf{BL} &= \mathop{R}_{i=1}^{n}(\frac{\partial}{\partial c_i}C'\mathbf{BL} \to e\mathbf{BL}) \\
&= \mathop{R}_{i=1}^{3}(\frac{\partial\{rained\mathbf{BL},\ sprinkled\mathbf{BL},\ flooded\mathbf{BL}\}}{\partial c_i} \to e\mathbf{BL}) \\
&= \{\mathop{R}_{i=1}^{3}c_i\mathbf{BL}\ |\ [(\frac{\partial C'}{\partial rained\mathbf{BL}} \to wet\_road\mathbf{BL}),(\frac{\partial C'}{\partial sprinkled\mathbf{BL}} \to wet\_road\mathbf{BL}), \\
&\qquad (\frac{\partial C'}{\partial flooded\mathbf{BL}} \to wet\_road\mathbf{BL})]\} \\
&= \{\mathop{R}_{i=1}^{3}c_i\mathbf{BL}\ |\ [\mathbf{F,T,F}]\} \\
&= \{\varnothing,\ sprinkled\mathbf{BL},\ \varnothing\} \\
&= \{sprinkled\mathbf{BL}\}
\end{aligned}
\tag{31}
$$

Example 2: On the basis of Example 1, a set of concrete causations C'' **BL** can be further inferred according to Definition 22, based on $C'\mathbf{BL} = \{rained\mathbf{BL}, sprinkled\mathbf{BL}, flooded\mathbf{BL}\}$ with known factors that there was no rain and accident, as in Box 2, Equation 31.

The inference result by the qualitative causal deferential indicates that the concrete causation

set for the target effect as deduced from the valid causations is: $\{sprinkled\mathbf{BL}\} \vdash wet\_road\mathbf{BL}$.

Example 3: The following question is formulated by Lotfi A. Zadeh (BISC, 2010): After a significant promotion effort by three-month advertisements, Company A's annual sales of umbrellas have raised 10%. What are the qualitative and quantitative causations

Box 3. Equation 32

$$
\begin{aligned}
C'\mathbf{BL} &= \mathop{R}\limits_{i=1}^{n}\left(\frac{\partial}{\partial c_i}C\mathbf{BL} \to e\mathbf{BL}\right) \\
&= \mathop{R}\limits_{i=1}^{4}\left(\frac{\partial}{\partial c_i}\{persuasiveness\mathbf{BL}, competitive\_product\mathbf{BL}, alternative\_means\mathbf{BL}, weather\mathbf{BL}\} \to sale\_growth\mathbf{BL}\right) \\
&= \{\mathop{R}\limits_{i=1}^{4} c_i\mathbf{BL} \mid [[\frac{\partial C\mathbf{BL}}{\partial\, persuasiveness\mathbf{BL}} \to sale\_growth\mathbf{BL}), (\frac{\partial C\mathbf{BL}}{\partial\, competitive\_product\mathbf{BL}} \to sale\_growth\mathbf{BL}), \\
&\quad (\frac{\partial C\mathbf{BL}}{\partial\, alternative\_means\mathbf{BL}} \to sale\_growth\mathbf{BL}), (\frac{\partial C\mathbf{BL}}{\partial\, weather\mathbf{BL}} \to sale\_growth\mathbf{BL})]\} \\
&= \{\mathop{R}\limits_{i=1}^{4} c_i\mathbf{BL} \mid [\mathbf{T}, \mathbf{T}^c, \mathbf{T}^c, [\mathbf{T} \mid \mathbf{T}^c]]\} \to sale\_growth\mathbf{BL} \\
&= \{persuasiveness\mathbf{BL}, competitive\_product^c\mathbf{BL}, alternative\_means^c\mathbf{BL}, [weather\mathbf{BL} \mid weather^c\mathbf{BL}]\} \to sale\_growth\mathbf{BL}
\end{aligned}
$$
(32)

between the advertisements and the sale growth of Company A?

The above problem can be analyzed by causal differential of IA. First, a set of hypothetic causations may be identified based on known principles, empirical data, and expert knowledge as: $C\mathbf{BL}$ = {$persuasiveness\mathbf{BL}$, $competitive\_products\mathbf{BL}$, $alternative\_means\mathbf{BL}$, $weather\mathbf{BL}$}. Then, a qualitative causal differential on $C\mathbf{BL}$ with respect to the individual potential cause $c_i\mathbf{BL} \in C\mathbf{BL}$, will result in a set of valid causations $C'\mathbf{BL}$ as in Box 3, Equation 32.

Based on the valid causation set of $C\mathbf{R}$ = {$persuasiveness\mathbf{R}$, $competitive\_products\mathbf{R}$, $alternative\_means\mathbf{R}$, $weather\mathbf{R}$} and the known analytic effect function $e\mathbf{R} = f(c_1\mathbf{R}, c_2\mathbf{R}, c_3\mathbf{R}, c_4\mathbf{R})$:

$$
\begin{aligned}
e\mathbf{R} &= sale\_growth\mathbf{R} = f(c_1\mathbf{R}, c_2\mathbf{R}, c_3\mathbf{R}, c_4\mathbf{R}) \\
&= \sum_{i=1}^{4} e_{c_i} \\
&= 60\% - 40\% - 30\% + 25\%
\end{aligned}
$$
(33)

where the set of concrete causations $C''\mathbf{R}$ can be further inferred using quantitative causal differ-

Box 4. Equation 34

$$
\begin{aligned}
C'\mathbf{R} &= \mathop{R}\limits_{i=1}^{n}\left(\frac{\partial}{\partial c_i}C\mathbf{R} \to e\mathbf{R}\right) \\
&= \mathop{R}\limits_{i=1}^{n}\left(\frac{\partial}{\partial c_i}\{c_1\mathbf{R}, c_2\mathbf{R}, ..., c_n\mathbf{R}\} \to e\mathbf{R}\right) \\
&= \{\mathop{R}\limits_{i=1}^{n} c_i\mathbf{R} \mid \frac{\partial}{\partial c_i}f(C\mathbf{R}) = \frac{\partial}{\partial c_i}e\mathbf{R} \neq 0\} \to e\mathbf{R} \\
&= \{\mathop{R}\limits_{i=1}^{4} c_i\mathbf{R} \mid \frac{\partial}{\partial c_i}f(persuasiveness\mathbf{R}, competitive\_product^c\mathbf{R}, alternative\_means^c\mathbf{R}, [weather\mathbf{BL} \mid weather^c\mathbf{BL}]) \neq 0\} \\
&\quad \to 10\%\_sale\_growth\mathbf{R} \\
&= \{\mathop{R}\limits_{i=1}^{4} c_i\mathbf{R} \mid [60\%, -40\%, -30\%, 20\%]\} \to 10\%\_sale\_growth\mathbf{R} \\
&= \{persuasiveness\mathbf{BL}(60\%), competitive\_product\mathbf{BL}(-40\%), alternative\_means\mathbf{BL}(-30\%), weather\mathbf{BL}(20\%)\} \\
&\quad \to 10\%\_sale\_growth\mathbf{R}
\end{aligned}
$$
(34)

ential as modeled in Definition 23, as see in Box 4, Equation 34.

The second part of this work, Sections 4 through 7, on the inference operators of IA as well as their extensions and applications will be followed in IJCINI 6(1) (Wang, 2012).

7. CONCLUSION

This paper has presented a novel denotational mathematics known as *inference algebra* (IA), which established a theory and a framework of formal causal inferences. Two gifted privileges of human intelligence, the causal sense and inference, have been formally investigated. The taxonomy and properties of causal inferences have been systematically explored. Cognitive models of causations and causal inferences have been rigorously described in three categories such as *logical inferences* on Boolean, fuzzy, and logical function causations, *analytic inferences* on general functional, correlative, linear regression, and nonlinear regression causations, and *hybrid inferences* on qualification and quantification causations. As that of Boolean algebra for explicit logical reasoning and fuzzy logic for approximate and uncertainty reasoning, IA has introduced a denotational mathematical structure with a set of algebraic operators on a set of formal causations for logical, analytic, and hybrid inferences. This work has demonstrated that highly complicated inference mechanisms and their mathematical rules can be rigorously treated and modeled by IA. The formal causation models and inference rules of IA have enabled artificial and computational intelligent systems to mimic and implement similar inference abilities of the brain in cognitive computing. A wide range of applications of IA have been identified and demonstrated in cognitive informatics, computational intelligence, and cognitive computers.

ACKNOWLEDGMENT

A number of notions in this work have been inspired by Prof. Lotfi A. Zadeh during my sabbatical leave at BISC, UC Berkeley and Stanford University as a visiting professor in 2008. I am grateful to Prof. Zadeh for his vision, insight, and kind support. The author acknowledges the Natural Science and Engineering Council of Canada (NSERC) for its partial support to this work. The author would like to thank the anonymous reviewers for their valuable comments and suggestions.

REFERENCES

Aristotle,. (1989). *Prior analytics* (Smith, R., Trans.). Cambridge, MA: Hackett.

Beebee, H., Hitchcock, C., & Menzies, P. P. (Eds.). (2009). *The Oxford handbook of causation*. Oxford, UK: Oxford University Press.

Bender, E. A. (1996). *Mathematical methods in artificial intelligence*. Los Alamitos, CA: IEEE Press.

BISC. (2010). *Internal communications*. Berkeley, CA: University of California.

Boole, G. (1854). *The laws of thought*. New York, NY: Prometheus Books.

Hurley, P. J. (1997). *A concise introduction to logic* (6th ed.). London, UK: Wadsworth.

Johnson, R. A., & Bhattacharyya, G. K. (1996). *Statistics: Principles and methods* (3rd ed.). New York, NY: John Wiley & Sons.

Leahey, T. H. (1997). *A history of psychology: Main currents in psychological thought* (4th ed.). Upper Saddle River, NJ: Prentice Hall.

Lewis, D. (1973). Causation. *The Journal of Philosophy*, 70, 556–567. doi:10.2307/2025310

McDermid, J. (Ed.). (1991). *Software engineer's reference book*. Oxford, UK: Butterworth Heinemann.

Mellor, D. H. (1995). *The facts of causation*. London, UK: Routledge. doi:10.4324/9780203302682

Mill, J. S. (1874). *A system of logic*. New York, NY: Harper & Brothers.

Newton, I. (1687). *The mathematical principles of natural philosophy* (1st ed.). Berkeley, CA: University of California Press.

Pearl, J. (2009). *Causality: Models, reasoning, and inference*. Berkeley, CA: California University Press.

Ross, T. J. (1995). *Fuzzy logic with engineering applications*. New York, NY: McGraw-Hill.

Russel, B. (1903). *The principles of mathematics*. London, UK: George Allen & Unwin.

Schoning, U. (1989). *Logic for computer scientists*. Boston, MA: Birkhauser.

Sperschneider, V., & Antoniou, G. (1991). *Logic: A foundation for computer science*. Reading, MA: Addison-Wesley.

Tomassi, P. (1999). *Logic*. London, UK: Routledge. doi:10.4324/9780203197035

Turner, S. (2010). *Causality*. Thousand Oaks, CA: Sage.

Wang, Y. (2002). The Real-Time Process Algebra (RTPA). *Annals of Software Engineering*, (14): 235–274. doi:10.1023/A:1020561826073

Wang, Y. (2003). On cognitive informatics. *Brain and Mind: A Transdisciplinary Journal of Neuroscience and Neurophilosophy, 4*(3), 151-167.

Wang, Y. (2007a). The cognitive processes of formal inferences. *International Journal of Cognitive Informatics and Natural Intelligence, 1*(4), 75–86. doi:10.4018/jcini.2007100106

Wang, Y. (2007b). The theoretical framework of cognitive informatics. *International Journal of Cognitive Informatics and Natural Intelligence, 1*(1), 1–27. doi:10.4018/jcini.2007010101

Wang, Y. (2007c). *Software engineering foundations: A software science perspective* (*Vol. 2*). Boca Raton, FL: Auerbach.

Wang, Y. (2008a). On contemporary denotational mathematics for computational intelligence. *Transactions of Computational Science, 2*, 6–29. doi:10.1007/978-3-540-87563-5_2

Wang, Y. (2008b). RTPA: A denotational mathematics for manipulating intelligent and computing behaviors. *International Journal of Cognitive Informatics and Natural Intelligence, 2*(2), 44–62. doi:10.4018/jcini.2008040103

Wang, Y. (2008c). On concept algebra: A denotational mathematical structure for knowledge and software modeling. *International Journal of Cognitive Informatics and Natural Intelligence, 2*(2), 1–19. doi:10.4018/jcini.2008040101

Wang, Y. (2009a). On abstract intelligence: Toward a unified theory of natural, artificial, machinable, and computational intelligence. *International Journal of Software Science and Computational Intelligence, 1*(1), 1–18. doi:10.4018/jssci.2009010101

Wang, Y. (2009b). On cognitive computing. *International Journal of Software Science and Computational Intelligence, 1*(3), 1–15. doi:10.4018/jssci.2009070101

Wang, Y. (2009c, June). Fuzzy inferences methodologies for cognitive informatics and computational intelligence. In *Proceedings of the 8th IEEE International Conference on Cognitive Informatics*, Hong Kong (pp. 241-248).

Wang, Y. (2010a). Cognitive robots: A reference model towards intelligent authentication. *IEEE Robotics and Automation, 17*(4), 54–62. doi:10.1109/MRA.2010.938842

Wang, Y. (2010b). On formal and cognitive semantics for semantic computing. *International Journal of Semantic Computing, 4*(2), 203–237. doi:10.1142/S1793351X10000833

Wang, Y. (2011a). On cognitive models of causal inferences and causation networks. *International Journal of Software Science and Computational Intelligence, 3*(1), 50–60. doi:10.4018/jssci.2011010104

Wang, Y. (2011b). On concept algebra for Computing with Words (CWW). *International Journal of Semantic Computing, 4*(3), 331–356. doi:10.1142/S1793351X10001061

Wang, Y. (2012). Inference Algebra (IA): A denotational mathematics for cognitive computing and machine reasoning (II). *International Journal of Cognitive Informatics and Natural Intelligence, 6*(1).

Wang, Y., Kinsner, W., & Zhang, D. (2009a). Contemporary cybernetics and its faces of cognitive informatics and computational intelligence. *IEEE Transactions on System, Man, and Cybernetics. Part B, 39*(4), 1–11.

Wang, Y., Wang, Y., Patel, S., & Patel, D. (2006). A layered reference model of the brain (LRMB). *IEEE Transactions on Systems, Man, and Cybernetics. Part C, 36*(2), 124–133.

Wang, Y., Zadeh, L. A., & Yao, Y. (2009). On the system algebra foundations for granular computing. *International Journal of Software Science and Computational Intelligence, 1*(1), 1–17. doi:10.4018/jssci.2009010101

Wang, Y., Zhang, D., & Kinsner, W. (Eds.). (2010). *Advances in cognitive informatics and cognitive computing.* Berlin, Germany: Springer-Verlag.

Wilson, R. A., & Keil, F. C. (2001). *The MIT encyclopaedia of the cognitive sciences.* Cambridge, MA: MIT Press.

Zadeh, L. A. (1965). Fuzzy sets. *Information and Control, 8,* 338–353. doi:10.1016/S0019-9958(65)90241-X

Zadeh, L. A. (1975). Fuzzy logic and approximate reasoning. *Syntheses, 30,* 407–428. doi:10.1007/BF00485052

Zadeh, L. A. (1999). From computing with numbers to computing with words – from manipulation of measurements to manipulation of perception. *IEEE Transactions on Circuits and Systems, 45,* 105–119.

Zadeh, L. A. (2004). Precisiated Natural Language (PNL). *AI Magazine, 25*(3), 74–91.

Zadeh, L. A. (2008). Is there a need for fuzzy logic? *Information Sciences, 178,* 2751–2779. doi:10.1016/j.ins.2008.02.012

This work was previously published in the International Journal of Cognitive Informatics and Natural Intelligence, Volume 5, Issue 4, edited by Yingxu Wang, pp. 61-82, copyright 2011 by IGI Publishing (an imprint of IGI Global).

Chapter 12
Human Centricity and Perception–Based Perspective and Their Centrality to the Agenda of Granular Computing

Witold Pedrycz
University of Alberta, Canada, & Polish Academy of Sciences, Poland

ABSTRACT

In spite of their striking diversity, numerous tasks and architectures of intelligent systems such as those permeating multivariable data analysis, decision-making processes along with their underlying models, recommender systems and others exhibit two evident commonalities. They promote (a) human centricity and (b) vigorously engage perceptions (rather than plain numeric entities) in the realization of the systems and their further usage. Information granules play a pivotal role in such settings. Granular Computing delivers a cohesive framework supporting a formation of information granules and facilitating their processing. The author exploits two essential concepts of Granular Computing. The first one deals with the construction of information granules. The second one helps endow constructs of intelligent systems with a much needed conceptual and modeling flexibility. The study elaborates in detail on the three representative studies. In the first study being focused on the Analytic Hierarchy Process (AHP) used in decision-making, the author shows how an optimal allocation of granularity helps improve the quality of the solution and facilitate collaborative activities in models of group decision-making. The second study is concerned with a granular interpretation of temporal data where the role of information granularity is profoundly visible when effectively supporting human centric description of relationships existing in data. The third study concerns a formation of granular logic descriptors on a basis of a family of logic descriptors.

DOI: 10.4018/978-1-4666-2476-4.ch012

1. INTRODUCTION

Let us consider a system (process) for which constructed is a family of models. The system can be perceived from different points of view, observed over some time periods and analyzed at different levels of detail. Subsequently, the resulting models are built with various objectives in mind. They offer some particular, albeit useful views at the system. We are interested in forming a *holistic* model of the system by taking advantage of the individual sources of knowledge – models, which have been constructed so far. When doing this, we are obviously aware that the sources of knowledge exhibit diversity and hence this diversity has to be taken into consideration and carefully quantified. No matter what the local models may look like, it is legitimate to anticipate that the global model (say, the one formed at the higher level of hierarchy) is more general, abstract. Another point of interest is to engage the sources of knowledge in intensive and carefully orchestrated procedures of knowledge reconciliation and consensus building. Granularity of information (Bargiela & Pedrycz, 2003, 2005, 2008, 2009; Bezdek, 1981; Pedrycz & Rai, 2008; Pedrycz & Hirota, 2008; Zadeh, 1997, 1999) becomes of paramount importance, both from the conceptual as well as algorithmic perspective, in the realization of granular fuzzy models. Subsequently, processing realized at the level of information granules gives rise to the discipline of Granular Computing (Bargiela & Pedrycz, 2003). We envision here a vast suite of formal approaches of fuzzy sets (Zadeh, 1997) rough sets (Pawlak, 1991, 1985; Pawlak & Skowron, 2007), shadowed sets (Pedrycz, 1998, 1999, 2005), probabilistic sets (Hirota, 1981; Hirota & Pedrycz, 1984) and alike. Along with the conceptual setups, we also encounter a great deal of interesting and relevant ideas supporting processing of information granules. For instance, in the realm of rough sets we can refer to Pawlak (1991). From the algorithmic perspective, fuzzy clustering (Bezdek, 1981), rough clustering, and clustering are regarded as fundamental development frameworks in which information granules are constructed.

The objective of this study is to introduce two conceptual and algorithmic underpinnings of Granular Computing overarching the entire domain, namely a principle of justifiable granularity and an optimal allocation of information granularity (Section 2 and 3). In Section 4, we focus on the role of second principle in the AHP models of decision-making including their group alternatives. Granular time series are investigated in Section 5; here we demonstrate a pivotal role of the principle of justifiable granularity. The ensuing optimization problems are formulated. In Section 6, we discuss an idea of granular logic and discuss how it emerges as a result of a global view at a collection of local logic descriptors. Conclusions of the study are covered in Section 7.

2. THE PRINCIPLE OF JUSTIFIABLE GRANULARITY

Here we are concerned with the formation of a single information granule Ω based on some experimental evidence being a set of a single-dimensional (scalar) numeric data, $\mathbf{D} = \{x_1, x_2, \ldots, x_N\}$. The information granule itself could be expressed in a certain formal framework of Granular Computing. The principle of justifiable granularity (Pedrycz & Gomide, 2007) is concerned with a formation of a meaningful information granule based on available experimental evidence. In its formation, such a construct has to adhere to the two intuitively compelling requirements:

1. The numeric evidence accumulated within the bounds of Ω has to be as *high* as possible. By doing so, we anticipate that the existence of the information granule is well motivated (justified) as being reflective of the existing experimental data.

2. At the same time, the information granule should be as *specific* as possible meaning that it comes with a well-defined semantics. In other words, we would like to have Ω as detailed (specific) as possible.

While these two requirements are appealing, they have to be translated into some operational framework where the formation of the information granule can be realized. This framework depends upon the accepted formalism of information granulation, viz. a way in which information granules are described as sets, fuzzy sets, shadowed sets, rough sets, probabilistic granules and alike. To start with a simple and convincing constructs, let us treat Ω as an interval to be constructed. The first requirement is quantified by counting the number of data falling within the bounds of Ω. In the simplest scenario, we can look at the cardinality of Ω, namely card$\{x_k \in \Omega\}$. More generally, we may consider an increasing functional of the cardinality, say $f_1(\text{card}\{x_k \in \Omega\})$ where f_1 is an increasing function of its argument. The simplest case concerns an identity functional, $f_1(u) = u$. The specificity of the information granule can be quantified by taking into account its size. The inverse of the length of the interval Ω or more generally a decreasing functional of the length, f_2, can serve as a sound measure of specificity. The lower the value of $f_2(\text{length}(\Omega))$, the higher the specificity is and the more appealing the resulting information granule is in light of the second criterion. It is apparent that these two requirements discussed are in conflict: the increase of the criterion of experimental evidence comes with a deterioration of the specificity of the information granule. As usual, we are interested in forming a sound compromise.

Having these two criteria in mind, let us proceed with the detailed construct of interval information granules. We start with a determination of the numeric representative of the set of data $\mathbf{D}$. A sound representative is its median, med($\mathbf{D}$), as it is robust estimator of the sample and typically comes as one of the elements of $\mathbf{D}$. An information granule Ω is formed by forming its lower and upper bound, denoted here by "a" and "b", respectively.

The determination of the bounds is realized independently. In this way, we can concentrate on the optimization of the upper bound (b). The calculations of the lower bound (a) are carried out in an analogous fashion. The length of Ω, which quantifies the specificity of the information granule is given now as |med($\mathbf{D}$)-b|. More generally, in the realization of the second criterion we use $f_2(|\text{med}(D)-b|)$. In the calculations of the cardinality of the information granule, we take into account the elements of $\mathbf{D}$ positioned to the right from the median, say card $\{x_k \in \Omega, x_k > \text{med}(\mathbf{D})\}$. Again, in general, we compute $f_1(\text{card } \{x_k \in \Omega, x_k > \text{med}(\mathbf{D})\})$. As the requirements of experimental evidence and specificity are in conflict, we can either resort ourselves to a certain version of multiobjective optimization or consider a maximization of the product $V = f_1 * f_2$ whose optimization is to be realized with respect to the upper bound of the information granule, that is $V(b_{opt}) = \max_{b > \text{med}(\mathbf{D})} V(b)$. In the same way, constructed is the lower bound of the information granule, a_{opt}; $V(a_{opt}) = \max_{a < \text{med}(\mathbf{D})} V(a)$. One among possible design alternatives, we can consider the functionals f_1 and f_2 assuming the following form $f_1(u) = u$ and $f_2(u) = \exp(-\alpha u)$ where α is a positive parameter offering some flexibility in the produced information granule Ω. Its role is to calibrate an impact of the specificity criterion on the constructed information granule.

We include some illustrative examples, see Figure 1, by showing how the intervals information granules are constructed in presence of data governed by some probability density function (pdf). We show the optimization of the upper bound only. The plots show the optimized performance index viewed as a function of the upper bound (b) for selected values of α.

Figure 1. Plots of V treated as functions of b: (a) triangular pdf over [0,1] and α=0.6, (b) triangular pdf over [0,1] and α=2.0, (c) Gaussian pdf N(0, 1) and α=0.6, (d) Gaussian pdf N(0, 1) and α=2.0, (e) exponential pdf, α =0.6, (f) exponential pdf, α =2.0; solid line – cumulative probability, dotted line exp(-αb), thick solid line- optimized objective function V

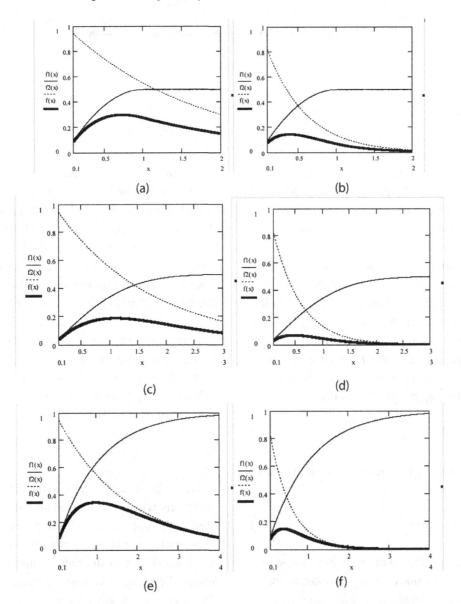

Alluding to the format of the maximized multiplicative objective function f_1f_2, it is helpful to elaborate on the choice of the maximal value of α. Note that for α = 0.0, we have f_2 equal identically to 0 and only the first component of V is used in the formation of the information granule; naturally the interval includes all experimental data. We observe the following two relationships, refer to the notation displayed in Figure 2,

Figure 2. A determination of a maximal value of a, a_{max}; see details in the text

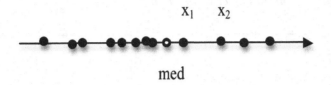

med

- If the upper bound a set to x_1 returns card(A) =1 and $\exp(-\alpha|x_1\text{-med}|)$.
- If the upper bound a is specified in such a way that embraces two the elements of the data closest to the median, $\{x_1, x_2\}$, we have card(A) = 2 and $\exp(-\alpha|x_2\text{-med}|)$.

Here the median, med, is treated as a numeric representative of the experimental data. We also consider only the upper bound of the information granule; the construction in case of the lower bound is realized in the same manner as discussed here.

Let us request that the bound located at x_1 maximizes the performance index. This means that the following inequality is satisfied:

$$\exp(-\alpha|x_1\text{-med}|)>2\exp(-\alpha|x_2\text{-med}|) \tag{1}$$

Introducing the notation $d_1 = |x_1\text{-med}|$ and $d_2 =|x_2\text{-med}|$ we obtain

$$\exp(-\alpha b_1)>2\exp(-\alpha b_2) \tag{2}$$

Its solution comes in the form $\exp(\alpha(d_2\text{-}d_1)) > 2$. The maximal value of α, α_{max} is taken as $\ln(2)/(d_2\text{-}d_1)$. It is associated with the maximal value of the membership grade that is made close to 1, say 1-ε where e is a small positive value, say 0.01. Practically, we can consider this value being equal to 0. In this way we map the values of α used in the principle of justifiable granularity onto the [0,1] interval of membership values

$$[0, \alpha_{max}] \to [0,1] \tag{3}$$

In virtue of the monotonicity of the interval information granules with regard to the values of α, each information granule can be associated with the corresponding value in [0,1]. Interestingly, a family of intervals indexed by the normalized values of α can be treated as a collection of α-cuts. Subsequently, by varying the values of α we form the corresponding α-cut and as a result, by making use of the representation theorem (Pedrycz & Gomide, 2007) arrive at an information granule in the form of a fuzzy set.

In this way, we map the values of α used in the principle of justifiable granularity onto the [0,1] interval of membership values

$$[0, \alpha_{max}] \to [0,1] \tag{4}$$

The above construct can be augmented to situations where the individual data are associated with some weights (which could quantify their quality which may vary from one element to another). Given the data in the form (x_i, w_i) where the weights w_i assume values located in the [0,1] interval, we reformulate the maximized performance index to be in the form

$$V(a) = f_1(\sum_{k=1}^{N} w_i)*f_2(|a\text{-med}|) \tag{5}$$

where "m" is a weighted median whose computing uses the weighted data.

It is worth stressing that the principle of justifiable granularity covers a broad spectrum of scenarios- all of them can be arranged along two main coordinates. The first one is concerned with the formal environment of information granules. The second one points at the nature of available experimental evidence (sets, fuzzy sets, etc.)

In our considerations we have focused on the use of numeric evidence while constructed were interval like information granules and those expressed by fuzzy sets. The developed principle covers other cases of formal approaches to information granules with some modifications to the criteria that are pertinent to the specificity of the contemplated realization.

The realized information granule (either in a form of an interval or a fuzzy set) concerns one-dimensional numeric data. The extension to the multidimensional case is straightforward by constructing a Cartesian product of the information granules formed for the individual variables. For instance, given an information granule A defined in X_1, B arising at X_2 and C at X_3, the result is in the form $A \times B \times C$.

3. OPTIMAL ALLOCATION OF INFORMATION GRANULARITY

Information granularity is an important design asset. Information granularity allocated to the original numeric construct elevates a level of abstraction (generalizes) of the original construct developed at the numeric level. A way in which such an asset is going to be distributed throughout the construct or a collection of constructs to make the abstraction more efficient, is a subject to optimization.

Consider a certain mapping $y = f(\mathbf{x}, \mathbf{a})$ with $\mathbf{a}$ being a vector of parameters of the mapping. The mapping can be sought in a general way. One may think of a fuzzy model, neural network, polynomial, differential equation, linear regression, etc. The granulation mechanism G is applied to $\mathbf{a}$ giv-

ing rise to its granular counterpart, $\mathbf{A} = G(\mathbf{a})$ and subsequently producing a granular mapping, $Y = G(f(\mathbf{x}, \mathbf{a})) = f(\mathbf{x}, G(\mathbf{a})) = f(\mathbf{x}, \mathbf{A})$. Given the diversity of the underlying constructs as well as a variety of ways information granules can be formalized, we arrive at a suite of interesting constructs such as granular neural networks, say interval neural networks, fuzzy neural networks, probabilistic neural networks, etc.

There are a number of well-justified and convincing arguments behind elevating the level of abstraction of the existing constructs. Those include: an ability to realize various mechanisms of collaboration, quantification of variability of sources of knowledge considered, better modelling rapport with systems when dealing with nonstationary environments. In what follows, we will elaborate on the general categories of problems in which information granularity plays a pivotal role.

Information granularity provided to form a granular construct is a design asset whose allocation throughout the mapping can be guided by certain optimization criteria. Let us discuss the underlying optimization problem in more detail. In addition to the mapping itself, we are provided with some experimental evidence in the form of input-output pairs $(\mathbf{x}_k, t_k)$, $k = 1, 2, \ldots, M$. Given is a level of information granularity ε, $\varepsilon \in [0,1]$. We allocate the available level ε to the parameters of the mapping, $\dim(\mathbf{a}) = h$, so that the some optimization criteria are satisfied while the allocation of granularity satisfies the following balance $\varepsilon = \sum_{i=1}^{h} \varepsilon_i$ where ε_i is a level of information granularity associated with the i-th parameter of the mapping. All of the individual allocations are organized in a vector format $[\varepsilon_1 \ \varepsilon_2 \ldots \ \varepsilon_h]^T$. There are two optimization criteria to be considered in the optimization. The first one is concerned with the coverage of data t_k. For $\mathbf{x}_k$ we compute $Y_k = f(\mathbf{x}_k, G(\mathbf{a}))$ and determine a degree of inclusion of t_k in information granule Y_k, $incl(t_k, Y_k) = t_k \subset Y_k$. Then we compute an average sum of the degrees of inclusion taken over all the data, that is

$\frac{1}{M}\sum_{k=1}^{M}\text{incl}(t_k,Y_k)$ Depending upon the formalism of information granulation, the inclusion returns a Boolean value in case of intervals (sets) or a certain degree of inclusion in case of fuzzy sets. The second criterion is focused on the specificity of Y_k - we want it to make it as high as possible. The specificity could be viewed as a decreasing function of the length of the interval in case of set –based information granulation. For instance, one can consider the inverse of the length of Y_k, say $1/\text{length}(Y_k)$, $\exp(-\text{length}(Y_k))$, etc. In case of fuzzy sets, one consider the specificity involving the membership grades. The length of the fuzzy set Y_k is computed by integrating the lengths of the β-cuts, $\int_0^1 \text{length}(Y_k^\beta)\beta d\beta$. More formally, the two-objective optimization problem is formulated as follows: Distribute (allocate) a given level of information granularity e so that the following two criteria are maximized

- Maximize $\frac{1}{M}\sum_{k=1}^{M}\text{incl}(t_k,Y_k)$

- Maximize $g(\text{length}(Y_k))$ (where g is a decreasing function of its argument) subject to $\varepsilon = \sum_{i=1}^{h}\varepsilon_i$

A simpler, optimization scenario involves a single coverage criterion regarded as a single most essential criterion considered in the problem

$$\text{Maximize } Q = \frac{1}{M}\sum_{k=1}^{M}\text{incl}(t_k,Y_k) \text{ subject to } \varepsilon = \sum_{i=1}^{h}\varepsilon_i \qquad (6)$$

There is an interesting monotonicity property: higher values of ε lead to higher values of the maximized objective function. By taking account the nature of the above relationship, we can arrive at some global view at the relationship that is independent from a specific value of ε by taking

an area under curve (AUC) computed as AUC = $\int_0^1 Q(\varepsilon)d\varepsilon$. The higher the value of the AUC, the higher the performance of the granular version of the mapping.

Information granularity can be realized in the setting of a certain information allocation protocol. Several main categories of such protocols can be envisioned:

P_1: Uniform allocation of information granularity. This process is the simplest one and in essence does not call for any optimization mechanism. The numeric parameter "a" of the mapping is replaced by the information granule $G(a)$, which is the same in terms of the size and the distribution around a. If the formal setup of G concerns intervals then the numeric parameters of the mapping are replaced by intervals of the same length (ε) and distributed symmetrically around the parameters of the mapping.

P_2: Uniform allocation of information granularity with asymmetric position of intervals.

P_3: Non-uniform allocation of information granularity with symmetrically distributed intervals of information granules.

P_4: Non-uniform allocation of information granularity with asymmetrically distributed intervals of information granules.

P_5: An interesting point of reference, which is helpful in assessing a relative performance of the above methods, is to consider a random allocation of granularity. By doing this, one can quantify how the optimized and carefully thought out process of granularity allocation is superior over a purely random allocation process.

Depending upon the formalism of information granularity, the protocols can be made more specific.

The quality of the resulting granular mappings produced through invoking different granularity allocation protocols can be assessed by computing the resulting value of the AUC. In this way, we can establish a liner order within a collection of the protocols. In virtue of the increasing generality of the protocols, we can envision the following ordering $P_1 \prec P_2 \prec P_3 \prec P_4$ where the notation $P_i \prec P_j$ denotes that P_j is preferred over P_i as producing higher values of the AUC.

We can think of fuzzy sets built around numeric values of the parameters where depending upon a certain the membership functions may exhibit symmetric or asymmetric character as well as come with various supports. In case of probabilistic information granules, one may talk about symmetric and asymmetric probability density functions with the modal values allocated to the numeric values of the parameters and standard deviations whose values vary from parameter to parameter. In total, we require a sum of the standard deviations to satisfy the predefined level of granularity that is $\sigma = \sum_{i=1}^{h} \sigma_i$.

4. GRANULAR ANALYTIC HIERARCHY PROCESS (AHP)

This model serves as a simple yet a convincing example in which the idea of granularity allocation can be used effectively in improving the quality of a solution both in case of a individual decision-making as well as its group version. Let us recall that the Analytic Hierarchy Process (AHP) is aimed at forming a vector of preferences for a finite set of n alternatives. These preferences are formed on a basis of a reciprocal matrix R, $R=[r_{ij}]$, i, j=1, 2, ..., n whose entries are a result of pairwise comparisons of alternatives provided by a decision-maker. The quality of the result (reflecting the consistency of the judgment of the decision-maker) is expressed in terms of the following inconsistency index

$$\nu = \frac{\lambda_{max} - n}{n - 1} \tag{7}$$

where λ_{max} is the largest eigenvalue associated with the reciprocal matrix. The larger the value of this index is, the more significant level of inconsistency is associated with the preferences collected in the reciprocal matrix.

We distinguish here two main categories of design scenarios: a single decision-maker is involved or we are concerned with a group decision-making where there is a collection of reciprocal matrices provided by each of the member of the group.

4.1 A Single Decision-Maker Scenario

The results of pairwise comparisons usually exhibit a certain level of inconsistency. The inconsistency index presented above quantifies this effect. We generalize the numeric reciprocal matrix R by forming its granular counterpart and allocating the admissible level of granularity to the individual entries of the matrix. Formally, the process can be schematically described in the following form

$$R \xrightarrow{\varepsilon} G(R) \tag{8}$$

where $G(R)$ stands for the granular version of the reciprocal matrix. A certain predetermined level of information granularity e is distributed among elements of the reciprocal matrix R. More specifically, we look at the entries of the reciprocal matrix, which are below 1 and form information granules around those. Confining ourselves to intervals (for illustrative purposes), formed are intervals around the corresponding entries of the matrix whose total length satisfies the constraint $\sum_{i,j} \varepsilon_{ij} = p\varepsilon$ where "p" stands for the number of elements of R assuming values below 1. Thus the original entry r_{ij} is replaced by the interval whose lower and upper bound are expressed as max(1/9,

$r_{ij}-\varepsilon_{ij}(8/9))$ and $\min(1, r_{ij}+\varepsilon_{ij}(8/9))$. Here the number 9 reflects the largest length of the scale used in the realization of pairwise comparisons. For the reciprocal entry of the matrix, we compute the inverse of the lower and upper bound of the interval, round off the results to the closest integers (here we use the integers from 1 to 9) and map the results to the interval of the reciprocals. In this way an original numeric entry r_{ij} and $1/r_{ij}$ are made granular. The same process is completed for the remaining entries of the reciprocal matrix.

As an illustration, let us show the calculations in case where $r_{ij} = 1/3$ and $\varepsilon_{ij} = 0.10$. The numeric value is replaced by the bounds 0.24 and 0.42. The inverse produces the integers (after rounding off) being equal to 4 and 2. Mapping them again by computing the inverse produces the entry of the reciprocal matrix equal to $[1/4, 1/2]$. Summarizing, through an allocation of granularity, the original entries 1/3 and 3 were replaced by their granular (interval) counterparts of $[1/4, 1/2]$ and $[2, 4]$. The resulting information granule depends upon the assumed level of granularity as well as the protocol of granularity allocation. For instance, for asymmetric allocation of granularity with $\varepsilon_{ij-} = 0.1$ and $\varepsilon_{ij+} = 0.2$, we arrive at the intervals $[1/4, 1]$ and $[1, 4]$, respectively.

The granular (interval-valued) reciprocal matrix $P(R)$ manifest in a numeric fashion in a variety of ways. To realize such manifestation, we randomly pick up the numeric values from the corresponding intervals (maintaining the reciprocality condition, that is when a value of r_{ij} has been selected from the range $[r_{ij-}, r_{ij+}]$ the value of r_{ji} is computed as the inverse of the one being already selected). For the matrix obtained in this way computed is its inconsistency index. The overall process is repeated a number of times and determined is the average of the corresponding values of the inconsistency index of the matrices. Denote the average by $E(v)$ This average quantifies the quality of the granular reciprocal matrix being a result of allocation (distribution) of the level of information granularity ε. The goal of optimization is to minimize $E(v)$ by determining ε_{ij} so that $\min_{\varepsilon_{ij}} E(v)$ subject to constraints $\sum_{i,j} \varepsilon_{ij} = p\varepsilon$.

4.2 Group Decision Making

In this situation, we are concerned with a group of reciprocal matrices R[1], R[2],..., R[c] along with the preferences (preference vectors), **e**[1], **e**[2],..., **e**[c] obtained by running the AHP for the corresponding reciprocal matrices. Furthermore the quality of preference vectors is quantified by the associated inconsistency index v[i]. First, in the optimization problem, we bring all preferences close to each other and this goal is realized by adjusting the reciprocal matrices within the bounds offered by the admissible level of granularity provided to each decision-maker.

$$Q_1 = \sum_{i=1}^{c} (1 - \nu_i) \| e[i] - \hat{\mathbf{e}} \|^2 \qquad (9)$$

where $\hat{\mathbf{e}}$ stands for the vector of preferences which minimizes the weighted sum of differences $\|.\|$ between $e[i]$ and $\hat{\mathbf{e}}$. Second, we increase the consistency of the reciprocal matrices and this improvement is realized at the level of individual decision-maker. The following performance index quantifies this aspect of collaboration

$$Q_2 = \sum_{i=1}^{c} \nu_i \qquad (10)$$

These are the two objectives to be minimized. If we consider the scalar version of the optimization problem, it can arise in the following additive format $Q = gQ_1 + Q_2$ where $g \geq 0$. The overall optimization problem with constraint reads now as follows

$$\min_{R[1], R[2]...,R[c] \in G(R)} Q \qquad (11)$$

subject to predetermined level of granularity ε where $G(R)$ stands for the granular version of the of the reciprocal matrix. We require that the overall balance of the predefined level of granularity given in advance ε is retained and allocated throughout all reciprocal matrices (Pedrycz & Song, 2011).

5. GRANULAR TIME SERIES–TOWARDS INTERPRETABLE MODELS

Time series are those essential and commonly encountered data structures, which are often perceived visually and interpreted in a qualitative fashion by humans. Decisions made by humans are formed on a basis of Information granularity is inherently present in the perception and interpretation of time signals: information granules are entities central to a description of signals. In one way or another, granularity of information becomes essential to all mechanisms of data mining of temporal information. As usual, describing time series through a vocabulary of information granules makes the interpretation of data easier and more transparent as well as help navigate through various levels of abstraction/specificity by adjusting sizes of information granules used in the description. The flexibility offered by information granules with this regard is of importance.

Let us consider a finite time series (consisting of a sequence of samples of a certain signal coming in successive discrete time moments). There are a number of different ways of looking at the signal and coming up with its interpretation. We discuss three of them alternatives starting with the simplest one. We stress that in spite of the visible diversity, the principle of justifiable granularity permeates all the constructs by offering an intuitively appealing description of the time series as well contributes to the formulation of the performance index under optimization

5.1 Representation of Time Series as a Sequence of its Granular Magnitudes

Consider a discrete time series shown in Figure 1 (a collection of dots). Perhaps the simplest yet still a convincing description is realized as a sequence of information granules of magnitude of the time series where each information granule spreads over some time interval (time slice) T_1, T_2, ..., T_p where "p" denotes a number of time slices predefined in advance. For all samples of the time series falling within the temporal window T_i, we form an information granule of the magnitudes of the series by invoking the principle of justifiable granularity, see Figure 1. More formally, we obtain $A_i(\alpha) = G(\{x_1, x_2, ..., x_N\}, T_i, \alpha)$ with the samples of the time series $\{x_1, x_2, .., x_N\}$ falling within the window T_i where α stands for the normalized parameter used in the construction of the information granules, refer to Section 2. The result is a collection of α-cuts $A_i(\alpha)$ of a fuzzy set describing the information granule of the magnitude.

Denote by Ω_i a set of samples falling within the temporal window T_i, $\Omega_i = \{x_{i1}, x_{i2} ... x_{iNi}\}$. We define a volume of the information granule $G(\Omega_i)$ as the following integral

$$\text{Vol}(G(\Omega_i)) = \int_0^1 T_i \|G(\Omega_i)(\alpha)\| d\alpha \qquad (12)$$

where the integration above is completed over α-cuts of the fuzzy set $G(\Omega_i)$. The symbol $\|.\|$ denotes a length (size) of the corresponding α-cut. Our intent is to arrive at information granules $G(\Omega_1)$, $G(\Omega_2)$,..., $G(\Omega_p)$ that are the most "informative" (compact) so that they carry a clearly articulated semantics. An overall measure with this regard, which articulates a quality of the granular description of the time series, is obtained by summing volumes of the information granules, namely

Figure 3. A discrete time series and its description as a sequence of information granules of magnitude (shaded areas) formed over temporal windows

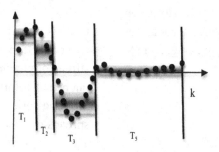

$$Q = Vol(G(W_1)) + Vol(G(W_2)) + ... + Vol(G(W_p)) \quad (13)$$

It is noticeable that the values assumed by Q depend upon a way in which the temporal windows are formed. This gives rise to the optimization problem of the form

$$\text{Min } Q \text{ with respect to } \{T_1, T_2, .., T_p\} \quad (14)$$

Apparently, the optimization here is of combinatorial character and as such calls for the use of methods of global optimization such as ant colonies, genetic algorithms, particle swarm optimization or alike.

5.2 Representation of Time Series as a Sequence of its Granular Magnitudes and Magnitude Changes

This approach comes as a direct extension of the previous way of representing time series. In addition to the original time series, we take into account the sequence of changes of the magnitude, $x_{t+1} - x_t$. These changes capture the dynamics of time series. Denote it by $d\Omega_i$ and for each time window T_i, consider volumes of the information granule $G(\Omega_i)$ and $G(d\Omega_i)$. These two combined describe an overall volume of the granular descriptors of

Ω_i, $vol(G(\Omega_i) + vol(G(d\Omega_i))$ or a weighted sum of these components. The optimization problem of the formation of the temporal windows is practically the same as shown in (7) however Q is taken as the sum computed over the volumes of the Cartesian products

$$Q = \sum_{i=1}^{p} [vol(G(\Omega_i \times d\Omega_i))] \quad (15)$$

5.3 Representation of Time Series Based on Errors of Linear Segments

The temporal segments of the data can be approximated through piecewise linear curve fitting. For each time window we form a linear approximation L_i and then compute errors between the approximation and the samples of the time series, that is $\varphi_i = x_i - L(i)$. Instead of the elements of the series falling within T_i, we consider the series $\{\phi_{i1}, \phi_{i2} ... \phi_{iNi}\}$ and for this collection of data we form the information granule of the representation error. Its construction is realized in the same way as discussed so far. As before the optimization problem is concerned with the determination of the temporal windows.

6. THE DEVELOPMENT OF GRANULAR LOGIC: A HOLISTIC VIEW AT A COLLECTION OF LOGIC DESCRIPTORS

We consider a collection of logic descriptors describing some local relationships. We intend to view them at a global level to arrive at a generalized description in the form of a *granular* logic descriptor. Information granularity arises here as a result of inherent diversity within a family of individual descriptors. It also helps quantify the existing diversity as well as support some mechanisms of successive reconciliations of the sources of knowledge (logic descriptors).

Let us define a logic descriptor as a certain quantified logic expression coming in a conjunctive or disjunctive form. Given is a collection of "c" logic descriptors involving "n" variables in either a conjunctive form

$$L_{ii}: y[ii] = (w_1[ii]\text{or } x_1) \text{ and } (w_2[ii]\text{or } x_2) \text{ and .. and } (w_n[ii]\text{or } x_n) \qquad (16)$$

or a disjunctive form

$$L_{ii}: y[ii] = (w_1[ii]\text{and } x_1) \text{ or } (w_2[ii]\text{and } x_2) \text{ or .. or } (w_n[ii]\text{and } x_n) \qquad (17)$$

$ii=1, 2,..., c$. In the above logic descriptors, x_1, $x_2,..., x_n$ are the input variables assuming values in the unit interval and $w_j[ii]$ are the weights calibrating a level of contribution to the individual inputs, $\mathbf{w}[ii] = [w_1[ii] \; w_2[ii]... \; w_n[ii]]^T$. Each logic descriptor (16)-(17) denoted here briefly as $L_1, L_2, ..., L_c$ is a logic mapping from $[0,1]^n$ to $[0,1]$. We assume that all of them are disjunctive or conjunctive (if this does not hold, all are transformed to either of these formats). As the logic connectives are modeled by t-norms or t-conorms, the expressions shown above read as

$$L_{ii}: y[ii] = (w_1[ii]\text{s } x_1) \text{ t } (w_2[ii]\text{s } x_2) \text{ t .. t } (w_n[ii]\text{s } x_n) \qquad (18)$$

or

$$L_{ii}: y[ii] = (w_1[ii]\text{t } x_1) \text{ s } (w_2[ii]\text{t } x_2) \text{ s .. s } (w_n[ii]\text{t } x_n) \qquad (19)$$

We form a unified, holistic view at all of them by forming a certain granular abstraction of $\{L_1, L_2, ...L_c\}$, denoted here as $GL(\mathbf{x})$, refer also to Figure 4.

It reads as follows

$$L: y= (W_1\text{s } x_1) \text{ t } (W_2 \text{ s } x_2) \text{ t .. t } (W_n \text{ s } x_n) \qquad (20)$$

or

$$L: y= (W_1\text{t } x_1) \text{ s } (W_2\text{t } x_2) \text{ s .. s } (W_n\text{t } x_n) \qquad (21)$$

where W_j is a granular weight coming as a result of the granulation of the family of the weights (connections) present in the individual descriptors, namely $W_j = G(\{w_j[1], w_j[2],..., w_j[c]\})$.

Figure 4. From local logic descriptors to its global granular description G(L)

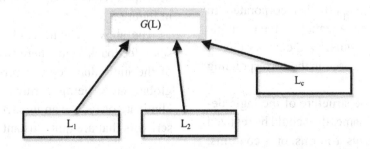

The principle of justifiable granularity is used as a means to construct the granular counterpart of the local descriptors. The area under curve characteristics serves as a global descriptor of the granular construct formed in this way.

One has to note that when forming the granular construct, it is assumed that the corresponding counterparts involve the logic structures, which are structurally the same, that is they come with the same variables and in their realization use the same t-norms and t-conorms. These assumptions might not be always satisfied. To make necessary adjustments, some additional processing is considered:

6.1 Missing Input Variables

If a certain variable x_j is not present in the logic descriptor, it is included in the augmented description by associating it with the connection that is either equal to 0 or 1. More specifically, we proceed with the following structural construct:

A. For the conjunctive descriptor, we include the expression x_j s 1.0 so in fact the associated connection is equal to 1, which formally incorporates this variable as a part of the description but in essence given the value of the connection, the variable does not impact the output. Nevertheless the value of the connection is now considered when running the principle of justifiable granularity to construct the information granule of the j-th variable. For the disjunctive descriptor, we follow the same way as described above but in virtue of the boundary condition of the t-norm, the term x_j t 0.0 is incorporated. In the consecutive realization of the principle of justifiable granularity, the zero value of the connection is used in the corresponding computing.

Even though the structure of the logic descriptors is the same, they could be realized by using various t-norms or t-conorms.

Because of this the values of the weights (connections) cannot be used explicitly in the formation of the granular counterpart but instead they call for some calibration whose essence can be outlined as follows. The current realization involves the expression x_i t w_i while we are interested in the realization invoking another t-norm, say t'. We need some modification to the original weight, call it w_i' so that the approximate equality x_i t $w_i \approx x_i$ t' w_i' holds. The satisfaction of this relationship is anticipated to be satisfied for some input values assumed by x_i. Let the set of such "P" values is denoted by F; $F = \{x_i\{p\}\}$, p=1, 2, ..., P. Then we require the approximate satisfaction of the following set of relationships

$$x_i(p) \text{ t } w_i \approx x_i(p) \text{ t' } w_i' \quad p=1, 2, ..., P \quad (22)$$

Denote the left-hand side of the expression by c(p). To achieve the highest approximate equality of these two sides (22) we consider the optimization problem

$$Q = \sum_F (c(p) - x_i(p)t'w_i')^2 \quad (23)$$

whose minimization is realized with respect to the unknown connection w_i'. min Q and $w_{i\ opt}' =$ arg Min Q. In the sequel the optimal connection is used when running the principle of justifiable granularity.

An overall look at the formation of the granular descriptor realized at the global level is illustrated in Figure 5.

The above scheme could be augmented by some feedback loop where we assess the quality of the individual logic descriptor vis-à-vis the global view being formed. This could help eliminate or reduce an impact of some logic descriptors that are quite distant from the rest. For

Figure 5. A schematic view at the formation of the granular logic descriptor G(L)

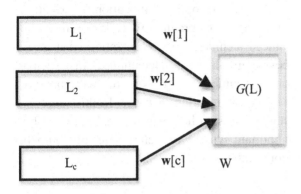

calculating the ratio $g_i = card(k| L_i(\mathbf{x}(k)) \in Y_k)/N$. This index can serve as a consistency measure of the i-th logic descriptor when assessed in light of the global descriptor $G(L)$. The higher the value of g_i, the better the consistency of the logic descriptor with the global one already constructed through the principle of justifiable granularity. The values of g_i, i=1,2, ..., c are incorporated into the determination of the granular connections in the sense that now g_is are used to form the corresponding granules. More specifically, the optimized information granule minimizes the objective function of the form

the realization of the feedback loop we consider that there is a validation set available which comes in the form of some data $\mathbf{D} = \{(\mathbf{x}(k)\}, k=1, ...,N$. The granular descriptor forms the granular output for every vector $\mathbf{x}(k)$ from $\mathbf{D}$, that is $Y_k = G(L(\mathbf{x}(k)))$. Then we assess the performance of each local descriptor L_i by counting the number of cases the result $L_i(\mathbf{x}(k))$ is included in Y_k and

$$V(w^+[ii]) = \sum_{w_i[ii]>med[ii]} g_i exp(-a|med[ii] - w^+[ii]|$$

(24)

where $w^+[ii]$ stands for the upper bound of the granular weight and the median, $med[ii]$. Is taken as a numeric representative of the collection of the weights.

Figure 6. The design of the granular descriptor G(L): an iterative development scheme

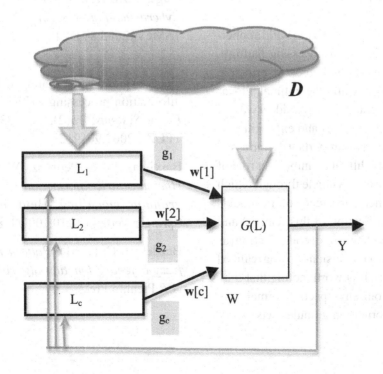

Where the weights f_i help differentiate contributions coming from the individual logic descriptors. Note that the information granule is constructed around some numeric representative, which could e.g., a weighted median. The essence of this process is captured in Figure 6.

The overall optimization can be realized iteratively so we start with equally weighted logic descriptors to form a granular generalization $G(L)$, then assess the quality (consistency) of each logic descriptor in terms of g_i; revisit the granular weights of $G(L)$, again assess the consistency of L_i in light of the revisited $G(L)$, modify the granular weights of the next version of $G(L)$ and proceed with the next iteration.

Note that the overall scheme is invoked in the context of some data D so a change in the data might result in a different granular logic descriptor. This points at the role of the data in the overall construct and underlines a need for its representativeness in describing the characteristics of the logic descriptors.

7. CONCLUSION

In Granular Computing, we strive to build a coherent and algorithmically sound processing platform. The two mechanisms studied here, the principle of justifiable granularity and the optimal allocation of information granularity provide a way of forming information granules and exploiting information granularity as an important design asset. In this context, we highlight an important role of information granules as a vehicle through which we can achieve higher consistency of the models (along with a quantification of this feature) and facilitate various mechanisms of collaboration (as exemplified in the group decision-making realized via the AHP model). It is worth noting that they are independent from any specific formal way of representing information granules sets, fuzzy sets, rough sets, etc.) and in this manner, the two schemes are of a general character.

The idea of optimal allocation (distribution) of information granularity calls for more advanced techniques of optimization (that go far beyond gradient-based techniques). In particular, one can anticipate the usage of evolutionary of swarm optimization methods. In this sense, we start witnessing here yet another example of an important synergy of technologies of Computational Intelligence.

REFERENCES

Bargiela, A., & Pedrycz, W. (2003). *Granular computing: An introduction*. Dordrecht, The Netherlands: Kluwer Academic.

Bargiela, A., & Pedrycz, W. (2005). A model of granular data: a design problem with the Tchebyschev FCM. *Soft Computing, 9*, 155–163. doi:10.1007/s00500-003-0339-2

Bargiela, A., & Pedrycz, W. (2005). Granular mappings. *IEEE Transactions on Systems, Man, and Cybernetics-Part A, 35*(2), 292–297. doi:10.1109/TSMCA.2005.843381

Bargiela, A., & Pedrycz, W. (2008). Toward a theory of granular computing for human-centered information processing. *IEEE Transactions on Fuzzy Systems, 16*(2), 320–330. doi:10.1109/TFUZZ.2007.905912

Bargiela, A., & Pedrycz, W. (Eds.). (2009). *Human-centric information processing through granular modelling*. Heidelberg, Germany: Springer-Verlag. doi:10.1007/978-3-540-92916-1

Bezdek, J. C. (1981). *Pattern recognition with fuzzy objective function algorithms*. New York, NY: Plenum Press.

Hirota, K. (1981). Concepts of probabilistic sets. *Fuzzy Sets and Systems, 5*(1), 31–46. doi:10.1016/0165-0114(81)90032-4

Hirota, K., & Pedrycz, W. (1984). Characterization of fuzzy clustering algorithms in terms of entropy of probabilistic sets. *Pattern Recognition Letters, 2*(4), 213–216. doi:10.1016/0167-8655(84)90027-8

Pawlak, Z. (1985). Rough sets and fuzzy sets. *Fuzzy Sets and Systems, 17*(1), 99–102. doi:10.1016/S0165-0114(85)80029-4

Pawlak, Z. (1991). *Rough sets: Theoretical aspects of reasoning about data, system theory.* Dordrecht, The Netherlands: Kluwer Academic.

Pawlak, Z., & Skowron, A. (2007). Rudiments of rough sets. *Information Sciences, 177*(1), 1 3-27.

Pedrycz, W. (1998). Shadowed sets: representing and processing fuzzy sets. *IEEE Transactions on Systems, Man, and Cybernetics. Part B, 28,* 103–109.

Pedrycz, W. (1999). Shadowed sets: bridging fuzzy and rough sets. In Pal, S. K., & Skowron, A. (Eds.), *Rough fuzzy hybridization: A new trend in decision-making* (pp. 179–199). Berlin, Germany: Springer-Verlag.

Pedrycz, W. (2005). Interpretation of clusters in the framework of shadowed sets. *Pattern Recognition Letters, 26*(15), 2439–2449. doi:10.1016/j.patrec.2005.05.001

Pedrycz, W., & Gomide, F. (2007). *Fuzzy systems engineering: Toward human-centric computing.* Hoboken, NJ: John Wiley & Sons.

Pedrycz, W., & Hirota, K. (2008). A consensus-driven clustering. *Pattern Recognition Letters, 29,* 1333–1343. doi:10.1016/j.patrec.2008.02.015

Pedrycz, W., & Rai, P. (2008). Collaborative clustering with the use of Fuzzy C-Means and its quantification. *Fuzzy Sets and Systems, 159*(18), 2399–2427. doi:10.1016/j.fss.2007.12.030

Pedrycz, W., & Song, M. (2011). Analytic Hierarchy Process (AHP) in group decision making and its optimization with an allocation of information granularity. *IEEE Transactions on Fuzzy Systems, 19*(3), 527–539. doi:10.1109/TFUZZ.2011.2116029

Zadeh, L. A. (1997). Towards a theory of fuzzy information granulation and its centrality in human reasoning and fuzzy logic. *Fuzzy Sets and Systems, 90,* 111–117. doi:10.1016/S0165-0114(97)00077-8

Zadeh, L. A. (1999). From computing with numbers to computing with words-from manipulation of measurements to manipulation of perceptions. *IEEE Transactions on Circuits and Systems, 45,* 105–119.

This work was previously published in the International Journal of Cognitive Informatics and Natural Intelligence, Volume 5, Issue 4, edited by Yingxu Wang, pp. 44-60, copyright 2011 by IGI Publishing (an imprint of IGI Global).

Chapter 13
Semantic Manipulations and Formal Ontology for Machine Learning Based on Concept Algebra

Yingxu Wang
University of Calgary, Canada

Yousheng Tian
University of Calgary, Canada

Kendal Hu
University of Calgary, Canada

ABSTRACT

Towards the formalization of ontological methodologies for dynamic machine learning and semantic analyses, a new form of denotational mathematics known as concept algebra is introduced. Concept Algebra (CA) is a denotational mathematical structure for formal knowledge representation and manipulation in machine learning and cognitive computing. CA provides a rigorous knowledge modeling and processing tool, which extends the informal, static, and application-specific ontological technologies to a formal, dynamic, and general mathematical means. An operational semantics for the calculus of CA is formally elaborated using a set of computational processes in real-time process algebra (RTPA). A case study is presented on how machines, cognitive robots, and software agents may mimic the key ability of human beings to autonomously manipulate knowledge in generic learning using CA. This work demonstrates the expressive power and a wide range of applications of CA for both humans and machines in cognitive computing, semantic computing, machine learning, and computational intelligence.

DOI: 10.4018/978-1-4666-2476-4.ch013

1. INTRODUCTION

Concepts are the basic unit of semantic cognition that carries certain meanings in expression, thinking, reasoning, and system modeling (Smith & Medin, 1981; Wille, 1982; Medin & Shoben, 1988; Murphy, 1993; Codin et al., 1995; Zadeh, 1999; Ganter & Wille, 1999; Wilson & Keil, 2001). In *denotational mathematics* (Wang, 2007a, 2008d), a concept is formally modeled as an abstract and dynamic mathematical structure that encapsulates a coherent hierarchy of attributes, objects, and relations (Wang, 2010b). The formal methodology for manipulating knowledge by *concept algebra* is developed by Wang (2008c), which provides a generic and formal knowledge manipulation means for dealing with complex knowledge and language structures as well as their algebraic operations.

Knowledge and concepts may be represented by ontology, which is the branch of metaphysics dealing with the nature of being in philosophy (Wilson & Keil, 2001). However, in computing and AI, ontology (Cocchiarella, 1996; GOLD, 2010) is both a method for modeling a domain of knowledge and an entity that represents a part of knowledge in knowledge engineering. Ontological engineering is a method of knowledge engineering. Typical ontological systems are WordNet – a lexical knowledgebase (Miller et al., 1990; Miller, 1995; Vossen, 1998), Dublin Core – an ontology for documents and publishing as standardized in ISO 15836 (ISO, 2011), and GOLD – a general ontology for linguistic description (GOLD, 2010). However, ontological technologies may only represent a set of static knowledge and are highly application-specific and subjective, which may not allow machines to mimic the process of human ontology building. In order to solve his problem, a mathematical model of general concept is formally elicited as an abstract and dynamic mathematical structure that denotes a concept as a triple of sets of attributes, objects, and relations (Wang, 2007c).

Based on the mathematical model of concepts, a formal methodology for manipulating knowledge is developed known as concept algebra (Wang, 2008c, 2010b), which provides a generic and formal knowledge manipulation means for dealing with complex knowledge and natural language structures as well as their algebraic operations. In concept algebra, the formal methodology for visualizing knowledge as a concept network is enabled for knowledge engineering.

It is recognized that new types of problems require new forms of mathematics. The maturity of a discipline is characterized by the maturity of its *mathematical means*. The family of mathematics may be classified into *analytic, numerical*, and *denotational mathematics* (Bender, 2000; Wang, 2008d). Analytic mathematics deals with mathematical entities on $\mathbb{R}$ (real numbers) with static relations and deterministic functions. Numerical mathematics deals with mathematical entities on $\mathbb{R}$ or $\mathbb{B}$ (bits) with discrete and recursively approximate functions. However, the domain of problems in machine learning and cognitive computing are *hyper-structures* ($\mathbb{HS}$) beyond that of pure numbers on $\mathbb{R}$. Therefore, the requirement for reduction of complex knowledge onto the *low-level data objects* in conventional computing technologies and their associated analytic mathematical means has greatly *constrained* the inference and computing ability toward the development of intelligent knowledge processors known as *cognitive computers* (Wang, 2009c). This observation (Bender, 2000; Wang, 2008d) has triggered the current transdisciplinary investigation into *new mathematical structures* for machine learning and cognitive computing collectively known as *denotational mathematics*.

Definition 1: A *hype-structure,* $\mathbb{HS}$, is a type of mathematical entities that is a complex *n*-tuple with multiple fields of attributes and constraints, as well as their interrelations, i.e.:

$$HS \triangleq \mathop{R}_{i-1}^{n}(A_i \mid \forall e \in A_i, p_i(e)) \qquad (1)$$

where $\mathop{R}\limits_{i=1}^{n}$ is the big-R notation of Real-Time Process Algebra (RTPA) (Wang, 2002, 2008a) that denotes a repetitive structure or operation, and A_i, $1 \le i \le n$, is a set of attributes that is equivalent to a type in computing for elements e, in which all e share the property $p_i(e)$.

Typical mathematical entities with $\mathbb{HS}$, as shown in Figures 1, 2, 3, and 4, are such as abstract objects, complex relations, perceptual information, abstract concepts, formal knowledge, intelligent behaviors, behavioral processes, rational decisions, language semantics, visual semantics, causal inference, and generic systems (Wang, 2007a, 2008d).

Definition 2: *Denotational mathematics* is a category of expressive mathematical structures that deals with high-level mathematical entities with hyper-structures on $\mathbb{HS}$ beyond numbers on R with a series of embedded dynamic processes (functions).

Denotational mathematics extends the formal means for rigorous machine reasoning to a level closer to that of humans. Typical denotational mathematics are concept algebra (Wang, 2008c), inference algebra (Wang, 2011a, 2011b), system algebra (Wang, 2008e; Wang, Zadeh, & Yao,

2009), RTPA (Wang, 2002, 2008a), and visual semantic algebra (Wang, 2009d).

This paper presents the operational semantics of concept algebra, which is a denotational mathematical structure for formal knowledge representation and manipulations in cognitive computing, knowledge representation, and machine learning. In the remainder of this paper, concept algebra is introduced in Section 2 as a generic and dynamic knowledge representation and modeling means. The semantics of concept algebra, particularly those of its relational and compositional operators, are rigorously presented in Section 3 and illustrated by a set of computational processes in RTPA (Wang, 2002, 2008a). A case study on formal concept manipulations is presented in Section 4 for demonstrating the expressive power and usages of concept algebra in machine learning, cognitive informatics, cognitive computing, and computational intelligence.

2. RELATED WORK: ONTOLOGY AND CONCEPT ALGEBRA

Investigations into the cognitive models of information and knowledge representation in the brain are perceived to be one of the fundamental research areas that help to unveil the mechanisms of the brain (Wang, 2007b, 2010a; Wang et al., 2006; Wang, Zadeh, & Yao, 2009). This section introduces related work on ontology and formal

Figure 1. The structural model of the language knowledge base

$$LKB\mathbf{UDM} \triangleq \mathop{R}\limits_{i\mathbf{N}=1}^{SCB\mathbf{N}.\#Words\mathbf{N}} Word(i\mathbf{N})\mathbf{HS} ::$$

$$\{<\text{Word} : \mathbf{S} \mid Word\mathbf{S} \neq \varnothing>,$$
$$<\text{Index} : \mathbf{N} \mid Index\mathbf{N} := i\mathbf{N}>,$$
$$<\text{Hyponym} : \mathbf{SET} \mid HyponymRelation\,\mathbf{SET} \neq \varnothing>, \qquad // \text{ Subconcept}$$
$$<\text{Hypernym} : \mathbf{SET} \mid HypernymRelation\,\mathbf{SET} \neq \varnothing>, \qquad // \text{ Superconcept}$$
$$<\text{Synonym} : \mathbf{SET}>, \qquad // \text{ Equivalent concept}$$
$$<\text{Antonym} : \mathbf{SET}> \qquad // \text{ Opposite concept}$$
$$\}$$

Figure 2. The OAR model of logical memory architectures

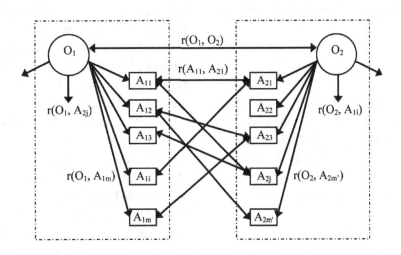

Figure 3. The structural model of a formal ontology of knowledge

$$\text{Knowledge}\textbf{UDM} \triangleq \overset{\#CurrentConcepts\textbf{N}}{\underset{i\textbf{N}=1}{R}} \quad \text{OAR}(i\textbf{N})\textbf{HS} ::$$

$$\{<c : \textbf{S} \mid 1 \leq |c\textbf{S}| \leq 255>, \qquad\qquad \text{// ID of the concept}$$
$$<A : \textbf{SET} \mid A\textbf{SET} = \{A_1\textbf{S}, A_2\textbf{S}, ..., A_n\textbf{S}\}>, \qquad \text{// Attribute}$$
$$<O : \textbf{SET} \mid O\textbf{SET} = \{O_1\textbf{S}, O_2\textbf{S}, ..., O_m\textbf{S}\}>, \qquad \text{// Object}$$
$$<RC : \textbf{SET} \mid RC\textbf{SET} = \{(O\textbf{SET} \times A\textbf{SET})\}\textbf{SET}>, \qquad \text{// Internal relation}$$
$$<RI : \textbf{SET} \mid RI\textbf{SET} = \{(C(i'\textbf{N})\textbf{HS} \times C(i\textbf{N})\textbf{HS})\}\textbf{SET}>, \qquad \text{// External input relation}$$
$$<RO : \textbf{SET} \mid RO\textbf{SET} = \{(C(i\textbf{N})\textbf{HS} \times C(i'\textbf{N})\textbf{HS})\}\textbf{SET}>, \qquad \text{// External output relation}$$
$$<RS : \textbf{SET} \mid \text{Equivalence}(C(i\textbf{N})\textbf{HS}, C(i'\textbf{N})\textbf{HS}) = 1>, \qquad \text{// Synonym relation}$$
$$<RA : \textbf{SET} \mid \text{Equivalence}(C(i\textbf{N})\textbf{HS}, \quad (i'\textbf{N})\textbf{HS}) = 1>, \qquad \text{// Antonym relation}$$
$$<\text{TimeStamp: }\textbf{yyyy:MM:dd:hh:mm:ss} \mid 2000:01:01:00:00:00 \leq$$
$$\text{TimeStamp } \textbf{yyyy:MM:dd:hh:mm:ss} \leq 2999:12:31:23:59:59>$$
$$\}$$

Figure 4. The structural model of a general concept

$$\text{Concept}\textbf{UDM} \triangleq \textbf{CHS} ::$$

$$\{<c : \textbf{S} \mid 1 \leq |c\textbf{S}| \leq 255>,$$
$$<A : \textbf{SET} \mid A\textbf{SET} = \{A_1\textbf{S}, A_2\textbf{S}, ..., A_n\textbf{S}\}>,$$
$$<O : \textbf{SET} \mid O\textbf{SET} = \{O_1\textbf{S}, O_2\textbf{S}, ..., O_m\textbf{S}\}>,$$
$$<RC : \textbf{SET} \mid RC\textbf{SET} = \{(O\textbf{SET} \times A\textbf{SET})\}>,$$
$$<RI : \textbf{SET} \mid RI\textbf{SET} = \{(\text{OAR}\textbf{HS} \times C\textbf{HS})\}>,$$
$$<RO : \textbf{SET} \mid RO\textbf{SET} = \{(C\textbf{HS} \times \text{OAR}\textbf{HS})\}>$$
$$\}$$

knowledge representation in order to explain the denotational mathematical structure of concept algebra.

2.1 Ontology

Ontology is an old branch of metaphysics dealing with the nature of being in philosophy (Wilson & Keil, 2001; Cocchiarella, 1996). The concept of ontology is recently extended to represent a method for modeling a domain or structure of knowledge in a knowledge system.

Definition 3: An *ontology* is a taxonomic hierarchy of knowledge, which uses lexical terms (words) and their associations to model and represent a framework of structured knowledge.

A widely used ontological knowledgebase, *WordNet*, was developed by the Cognitive Science Laboratory at Princeton University since 1985 (Miller et al., 1990). It is a general-purpose knowledge base of words, which covers most English nouns, adjectives, verbs, and adverbs. The structure of WordNet is a relational lexical network where each node stands for a specific 'sense' and is expressed by a lexical unit called 'synset' with a number of synonyms.

Parallel with WordNet, *ConceptNet* (Liu & Singh, 2004) is the largest commonsense knowledge base developed by the Media Laboratory at MIT. ConceptNet extends WordNet's concepts from purely lexical items (words and simple phrases) to complex concepts. ConceptNet also supply plenty of high-order concepts that compose verb with arguments such as events and processes. However, ConceptNet describes various relations between concepts in a planar way without distinguishing the difference between concept relations and attribute relations. There-

fore, knowledge derived from ConceptNet must be further processed and formalized in order to facilitate machine learning and causal reasoning.

On the basis of WordNet and ConceptNet, a language knowledge base, LKB**UDM**, can be formally modeled as shown in Figure 1, where **UDM** is a type suffix of RTPA (Wang, 2002) representing a *unified data model*. LKB**UDM** encompasses a set of structured words denoted by Word**HS**, which is refined by the field of Word**S**, Index**N**, HyponymRelation**SET**, HypernymRelation**SET**, Synonym**SET**, and Antonym**SET**. The field Word**S** models the name of the word as a string; the Index**N** denotes an index number of the word in the entire knowledgebase; the HyponymRelation**SET** and HypernymRelation**SET** denote subconcept and superconcept relations of the word with other words, respectively; and the fields of Synonym**SET** and Antonym**SET** represent equivalent or opposite concepts.

2.2 The OAR Model of Knowledge Representation

To rigorously model the hierarchical and dynamic neural cluster structures of memory at the neurological and physiological levels, a logical model of memory and internal knowledge representation is developed as given (Wang, 2007c).

Definition 4: The *Object-Attribute-Relation* (OAR) *model* of long-term memory can be described as a triple, i.e.:

$$OAR \triangleq (O, A, R) \tag{2}$$

where O is a finite set of objects identified by unique symbolic names, A is a finite set of attributes for characterizing the object, and R is a set of relations between the objects and attributes.

On the basis of OAR, the logical and mathematical models of concepts and knowledge may be rigorously derived in the following subsections. To a certain extent, the entire knowledge in the brain can be modeled as a global OAR model as illustrated in Figure 2. According to Figure 2, the human knowledge may be modeled by the interrelations and interactions among individual concepts in a concept network formally defined.

Definition 5: *Knowledge,* $\mathfrak{K}$, *as a concept network* (*CN*) is a hierarchical digraph of concepts c_i, $1 \leq i \leq n$, associated by a set of relational and compositional connections $\mathfrak{R}$ as defined in concept algebra, i.e.:

$$\mathfrak{K} \triangleq CN = \mathfrak{R} : \underset{i=1}{\overset{n}{X}} C_i \to \underset{i=1}{\overset{n}{X}} C_i \qquad (3)$$

where $\underset{i=1}{\overset{n}{X}}$ denote an *n* dimensional Cartesian product and $\mathfrak{R}$ is a set of 17 relational operators formally defined in RTPA (Wang, 2002).

The OAR model describes human memory, particularly the long-term memory (LTM), by using the *relational* metaphor, rather than the traditional *container* metaphor that used to be adopted in psychology, computing, and information science. The OAR model shows that human memory and knowledge are represented by relations, i.e., synaptic connections between neurons, rather than by the neurons themselves as the traditional container metaphor suggested.

On the basis of the OAR model and concept network, a formal ontology of knowledge, Knowledge**UDM**, can be refined as shown in Figure 3. Knowledge**UDM** repetitively defines a set of concepts and their relations by OAR(i**N**)**HS** on ℍ𝕊. Each concept in Knowledge**UDM** encompasses a set of fields represented by variable in a type and related constraints such as the concept ID in string type (c**S**), the attributes as a set of strings (A**SET**), and the objects as a set of strings (O**SET**), as well as any synonyms (RS**SET**) and antonyms

(RA**SET**). A time stamp is created for each concept in the knowledge model with the type suffix range from year to second (**yyyy:MM:dd:hh:mm:ss**), which is useful for time-specific manipulations on knowledge.

2.3 The Mathematical Model of Concepts and Knowledge

Definition 6: Let U denote a finite nonempty set of objects and M be a finite nonempty set of attributes, the *discourse* of the context or the *semantic environment*, Θ, is denoted as a triple, i.e.:

$$\begin{aligned} \Theta &= (U, M, R) \\ &= R : U \to U \,|\, U \to M \,|\, M \to U \,|\, M \to M \end{aligned} \qquad (4)$$

where R is a set of relations between U and M.

On the basis of the semantic environment Θ of human knowledge, a general concept is a composition of the above three elements as given below.

Definition 7: A *formal concept c* on Θ is a 5-tuple, i.e.:

$$C \triangleq (O, A, R^c, R^i, R^o) \qquad (5)$$

where

- O is a nonempty set of objects of the concept, $O = \{o_1, o_2, ..., o_m\} \subseteq \text{Þ}U$, where $\text{Þ}U$ denotes a power set of U.
- A is a nonempty set of attributes, $A = \{a_1, a_2, ..., a_n\} \subseteq \text{Þ}M$.
- $R^c \subseteq O \times A$ is a set of internal relations.
- $R^i \subseteq A' \times A$, $A' \subseteq M \wedge A' \not\subset A$, is a set of input relations with external concepts. For convenience, R^i may be simply denoted as $R^i = \mathfrak{K} \times C$ where $\mathfrak{K}$ denotes existing knowledge with all known concepts.
- $R^o = C \times \mathfrak{K}$ is a set of output relations.

The most important properties of the formal concept model, as defined in Equation (5), are the capture of a set of essential attributes as its *intension*; the classification of a set of instantiation objects as its *extension*; and the adaptive capability to autonomously interrelate itself to other concepts in existing knowledge.

According to Definition 7, the general schema of concepts, Concept**UDM**, can be uniquely modeled as a UDM as shown in Figure 4. The concept C**HS** is a hyper-structure with six fields where c**S** denotes the name of the concept; A**SET** a set of attributes of the concept; O**SET** a set of objects as instances of the concept; RC**SET** a set of internal relations between the objects and attributes; RI-**SET** and RO**SET** are sets of input/output relations between the entire OAR and the given concept or vice versa.

It is noteworthy that, although the semantics of words may be ambiguity, the semantics of concept is always unique and precise. For example, the word, "bank", is ambiguity because it may be a notion of a financial institution, a geographic location of raised ground of a river/

Figure 5. Formal concepts derived from the word "bank" in natural language

$b_o\textbf{HS} \triangleq (A_1, O_1, R^c_1, R^i_1, R^o_1)$ // A financial bank

$= (A_1 = \{$organization, company, financial business, money, deposit, withdraw, invest, exchange$\}$,

$O_1 = \{$international_bank, national_bank, local_bank, investment_bank, ATM$\}$,

$R^c_1 = O \times A$,

$R^i_1 = \mathfrak{K} \times b_o\textbf{HS}$,

$R^o_1 = b_o\textbf{HS} \times \mathfrak{K}$

)

$b_r\textbf{HS} \triangleq (A_2, O_2, R^c_2, R^i_2, R^o_2)$ // A river bank

$= (A_2 = \{$sides of a river, raised ground, a pile of earth, location$\}$,

$O_2 = \{$river_bank, lake_bank, canal_bank$\}$

$R^c_2 = O \times A$,

$R^i_2 = \mathfrak{K} \times b_r\textbf{HS}$,

$R^o_2 = b_r\textbf{HS} \times \mathfrak{K}$

)

$b_s\textbf{HS} \triangleq (A_3, O_3, R^c_3, R^i_3, R^o_3)$ // A storage bank

$= (A_3 = \{$storage, container, place, organization$\}$,

$O_3 = \{$information_bank, resource_bank, blood_bank$\}$

$R^c_3 = O \times A$,

$R^i_3 = \mathfrak{K} \times b_s\textbf{HS}$,

$R^o_3 = b_s\textbf{HS} \times \mathfrak{K}$

)

lake, and/or a storage of something. However, the three individual concepts denoted by bank, i.e., b_o = bank(organization), b_r = bank(river), and b_s = bank(storage), are precisely unique, which can be formally described according to Definition 7 and be illustrated by Figure 5.

2.4 The Mathematical Structure of Concept Algebra

Concept algebra is a denotational mathematical structure for the formal treatment of concepts as given in Definition 7 and their algebraic relations, operations, and associative rules (Wang, 2008c).

Definition 8: A *concept algebra*, *CA*, is a triple on the formal model of concepts *C* on the general discourse of the semantic environment Θ, i.e.:

$$CA \triangleq (C, OP, \Theta) = (\{O, A, R^c, R^i, R^o\}, \{\bullet_r, \bullet_c\}, \Theta)$$
(6)

where $OP = \{\bullet_r, \bullet_c\}$ are the sets of *relational* and *compositional* operators, respectively, on formal concepts.

The relational and compositional operators of CA will be rigorously defined in Section 3.1 and 3.2, respectively. An overview and summary of the compositional operators of CA are provided in Table 1 via Equations. 6.1, 6.2, 6.3, 6.4, 6.5, 6.6, 6.7, 6.8, and 6.9.

CA provides a rigorous means for algebraic manipulations of structured concepts. CA can be used to model, specify, and manipulate knowledge systems and models of domain ontologies for machine learning studied in cognitive informatics (Wang, 2007b; Wang & Wang, 2006), cognitive computing (Wang, 2009c), knowledge engineering, computational intelligence, and soft computing (Zadeh, 1998) for machine learning.

A comparative analysis of the relationship between CA and WordNet is shown in Table 2. Two categories of operators for concept formation and algebraic relation are contrasted in the table. It is obvious that CA provides a set of rigorous

definitions of the informally described semantics of concept structures and relational operations as adopted in WordNet. Furthermore, a set of compositional operators is provided in CA in order to algebraically manipulate complex concepts and the semantics of knowledge.

3. SEMANTIC OPERATIONS ON FORMAL CONCEPTS IN CONCEPT ALGEBRA

As described in the preceding section, CA is a set of algebraic operators defined on formal concepts as an algebraic structure. The algebraic operators of CA can be classified in two categories known as the relational and compositional operators. This section describes the operational semantics of CA and its usages in knowledge manipulations and machine learning.

3.1 Relational Manipulations of Concepts

Definition 9: The *relational operations* $\bullet_r$ of CA encompass eight comparative operators for manipulating the algebraic relations between concepts, i.e.:

$$\bullet_r \triangleq \{\leftrightarrow, \nleftrightarrow, \prec, \succ, =, \cong, \sim, \triangleq\}$$
(7)

where the relational operators stand for *related, independent, subconcept, superconcept, equivalent, consistent, comparison,* and *definition,* respectively.

Details of the relational operations on formal concepts may refer to Wang (2008c). CA provides a denotational mathematical means for algebraic manipulations of abstract concepts. CA can be used to model, specify, and manipulate generic *"to be"* type problems, particularly system architectures, knowledge bases, and detail-level system designs, in cognitive informatics, cognitive computing,

Table 1. Compositional operations on formal concepts in CA

No.	Operator	Description	Definition in CA
1	$\Rightarrow$	Inheritance (Eq. 6.1)	$c(O, A, R^c, R^i, R^o) \Rightarrow c_1(O_1, A_1, R^c_1, R^i_1, R^o_1)$ $\triangleq c_1(O_1, A_1, R^c_1, R^i_1, R^o_1 \mid O_1 = O, A_1 = A, R^c_1 = R^c, R^i_1 = R^i \cup (c, c_1), R^o_1 = R^o \cup (c_1, c))$ $\parallel c(O, A, R^c, R^{i'}, R^{o'} \mid R^{i'} = R^i \cup (c_1, c), R^{o'} = R^o \cup (c, c_1))$
2	$\stackrel{\approx}{\Rightarrow}$	Tailor (Eq. 6.2)	$c(O, A, R^c, R^i, R^o) \stackrel{\approx}{\Rightarrow} c_1(O_1, A_1, R^c_1, R^i_1, R^o_1)$ $\triangleq c_1(O_1, A_1, R^c_1, R^i_1, R^o_1 \mid O_1 = O \setminus O', A_1 = A \setminus A', R^c_1 = R^c \setminus (\{O \times A'\} \cup \{O' \times A\}), R^i_1 = R^i \cup (c, c_1), R^o_1 = R^o \cup (c_1, c))$ $\parallel c(O, A, R^c, R^{i'}, R^{o'} \mid R^{i'} = R^i \cup (c_1, c), R^{o'} = R^o \cup (c, c_1))$
3	$\stackrel{+}{\Rightarrow}$	Extension (Eq. 6.3)	$c(O, A, R^c, R^i, R^o) \stackrel{+}{\Rightarrow} c_1(O_1, A_1, R^c_1, R^i_1, R^o_1)$ $\triangleq c_1(O_1, A_1, R^c_1, R^i_1, R^o_1 \mid O_1 = O \cup O', A_1 = A \cup A', R^c_1 = R^c \cup \{O' \times A'\} \cup \{O' \times A\} \cup \{O \times A'\},$ $R^i_1 = R^i \cup (c, c_1), R^o_1 = R^o \cup (c_1, c))$ $\parallel c(O, A, R^c, R^{i'}, R^{o'} \mid R^{i'} = R^i \cup (c_1, c), R^{o'} = R^o \cup (c, c_1))$
4	$\stackrel{\sim}{\Rightarrow}$	Substitute (Eq. 6.4)	$c(O, A, R^c, R^i, R^o) \stackrel{\sim}{\Rightarrow} c_1(O_1, A_1, R^c_1, R^i_1, R^o_1)$ $\triangleq c_1(O_1, A_1, R^c_1, R^i_1, R^o_1 \mid O_1 = (O \setminus O') \cup O'', A_1 = (A \setminus A') \cup A'', R^c_1 = O_1 \times A_1,$ $R^i_1 = R^i \cup (c, c_1), R^o_1 = R^o \cup (c_1, c))$ $\parallel c(O, A, R^c, R^{i'}, R^{o'} \mid R^{i'} = R^i \cup (c_1, c), R^{o'} = R^o \cup (c, c_1))$
5	$\uplus$	Composition (Eq. 6.5)	$c(O, A, R^c, R^i, R^o) \triangleq \biguplus_{i=1}^{n} c_i(O_i, A_i, R^c_i, R^i_i, R^o_i)$ $= c(O, A, R^c, R^i, R^o \mid O = \bigcup_{i=1}^{n} O_i, A = \bigcup_{i=1}^{n} A_i, R^c = \bigcup_{i=1}^{n} (R^c_i \cup (O_i \times A \setminus A_i)), R^i = \bigcup_{i=1}^{n} (R^i_i \cup (c_i, c)), R^o = \bigcup_{i=1}^{n} (R^o_i \cup (c, c_i))$ $\parallel \mathop{R}_{i=1}^{n} c_i(O_i, A_i, R^c_i, R^{i'}_i, R^{o'}_i \mid R^{i'}_i = R^i_i \cup (c, c_i), R^{o'}_i = R^o_i \cup (c_i, c))$
6	$\mathbin{\uparrow\!\uparrow}$	Decomposition (Eq. 6.6)	$c(O, A, R^c, R^i, R^o) \mathbin{\uparrow\!\uparrow} \mathop{R}_{i=1}^{n} c_i(O_i \mid O = \bigcup_{i=1}^{n} O_i)$ $\triangleq \mathop{R}_{i=1}^{n} \{ c_i(O_i, A_i, R^c_i, R^i_i, R^o_i \mid O_i \subset O \mid O = \bigcup_{i=1}^{n} O_i, A_i = A \setminus \{(O_i \times A) \notin R^c\}, R^c_i = (O_i \times A_i), R^i_i = (c, c_i), R^o_i = (c_i, c)\}$ $\parallel c(O, A, R^c, R^{i'}, R^{o'} \mid R^{i'} = R^i \cup \mathop{R}_{i=1}^{n} (c_i, c), R^{o'} = R^o \cup \mathop{R}_{i=1}^{n} (c, c_i))$
7	$\Leftarrow$	Aggregation (Eq. 6.7)	$c(O, A, R^c, R^i, R^o) \Leftarrow \mathop{R}_{i=1}^{n} c_i$ $\triangleq c(O, A, R^c, R^i, R^o \mid O = \bigcup_{i=1}^{n} O_i, A = \bigcap_{i=1}^{n} A_i, R^c = O \times A, R^i = \mathop{R}_{i=1}^{n} (c_i, c), R^o = \mathop{R}_{i=1}^{n} (c, c_i))$ $\parallel \mathop{R}_{i=1}^{n} c_i(O_i, A_i, R^c_i, R^{i'}_i, R^{o'}_i \mid R^{i'}_i = R^i_i \cup (c, c_i), R^{o'}_i = R^o_i \cup (c_i, c))$
8	$\vdash$	Specification (Eq. 6.8)	$c(O, A, R^c, R^i, R^o) \vdash \mathop{R}_{i=1}^{n} c_i$ $\triangleq \mathop{R}_{i=1}^{n} c_i(O_i, A_i, R^c_i, R^i_i, R^o_i \mid O_i = o_i \in O, A_i = A \setminus (o_i \times A) \notin R^c, R^c_i = (O_i \times A_i), R^i_i = (c, c_i), R^o_i = (c_i, c)\}$ $\parallel c(O, A, R^c, R^{i'}, R^{o'} \mid R^{i'} = R^i \cup \mathop{R}_{i=1}^{n} (c_i, c), R^{o'} = R^o \cup \mathop{R}_{i=1}^{n} (c, c_i))$
9	$\mapsto$	Instantiation (Eq. 6.9)	$c(O, A, R^c, R^i, R^o) \mapsto o(A_o, R^c_o, R^i_o)$ $\triangleq o(A_o, R^c_o, R^{i'}_o \mid o \subset O, A_o = A, R^c_o = o \times A_o, R^{i'}_o = R^i_o \cup \{(c, o)\})$ $\parallel c(O, A, R^c, R^{i'}, R^{o'} \mid R^{i'} = R^i \cup \{(o, c)\}, R^{o'} = R^o \cup \{(c, o)\})$

Table 2. Relationship between concept algebra and WordNet

Category	Concept Algebra		WordNet		Definition
	Operation	Symbol	Operation	Symbol	
Concept formation	Definition (Identification)	$\triangleq$			$c \triangleq (O, A, R^c, R^i, R^o)$ $c = (O, A, R^c, R^i, R^o \mid O \subset U, A \subset M,$ $R^c = O \times A, R^i = \varnothing, R^o = \varnothing)$
	Attribute (Qualification)	$\forall$	Meronym/ Attribute	%/=	$\forall c = \forall c(O, A, R^c, R^i, R^o) = A = \{a_1, a_2, ..., a_n \mid a_i \in \overset{m}{\underset{i=1}{R}} c.o_i\}$
	Object (Elicition)	*	Hyponym/ Synset	~ / *	$c = c_1(O_1, A_1, R^c_1, R^i_1, R^o_1) * c_2(O_2, A_2, R^c_2, R^i_2, R^o_2)$ $= c(O, A, R^c, R^i, R^o \mid O = O_1 \cup O_2, A = A_1 \cap A_2,$ $R^c = O \times A, R^i = \{(c_1, c), (c_2, c)\}, R^o = \{(c, c_1), (c, c_2)\})$
Relations	Synonym (Equivalent)	=	Synonym/Synset	=	$A_1 = A_2 \wedge O_1 = O_2 \wedge R_1 = R_2 \Rightarrow c_1 = c_2$
	Antonum	!	Antonym	!	-
	Hyponym (Subconcept)	$\prec$	Hyponym	~	$A_1 \subset A_2 \Rightarrow c_1 \prec c_2$
	Holonym (Super concept)	$\succ$	Holonym	#	$A_1 \supset A_2 \Rightarrow c_1 \succ c_2$
	Related	$\leftrightarrow$			$A_1 \cap A_2 \neq \varnothing \Rightarrow c_1 \leftrightarrow c_2$
	Independent	$\nleftrightarrow$			$A_1 \cap A_2 = \varnothing \Rightarrow c_1 \nleftrightarrow c_2$
	Consistent	$\cong$			a) $c_1, c_2 \in \Theta$; b) $c_1 \leftrightarrow c_2$; c) $(c_1 \succ ... \succ c_2) \vee (c_1 \prec ... \prec c_2)$

software engineering, computational intelligence, and soft computing.

3.1.1 Concept Relations

Definition 10: The *related* and *independent concepts* c_1 and c_2 on Θ, denoted by $c_1 \leftrightarrow c_2$ and $c_1 \nleftrightarrow c_2$, respectively, are a pair of relational operations determined by if they have shared attributes in their intensions A_1 and A_2, i.e.:

$$\begin{cases} \text{Related concepts:} & c_1 \leftrightarrow c_2 \triangleq A_1 \cap A_2 \neq \varnothing \\ \text{Independent concepts:} & c_1 \nleftrightarrow c_2 \triangleq A_1 \cap A_2 = \varnothing \end{cases} \tag{8}$$

Definition 11: The *subconcept* and *superconcept* between concepts c_1 and c_2 on Θ, denoted by $c_1 \prec c_2$ and $c_1 \succ c_2$, are a pair of relational operations determined by if their intensions A_1 and A_2 are related and mutually inclusive, i.e.:

$$\begin{cases} \text{Subconcept } c_1: & c_1 \prec c_2 \triangleq A_1 \supset A_2 \\ \text{Superconcept } c_2: & c_2 \succ c_1 \triangleq A_2 \subset A_1 \end{cases} \tag{9}$$

where the intension A_2 of superconcept c_2 includes fewer attributes than A_1 because a superconcept is more general than a specific concept c_1.

Definition 12: The *equal concepts* c_1 and c_2 on Θ, denoted by $c_1 = c_2$, are a relational operation that determines if two given concepts are the same with identical intensions A_1 and A_2, i.e.:

$$c_1 = c_2 \triangleq A_1 = A_2 \tag{10}$$

The operational semantics of related/independent, equal, and sub/super concept operations is illustrated in Figure 6 by the process ConceptRelations**PC** in RTPA according to Equations 8, 9, and 10. In ConceptRelations**PC**, the UDM represents global architectural models as hyperstructures. The type suffixes **PC, HS, S, SET, N**, and

Figure 6. The operational semantics of relational operators in CA

$$\text{ConceptRelations}\textbf{PC}(<\textbf{I::} \text{ ConceptX}\textbf{HS}, \text{ConceptY}\textbf{HS}>; <\textbf{O::} \text{ Relation}\textbf{S}>;$$
$$<\textbf{UDM::} \text{ Concept}\textbf{HS}, \text{OAR}\textbf{HS}>) \triangleq$$

$\{A_X\textbf{SET} := \text{ConceptX}\textbf{HS}.A\textbf{SET}$

$\rightarrow A_Y\textbf{SET} := \text{ConceptY}\textbf{HS}.A\textbf{SET}$

$\rightarrow (\blacklozenge A_X\textbf{SET} \cap A_Y\textbf{SET} = \varnothing$

$\quad\quad \rightarrow \text{Relation}\textbf{S} := \text{'Independent'}$ $/\!/ \text{ X} \leftrightarrow \text{Y}$

$|\blacklozenge \sim$ $/\!/ \text{ X} \leftrightarrow \text{Y}$

$\quad\quad \rightarrow (\blacklozenge A_X\textbf{SET} = A_Y\textbf{SET}$

$\quad\quad\quad \rightarrow \text{Relation}\textbf{S} := \text{'Equal'}$ $/\!/ \text{ X} = \text{Y}$

$\quad\quad | \blacklozenge A_X\textbf{SET} \supset A_Y\textbf{SET}$

$\quad\quad\quad \rightarrow \text{Relation}\textbf{S} := \text{'Subconcept'}$ $/\!/ \text{ A} \prec \text{B}$

$\quad\quad | \blacklozenge A_X\textbf{SET} \subset A_Y\textbf{SET}$

$\quad\quad\quad \rightarrow \text{Relation}\textbf{S} := \text{'Superconcept'}$ $/\!/ \text{ A} \succ \text{B}$

$\quad\quad)$

$\quad)$

$\}$

BL denote that an entity is in the type of process, hyper structure, string, set, natural number, and Boolean variable, respectively (Wang, 2002).

3.1.2 Concept Consistency

Definition 13: The *concept consistency* operation between concepts c_1 and c_2 on Θ, denoted by $c_1 @ c_2$, determines if the two concepts are consistent by checking their intensions as being either a subconcept or superconcept, i.e.:

$$c_1 \cong c_2 \triangleq A_1 \supseteq A_2 \vee A_1 \subseteq A_2 \quad\quad (11)$$

The operational semantics of consistent concept operation is illustrated in Figure 7 by the process ConceptConsistency**PC** according to Equation 11. ConceptConsistency**PC** tests the consistency of a pair of given concepts by invoking the predefined process ConceptRelations**PC**.

3.1.3 Concept Equivalence

Definition 14: The *concept equivalence* operation between concepts c_1 and c_2 on Θ, denoted by $c_1 \sim c_2$, compares the extend of equivalency or similarity between the intensions of the two given concepts A_1 and A_2, i.e.:

$$c_1 \sim c_2 \triangleq \frac{|A_1 \cap A_2|}{|A_1 \cup A_2|} \quad\quad (12)$$

The operational semantics of concept equivalence operation is illustrated in Figure 8 by the process ConceptEquivalence**PC** according to Equation 12. ConceptEquivalence**PC** analyzes the extent of concept equivalency in the scope of [0, 1] where $c_1 \sim c_2 = 1$ indicates an identical concept, i.e., $c_1 = c_2$.

Figure 7. The operational semantics of concept consistency in CA

$$\text{ConceptConsistency}\textbf{PC}(<\textbf{I}:: \text{ConceptX}\textbf{HS}, \text{ConceptY}\textbf{HS}>; <\textbf{O}:: \text{Consistency}\textbf{BL}>;$$
$$<\textbf{UDM}:: \text{Concept}\textbf{HS}, \text{OAR}\textbf{HS}>) \triangleq$$
$$\{ \mapsto \text{ConceptRelations}\textbf{PC}(<\textbf{I}:: \text{ConceptX}\textbf{HS}, \text{ConceptY}\textbf{HS}>; <\textbf{O}:: \text{Relation}\textbf{S}>;$$
$$<\textbf{UDM}:: \text{Concept}\textbf{HS}, \text{OAR}\textbf{HS}>)$$
$$\rightarrow (\quad \blacklozenge \ \text{Relation}\textbf{S} = \text{'Subconcept'} \lor \text{Relation}\textbf{S} = \text{'Superconcept'}$$
$$\rightarrow \text{Consistency}\textbf{BL} := \textbf{T} \qquad\qquad\qquad // \ X \cong Y$$
$$| \ \blacklozenge \sim$$
$$\rightarrow \text{Consistency}\textbf{BL} := \textbf{F}$$
$$)$$
$$\}$$

Figure 8. The operational semantics of concept equivalence in CA

$$\text{ConceptEquivalence}\textbf{PC} (<\textbf{I}:: \text{ConceptX}\textbf{HS}, \text{ConceptY}\textbf{HS}>; <\textbf{O}:: \text{Equivalence}\textbf{R}>;$$
$$<\textbf{UDM}:: \text{Concept}\textbf{HS}, \text{OAR}\textbf{HS}>) \triangleq$$
$$\{ \rightarrow A_X\textbf{SET} := \text{ConceptX}\textbf{HS}.\textbf{ASET}$$
$$\rightarrow A_Y\textbf{SET} := \text{ConceptY}\textbf{HS}.\textbf{ASET}$$
$$\rightarrow \text{Equivalence}\textbf{R} := \frac{\#(A_X\textbf{SET} \cap A_Y\textbf{SET})}{\#(A_X\textbf{SET} \cup A_Y\textbf{SET})} \qquad\qquad // \ X \sim Y$$
$$\}$$

3.2 Compositional Manipulations of Concepts

Definition 15: The *compositional operations* $\bullet_c$ of CA encompass nine algebraic operators for manipulating concept reproduction, composition, and decomposition, i.e.:

$$\bullet_c \triangleq \{\Rightarrow, \overset{-}{\Rightarrow}, \overset{+}{\Rightarrow}, \overset{\sim}{\Rightarrow}, \uplus, \pitchfork, \Leftarrow, \vdash, \mapsto\} \qquad (13)$$

where the operators stand for *inheritance, tailor, extension, substitute, composition, decomposition, aggregation, specification,* and *instantiation,* respectively.

Detailed definitions of the compositional operators in CA may refer to Table 1 and Wang

(2008c). The following subsections describe the semantics of the nine compositional operators of CA using RTPA (Wang, 2002, 2008a, 2008b).

3.2.1 Concept Inheritance

Definition 16: The *concept inheritance* operation, $c \Rightarrow c_1$, derives a new concept (*synonym*) c_1 by reproducing c and creates a pair of new associations between them as given in Equation 6.1 in Table 1.

The semantics of concept inheritance is illustrated in Figure 9 by the process Concept-Inheritance**PC** in RTPA. ConceptInheritance**PC** derives a synonym, ConceptY**HS** from the given concept ConceptX**HS** as follows: a) ConceptY**HS** reproduces the same structure and intension/exten-

Figure 9. The semantics of concept inheritance operation

ConceptInheritance**PC**(<**I:**: ConceptX**HS**>; <**O**:: ConceptY**HS**>;

        <**UDM**:: Concept**HS**, OAR**HS**>) ⊠

{ // *CA definition:* X ⇒ Y

 → ConceptY**HS** := ConceptX**HS**   // Form ConceptY

 → ConceptY**HS**.RI**SET** := ConceptY**HS**.RI**SET**⊠ (ConceptX**HS**, ConceptY**HS**)

 → ConceptY**HS**.RO**SET** := ConceptY**HS**.RO**SET** ⊠ (ConceptY**HS**, ConceptX**HS**)

 // *Update ConceptX*

 → ConceptX**HS**.RI**SET** := ConceptX**HS**.RI**SET**⊠ (ConceptY**HS**, ConceptX**HS**)

 → ConceptX**HS**.RO**SET** := ConceptX**HS**.RO**SET** ⊠ (ConceptX**HS**, ConceptY**HS**)

}

sion of ConceptX**HS**; and b) The input and output relation sets of ConceptY**HS**, i.e., ConceptY**HS**.RI**SET** and ConceptY**HS**.RO**SET**, are updated by the newly established relations (ConceptX**HS**, ConceptY**HS**) and (ConceptY**HS**, ConceptX**HS**), respectively. As a reflexive consequence, the input and output relation sets of ConceptX**HS**, i.e., ConceptX**HS**.RI**SET** and ConceptX**HS**.RO**SET**, are also extended by (ConceptY**HS**, ConceptX**HS**) and (ConceptX**HS**, ConceptY**HS**), respectively, which represent the newly created associations between the given concept and its inherited counterpart in the dynamic concept network.

The semantics of *multiple inheritance* of concepts, $c \Rightarrow \overset{n}{\underset{i=1}{R}} c_i$, can be denoted in a similar way as that of single inheritance, where $\overset{n}{\underset{i=1}{R}} c_i$ is known as the *big-R* notation (Wang, 2007a) that denotes a repetitive behavior or a *recurrent* structure in system modeling.

3.2.2 Concept Tailoring

Definition 17: The *concept tailor* operation, $c \overset{\Rightarrow}{\Rightarrow} c_1$, is a special inheritance that derives a new concept c_1 from an existing concept c where its objects and attributes are reduced by specific subsets of objects and attributes, O' and A', respectively, as given in Equation 6.2 in Table 1.

The semantics of concept tailor is illustrated in Figure 10 by the process ConceptTailor**PC**. ConceptTailor**PC** generates a tailored concept ConceptY**HS** based on the given concept ConceptX**HS** as follows: (a) ConceptY**HS** reproduces the same structure and contents of ConceptX**HS**; (b) Each of the target objects as specified in ObjectForTailor**SET** is iteratively removed from ConceptY**HS**.O**SET** after it is been confirmed belonging to ConceptX**HS**; (c) Each of the target attributes as specified in AttributesForTailor**SET** is removed from ConceptY**HS**.A**SET**; (d) The internal relation set ConceptY**HS**.RC**SET** is updated by removing the relations between the tailored object and all remaining attributes as well as those between all remaining objects and the tailored attributes; and (e) The input and output relation sets of ConceptY**HS**, i.e., ConceptY**HS**.RI**SET** and ConceptY**HS**.RO**SET**, are updated by the newly established relations (ConceptX**HS**, ConceptY**HS**) and (ConceptY**HS**, ConceptX**HS**), respectively. As a reflexive consequence, the input and output relation sets of ConceptX**HS**, i.e., ConceptX**HS**.RI**SET** and ConceptX**HS**.RO**SET**, arc also extended by (ConceptY**HS**, ConceptX**HS**) and (ConceptX**HS**, ConceptY**HS**), respectively, which represent the newly created associations between the given concept and its tailored counterpart in the dynamic concept network.

Figure 10. The semantics of concept tailor operation

ConceptTailor**PC**(<**I::** ConceptX**HS**, ObjectsForTailor**SET**, AttributesForTailor**SET**>;

<**O::** ConceptY**HS**>; <**UDM::** Concept**HS**, OAR**HS**>) ≜

{ // *CA definition:* $X \overset{\rightarrow}{\Rightarrow} Y$

→ ConceptY**HS** := ConceptX**HS**   // Form ConceptY

$$\overset{\#(\text{ObjectsForTailor\textbf{SET}})}{\underset{i\textbf{N}=1}{R}} \quad (\; \blacklozenge \; \text{ObjectsForTailor(i\textbf{N})\textbf{SET}} \in \text{ConceptX\textbf{HS}.O\textbf{SET}}$$

→ ConceptY**HS**.O**SET** := ConceptY**HS**.O**SET** –
ObjectsForTailor(i**N**)**SET**

)

$$\overset{\#(\text{AtributesForTailor\textbf{SET}})}{\underset{i\textbf{N}=1}{R}} \quad (\; \blacklozenge \; \text{AttributesForTailor(i\textbf{N})\textbf{SET}} \in \text{ConceptX\textbf{HS}.A\textbf{SET}}$$

→ ConceptY**HS**.A**SET** := ConceptY**HS**.A**SET** –
AttributesForTailor(i**N**)**SET**

)

→ ConceptY**HS**.RC**SET** := ConceptY**HS**.RC**SET** –
{(ObjectForTailor**SET** × ConceptY**HS**.A**SET**),
(ConceptY**HS**.O**SET** × AttributesForTailor**SET**)}

→ ConceptY**HS**.RI**SET** := ConceptY**HS**.RI**SET** ∪ (ConceptX**HS**, ConceptY**HS**)

→ ConceptY**HS**.RO**SET** := ConceptY**HS**.RO**SET** ∪ (ConceptY**HS**, ConceptX**HS**)

// *Update ConceptX*

→ ConceptX**HS**.RI**SET** := ConceptX**HS**.RI**SET** ∪ (ConceptY**HS**, Concept**NS**)

→ ConceptX**HS**.RO**SET** := ConceptX**HS**.RO**SET** ∪ (ConceptX**HS**, ConceptY**HS**)

}

3.2.3 Concept Extension

Definition 18: The *concept extension* operation, $c \overset{+}{\Rightarrow} c_1$, is a special inheritance that derives a new concept c_1 from an existing concept c where its objects and attributes are extended by specific sets of objects O' and attributes A', respectively, as given in Equation 6.3 in Table 1.

The semantics of concept extension is illustrated in Figure 11 by the process ConceptExtension**PC**. ConceptExtension**PC** generates an extended concept ConceptY**HS** based on the given concept Con-

ceptX**HS** as follows: (a) ConceptY**HS** reproduces the same structure and contents of ConceptX**HS**; (b) Each of the target objects as specified in Ob-jectForExtension**SET** is iteratively added in ConceptY**HS**.O**SET** after it is confirmed as a new object for ConceptY**HS**; (c) Each of the target attributes as specified in AttributesForExtension**SET** is added in ConceptY**HS**.A**SET**; (d) The internal relation set ConceptY**HS**.RC**SET** is updated by including the newly created relations between the extended objects and attributes, the extended objects and all remaining attributes, and the remaining objects and the extended attributes; and (e) The input and output relation sets of ConceptY**HS**, i.e., ConceptY**HS**.RI**SET** and ConceptY**HS**.RO**SET**, are updated

Figure 11. The semantics of concept extension operation

$$
\begin{aligned}
&\text{ConceptExtension}\mathbf{PC}(<\mathbf{I}:: \text{ConceptX}\mathbf{HS}, \text{ObjectForExtension}\mathbf{SET}, \\
&\qquad\qquad\qquad \text{AttributesForExtension}\mathbf{SET}>; <\mathbf{O}:: \text{ConceptY}\mathbf{HS}>; \\
&\qquad\qquad\qquad <\mathbf{UDM}:: \text{Concept}\mathbf{HS}, \text{OAR}\mathbf{HS}>) \triangleq
\end{aligned}
$$

{ // *CA definition:* $X \overset{+}{\Rightarrow} Y$

$\rightarrow$ ConceptY**HS** := ConceptX**HS**   // Form ConceptY

$\qquad$ #(ObjectsForExtension**SET**)

$\rightarrow \qquad \underset{iN=1}{R} \qquad (\blacklozenge \text{ObjectsForTailor(i}\mathbf{N})\mathbf{SET} \notin \text{ConceptX}\mathbf{HS}.\mathbf{O}\mathbf{SET}$

$\qquad\qquad\qquad\qquad \rightarrow \text{ConceptY}\mathbf{HS}.\mathbf{O}\mathbf{SET} := \text{ConceptY}\mathbf{HS}.\mathbf{O}\mathbf{SET} \cup$
$\qquad\qquad\qquad\qquad\qquad \text{ObjectsForExtension(i}\mathbf{N})\mathbf{SET}$

$\qquad\qquad\qquad)$

$\qquad$ #(AtributesForExtension**SET**)

$\rightarrow \qquad \underset{iN=1}{R} \qquad (\blacklozenge \text{AttributesForTailor(i}\mathbf{N})\mathbf{SET} \notin \text{ConceptX}\mathbf{HS}.\mathbf{A}\mathbf{SET}$

$\qquad\qquad\qquad\qquad \rightarrow \text{ConceptY}\mathbf{HS}.\mathbf{A}\mathbf{SET} := \text{ConceptY}\mathbf{HS}.\mathbf{A}\mathbf{SET} \cup$
$\qquad\qquad\qquad\qquad\qquad \text{AttributesForExtension(i}\mathbf{N})\mathbf{SET}$

$\qquad\qquad\qquad)$

$\rightarrow \text{ConceptY}\mathbf{HS}.\mathbf{RC}\mathbf{SET} := \text{ConceptY}\mathbf{HS}.\mathbf{RC}\mathbf{SET} \cup$
$\qquad\qquad\qquad \{(\text{ObjectForExtension}\mathbf{SET} \times \text{AttributeForExtension}\mathbf{SET}) \cup$
$\qquad\qquad\qquad (\text{ObjectForExtension}\mathbf{SET} \times \text{ConceptY}\mathbf{ST}.\mathbf{A}\mathbf{SET}) \cup$
$\qquad\qquad\qquad (\text{ConceptY}\mathbf{HS}.\mathbf{O}\mathbf{SET} \times \text{AttributesForTailor}\mathbf{SET})\}$

$\rightarrow \text{ConceptY}\mathbf{HS}.\mathbf{RI}\mathbf{SET} := \text{ConceptY}\mathbf{HS}.\mathbf{RI}\mathbf{SET} \cup (\text{ConceptX}\mathbf{HS}, \text{ConceptY}\mathbf{HS})$

$\rightarrow \text{ConceptY}\mathbf{HS}.\mathbf{RO}\mathbf{SET} := \text{ConceptY}\mathbf{HS}.\mathbf{RO}\mathbf{SET} \cup (\text{ConceptY}\mathbf{HS}, \text{ConceptX}\mathbf{HS})$

// *Update ConceptX*

$\rightarrow \text{ConceptX}\mathbf{HS}.\mathbf{RI}\mathbf{SET} := \text{ConceptX}\mathbf{HS}.\mathbf{RI}\mathbf{SET} \cup (\text{ConceptY}\mathbf{HS}, \text{ConceptX}\mathbf{HS})$

$\rightarrow \text{ConceptX}\mathbf{HS}.\mathbf{RO}\mathbf{SET} := \text{ConceptX}\mathbf{HS}.\mathbf{RO}\mathbf{SET} \cup (\text{ConceptX}\mathbf{HS}, \text{ConceptY}\mathbf{HS})$

}

by the newly established relations (ConceptX**HS**, ConceptY**HS**) and (ConceptY**HS**, ConceptX**HS**), respectively. As a reflexive consequence, the input and output relation sets of ConceptX**HS**, i.e., ConceptX**HS**.RI**SET** and ConceptX**HS**.RO**SET**, are also extended by (ConceptY**HS**, ConceptX**HS**) and (ConceptX**HS**, ConceptY**HS**), respectively, which represent the newly created associations between the given concept and its extended counterpart in the dynamic concept network.

3.2.4 Concept Substitute

Definition 19: The *concept substitution* operation, $c \overset{\sim}{\Rightarrow} c_1$, is a special inheritance that derives a new concept c_1 from an existing concept c where its objects and attributes are extended by specific sets of objects O' and attributes A', respectively, as given in Equation 6.4 in Table 1.

The semantics of concept substitute is illustrated in Figure 12 by the process ConceptSubstitute**PC**. ConceptSubstitute**PC** generates a substituted concept ConceptY**HS** based on the given concept ConceptX**HS** as follows: a) ConceptY**HS** reproduces the same structure and contents of ConceptX**HS**; b) If the numbers of the target and substitute objects and attributes meet, respectively, the target objects and attributes are removed by calling the process ConceptTailor**PC**; c) The substitute objects and attributes are added in ConceptY**HS** by calling the process ConceptEx-

Figure 12. The semantics of concept substitution operation

```
ConceptSubstitutePC(<I:: ConceptXHS, (TargetObjectsSET, SubstituteObjectsSET),
                     (TargetAttributesSET, SubstituteAttributesSET)>;
                     <O:: ConceptYHS>; <UDM:: ConceptHS, OARHS>) ≜
{ // CA definition: X ⇒ Y
  → ( ◆ #(TargetObjectsSET) = #(SubstituteObjectsSET ) ∧
      #(TargetAttributesSET) = #(SubstituteAttributesSET)
      → ObjectForTailorSET := TargetObjectsSET          // Remove target O/A
      → AttributesForTailorSET := TargetAttributesSET
      ↦ ConceptTailorPC(<I:: ConceptXHS, ObjectForTailorSET,
                        AttributesForTailorSET>; <O:: ConceptZHS>;
                        <UDM:: ConceptHS, OARHS>)
      → ObjectForExtensionSET := SubstituteObjectsSET    // Add substitute O/A
      → AttributesForExtensionSET := SubstituteAttributesSET
      ↦ ConceptExtensionPC(<I:: ConceptZHS, ObjectForExtensionSET,
                           AttributesForExtensionSET>; <O:: ConceptYHS>;
                           <UDM:: ConceptHS, OARHS>)
      // Update ConceptY
      → ConceptYHS.RISET := ConceptYHS.RISET − (ConceptZHS, ConceptYHS)
      → ConceptYHS.RISET := ConceptYHS.RISET ∪ (ConceptXHS, ConceptYHS)
      → ConceptYHS.ROSET := ConceptYHS.ROSET − (ConceptYHS, ConceptZHS)
      → ConceptYHS.ROSET := ConceptXHS.ROSET ∪ (ConceptYHS, ConceptXHS)
      // Update ConceptX
      → ConceptXHS.RISET := ConceptXHS.RISET − (ConceptYHS, ConceptZHS)
      → ConceptXHS.RISET := ConceptXHS.RISET ∪ (ConceptYHS, ConceptXHS)
      → ConceptXHS.ROSET := ConceptXHS.ROSET − (ConceptZHS, ConceptYHS)
      → ConceptXHS.ROSET := ConceptXHS.ROSET ∪ (ConceptXHS, ConceptYHS)
  | ◆~
      → !(Substitute pairs are not equal.)
  )
}
```

tensionPC; and d) The input and output relation sets of ConceptYHS, i.e., ConceptYHS.RISET and ConceptYHS.ROSET, are updated by the newly established relations (ConceptXHS, ConceptYHS) and (ConceptYHS, ConceptXHS), respectively. At the same time, the intermediate relations, (ConceptZHS, ConceptYHS) and (ConceptYHS, ConceptZHS), are removed from ConceptYHS. RISET and ConceptYHS.ROSET, respectively. As a reflexive consequence, the input and output relation sets of ConceptXHS, i.e., ConceptXST. RISET and ConceptXHS.ROSET, are also updated by the newly established relations (ConceptYHS, ConceptXHS) and (ConceptXHS, ConceptYHS),

respectively. At the same time, the intermediate relations, (ConceptYHS, ConceptZHS) and (ConceptZHS, ConceptYHS), are removed from ConceptXHS.RISET and ConceptXHS.ROSET, respectively.

3.2.5 Concept Composition

Definition 20: The *concept composition* operation, $c(O, A, R^c, R^i, R^o) \uplus \overset{n}{\underset{i=1}{R}} c_i$, is a generation of a superconcept c by integrating a set of subconcepts c_i, as given in Equation 6.5 in Table 1.

Figure 13. The semantics of concept composition operation

```
ConceptCompositionPC(<I:: ConceptXHS, ConceptYHS>; <O:: ConceptZHS>;
                     <UDM:: ConceptHS, OARHS>) ≙
{ // CA definition: ConceptZHS = ConceptXHS ⊎ ConceptYHS
   // Form ConceptZHS
   → ConceptZHS.OSET := ConceptXHS.OSET ∪ ConceptYHS.OSET
   → ConceptZHS.ASET := ConceptXHS.ASET ∪ ConceptYHS.ASET
   → ConceptZHS.RCSET := ConceptXHS.RCSET ∪ ConceptYHS.RCSET ∪
                         (ConceptXHS.OSET × ConceptYHS.ASET) ∪
                         (ConceptYHS.OSET × ConceptXHS.ASET)
   → ConceptZHS.RISET := ConceptXHS.RISET ∪ ConceptYHS.RISET ∪
                         (ConceptXHS, ConceptZHS) ∪
                         (ConceptYHS, ConceptZHS)
   → ConceptZHS.ROSET := ConceptXHS.ROSET ∪ ConceptYHS.ROSET ∪
                         (ConceptZHS, ConceptXHS) ∪
                         (ConceptZHS, ConceptYHS)
   // Update ConceptX
   → ConceptXHS.RISET := ConceptXHS.RISET ∪ (ConceptZHS, ConceptXHS)
   → ConceptXHS.ROSET := ConceptXHS.ROSET ∪ (ConceptXHS, ConceptZHS)
   // Update ConceptY
   → ConceptYHS.RISET := ConceptYHS.RISET ∪ (ConceptZHS, ConceptYHS)
   → ConceptYHS.ROSET := ConceptYHS.ROSET ∪ (ConceptYHS, ConceptZHS)
}
```

The semantics of concept composition is illustrated in Figure 13 by the process ConceptComposition**PC**. ConceptComposition**PC** integrates both given subconcepts ConceptX**HS** and ConceptY**HS** into a superconcept as follows: a) The object set ConceptZ**HS**.O**SET** is formed by a conjunction of the counterparts of ConceptX**HS** and ConceptY**HS**; b) The attribute set ConceptZ**HS**.A**SET** is also a conjunction of the counterparts of ConceptX**HS** and ConceptY**HS**; c) The internal relation set ConceptZ**HS**.RC**SET** is formed by the union of both ConceptX**HS**.RC**SET** and ConceptY**HS**.RC**SET**, as well as the Cartesian products (ConceptX**HS**.O**SET** × ConceptY**HS**. A**SET**) and (ConceptY**HS**.O**SET** × ConceptX**HS**. A**SET**); d) The input relation set of the concept, ConceptZ**HS**.RI**SET**, is generated by the union of relations ConceptX**HS**.RI**SET**, ConceptY**HS**.RI**SET**, and both pairs of (ConceptX**HS**, ConceptZ**HS**) and

(ConceptY**HS**, ConceptZ**HS**); and e) The output relation set of the concept, ConceptZ**HS**.RO**SET**, is generated by the union of relations ConceptX**HS**. RO**SET**, ConceptY**HS**.RO**SET**, and both pairs of (ConceptZ**HS**, ConceptX**HS**) and (ConceptZ**HS**, ConceptY**HS**). As a reflexive consequence, the input and output relation sets of ConceptX**HS**, i.e., ConceptX**HS**.RI**SET** and ConceptX**HS**.RO**SET**, are also updated by the newly established relations (ConceptZ**HS**, ConceptX**HS**) and (ConceptX**HS**, ConceptZ**HS**), respectively. So do ConceptY**HS**. RI**SET** and ConceptY**HS**.RO**SET**.

3.2.6 Concept Decomposition

Concept decomposition is an inverse operation of concept composition.

Figure 14. The semantics of concept decomposition operation

```
ConceptDecompositionPC(<I:: ConceptZHS, OxSET ∪ OySET = OzSET>; <O:: ConceptXHS, ConceptYST>;
                    <UDM:: ConceptHS, OARST>) ≜
{ // CA definition: ConceptZHS ⫛ {ConceptXHS, ConceptYHS}
  → ConceptXHS.OSET := OxSET                                                   // Form SubConceptXST
  → ConceptXHS.RCSET := ∅
       #(ConceptZHS.OSET)  #(ConceptXHS.OSET)
  →        R                  R           (◆ ConceptZHS.O(iN)SET = ConceptXHS.O(jN)SET
         iN=1               jN=1
                                  #(ConceptZHS.ASET)
                              →        R          (◆ (ConceptZHS.O(iN)SET, ConceptZHS.A(kN)SET) ∈ ConceptZHS.RCSET
                                     kN=1
                                            → ConceptXHS.ASET := ConceptXHS.ASET ∪ ConceptZHS.A(kN)SET
                                            → ConceptXHS.RCSET := ConceptXHS.RCSET ∪
                                              (ConceptXHS.O(jN)SET, ConceptZHS.A(kN)SET)
                                          )
                       )
  → ConceptXHS.RISET := (ConceptZHS, ConceptXST)
  → ConceptXHS.ROSET := (ConceptXHS, ConceptZST)
  → ConceptYHS.OSET := OySET                                                   // Form SubConceptYHS
  → ConceptYHS.RCSET := ∅
       #(ConceptZHS.OSET)  #(ConceptYHS.OSET)
  →        R                  R           (◆ ConceptZHS.O(iN)SET = ConceptYHS.O(jN)SET
         iN=1               jN=1
                                  #(ConceptZHS.ASET)
                              →        R          (◆ (ConceptZHS.O(iN)SET, ConceptZHS.A(kN)SET) ∈ ConceptZHS.RCSET
                                     kN=1
                                            → ConceptYHS.ASET := ConceptYHS.ASET ∪ ConceptZHS.A(kN)SET
                                            → ConceptYHS.RCSET := ConceptYHS.RCSET ∪
                                              (ConceptYHS.O(jN)SET, ConceptZHS.A(kN)SET)
                                          )
                       )
  → ConceptYHS.RISET := (ConceptZHS, ConceptYHS)
  → ConceptYHS.ROSET := (ConceptYHS, ConceptZHS)
  // Update ConceptZHS
  → ConceptZHS.RISET := ConceptZHS.RISET ∪ (Concept(XN)HS, Concept(ZN)HS) ∪ (Concept(YN)HS, Concept(ZN)HS)
  → ConceptZHS.ROSET := ConceptZHS.ROSET ∪ (Concept(ZN)HS, Concept(XN)HS) ∪ (Concept(ZN)HS, Concept(YN)HS)
}
```

Definition 21: The *concept decomposition* operation, $c(O, A, R^c, R^i, R^o) \pitchfork \underset{i=1}{\overset{n}{R}} c_i$, is a partition of a superconcept c into a set of subconcepts c_i, as given in Equation 6.6 in Table 1.

The semantics of concept decomposition is illustrated in Figure 14 by the process ConceptDecomposition**PC**. Conceptdecomposition**PC** separates the superconcept ConceptZ**HS** according to the given partitions O_x**SET** and O_y**SET** as follows: (a) The object set of ConceptZ**HS**.O**SET** is decomposed into two sets ConceptX**HS**.O**SET**:= O_x**SET** and ConceptY**HS**.O**SET**:= O_y**SET**; (b) Each of the attributes related to any object in ConceptZ**HS**.O(i**N**)**SET** is included in ConceptX**HS**.A**SET**; (c) Each of the internal relations associated with (ConceptX**HS**.O**SET**, ConceptZ**HS**.A**SET**) in ConceptZ**HS**.RC**SET** is added to ConceptX**HS**.RC**SET**; (d) The input relation set ConceptX**HS**.RI**SET** is generated by the relation (ConceptZ**HS**, ConceptX**HS**); (e) The output relation set ConceptX**HS**.RO**SET** is generated by the relation (ConceptX**HS**, ConceptZ**HS**); and f) So does ConceptY**HS** for deriving the second partition. As a reflexive consequence, the input and output relation sets of ConceptZ**HS**, i.e., ConceptZ**HS**.RI**SET** and ConceptZ**HS**.RO**SET**, are also updated by the newly established relations (ConceptX**HS**, ConceptZ**HS**) ∪ (ConceptY**HS**, ConceptZ**HS**), as well as (ConceptZ**HS**, ConceptX**HS**) ∪ (ConceptZ**HS**, ConceptY**HS**), respectively.

3.2.7 Concept Aggregation

Definition 22: The *concept aggregation* operation,

$$c(O, A, R^c, R^i, R^o) \Leftarrow \mathop{R}_{i=1}^{n} c_i \text{, is a creation}$$

of a superconcept c with a set of subconcepts c_i by eliciting their shared attributes and representative objects as given in Equation 6.7 in Table 1.

The semantics of concept aggregation is illustrated in Figure 15 by the process ConceptAggregation**PC**. ConceptAggregation**PC** creates the superconcept, ConceptZ**HS**, using the materials of both subconcepts ConceptX**HS** and ConceptY**HS** as follows: (a) The object set of the concept, ConceptZ**HS**.O**SET**, is aggregated by a conjunction of the counterparts of ConceptX**HS** and ConceptY**HS**; (b) The attribute set of the concept, ConceptZ**HS**. A**SET**, is aggregated by a disjunction of the counterparts of ConceptX**HS** and ConceptY**HS**; (c) The internal relation set of the concept, ConceptZ**HS**. RC**SET**, is formed by a Cartesian product Con-

ceptZ**HS**.O**SET** × ConceptZ**HS**.A**SET**; (d) The input relation set of the concept, ConceptZ**HS**.RI**SET**, is generated by the relations between both pairs of (ConceptX**HS**, ConceptZ**HS**) and (ConceptY**HS**, ConceptZ**HS**); and (e) The output relation set of the concept, ConceptZ**HS**.RO**SET**, is generated by the relations between both pairs of (ConceptZ**HS**, ConceptX**HS**) and (ConceptZ**HS**, ConceptY**HS**). As a reflexive consequence, the material subconcepts must be updated, that is: (a) The input and output relations of ConceptX**HS**, are extended by the pairs (ConceptZ**HS**, ConceptX**HS**) and (ConceptX**HS**, ConceptZ**HS**), respectively; and (b) So does ConceptY**HS**.

The differences between concept composition and aggregation are that the former is an integration of multiple subconcepts in order to form a more comprehensive superconcept with conjoined attributes contributed by all subconcepts; while the latter is an elicitation of the materials from multiple subconcepts in order to form a more abstract superconcept with only commonly shared attributes by all objects.

Figure 15. The semantics of concept aggregation operation

ConceptAggregation**PC**(<I:: ConceptX**HS**, ConceptY**HS**>; <O:: ConceptZ**HS**>;
<UDM:: Concept**HS**, OAR**HS**>) ≜
{ // *CA definition:* ConceptZ**HS** ⇐ {ConceptX**HS**, ConceptY**HS**}
  // *Form ConceptZ**HS***
  → ConceptZ**HS**.O**SET** := ConceptX**HS**.O**SET** ∪ ConceptY**HS**.O**SET**
  → ConceptZ**HS**.A**SET** := ConceptX**ST**.A**SET** ∩ ConceptY**HS**.A**SET**
  → ConceptZ**HS**.RC**SET** := ConceptZ**ST**.O**SET** × ConceptZ**HS**.A**SET**
  → ConceptZ**HS**.RI**SET** := (ConceptX**HS**, ConceptZ**HS**) ∪
                (ConceptY**HS**, ConceptZ**HS**)
  → ConceptZ**HS**.RO**SET** := (ConceptZ**HS**, ConceptX**HS**) ∪
                (ConceptZ**HS**, ConceptY**HS**)
  // *Update ConceptX**HS** and ConceptY**HS***
  → ConceptX**HS**.RI**SET** := ConceptX**HS**.RI**SET** ∪ (ConceptZ**HS**, ConceptX**HS**)
  → ConceptX**HS**.RO**SET** := ConceptX**HS**.RO**SET** ∪ (ConceptX**HS**, ConceptZ**HS**)
  → ConceptY**HS**.RI**SET** := ConceptY**HS**.RI**SET** ∪ (ConceptZ**HS**, ConceptY**HS**)
  → ConceptY**HS**.RO**SET** := ConceptY**HS**.RO**SET** ∪ (ConceptY**HS**, ConceptZ**HS**)
}

3.2.8 Concept Specification

Concept specification is an inverse operation of concept aggregation.

Definition 23: The *concept specification* operation, $c(O, A, R^c, R^i, R^o) \vdash \underset{i=1}{\overset{n}{R}} c_i$, is a deductive refinement of a set of subconcepts c_i based on each of the individual objects of c with an extended intension of more specific attributes as given in Equation 6.8 in Table 1.

The semantics of concept specification is illustrated in Figure 16 by the process ConceptSpecification**PC**. ConceptSpecification**PC** iteratively derives a set of more precise and specific subconcepts from the given superconcept ConceptX**HS** as follows: (a) The object set of the specified concept, ConceptX$_i$**HS**.O**SET**, is a single object elicited from the ith object of ConceptX**HS**; (b) The attribute set of the specified concept, ConceptX$_i$**HS**.A**SET**, is selected from those of ConceptX**HS**.A**SET** in which who is associated to ConceptX**HS**.O(i**N**)**SET**; (c) The internal relation set of the specified concept, ConceptX$_i$**HS**.RC**SET**, is formed by a Cartesian product ConceptX$_i$**HS**.

Figure 16. The semantics of concept specification operation

O**SET** × ConceptX$_i$**HS**.A**SET**; d) The input relation set of the specified concept, ConceptX$_i$**HS**.RI**SET**, is generated by the relation (ConceptX**HS**, ConceptX$_i$**HS**); and e) The output relation set of the specified concept, ConceptX$_i$**HS**.RO**SET**, is generated by the relations (ConceptX$_i$**HS**, ConceptX**HS**). As a reflexive consequence, the superconcept must be updated as a result of the specification operation, that is: (a) The input relation set of ConceptX**HS** is extended by the new relation (ConceptX$_i$**HS**, ConceptX**HS**); and (b) The output relation set of ConceptX**HS** is extended by the new relation (ConceptX**HS**, ConceptX$_i$**HS**).

It is noteworthy that concept decomposition resolves a superconcept into n partitioned subconcepts where n is determined by the number of parts in the expected partitions; while concept specification resolves a superconcept into a set of m individual subconcepts where m is determined by the number of objects in the superconcept. It is obvious that $n \leq m$.

3.2.9 Concept Instantiation

Definition 24: The *concept instantiation* operation, $c(O, A, R^c, R^i, R^o) \mapsto o,\ o \in O$, is a derivation of a specific object o from the given concept c as given in Equation 6.9 in Table 1.

The semantics of concept instantiation is illustrated in Figure 17 by the process ConceptInstan-

Figure 17. The semantics of concept instantiation operation

tiation$\mathbf{PC}$. ConceptInstantiation$\mathbf{PC}$ derives a set of objects associated to ConceptX$\mathbf{HS}$ as follows: (a) Each of the derived object, ObjectX$_i\mathbf{HS}$, is a reproduction of the parent concept and its set of objects is a single object formed by ConceptX$\mathbf{HS}$.O(i$\mathbf{N}$)$\mathbf{SET}$; (b) The attribute set of the object, ObjectX$_i\mathbf{HS}$.A$\mathbf{SET}$, is elicited from all associated attributes to the specific object in ConceptX$\mathbf{HS}$; (c) The internal relation set of the object, ObjectX$_i\mathbf{HS}$.RC$\mathbf{SET}$, is generated by all relations between the object and associated attributes; (d) The input relation set of the object, ObjectX$_i\mathbf{HS}$.RI$\mathbf{SET}$, is generated by the relation (ConceptX$\mathbf{HS}$, ObjectX$_i\mathbf{HS}$); and (e) The output relation set of the object, ObjectX$_i\mathbf{HS}$.RO$\mathbf{SET}$, is generated by the relation (ObjectX$_i\mathbf{HS}$, ConceptX$\mathbf{HS}$). As a reflexive consequence, the input and output relations of ConceptX$\mathbf{HS}$, i.e., ConceptX$\mathbf{HS}$.RI$\mathbf{SET}$ and ConceptX$\mathbf{HS}$.RO$\mathbf{SET}$, are extended by the pairs (ObjectX$_i\mathbf{HS}$, ConceptX$\mathbf{HS}$) and (ConceptX$\mathbf{HS}$, ObjectX$_i\mathbf{HS}$), respectively.

4. APPLICATIONS OF CONCEPT ALGEBRA IN KNOWLEDGE REPRESENTATION AND MACHINE LEARNING

The general mathematical model of concepts and the relational and compositional operators of CA described in preceding sections enable the formal representation of both abstract and concrete concepts as well as knowledge systems and their manipulations. On the basis of CA as well as its operational semantics and related RTPA algorithms, this section presents some examples and case studies on CA applications.

Example 1: Three concrete concepts, Pen$\mathbf{C}$, Printer$\mathbf{C}$, and Computer$\mathbf{C}$, can be formally defined in CA as instances of the formal concept model according to Definition 7 as follows:

$$\text{Pen}\mathbf{C} \triangleq C_1\mathbf{HS}(A_1, O_1, R^c_1, R^i_1, R^o_1)$$
$$= \begin{cases} A_1 = \{a_1,a_2,a_3,a_4,a_5,a_6\} = \{\text{a\_writing\_tool, using\_ink,} \\ \qquad \text{having\_a\_nib, office, paper, ink\_container}\} \\ O_1 = \{o_1,o_2,o_3\} = \{\text{ballpoint, fountain, brush}\} \\ R^c_1 = \{(o_1,a_1),(o_1,a_2),(o_1,a_3),(o_1,a_4),(o_1,a_5),(o_1,a_6)\} \cup \\ \qquad \{(o_2,a_1),(o_2,a_2),(o_2,a_3),(o_2,a_4),(o_2,a_5),(o_2,a_6)\} \cup \\ \qquad \{(o_3,a_1),(o_3,a_2),(o_3,a_3),(o_3,a_4),(o_2,a_5)\} \\ R^i_1 = \mathfrak{K} \times C_1\mathbf{HS} \\ R^o_1 = C_1\mathbf{HS} \times \mathfrak{K} \end{cases}$$

$$(14)$$

$$\text{Printer}\mathbf{C} \triangleq C_2\mathbf{HS}(A_2, O_2, R^c_2, R^i_2, R^o_2)$$
$$= \begin{cases} A_2 = \{a_1,a_2,a_3,a_4,a_5,a_6\} = \{\text{a\_printing\_tool, office,} \\ \qquad \text{using\_ink, paper, ink\_container, toner\_cartridge}\} \\ O_2 = \{o_1,o_2\} = \{\text{ink\_jet\_printer, laser\_printer}\} \\ R^c_2 = \{(o_1,a_1),(o_1,a_2),(o_1,a_3),(o_1,a_4),(o_1,a_5),(o_1,a_6)\} \cup \\ \qquad \{(o_2,a_1),(o_2,a_2),(o_2,a_3),(o_2,a_4),(o_2,a_5),(o_2,a_6)\} \\ R^i_2 = \mathfrak{K} \times C_2\mathbf{HS} \\ R^o_2 = C_2\mathbf{HS} \times \mathfrak{K} \end{cases}$$

$$(15)$$

$$\text{Computer}\mathbf{C} \triangleq C_3\mathbf{HS}(A_3, O_3, R^c_3, R^i_3, R^o_3)$$
$$= \begin{cases} A_3 = \{a_1,a_2,a_3,a_4,a_5,a_6\} = \{\text{computing\_tool,} \\ \qquad \text{CPU, memory, I/O, instructions, data}\} \\ O_3 = \{o_1,o_2,o_3\} = \{\text{main\_frame, PC, laptop}\} \\ R^c_3 = O_3 \times A_3 \\ R^i_3 = \mathfrak{K} \times C_3\mathbf{HS} \\ R^o_3 = C_3\mathbf{HS} \times \mathfrak{K} \end{cases}$$

$$(16)$$

Example 2: A more abstract superconcept, Stationery$\mathbf{C}$, can be formally derived by concept *aggregation* ($\Leftarrow$) in CA based on the given concrete concepts Pen$\mathbf{C}$ and Printer$\mathbf{C}$ in Example 1 as follows:

$$\text{Stationery}\mathbf{C} \triangleq C_4\mathbf{HS}(A_4, O_4, R^c{}_4, R^i{}_4, R^o{}_4)$$
$$\Leftarrow \{\text{Pen}\mathbf{C}, \text{Printer}\mathbf{C}\}$$

$$= \begin{cases} A_4 = C_1\mathbf{HS}.A_1 \cap C_2\mathbf{HS}.A_2 = C_4\mathbf{HS}.\{a_1, a_2, a_3\} \\ \qquad = \{\text{office, paper, using\_ink}\} \\ O_4 = C_1\mathbf{HS}.O_1 \cup C_2\mathbf{HS}.O_2 = C_1\mathbf{HS}.\{o_1, o_2, o_3\} \\ \qquad\qquad\qquad\qquad\quad \cup C_2\mathbf{HS}.\{o_1, o_2\} \\ R^c{}_4 = O_4 \times A_4 \\ R^i{}_4 = \mathfrak{K} \times C_4\mathbf{HS} \\ R^o{}_4 = C_4\mathbf{HS} \times \mathfrak{K} \end{cases}$$

$$(17)$$

where the abstract concept may be further developed by multiple aggregations according to Defined 22 when more subconcepts of stationery are provided.

Example 3: A comprehensive superconcept, DeskComputerSys**C**, can be formally derived by concept *composition* (⊎) according to Definition 20 in CA based on the given subconcepts Computer**C** and Printer**C** as follows:

$$\text{DeskComputerSys}\mathbf{C} \triangleq C_5\mathbf{HS}(A_5, O_5, R^c{}_5, R^i{}_5, R^o{}_5)$$
$$= (\text{Computer}\mathbf{C} \Rightarrow \text{DeskComputer}\mathbf{C}) \uplus \text{Printer}\mathbf{C}$$

$$= \begin{cases} A_5 = C'_3\mathbf{HS}.A_3 \cup C_2\mathbf{HS}.A_2 = C_3\mathbf{HS}.\{a_1, a_2, a_3, a_4, a_5, a_6\} \\ \qquad\qquad\qquad\qquad\qquad \cup C_2\mathbf{HS}.\{a_1, a_2, a_3, a_4, a_5, a_6\} \\ O_5 = C'_3\mathbf{HS}.O_3 \cup C_2\mathbf{HS}.O_2 = C_3\mathbf{HS}.\{o_1, o_2, o_3\} \setminus o_1 \\ \qquad\qquad\qquad\qquad\qquad \cup C_2\mathbf{HS}.\{o_1, o_2\} \\ R^c{}_5 = O_5 \times A_5 \\ R^i{}_5 = \mathfrak{K} \times C_5\mathbf{HS} \\ R^o{}_5 = C_5\mathbf{HS} \times \mathfrak{K} \end{cases}$$

$$(18)$$

where DeskComputer**C** (C'$_3$**HS**) is first derived by tailoring Computer**C** (C$_3$**HS**), i.e., $C_3\mathbf{HS}.O \Rightarrow C'_3\mathbf{HS}.O = C_3\mathbf{HS}.O \setminus o_1$.

It is noteworthy that the CA operators for concept composition and aggregation are different from using the component materials of given subconcepts in the process of superconcept construc-tion. Concept composition assembles multiple subconcepts as a whole into the superconcept; while concept aggregation assembles selected elements of individual subconcepts into the super-concept. In other words, concept composition is a direct assembly of multiple concepts operating at the level of the whole subconcept as is; while concept aggregation is a selective assembly of tailored concepts operating at the level of partial elements of the given subconcepts.

A screenshot of the formal knowledge representation and learning system tool for concept manipulations powered by CA is shown in Figure 18. In Figure 18, the operation of concept composition in CA is autonomously carried out with two existing concepts of the machine. As a result, the superconcept, DeskComputerSys**C**, is created based on the existing subconcepts Computer**C** and Printer**C**. The above system implements the fundamental capability for knowledge representation, learning, and reasoning in cognitive computing on the basis of CA.

The case studies provided in this section demonstrate that CA is not only an expressive conceptual modeling methodology for knowledge representation, but also a rigorous computational methodology for machine learning. A wide range of applications of CA have been reported in real-world projects on cognitive learning (Wang, 2007d; Tian et al., in press), knowledge representation (Wang, 2009a, 2009b; Tian et al., 2009), cognitive robots (Wang, 2010a), formal ontological system modeling and semantic computing (Wang, 2010b), as well as web knowledge extraction and web text processing (Hu et al., 2010).

5. CONCLUSION

It has been recognized that concepts are the basic unit of human cognition that carries certain meanings in expression, thinking, reasoning, and system modeling. A formal concept model has been created as a new mathematical structure

Figure 18. A screenshot of concept composition automatically operated by the formal knowledge representation and learning system

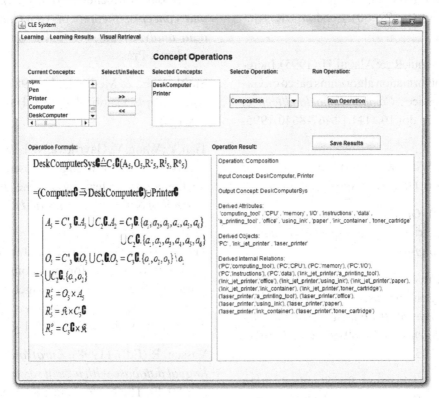

from the commonly shared properties of concepts in the real world and human cognitive processes. The operational semantics of Concept Algebra (CA) has been elaborated for formal knowledge representation and manipulations, which extends conventional ontology to more rigorous algebraic operations on formal concepts. This paper has demonstrated that the denotational mathematical operators of CA can be rigorously implemented by RTPA algorithms and processes, which enable computational simulations and implementations in cognitive computing. This capability has been demonstrated as not only necessary in autonomous machine learning, but also highly demanded in cognitive informatics, cognitive computing, knowledge engineering, and computational intelligence. A wide range of applications of CA have been reported in real-world projects on knowledge representation, formal ontological system modeling, web knowledge extraction, and cognitive learning systems.

ACKNOWLEDGMENT

The authors would like to acknowledge the Natural Science and Engineering Council of Canada (NSERC) for its partial support to this work. We would like to thank the anonymous reviewers for their valuable comments and suggestions.

REFERENCES

Bender, E. A. (2000). *Mathematical methods in artificial intelligence*. Los Alamitos, CA: IEEE Press.

Cocchiarella, N. (1996). Conceptual realism as a formal ontology. In Poli, R., & Simons, P. (Eds.), *Formal ontology* (pp. 27–60). London, UK: Kluwer Academic.

Codin, R., Missaoui, R., & Alaoui, H. (1995). Incremental concept formation algorithms based on Galois (concept) lattices. *Computational Intelligence, 11*(2), 246–267. doi:10.1111/j.1467-8640.1995.tb00031.x

Ganter, B., & Wille, R. (1999). *Formal concept analysis*. Berlin, Germany: Springer-Verlag.

GOLD. (2010). *General ontology for linguistic description*. Retrieved from http://www.linguistics-ontology.org/gold/

Hu, K., Wang, Y., & Tian, Y. (2010). A web knowledge discovery engine based on concept algebra. *International Journal of Cognitive Informatics and Natural Intelligence, 4*(1), 80–97. doi:10.4018/jcini.2010010105

International Organization for Standardization (ISO). (2011). *ISO Standard 15836*. Geneva, Switzerland: ISO.

Liu, H., & Singh, P. (2004). ConceptNet - A practical commonsense reasoning toolkit. *BT Technology Journal, 22*(4), 211–225. doi:10.1023/B:BTTJ.0000047600.45421.6d

Medin, D. L., & Shoben, E. J. (1988). Context and structure in conceptual combination. *Cognitive Psychology, 20*, 158–190. doi:10.1016/0010-0285(88)90018-7

Miller, G. A. (1995). WordNet: A lexical database for English. *Communications of the ACM, 38*(11), 39–41. doi:10.1145/219717.219748

Miller, G. A., Beckwith, R., Fellbaum, C. D., Gross, D., & Miller, K. (1990). WordNet: An online lexical database. *International Journal of Lexicograph, 3*(4), 235–244. doi:10.1093/ijl/3.4.235

Murphy, G. L. (1993). Theories and concept formation. In Mechelen, I. V. (Ed.), *Categories and concepts, theoretical views and inductive data analysis* (pp. 173–200). New York, NY: Academic Press.

Smith, E. E., & Medin, D. L. (1981). *Categories and concepts*. Cambridge, MA: Harvard University Press.

Tian, Y., Wang, Y., Gavrilova, M. L., & Ruhe, G. (in press). A formal knowledge representation system for the cognitive learning engine. *International Journal of Software Science and Computational Intelligence, 3*(4).

Tian, Y., Wang, Y., & Hu, K. (2009). A knowledge representation tool for autonomous machine learning based on concept algebra. *Transactions of Computational Science, 5*, 143–160. doi:10.1007/978-3-642-02097-1_8

Vossen, P. (Ed.). (1998). *EuroWordNet: A multilingual database with lexical semantic networks*. Dordrecht, The Netherlands: Kluwer Academic.

Wang, Y. (2002). The real-time process algebra (RTPA). *International Journal of Annals of Software Engineering, 14*, 235–274. doi:10.1023/A:1020561826073

Wang, Y. (2007a). *Software engineering foundations: A software science perspective* (CRC Series in Software Engineering, Vol. 2). Boca Raton, FL: CRC Press.

Wang, Y. (2007b). The theoretical framework of cognitive informatics. *International Journal of Cognitive Informatics and Natural Intelligence, 1*(1), 1–27. doi:10.4018/jcini.2007010101

Wang, Y. (2007c). The OAR model of neural informatics for internal knowledge representation in the brain. *International Journal of Cognitive Informatics and Natural Intelligence, 1*(3), 64–75. doi:10.4018/jcini.2007070105

Wang, Y. (2007d, August). The theoretical framework and cognitive process of learning. In *Proceedings of the 6th IEEE International Conference on Cognitive Informatics* (pp. 470-479).

Wang, Y. (2008a). RTPA: A denotational mathematics for manipulating intelligent and computational behaviors. *International Journal of Cognitive Informatics and Natural Intelligence*, *2*(2), 44–62. doi:10.4018/jcini.2008040103

Wang, Y. (2008b). Deductive semantics of RTPA. *International Journal of Cognitive Informatics and Natural Intelligence*, *2*(2), 95–121. doi:10.4018/jcini.2008040106

Wang, Y. (2008c). On concept algebra: A denotational mathematical structure for knowledge and software modeling. *International Journal of Cognitive Informatics and Natural Intelligence*, *2*(2), 1–19. doi:10.4018/jcini.2008040101

Wang, Y. (2008d). On contemporary denotational mathematics for computational intelligence. *Transactions of Computational Science*, *2*, 6–29. doi:10.1007/978-3-540-87563-5_2

Wang, Y. (2008e). On system algebra: A denotational mathematical structure for abstract system modeling. *International Journal of Cognitive Informatics and Natural Intelligence*, *2*(2), 20–42. doi:10.4018/jcini.2008040102

Wang, Y. (2009a). Toward a formal knowledge system theory and its cognitive informatics foundations. *Transactions of Computational Science*, *5*, 1–19. doi:10.1007/978-3-642-02097-1_1

Wang, Y. (2009b). A formal syntax of natural languages and the deductive grammar. *Fundamenta Informaticae*, *90*(4), 353–368.

Wang, Y. (2009c). On cognitive computing. *International Journal of Software Science and Computational Intelligence*, *1*(3), 1–15. doi:10.4018/jssci.2009070101

Wang, Y. (2009d). On visual semantic algebra (VSA): A denotational mathematical structure for modeling and manipulating visual objects and patterns. *International Journal of Software Science and Computational Intelligence*, *1*(4), 1–16. doi:10.4018/jssci.2009062501

Wang, Y. (2010a). Cognitive robots: A reference model towards intelligent authentication. *IEEE Robotics and Automation*, *17*(4), 54–62. doi:10.1109/MRA.2010.938842

Wang, Y. (2010b). On concept algebra for computing with words (CWW). *International Journal of Semantic Computing*, *4*(3), 331–356. doi:10.1142/S1793351X10001061

Wang, Y. (2011a). On cognitive models of causal inferences and causation networks. *International Journal of Software Science and Computational Intelligence*, *3*(1), 50–60.

Wang, Y. (2011b, August 18-20). Keynote: On inference algebra: A formal means for machine reasoning and cognitive computing. In *Proceedings of the 10th IEEE International Conference on Cognitive Informatics and Cognitive Computing*, Banff, AB, Canada (pp. 4-6).

Wang, Y., Kinsner, W., & Zhang, D. (2009). Contemporary cybernetics and its faces of cognitive informatics and computational intelligence. [B]. *IEEE Transactions on Systems, Man, and Cybernetics*, *39*(4), 823–833. doi:10.1109/TSMCB.2009.2013721

Wang, Y., & Wang, Y. (2006). On cognitive informatics models of the brain. *IEEE Transactions on Systems, Man and Cybernetics. Part C, Applications and Reviews*, *36*(2), 203–207. doi:10.1109/TSMCC.2006.871151

Wang, Y., Zadeh, L. A., & Yao, Y. (2009). On the system algebra foundations for granular computing. *International Journal of Software Science and Computational Intelligence*, *1*(1), 64–86. doi:10.4018/jssci.2009010105

Wille, R. (1982). Restructuring lattice theory: An approach based on hierarchies of concepts. In Rival, I. (Ed.), *Ordered sets* (pp. 445–470). Dordrecht, The Netherlands: Reidel.

Wilson, R. A., & Keil, F. C. (Eds.). (2001). *The MIT encyclopedia of the cognitive sciences*. Cambridge, MA: MIT Press.

Zadeh, L. A. (1998). Some reflections on soft computing, granular computing and their roles in the conception, design and utilization of information/intelligent systems. *Soft Computing*, (2): 23–25. doi:10.1007/s005000050030

Zadeh, L. A. (1999). From computing with numbers to computing with words – from manipulation of measurements to manipulation of perception. *IEEE Transactions on Circuits and Systems I*, *45*(1), 105–119. doi:10.1109/81.739259

This work was previously published in the International Journal of Cognitive Informatics and Natural Intelligence, Volume 5, Issue 3, edited by Yingxu Wang, pp. 1-29, copyright 2011 by IGI Publishing (an imprint of IGI Global).

Chapter 14
Text Semantic Mining Model Based on the Algebra of Human Concept Learning

Jun Zhang
Shanghai University, China

Xiangfeng Luo
Shanghai University, China

Xiang He
Shanghai University, China

Chuanliang Cai
Shanghai University, China

ABSTRACT

Dealing with the large-scale text knowledge on the Web has become increasingly important with the development of the Web, yet it confronts with several challenges, one of which is to find out as much semantics as possible to represent text knowledge. As the text semantic mining process is also the knowledge representation process of text, this paper proposes a text knowledge representation model called text semantic mining model (TSMM) based on the algebra of human concept learning, which both carries rich semantics and is constructed automatically with a lower complexity. Herein, the algebra of human concept learning is introduced, which enables TSMM containing rich semantics. Then the formalization and the construction process of TSMM are discussed. Moreover, three types of reasoning rules based on TSMM are proposed. Lastly, experiments and the comparison with current text representation models show that the given model performs better than others.

DOI: 10.4018/978-1-4666-2476-4.ch014

INTRODUCTION

With the rapid growth of the Web, how to represent and organize the large-scale texts have drawn a lot of attentions. One of the most important works on text knowledge representation is to find out the semantics in texts. Plenty of scholars focus on many kinds of models that are used to represent text knowledge through various text analyzing methods. Such models are always expected to contain rich semantics, to obtain a robust reasoning ability and to be automatically constructed.

Currently, models referring to represent text knowledge can be mainly divided into four types. (1) Statistics models, which are generated by statistical methods. The typical ones include vector space model (VSM) (Salton & Wong, 1975) and latent semantic analysis (LSA) (Landauer & Foltz, 1998). VSM uses some words extracted from a text and their weights to represent the text semantics, but it doesn't take the relations between the words into account. Thus VSM is only able to express a little semantics in the text while much more semantics has been lost. On the contrary, the LSA model carries more semantics than the former one but its complexity is high because the construction of LSA is involved with the operation of singular value decomposition, whose complexity goes very high. (2) Cognition based models, whose basic idea is inspired by cognitive theories. Element fuzzy cognitive map (EFCM) (Luo & Xu, 2008) is one of the typical models. It obtains more semantics than VSM and a lower computation complexity than LSA. Meanwhile, it can be applied to large-scale text collections since it is constructed automatically. (3) Probability topic models, such as author-topic model (ATM) (Michal & Thomas, 2004), author-recipient-topic model (ART) (McCallum & Corrada-Emmanuel, 2004) and correlated topic models (CTM) (Blei & Lafferty, 2006). These models always need a lot of complex computations, which make probability topic models unsuitable to be used in large-scale text collections. (4) Ontology based models, which

are based on ontology languages and most of them are semi-automatically constructed. Ontology inference layer (OIL) (Horrocks & Fensel, 2000), web ontology language (OWL) (McGuinness & Harmelen, 2004) and simple html ontology extensions (SHOE) (Heflin & Hendler, 1999) are typical ontology based models. Since possessing a lot of semantics, ontology based models attracts plenty of researches on them. However, they can only be applied to special areas that contain a lot of human experiential knowledge, as the generation of ontology based models needs a mass of manual work. Thus, up to now, ontology based models still cannot be applied to automatically process large-scale text collections.

Consequently, according to the discussions above, we can see that some models are carrying abundant semantics but cannot be constructed automatically (e.g. OWL); some ones are both allowed to be automatically established and carrying a lot of semantics but still can't be applied to large-scale collections for their high complexities (e.g. CTM and ATM); some ones can be set up automatically with a lower complexity but carry little semantics (e.g. VSM). As a result, through the analysis of those models, we consider that a good text knowledge representation model should satisfy the two conditions listed below.

1. Contain rich text semantics;
2. Construct automatically with a lower complexity;

According to the two conditions, this paper proposes a text knowledge representation model called text semantic mining model (TSMM) based on the algebra of human concept learning. According to Cognitive Informatics (Wang, 2002, 2007a), a concept is defined as a cognitive unit to identify and/or model a real-world concrete entity and a perceived-world abstract object whereas the formal treatment of concepts and a new mathematical structure known are defined as Concept Algebra (Wang, 2006). Moreover, in consideration of the Object-Attribute-Relation model (Wang, 2003, 2007a, 2007b), a text can be regarded as a concept. Sentences in the text can be regarded as objects belonging to the concept and the keywords in sentences are the attributes of these objects (Fang & Luo, 2009). Furthermore, inspired by the algebra of human concept learning (Feldman, 2006), we propose TSMM model, which carries more semantics than statistics models and can be constructed automatically with a lower complexity. Some cases and comparisons have been presented to validate the performance of this model. TSMM takes an important part in plenty of areas, such as Machine Learning (Mirza & Sommers, 2010), e-Learning (Luo & Wei, 2010), Information Retrieval (Bansal & Garg, 2010), Semantic Link Network (Zhuge & Sun, 2010) and etc. Moreover, our on-going work, Association Link Network (Luo et al., in press), will also be improved by TSMM.

THE ALGEBRA OF HUMAN CONCEPT LEARNING

As Feldman (2006) said, any set of data exhibits many patterns, some of which will be simple or coarse, in the sense that they may be specified with reference to only a small number of features, while others may be complex and detailed, in the sense that they may only be described by reference to a larger number of features. But the key is how to capture combinations of component patterns, including simple components, complex components and those in between. This section introduces a description language called concept algebra consisting of a certain atomic set of rules based on the fundamental nature using a simple or expressively impoverished language to describe patterns. A simple example (Feldman, 2006), as shown in Figure 1, is given out to specify this language.

The "world" W in Figure 1 is composed of five amoeba-like objects. A simple language denoted as a list $\sum=\{\sigma_1, \sigma_2..., \sigma_d\}$ of abstract property tags is chosen to express the structure of this "world", which can be regarded as a concept. Herein, the features are assumed to be Boolean ones. According to the "world" in Figure 1, appropriate language might be:

$\sum=\{$ blob_shaped, shaded, has_nucleus, has_dotted_membrane, large$\}$,

which is abbreviated to $\sum=\{a, b, c, d, e\}$ under the assignment,

a = blob_shaped,
b = shaded,
c = has_nucleus,
d = has_dotted_membrane,
e = large.

Thus, the "world" W in Figure 1 can be represented as $W|_\Sigma=\{abc'd'e', ab'cde, ab'cde', abcde, ab'c'd'e'\}$, where a' means the opposite of a. However, such a representation is hard to be well understood by people. To deal with that, we introduce a more comprehensive description language in the following, whose basic idea is to pick out a certain atomic set of rules to be used as the building blocks of concepts or patterns. These building blocks together with the rules for com-

Figure 1. A "world" W containing amoeba-like objects (Feldman, 2006)

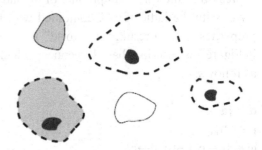

bining them constitute the description language, concept algebra (Feldman, 2006), which considers two simple kinds of rules defined as follows.

Definition 1: Implication

Implication is a very obvious kind of rule-implicit in the notion of "causality"-meaning when one property's value determines another's. This causality will be reflected proximally in a pattern among observed variables, namely a logical implication, $\sigma_1 \rightarrow \sigma_2$, which meaning that σ_1 is never true until σ_2 is also.

Definition 2: Affirmation

When one property holds consistently in all observed objects, i.e. a rule with the very simple form, σ, which is defined as affirmation.

For example, the property *blob_shaped* denoted as a in Figure 1 is one of affirmations in the "world" W.

Therefore, through the two kinds of rules defined above, the conceptual structure will be reduced to the following algebraic form:

$$\alpha \cdot [\omega \times \beta], \tag{1}$$

where α is the set of constant properties corresponding to affirmations, ω is the set of paired implications and β is the set of the rest properties. The elements in set β neither belong to the set of affirmations nor the set of implications but appear in $\sum$.

That is to say every possible "world" can be expressed as the Cartesian product of the lattice for ω with the lattice for β, conjoined with the properties α. As a result, the "world" W shown in Figure 1 is described based on concept algebra as follows.

$\alpha = \{a\}$,
$\beta = \{b\}$,
$\omega = \{e \rightarrow c, c \rightarrow d, d \rightarrow c\}$,
$W|_{\sum} \subseteq \alpha \cdot [\omega \times \beta]$.

From this description, much more information referring to this "world" will be learned by people, such as: objects are all *blob-shaped*; if one object has the nucleus then it must have the dotted membrane. Thus, concept algebra is consistent with human concept learning process.

TEXT SEMANTIC MINING MODEL BASED ON THE ALGEBRA OF HUMAN CONCEPT LEARNING

In view of Cognitive Informatics (Wang, 2002, 2007a), we regard each text as a concept in this paper. The sentences in a text are seen as the objects in the corresponding concept while the words are considered as attributes in the objects. Thus, a text can be structured as Figure 2. Based on this idea, the reading or understanding process of a text is able to be considered as a human concept learning process. However, directly processing the structure of a concept/text illustrated in Figure 2 is difficult and bears a high complexity since there are lots of discrete properties. However, the reduced conceptual structure formalized as Equation (1) is possible to well address such a problem. As a result, according to the idea of the algebra of human concept learning, we propose text semantic mining model. And the following definitions are given before formalizing the TSMM in this paper.

Definition 3: Text property set

Text property set, denoted as $\sum$, is defined as a set of text properties, which refers to the nouns or noun phrases in a text. In another word, all of the nouns and noun phrases in a text form the set of text properties.

The reason is that nouns or noun phrases have more expressive power than other words (Zheng & Kang, 2009). Therefore, it is reasonable to use nouns and noun phrases to describe the semantics in a text.

Figure 2. The structure of a text regarded as a concept

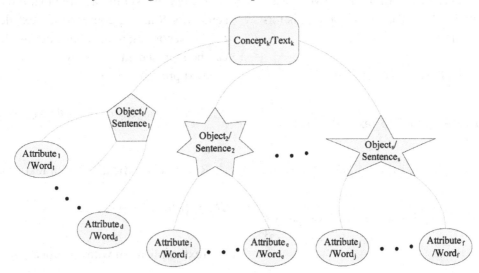

Definition 4: Relation specification set of text properties

Relation specification set of text properties, denoted as *Rs*, is a set of the verbs in a text, which can be used to represent the specific relations between text properties.

Definition 5: Text implication with specification

A text implication with specification consists of two parts: one is text implication and the other is its relation specification set.

A text implication with specification is formalized as follow.

$$tis_i : a \xrightarrow{\; rs(tis_i)\;} b, \qquad (2)$$

where, *tis*$_i$ refers to a text implication with specification, either *a* or *b* belongs to text property set $\sum$ and *rs(tis$_i$)* is a subset of relation specification set *Rs*.

Through such a text implication with specification, we are able to obtain more specific relations between text properties (e.g. $Pres.Hu \xrightarrow{\;\{visit\}\;} U.S.$). By contrast, rela-

tions in ontology based models are more abstract (e.g. *"a kind of"* and *"sub class of"*, etc) and need a lot of manual work to build them whereas our method appear more concrete and can be mined by machine automatically.

Definition 6: Text affirmation set

Text affirmation set, denoted as *Ta*, refers to a set of text affirmations, which represents the common sense properties in a text. In other words, it is a subset of text property set, denoted as *Ta*$\subseteq\sum$, which only containing the common sense properties in text.

Definition 7: Text weak property set

Text weak property set, denoted as *Tw*, is a set of text weak properties appearing neither in text affirmation set nor in the set of properties contained in text implications with specification. Nonetheless, the elements in *Tw* all belong to text property set, that is *Tw*$\subseteq\sum$.

Through text weak property set, people will have a deep understanding of texts and will be easy to distinguish the differences among texts.

Based on the above definitions, TSMM based on the algebra of human concept learning is formalized as follows.

$$T_s \mid_{\Sigma} = Ta \mid_{\Sigma} \cdot \left[Tis \mid_{\Sigma} \times Tw \mid_{\Sigma} \right],$$

$$Ta \mid_{\Sigma} = \left\{ x \mid x \in \Sigma \wedge x \text{ is a text affirmation} \right\},$$

$$Tis \mid_{\Sigma} = \left\{ x \xrightarrow{rs} y \mid x, y \in \Sigma \wedge rs \subseteq Rs \right\},$$

$$Tw \mid_{\Sigma} = \left\{ x \mid x \in \Sigma \wedge x \text{ is a text weak property} \right\}$$

(3)

where $T_s \mid_{\Sigma}$ indicates a text seen as a concept or a "world", $Ta \mid_{\Sigma}$ and $Tw \mid_{\Sigma}$ are respectively the text affirmation set and text weak property set lying in the text property set Σ. Whereas $Tis \mid_{\Sigma}$ means the set of text implications with specification.

Based on the TSMM developed by the algebra of human concept learning, people are able to have a better understanding of a text with a low complexity as well as the computers are. Because TSMM not only accords with human concept learning process but also can be automatically constructed by machine.

To better understand this model, a simple instance has been given as follows:

S_1: The **university** <u>offers</u> excellent **students** several kinds of **scholarships** every **year**.
S_2: **John** <u>failed</u> to <u>obtain</u> the **scholarship** last **year**.

Herein, we simply assume that only the two sentences, S_1 and S_2, compose of a text, denoted as $t_{example}$. According to the above definitions, $t_{example}$ can be represented as follows.

Text property set:

$$\Sigma = \left\{ university, student, scholarship, year, john \right\}$$

Relation specification set of text properties:

$$Rs = \left\{ offer, fail, obtain \right\}$$

Text implication with specification:

$$Tis \mid_{\Sigma} = \left\{ unversity \xrightarrow{\{offer\}} scholarship, john \xrightarrow{\{fail, obtain\}} scholaship \right\}$$

Text affirmation set:

$$Ta \mid_{\Sigma} = \left\{ scholarship, year \right\}$$

Text weak property set:

$$Tw \mid_{\Sigma} = \left\{ student \right\}$$

Consequently, the example text can be represented as shown in Box 1.

Trom the text $t_{example}$ the main idea can be learned that John failed to obtain the scholarship offered by the university last year. Meanwhile, from our model $t_{example} \mid_{\Sigma}$, we are supposed to learn that the text focuses on the topic of the scholarship, while the details are likely to tell about the two events:

Box 1.

$$t_{example} \mid_{\Sigma} =$$
$$\left\{ scholarship, year \right\} \cdot \left[\left\{ unversity \xrightarrow{\{offer\}} scholarship, john \xrightarrow{\{fail, obtain\}} scholaship \right\} \times \left\{ student \right\} \right]$$

(1) the university offers the scholarship and (2) John fails to obtain the scholarship. As a result, it is apparent that our model has well expressed the main idea of this text.

Although TSMM is proposed based on the idea of the algebra of human concept learning, there are several differences between them. Firstly, the algebra of human concept learning deals with concepts while TSMM processes texts. Secondly, the establishment of algebra of human concept learning depends to a great extent on human work whereas TSMM can be automatically constructed, which will be discussed in the following section. However, the former possesses more accurate semantics than TSMM for it is constructed by humans. Thirdly, comparing with algebra of human concept learning, TSMM offers more specific relations between text properties. For example, the relation between *"university"* and *"scholarship"* as shown above is described as *"offer"* in our model, which leads us to suppose that *"the university offers the scholarship"*. The differences between TSMM and the algebra of human concept learning are concluded in Table 1.

With the help of TSMM, the content of this text has been well expressed and is easy to be understood. Even if we have never read this text, we shall also know what the text discusses according to this representation.

Table 1. Differences between TSMM and algebra of human concept learning

	TSMM	Algebra of Human Concept Learning
Processing Object	Text	Concept
Construction	Automatic	Manual
Accuracy[1]	General	High
Relationship	Specific	Coarse

ESTABLISHMENT OF TEXT SEMANTIC MINING MODEL BASED ON THE ALGEBRA OF HUMAN CONCEPT LEARNING

The above section has discussed the formalization of text semantic mining model based on the algebra of human concept learning while this section will focus on its establishment method, which is also an important issue. According to formalization (3), the main components of TSMM include three parts: (1) the text affirmation set, (2) the set of text implications with specification and (3) the text weak property set. As text weak property set can be acquired with the help of text affirmation set and the set of text implications with specification, herein, we consider the establishment is composed of the following two parts:

1. Text affirmation set mining;
2. Text implication with specification mining.

Text Affirmation Set Mining

As given in Definition 6, text affirmations are the common sense properties, which refer to common sense nouns or noun phrases in a text. Accordingly, the whole process of text affirmation set mining is described as follows:

1. According to text properties' term frequencies, rank them in descending order;
2. Based the ranked list of text properties, select top four percent (Salton & Wong, 1975) of them as text affirmations.

To have a further explanation, an example is given:

[obama: 43, president: 35, campaign: 22, candidate: 18, economy: 17, crisis: 14, tax: 14, bill: 12, government: 10, country: 7,...] (Assuming the length of this list=100).

Each element in the above list consists of two components: the former is text property while the latter is its term frequency. This list has been sorted by each property's term frequency. As aforementioned, the top four percent properties of this list will be regarded as text affirmations. Therefore, in this example, *obama*, *president*, *campaign* and *candidate* are considered as text affirmations.

Text Implication with Specification Mining

Text implications with specification are mined to explain the specific relations between text properties. The mining process is divided into two steps in this paper. The first is to discover the atomic semantic relations while the second is to mine the specification set for each atomic semantic relation. And the definition of atomic semantic relation is given as follows.

Definition 8: Atomic semantic relation

Atomic semantic relation is defined as the most basic relation between text properties, which tells and only tells us that there exists a relation between text properties.

In this paper, we consider an association rule as an atomic semantic relation since it not only accords with the definition but also can be obtained by the machine.

As a result, in this paper the process of atomic semantic relation mining is treated as the process of association rule mining. Meanwhile, the algorithm of Apriori (Agrawal & Srikant, 1994) has been chosen to mine association rules and only the ones containing just one antecedent will be remained. Herein, each sentence in a text will be considered as a transaction.

After atomic semantic relation mining, pairs of text properties have been explored, between which there may exist a certain relation. Based on that, relation specification set mining is brought out

to enrich the atomic semantic relations, to make them more specific.

Generally, when two properties, between which an atomic semantic relation exists, appear in a sentence, the verbs in the same sentence are probably taking a very important part in describing the relation. For example,

The general **task** of <u>finding</u> an **exception set** can be NP-hard.

From the above sentence, we may find that between the properties "*task*" and "*exception set*" there is a verb "*finding*", which seems reasonable to be used to express the relation between "*task*" and "*exception set*".

Therefore, based on the above analyzing, relation specification set mining process can be concluded in two steps.

1. For each atomic semantic relation, find out the verbs that may relate to it among the sentences and treat those verbs as its relation specification candidate set.
2. Pick out the verbs that are really reasonable to describe the relations in the candidate set generated in first step. In other words, it is a process of removing noises in the candidate set. The basic and simple idea is first to count the co-appearances between an atomic semantic relation and the verbs in candidate set, and then to pick out the verbs possessing high co-appearances.

For example, given an association rule (regarded as an atomic semantic relation in this paper) mined from a text, *tree→apple*, we'll first find out all the verbs appearing in the same sentence with this association rule as its specification candidate set, which is shown as follows:

{need: 1, affect: 1, have: 2, change: 1, plant: 2, grow: 4, expect: 2, live: 3, start: 1, produce: 8}.

The number after colon means the times of the former verb and *tree→apple* appearing in the same sentence. Based on that, we'll pick out the top n verbs to compose the atomic semantic relation's specification set. Herein, assuming n=2, then *produce* and *grow* will be selected to describe the specific relation between *tree* and *apple*, denoted as $tree \xrightarrow{\{produce,grow\}} apple$.

According to the above, the whole process of TSMM construction is illustrated in Figure 3. As it shows, the first step of TSMM construction is to pre-process a text, including part-of-speech tagging, stop words removing, etc. After text pre-processing, only nouns/noun phrases and verbs will be remained. Because in this paper we just take these two types of words into consideration. Nouns/noun phrases are used to form the set of text properties while verbs are used to describe the specific relations between text properties. After that, the second step is to compute the term frequency of each word, according to which, find out the text affirmations. Meanwhile, taking sentences as transactions, mine association rules, which are regarded as atomic semantic relations in this paper. Then, for each atomic semantic relation, find out its specification set. Finally, based on the text semantics mined in former steps, construct the TSMM.

As a result, according to the whole process of TSMM construction, it can be learnt that TSMM is constructed automatically with a lower complex and accords with human concept learning. Moreover, as TSMM to a large extent depends on statistical method and has not taken the diversities among different languages into account, the whole process of TSMM construction is irrelevant with languages in this paper. However, we'll have a further research on language oriented TSMM with the help of syntax analysis in the future.

REASONING BASED ON TEXT SEMANTIC MINING MODEL

With the growth of the Web, dealing with large-scale text knowledge is meaningful and has become increasingly important. Therefore, how to organize the texts represented by TSMM is a burning question to be addressed. This section proposes three basic reasoning rules based on TSMM, including intersection operation, union operation and subtraction operation. Intersection can be used to discover the common semantic information among the texts while the union of texts will provide background knowledge. Furthermore, differences between texts are possible

Figure 3. The whole process of TSMM establishment based on the algebra of human concept learning

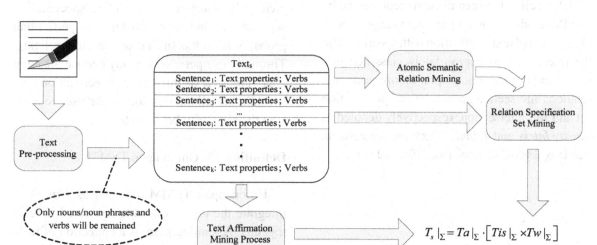

229

to be found out by subtraction. Meanwhile, latent semantics, which cannot be explored using general mining methods, are also probably discovered by the three basic operations. From this view, our model performs better than other models. VSM doesn't possess the ability of reasoning since it only contains keywords that are not sufficient to support reasoning. EFCM carries on a simple reasoning for it contains association rules. Ontology based models can express rich semantics, but the reasoning rules of these models are too strict. Thus such models cannot meet with the diversity and irregularity of web resources. Probability topic models are based on probability and statistic, with many given mathematical assumptions. But those assumptions are not suitable as the web resources are massive, dynamic and uncertain. However, our model obtains the capability of dealing with such a kind of resources for it is constructed automatically under a lower complexity. And its reasoning rules are discussed as below.

Assuming there are two texts, *Text_i* and *Text_j*, represented by TSMM.

$$T_i \mid_{\Sigma_i} = Ta \mid_{\Sigma_i} \cdot \left[Tis \mid_{\Sigma_i} \times Tw \mid_{\Sigma_i} \right],$$
$$T_j \mid_{\Sigma_j} = Ta \mid_{\Sigma_j} \cdot \left[Tis \mid_{\Sigma_j} \times Tw \mid_{\Sigma_j} \right].$$

Definition 9: Intersection upon TSMMs

Intersection between the texts represented by TSMM stands for the common knowledge of them. On account of text implication with specification, the intersection is appropriately classified into three kinds, intersection upon atomic semantic relation, intersection upon relation specification set and full intersection, respectively denoted as *Intasr*, *Intrss* and *Intfull*, which are formalized. See Box 2 for Equations (9), (10), and (11).

$$Intasr\left(T_i \mid_{\Sigma_i}, T_i \mid_{\Sigma_j} \right) = \left(Ta \mid_{\Sigma_i} \cap Ta \mid_{\Sigma_j} \right)$$
$$\cdot \left[Intasr\left(Tis \mid_{\Sigma_i}, Tis \mid_{\Sigma_j} \right) \times \left(Tw \mid_{\Sigma_i} \cap Tw \mid_{\Sigma_j} \right) \right],$$
$$\tag{4}$$

$$Intrss\left(T_i \mid_{\Sigma_i}, T_i \mid_{\Sigma_j} \right) = \left(Ta \mid_{\Sigma_i} \cap Ta \mid_{\Sigma_j} \right)$$
$$\cdot \left[Intass\left(Tis \mid_{\Sigma_i}, Tis \mid_{\Sigma_j} \right) \times \left(Tw \mid_{\Sigma_i} \cap Tw \mid_{\Sigma_j} \right) \right],$$
$$\tag{5}$$

$$Intfull\left(T_i \mid_{\Sigma_i}, T_i \mid_{\Sigma_j} \right) = \left(Ta \mid_{\Sigma_i} \cap Ta \mid_{\Sigma_j} \right)$$
$$\cdot \left[Intfull\left(Tis \mid_{\Sigma_i}, Tis \mid_{\Sigma_j} \right) \times \left(Tw \mid_{\Sigma_i} \cap Tw \mid_{\Sigma_j} \right) \right],$$
$$\tag{6}$$

$$Ta \mid_{\Sigma_i} \cap Ta \mid_{\Sigma_j} = \left\{ x \mid x \in Ta \mid_{\Sigma_i} \wedge x \in Ta \mid_{\Sigma_j} \right\},$$
$$\tag{7}$$

$$Tw \mid_{\Sigma_i} \cap Tw \mid_{\Sigma_j} = \left\{ x \mid x \in Tw \mid_{\Sigma_i} \wedge x \in Tw \mid_{\Sigma_j} \right\}$$
$$\tag{8}$$

Different kinds of intersection based on TSMM have different abilities. From the definitions above, it can be concluded that intersection upon atomic semantic relation is capable of enhancing the strength of association between related text properties and enriching their relationship. By that, the two properties and their relationship will be more credible and be understood more completely. Intersection upon relation specification set is able to find out pairs of text properties that probably have the similar or the same relation. Through this operation we may deduce the main idea of those texts through the specification sets. In addition, full intersection can be used to measure the similarity between texts.

Definition 10: Union upon TSMM

Union upon TSMM, denoted as *Uni*, is to integrate the text knowledge contained in texts together. Assuming there is a collection of texts

Box 2.

$$Intasr\left(Tis\,|_{\Sigma_i}, Tis\,|_{\Sigma_j}\right) = \left\{ x_k \xrightarrow{\,rs_i \cup rs_j\,} y_k \,\middle|\, x_k \xrightarrow{\,rs_i\,} y_k \in Tis\,|_{\Sigma_i} \wedge x_k \xrightarrow{\,rs_j\,} y_k \in Tis\,|_{\Sigma_j} \right\} \tag{9}$$

$$Intrss\left(Tis\,|_{\Sigma_i}, Tis\,|_{\Sigma_j}\right) = \left\{ S : \left\{ x \xrightarrow{\,rs_k\,} y \,\middle|\, x \xrightarrow{\,rs_k\,} y \in Tis\,|_{\Sigma_i} \cup Tis\,|_{\Sigma_j} \right\} \middle| rs_k \subseteq Rs_i \cup Rs_j \wedge |S| > 1 \right\} \tag{10}$$

$$Intfull\left(Tis\,|_{\Sigma_i}, Tis\,|_{\Sigma_j}\right) = \left\{ x_k \xrightarrow{\,rs_k\,} y_k \,\middle|\, x_k \xrightarrow{\,rs_k\,} y_k \in Tis\,|_{\Sigma_i} \wedge x_k \xrightarrow{\,rs_k\,} y_k \in Tis\,|_{\Sigma_j} \right\} \tag{11}$$

that are similar to each other, if we obtain the union operation of those texts, the background knowledge of this collection will come out. Based on this background knowledge, much more upper applications or deeper text knowledge discoveries can be carried out. The followings formalize the union upon TSMM. See Box 3 for Equation (15).

$$Uni\left(T_i\,|_{\Sigma_i}, T_i\,|_{\Sigma_j}\right) = \left(Ta\,|_{\Sigma_i} \cup Ta\,|_{\Sigma_j}\right)$$
$$\cdot \left[Uni\left(Tis\,|_{\Sigma_i}, Tis\,|_{\Sigma_j}\right) \times \left(Tw\,|_{\Sigma_i} \cup Tw\,|_{\Sigma_j}\right) \right], \tag{12}$$

$$Ta\,|_{\Sigma_i} \cup Ta\,|_{\Sigma_j} = \left\{ x \mid x \in Ta\,|_{\Sigma_i} \vee x \in Ta\,|_{\Sigma_j} \right\}, \tag{13}$$

$$Tw\,|_{\Sigma_i} \cup Tw\,|_{\Sigma_j} = \left\{ x \mid x \in Tw\,|_{\Sigma_i} \vee x \in Tw\,|_{\Sigma_j} \right\} \tag{14}$$

Definition 11: Subtraction upon TSMM

Subtraction from *text$_a$* to *text$_b$*, denoted as *Sub(text$_a$, text$_b$)*, refers the text knowledge contained in *text$_a$*, but not contained in *text$_b$*. In consideration of the characteristics of text implications with specification, subtraction upon TSMM is divided into three aspects called respectively, subtraction upon atomic semantic relation, subtraction upon relation specification set and strict subtraction, all of which are denoted as *Subasr*, *Subrss* and *Substr* and formalized. See Box 4 for Equations (21), (22), and (23).

$$Subasr\left(T_i\,|_{\Sigma_i}, T_i\,|_{\Sigma_j}\right) = Sub\left(Ta\,|_{\Sigma_i}, Ta\,|_{\Sigma_j}\right)$$
$$\cdot \left[Subasr\left(Tis\,|_{\Sigma_i}, Tis\,|_{\Sigma_j}\right) \times Sub\left(Tw\,|_{\Sigma_i}, Tw\,|_{\Sigma_j}\right) \right] \tag{16}$$

Box 3.

$$Uni\left(Tis\,|_{\Sigma_i}, Tis\,|_{\Sigma_j}\right) = \left\{ x_k \xrightarrow{\,rs_i \cup rs_j\,} y_k \,\middle|\, x_k \xrightarrow{\,rs_i\,} y_k \in Tis\,|_{\Sigma_i} \vee x_k \xrightarrow{\,rs_j\,} y_k \in Tis\,|_{\Sigma_j} \right\} \tag{15}$$

$$Subrss\left(T_i\mid_{\Sigma_i}, T_i\mid_{\Sigma_j}\right) = Sub\left(Ta\mid_{\Sigma_i}, Ta\mid_{\Sigma_j}\right)$$
$$\cdot\left[Subrss\left(Tis\mid_{\Sigma_i}, Tis\mid_{\Sigma_j}\right) \times Sub\left(Tw\mid_{\Sigma_i}, Tw\mid_{\Sigma_j}\right)\right]$$
(17)

$$Substr\left(T_i\mid_{\Sigma_i}, T_i\mid_{\Sigma_j}\right) = Sub\left(Ta\mid_{\Sigma_i}, Ta\mid_{\Sigma_j}\right)$$
$$\cdot\left[Substr\left(Tis\mid_{\Sigma_i}, Tis\mid_{\Sigma_j}\right) \times Sub\left(Tw\mid_{\Sigma_i}, Tw\mid_{\Sigma_j}\right)\right]$$
(18)

$$Sub\left(Ta\mid_{\Sigma_i}, Ta\mid_{\Sigma_j}\right) = \left\{x\mid x\in Ta\mid_{\Sigma_i} \wedge x\notin Ta\mid_{\Sigma_j}\right\}$$
(19)

$$Sub\left(Tw\mid_{\Sigma_i}, Tw\mid_{\Sigma_j}\right) = \left\{x\mid x\in Tw\mid_{\Sigma_i} \wedge x\notin Tw\mid_{\Sigma_j}\right\}$$
(20)

As shown above, the text knowledge obtained by subtraction upon atomic semantic relation is the one exactly does not appear in the other text. It reflects the main difference from *text_a* to *text_b*. Subtraction upon relation specification set remains the same text implications but removes the ones having the same specifications. Strict subtraction only removes the ones that are completely same to each other.

Overall, the three types of reasoning, intersection operation, union operation and subtraction operation, are the basic components of the reasoning rules. Intersection reflects the common text knowledge of the texts. Union shows us the full view of the texts and provides their background knowledge. Subtraction presents the diversities among the texts and helps us discover more individual text knowledge.

COMPARISONS AND EXPERIMENTS

In previous sections, the basic idea and formalization of text semantic mining model based on the algebra of human concept learning have been discussed, from which we shall learn that TSMM is in accordance with the process of human concept learning. Meanwhile, the implementation of TSMM is fully automatic under a lower complexity according to its establishing process. However, this section will give a validation for text semantic mining model through two different aspects: summary comparison and experimental verification.

Box 4.

$$Subasr\left(Tis\mid_{\Sigma_i}, Tis\mid_{\Sigma_j}\right) = \left\{x_k\xrightarrow{rs_k}y_k\mid x_k\xrightarrow{rs_k}y_k\in Tis\mid_{\Sigma_i} \wedge x_k\xrightarrow{\forall(rs_j\subseteq Rs_j)}y_k\notin Tis\mid_{\Sigma_j}\right\}$$ (21)

$$Subrss\left(Tis\mid_{\Sigma_i}, Tis\mid_{\Sigma_j}\right) = \left\{x_k\xrightarrow{rs_k}y_k\mid x_k\xrightarrow{rs_k}y_k\in Tis\mid_{\Sigma_i} \wedge(\forall x,y\in\Sigma_j)\left(x\xrightarrow{rs_k}y\notin Tis\mid_{\Sigma_j}\right)\right\}$$ (22)

$$Substr\left(Tis\mid_{\Sigma_i}, Tis\mid_{\Sigma_j}\right) = \left\{x_k\xrightarrow{rs_k}y_k\mid x_k\xrightarrow{rs_k}y_k\in Tis\mid_{\Sigma_i} \wedge x_k\xrightarrow{rs_k}y_k\notin Tis\mid_{\Sigma_j}\right\}$$ (23)

Table 2. Comparisons on the abilities of reasoning, automatically construction, computation complexity and semantic richness

Representation models		VSM	EFCM	TSMM	PLSA, LDA	OWL, SHOE
Reasoning	Union	√	√	√	×	√
	Intersection	√	√	√	×	√
	Subtraction	√	√	√	×	√
	Inclusive	√	√	√	×	√
	Exclusive	√	√	√	×	√
Automatically construction		√	√	√	Manual	Semi-Automatic
Computation Complexity		Not	Easy	Easy	Difficult	Not
Semantic richness		Poor	Moderate	Rich	Rich	Very rich

A Summary Comparison with Current Models

Herein, we will have a summary comparison between our model and the other current models, which are respectively vector space model, element fuzzy cognitive model, probabilistic latent semantic analysis (PLSA) (Hoffman, 1999), latent dirichlet allocation (LDA) (Blei & Ng, 2003), web ontology language and simple html ontology extensions, focusing on the abilities of reasoning, automatically construction, computation complexity and semantic richness (Table 2).

From Table 2, it can be learnt that ontology based models possess the richest semantics and are capable of reasoning whereas the semi-automatically construction is their fatal defect since this it makes them unsuitable to be applied into large-scale web environment. Only VSM, EFCM and TSMM have the ability of automatically constructing, among which, our model contains the richest semantics. As a result, on account of the rapid growth of the Web, the following section will mainly discuss these three models: VSM, EFCM and TSMM.

Experimental Verification

In this section, we first obtained the main text contents of three web pages, denoted as $Text_a$, $Text_b$

and $Text_c$, and then represented them with VSM, EFCM and TSMM respectively (Table 3). To simply the representations, only ten text properties, which have the highest appearing frequencies, are selected and each text implication is specified by only one verb. These three texts seem to discuss a similar topic, but nevertheless, actually, only $Text_a$ and $Text_b$ have a certain overlapping discussions whereas $Text_c$ totally talks about another topic. The properties, "*sun*" and "*apple*", appear in all the three texts but what they are really meant to be is different. "*Sun*" and "*apple*" in $Text_a$ and $Text_b$ refer to the companies named as *Sun* and *Apple* respectively. However, in $Text_c$, they are meant to be the sun and a kind of fruit.

As given in Table 3, it is obvious that TSMM contains the richest semantics among the three models while EFCM is the second. Through the text implication with specification "*tree→apple: produce*", it is easy to conceive that the "*apple*" in $Text_c$ refers to the fruit "*apple*" instead of the company "*apple*". Moreover, the specific relation between "*tree*" and "*apple*" can also be obtained by "*produce*", which apparently is reasonable. When $Text_c$ is represented by EFCM, we'll only study that *tree* is related to *apple*, through which it is also possible for us to suppose that $Text_c$ tells something about apple trees. However, what VSM can offer is just a bag of words, which make us hardly to understand what $Text_c$ is talking about.

Table 3. Comparison between VSM, EFCM and TSMM

	Text$_a$: *Sun and Apple almost merged three times*
Source	http://www.theregister.co.uk/2006/01/12/sun_apple_snapple/
VSM	*{sun, apple, mcnealy, joy, ipod, ceo, company, time, steve, cell}*
EFCM	*{sun:0.21, apple:0.23, mcnealy:0.14, joy:0.14, ipod:0.07, ceo:0.04, company:0.04, time:0.04, steve:0.04, cel:0.04l}·{time→apple, apple→sun, ceo→apple, ceo→joy, ipod→ mcnealy, sun→apple, joy→apple}*
TSMM	*{sun, apple, mcnealy, joy}·[{ time→apple: merged, apple→sun: tried, ceo→apple: said, ceo→joy: said, ipod→ mcnealy: convince, sun→apple: tried, joy→apple:said }×{company,steve,cell}]*
	Text$_b$: *Skepticism surrounds Apple*
Source	http://news.cnet.com/Skepticism-surrounds-Apple,-Sun-report/2100-1033_3-202914.html
VSM	*{sun, apple, herwick, analyst, culture, deal, cubbage, company, internet, ratcliffe}*
EFCM	*{sun:0.25, apple:0.24, herwick:0.09, analyst:0.09, culture:0.06, deal:0.06, cubbage:0.06, company:0.06, internet:0.04, ratcliffe:0.04}·{apple→sun, internet→sun, culture→apple, culture→cubbage, culture→company, culture→sun, deal→herwick, deal→sun, cubbage→culture, cubbage→analyst, cubbage→company, cubbage→sun, analyst→apple, analyst→sun, ratcliffe→apple, ratcliffe→sun, company→apple, company→culture, company→cubbage, company→analyst, company→sun, sun→apple}*
TSMM	*{sun, apple, herwick, analyst}·[[apple→sun:said, internet→sun:commented, culture→apple:fueled, culture→cubbage:says, culture→company:says, culture→sun:fueled, deal→herwick:questioned, deal→sun:put, cubbage→culture:says, cubbage→analyst:said, cubbage→company:says, cubbage→sun:said, analyst→apple:said, analyst→sun:said, ratcliffe→apple:added, ratcliffe→sun:added, company→apple:said, company→culture:says, company→cubbage:says, company→analyst:said, company→sun:said, sun→apple:said}×{}]*
	Text$_c$: *Apple Trees: Where and How to Plant*
Source	http://www.doityourself.com/stry/plantappletree
VSM	*{tree, apple, soil, foot, root, m., year, size, variety, rootstock}*
EFCM	*{tree:0.33, apple:0.25, soil:0.08, foot:0.06, root:0.06, m.:0.05, year:0.04, size:0.04, variety:0.03, rootstock}·{tree→apple, year→tree, year→apple, foot→tree, size→tree, size→apple, apple→tree}*
TSMM	*{tree, apple, soil, foot, root}·[{tree→apple:produce, year→tree:live, year→apple:live, foot→tree:grow, size→tree:affects, size→apple:affects, apple→tree:produce}×{soil, root, m., variety, rootstock}]*

Accordingly, the semantics retrieved from *Text$_a$* and *Text$_b$* using different models gives similar results as *Text$_c$*. When represented by TSMM, "*time→apple:merged*" and "*apple→sun:said*" are able to guide us to think the two texts refer to the company "*Apple*". Yet "*time→apple*" and "*apple→sun*" in EFCM are harder to make us to distinguish the company "*Apple*" and the tree "*Apple*". As a result, TSMM obtains a better capability of discriminating the differences among the three texts for it possesses the richest semantics of the three models. Therefore, it can be concluded that TSMM can not only be constructed automatically so as to satisfy the rapid growth of the Web, but also carries various semantics to provide better services.

CONCLUSION

In this paper, we have proposed text semantic mining model based on the algebra of human concept learning. Actually, TSMM is a mapping from a concept to a text so that the process of understanding a text represented by TSMM is accordance with the process of human concept learning. By that, it is possible to be deduced that TSMM possesses a lot of semantics. Meanwhile, on account of the establishing process of TSMM, we shall learn that TSMM can be constructed automatically with a lower complexity. Furthermore, reasoning based on TSMM is also feasible and is able to provide a novel web service. The comparisons with current models and experimental results have shown the good performance of TSMM. In future, our research group will put more efforts in dealing with the specification sets.

ACKNOWLEDGMENT

This work was supported in part by the Shanghai Science and Technology Commission under Grant 09JC1406200, in part by the National Science Foundation of China under Grant 91024012, Grant 61071110, Grant 90818004, Grant 90612010, and in part by the Shanghai Leading Academic Discipline Project (J50103).

REFERENCES

Agrawal, R., & Srikant, R. (1994). Fast algorithms for mining association rules. In Proceedings of the International Conference on Very Large Data Bases, Santiago, Chile (pp. 487-499).

Bansal, S., & Garg, R. (2010). A novel probabilistic approach for efficient information retrieval. *International Journal of Computers and Applications*, *9*(2), 44–48. doi:10.5120/1354-1827

Blei, D. M., & Lafferty, J. D. (2006). Correlated topic models. *Advances in Neural Information Processing Systems*, 18.

Blei, D. M., Ng, A. Y., & Jordan, M. I. (2003). Latent dirichlet allocation. *Journal of Machine Learning Research*, 3, 993–1022. doi:10.1162/jmlr.2003.3.4-5.993

Fang, N., Luo, X. F., & Xu, W. M. (2009). Measuring textual context based on cognitive principles. *International Journal of Software Science and Computational Intelligence*, *1*(4), 61–89. doi:10.4018/jssci.2009062504

Feldman, J. (2006). An algebra of human concept learning. *Journal of Mathematical Psychology*, *50*(4), 339–368. doi:10.1016/j.jmp.2006.03.002

Heflin, J., Hendler, J., & Luke, S. (1999). SHOE: A knowledge representation language for internet applications (Tech. Rep. No. CS-TR-4078). Baltimore, MD: University of Maryland.

Hoffman, T. (1999). Probabilistic latent semantic analysis. In Proceedings of the Conference on Uncertainty in Artificial Intelligence.

Horrocks, I., Fensel, D., Broekstra, J., Decker, S., Erdmann, M., Goble, C., et al. (2000). OIL: The ontology inference layer (Tech. Rep. No. IR-479). Amsterdam, The Netherlands: Vrije Universiteit Amsterdam.

Landauer, T. K., Foltz, P. W., & Laham, D. (1998). An introduction to latent semantic analysis. *Discourse Processes*, *25*(2-3), 259–284. doi:10.1080/01638539809545028

Luo, X. F., Xu, Z., Yu, J., & Chen, X. (in press). Building association link network for semantic link on web resources. *IEEE Transactions on Automation Science and Engineering*.

Luo, X. F., Xu, Z., Yu, J., & Liu, F. F. (2008). Discovery of associated topics for the intelligent browsing. In Proceedings of the 1st IEEE International Conference on Ubi-Media Computing, Lanzhou, China (pp. 119-125).

McCallum, A., Corrada-Emmanuel, A., & Wang, X. (2004). *The author-recipient-topic model for topic and role discovery in social networks: Experiments with Enron and academic email*. Amherst, MA: University of Massachusetts.

McGuinness, D. L., & Harmelen, F. (2004). OWL web ontology language. Retrieved from http://www.w3.org/TR/owl-features/

Michal, R. Z., Thomas, G., Mark, S., & Padhraic, S. (2004). The author-topic model for authors and documents. In Proceedings of the 20th Conference on Uncertainty in Artificial Intelligence, Banff, AB, Canada (pp. 487-494).

Mirza, M., Sommers, J., Barford, P., & Zhu, X. J. (2010). A machine learning approach to TCP throughput prediction. *IEEE/ACM Transactions on Networking, 18*(4), 1026–1039. doi:10.1109/TNET.2009.2037812

Salton, G., Wong, A., & Yang, C. S. (1975). A vector space model for automatic indexing. *Communications of the ACM, 18*(11), 613–620. doi:10.1145/361219.361220

Wang, Y. (2002). Keynote: On cognitive informatics. In Proceedings of the 1st IEEE International Conference on Cognitive Informatics, Calgary, AB, Canada (pp. 34-42).

Wang, Y. (2003). On cognitive informatics. Brain and Mind: A Transdisciplinary Journal of Neuroscience and Neurophilosophy, 4(2), 151-167.

Wang, Y. (2006). On concept algebra and knowledge representation. In Proceedings of the 5th IEEE International Conference on Cognitive Informatics, Beijing, China (pp. 320-331).

Wang, Y. (2007a). The theoretical framework of cognitive informatics. *International Journal of Cognitive Informatics and Natural Intelligence, 1*(1), 1–27. doi:10.4018/jcini.2007010101

Wang, Y. (2007b). The OAR model of neural informatics for international knowledge representation in the brain. *International Journal of Cognitive Informatics and Natural Intelligence, 1*(3), 64–75. doi:10.4018/jcini.2007070105

Zheng, H. T., Kang, B. Y., & Kim, H. G. (2009). Exploiting noun phrases and semantic relationships for text document clustering. *Information Science, 179*(13), 2249–2262. doi:10.1016/j.ins.2009.02.019

Zhuge, H., & Sun, Y. C. (2010). The schema theory for semantic link network. *Future Generation Computer Systems, 26*(3), 408–420. doi:10.1016/j.future.2009.08.012

ENDNOTES

[1] Herein, accuracy is used to check the qualities of affirmations or implications that are retrieved from a concept or a text.

This work was previously published in the International Journal of Cognitive Informatics and Natural Intelligence, Volume 5, Issue 2, edited by Yingxu Wang, pp. 80-96, copyright 2011 by IGI Publishing (an imprint of IGI Global).

Chapter 15
In Search of Effective Granulization with DTRS for Ternary Classification

Bing Zhou
University of Regina, Canada

Yiyu Yao
University of Regina, Canada

ABSTRACT

Decision-Theoretic Rough Set (DTRS) model provides a three-way decision approach to classification problems, which allows a classifier to make a deferment decision on suspicious examples, rather than being forced to make an immediate determination. The deferred cases must be reexamined by collecting further information. Although the formulation of DTRS is intuitively appealing, a fundamental question that remains is how to determine the class of the deferment examples. In this paper, the authors introduce an adaptive learning method that automatically deals with the deferred examples by searching for effective granulization. A decision tree is constructed for classification. At each level, the authors sequentially choose the attributes that provide the most effective granulization. A subtree is added recursively if the conditional probability lies in between of the two given thresholds. A branch reaches its leaf node when the conditional probability is above or equal to the first threshold, or is below or equal to the second threshold, or the granule meets certain conditions. This learning process is illustrated by an example.

INTRODUCTION

The Decision-Theoretic Rough Set (DTRS) model, proposed by Yao et al. (Yao, Wang, & Lingras, 1990; Yao & Wang, 1992; Yao, 2010) in the early 1990s, is a meaningful and useful generalization of the probabilistic rough set model (Pawlak, 1991). In probabilistic rough set models, three probabilistic regions are defined by considering the degree of overlap between an equivalence class and a set to be approximated. A conditional probability is used to state the degree of overlap and a pair of

DOI: 10.4018/978-1-4666-2476-4.ch015

thresholds is used to define the three regions. An equivalence class is in the probabilistic positive region if its relative overlap with the set is above or equal to a threshold, is in the negative region if its relative overlap is below or equal to another threshold, and is in the boundary region if the relative overlap is between the two parameters. DTRS offers a solid foundation for probabilistic rough sets by systematically calculating the pair of thresholds based on the well-established Bayesian decision theory. Many real world problems can be solved with DTRS. For instance, DTRS provides a three-way decision approach to classification problems by allowing the possibility of indecision to suspicious examples, those examples in the boundary region must be re-examined by collecting additional information. A fundamental question that remains in DTRS is how to determine the classification of these deferred examples.

Cognitive science and cognitive informatics (Wang, 2007; Wang et al., 2009, 2011) study the human intelligence and its computational process. As an effective way of thinking, we typically focus on a particular level of abstraction and ignore irrelevant details. This not only enables us to identify differences between objects in the real world, but also helps us to view different objects as being the same, if low-level detail is ignored. Granular computing (GrC) (Bargiela & Pedrycz, 2002; Liang & Qian, 2008; Qian, Liang, & Dang, 2009; Yao, 2004b, 2007b, 2009) can be seen as a formal way of modeling this human thinking process. GrC is an area of study that explores different levels of granularity in human-centered perception, problem solving, and information processing, as well as their implications and applications in the design and implementation of knowledge intensive intelligent systems. Rough set theory is one of the concrete models of GrC for knowledge representation and data analysis.

In this paper, an adaptive learning method is introduced that classifies the deferred examples by adaptively searching for effective granulization. A decision tree is constructed for classification. At each level, we sequentially choose the attributes that provide the most suitable granulization. A subtree is added if the conditional probability lies in between of the two thresholds. A branch reaches its leaf node when the conditional probability is above or equal to the first threshold, or is below or equal to the second threshold.

The rest of the paper is organized as follows. We briefly review the basic ideas of DTRS. We introduce the interpretations of concepts based on GrC. A new adaptive learning algorithm is introduced for ternary classification. An illustrative example is given. We conclude the paper and explain the future work.

BRIEF INTRODUCTION TO DECISION-THEORETIC ROUGH SET MODEL

Bayesian decision theory is a fundamental statistical approach that makes decisions under uncertainty based on probabilities and costs associated with decisions. Following the discussions given in the book by Duda and Hart (1973), the decision theoretic rough set model is a straightforward application of the Bayesian decision theory.

With respect to the set C to be approximated, we have a set of two states $\Omega = \{C, C^C\}$ indicating that an object is in C or not in C, respectively. We use the same symbol to denote both a set C and the corresponding state. With respect to the three regions in the rough set theory, the set of actions is given by $\mathbf{A} = \{a_P, a_B, a_N\}$, where a_P, a_B and a_N represent the three actions in classifying an object x, namely, deciding $x \in \text{POS}(C)$, deciding $x \in \text{BND}(C)$, and deciding $x \in \text{NEG}(C)$, respectively. The loss function is given by Matrix 1.

In the matrix, λ_{pp}, λ_{BP} and λ_{NP} denote the losses incurred for taking actions a_P, a_B and a_N respectively, when an object belongs to C, and $\lambda_{PN}, \lambda_{BN}$, and λ_{NN} denote the losses incurred for taking these actions when the object does not belong to C.

Matrix 1.

	$C(P)$	$C^C(N)$
a_P	$\lambda_{PP} = \lambda(a_P \mid C)$	$\lambda_{PN} = \lambda(a_P \mid C^C)$
a_B	$\lambda_{BP} = \lambda(a_B \mid C)$	$\lambda_{BN} = \lambda(a_B \mid C^C)$
a_N	$\lambda_{NP} = \lambda(a_N \mid C)$	$\lambda_{NN} = \lambda(a_N \mid C^C)$

We use $\Pr(C|[x])$ to represent the conditional probability of an object belonging to C given that the object is described by its equivalence class $[x]$. The expected losses associated with taking different actions for objects in $[x]$ can be expressed as:

$$R(a_P \mid [x]) = \lambda_{PP} \Pr(C \mid [x]) + \lambda_{PN} \Pr(C^C \mid [x]),$$
$$R(a_B \mid [x]) = \lambda_{BP} \Pr(C \mid [x]) + \lambda_{BN} \Pr(C^C \mid [x]),$$
$$R(a_N \mid [x]) = \lambda_{NP} \Pr(C \mid [x]) + \lambda_{NN} \Pr(C^C \mid [x]).$$

The Bayesian decision procedure suggests the following minimum-risk decision rules:

(P) If
$R(a_P \mid [x]) \leq R(a_B \mid [x])$ & $R(a_P \mid [x]) \leq R(a_N \mid [x])$, decide $x \in POS(C)$;

(B) If
$R(a_B \mid [x]) \leq R(a_P \mid [x])$ & $R(a_B \mid [x]) \leq R(a_N \mid [x])$, decide $x \in BND(C)$;

(N) If
$R(a_N \mid [x]) \leq R(a_P \mid [x])$ & $R(a_N \mid [x]) \leq R(a_B \mid [x])$, decide $x \in NEG(C)$.

Tie-breaking criteria should be added so that each object is put into only one region.

Since $\Pr(C|[x]) + \Pr(C^C|[x]) = 1$, we can simplify the rules based only on the probabilities $\Pr(C|[x])$ and the loss function λ. Consider a special kind of loss functions with:

(c0). $\lambda_{PP} \leq \lambda_{BP} < \lambda_{NP,} \ \lambda_{NN} \leq \lambda_{BN} < \lambda_{PN}$.

That is, the loss of classifying an object x belonging to C into the positive region POS(C) is less than or equal to the loss of classifying x into the boundary region BND(C), and both of these losses are strictly less than the loss of classifying x into the negative region NEG(C). The reverse order of losses is used for classifying an object not in C. Under condition (c0), the decision rules (P)-(N) can be re-expressed as:

(P) If $\Pr(C \mid [x]) \geq \alpha$ & $\Pr(C \mid [x]) \geq \gamma$, decide $x \in POS(C)$;

(B) If $\Pr(C \mid [x]) \leq \alpha$ & $\Pr(C \mid [x]) \geq \beta$, decide $x \in BND(C)$;

(N) If $\Pr(C \mid [x]) \leq \beta$ & $\Pr(C \mid [x]) \leq \gamma$, decide $x \in NEG(C)$;

where the threshold values α, β and γ are given by:

$$\alpha = \frac{(\lambda_{PN} - \lambda_{BN})}{(\lambda_{PN} - \lambda_{BN}) + (\lambda_{BP} - \lambda_{PP})},$$

$$\beta = \frac{(\lambda_{BN} - \lambda_{NN})}{(\lambda_{BN} - \lambda_{NN}) + (\lambda_{NP} - \lambda_{BP})},$$

$$\gamma = \frac{(\lambda_{PN} - \lambda_{NN})}{(\lambda_{PN} - \lambda_{NN}) + (\lambda_{NP} - \lambda_{PP})}.$$

In other words, from a loss function one can systematically determine the required threshold values. When
$(\lambda_{PN} - \lambda_{BN})(\lambda_{NP} - \lambda_{BP}) > (\lambda_{BP} - \lambda_{PP})(\lambda_{BN} - \lambda_{NN})$, we have $\alpha > \beta$, and thus $\alpha > \gamma > \beta$, after tie-breaking, we obtain:

P1: If $\Pr(C \mid [x]) \geq \alpha$, decide $x \in POS(C)$;

B1: If $\beta < \Pr(C \mid [x]) < \alpha$, decide $x \in BND(C)$;

N1: If $\Pr(C \mid [x]) \leq \beta$, decide $x \in NEG(C)$;

The threshold value γ is no longer needed. From the rules (P1), (B1), and (N1), the (α, β)-probabilistic positive, negative and boundary regions are given, respectively, by:

$$POS_{(\alpha,\beta)}(C) = \{x \in U \mid \Pr(C \mid [x]) \geq \alpha\},$$

$$BND_{(\alpha,\beta)}(C) = \{x \in U \mid \beta < \Pr(C \mid [x]) < \alpha\},$$

$$NEG_{(\alpha,\beta)}(C) = \{x \in U \mid \Pr(C \mid [x]) \leq \beta\}.$$

They are referred to as the three probabilistic regions (Greco, Matarazzo, & Słowiński, 2009; Herbert & Yao, 2009; Pawlak, Wang, & Ziarko, 1988; Slezak & Ziarko, 2002, 2005; Yao, 2007a; Ziarko, 1993). Therefore, the decision-theoretic rough set model provides both a theoretical basis and a practical interpretation of the probabilistic rough sets. The threshold values can be systematically calculated from a loss function based on the Bayesian decision procedure. Other probabilistic rough set models, such as the 0.5-probabilistic rough sets (Pawlak, Wang, &Ziarko, 1988), and the variable precision rough set model (Ziarko, 1993), can be derived from this approach.

INTERPRETATIONS OF CONCEPTS WITH GRANULAR COMPUTING

Granular computing is an emerging field of study that attempts to formalize and explore methods and heuristics of human problem solving with multiple levels of granularity and abstraction (Bargiela & Pedrycz, 2002; Yao, 2004a, 2004b, 2007b; Zadeh, 1997). A fundamental issue of granular computing is the representation and utilization of granules and granular structures. In this paper, we explore a connection of granules and concepts (Yao, 2009) in classification and learning. Concepts are assumed to be the basic units of knowledge, which play an important role in the study of psychology, cognitive science, and inductive learning (Mitchell, 1982, 1997; Michalski, Carbonell, & Mitchell, 1983; Smith, 1989; Sowa, 1984; van Mechelen, Hampton, Michalski, & Theuns, 1993; Wang, 2007). Following the classical interpretation of a concept (Demri & Orlowska, 1997; Michalski, Carbonell, & Mitchell, 1983; Wille, 1992), we interpret a granule as a pair of a set of objects and a logic formula describing the granule (Zhou &Yao, 2008). The detailed formulations are introduced as follows.

With respect to a dataset, we can build a model based on an information table, in which a set of objects is described by a set of attributes (Pawlak, 1991):

$$S = (U, At, \{Va \mid a \in At\}, \{Ia \mid a \in At\}),$$

where U is a finite nonempty set of objects, At is a finite nonempty set of attributes, Va is a nonempty set of values of $a \in At$, and $Ia : U \rightarrow Va$ is an information function that maps an object in U to exactly one value in Va. In classification problems, we consider an information table of the forms $S = (U, At = A \cup \{D\}, \{Va\}, \{Ia\})$, where A is a set of condition attributes describing the objects, and D is a decision attribute that indicates the classes of objects.

With any $A \subseteq At$, there is an associated equivalence relation $IND(A)$:

$$IND(A) = \{(x, y) \in U \times U \mid \forall a \in A(Ia(x) = Ia(y))\}.$$

Two objects in U satisfy $IND(A)$ if and only if they have the same values on all attributes in A. The relation $IND(A)$ is called A-indiscernibility relation. The partition of U is a family of all equivalence classes of $IND(A)$ and is denoted by

$U= IND(A)$ (U/A). The equivalence classes of the A-indiscernibility relation are denoted as $[x]_A$. Different attribute subsets will give different equivalence classes. For example, Table 1 is a simple information table. The column labeled by Class denotes an expert's classification of the objects. In Table 1, if attribute $A = \{Eyes\}$ is chosen, we can obtain the following family of equivalence classes, or a partition of U:

$$[x]_{\{Eyes\}} = \{\{o_1, o_3, o_4, o_6, o_8\}, \{o_2, o_5, o_7, o_9, o_{10}, o_{11}\}\}.$$

If we consider attribute $A = \{Eyes, Weight\}$ the family of equivalence classes is:

$$[x] = \{\{o_1\}, \{o_2, o_5, o_7, o_9\}, \{o_3, o_8\}, \{o_4, o_6\}, \{o_{10}\}, \{o_{11}\}\}.$$

If we consider each equivalence class as a granule, by choosing different set of attributes from an information table, different granularity can be produced. For certain applications, we may only need to look at granularity of certain level.

Traditionally, a concept is interpreted as a pair of intension and extension. The intension of a concept is given by a set of properties. In order to formally define intensions of concepts, we adopt the decision logic language L used and studied by Pawlak (2010). Formulas of L are constructed recursively based on a set of atomic formulas corresponding to some basic concepts. An atomic formula is given by $a = v$, where $a \in At$ and $v \in Va$. For each atomic formula $a = v$, an object x satisfies it if $Ia(x) = v$. Otherwise, it does not satisfy $a = v$. From atomic formulas, we can construct other formulas by applying the logic connectives $\neg, \wedge, \vee, \rightarrow$, and $\leftrightarrow$. Each formula represents an intension of a concept. For two formulas ϕ and φ, we say that ϕ is more specific than φ, and φ is more general than ϕ, if and only if $\phi \rightarrow \varphi$, namely, φ logically follows from ϕ. In other words, the formula $\phi \rightarrow \varphi$ is satisfied by all objects with respect to any universe U and any information function Ia. If ϕ is more

Table 1. An information table

Object	Weight	Hair	Eyes	Class
o_1	normal	red	blue	+
o_2	Low	dark	brown	+
o_3	low	grey	blue	+
o_4	high	red	blue	+
o_5	low	blond	brown	-
o_6	high	dark	blue	-
o_7	low	red	brown	+
o_8	low	blond	blue	+
o_9	low	grey	brown	-
o_{10}	normal	dark	brown	+
o_{11}	high	dark	brown	-

specific than φ, we write $\phi \prec \varphi$, and call ϕ a sub-concept of φ, and φ a super-concept of ϕ.

In inductive learning and concept formation, extensions of concepts are normally defined with respect to a particular training set of examples. If ϕ is a formula, the set $m(\phi)$ is called the meaning of the formula ϕ in M. The meaning of a formula ϕ is therefore the set of all objects having the property expressed by the formula ϕ. In other words, ϕ can be viewed as the description of the set of objects $m(\phi)$. Thus, a connection between formulas and subsets of U is established. For example, in Table 1, if attribute $A = \{Eyes\}$ is chosen, the two equivalence classes can be written as:

$$m(Eyes = blue) = \{o_1, o_3, o_4, o_6, o_8\},$$

$$m(Eyes = brown) = \{o_2, o_5, o_7, o_9, o_{10}, o_{11}\},$$

where Eyes=blue and Eyes=brown are the intensions of the concepts described by the formulas of the language L.

With the introduction of language L, we have a formal description of concepts. A concept definable in a model S is a pair $(\phi, m(\phi))$, where $\phi \in L$. More specifically, ϕ is a description of $m(\phi)$ in S, the intension of concept $(\phi, m(\phi))$, and $m(\phi)$ is the set of objects satisfying ϕ, the extension of concept $(\phi, m(\phi))$. A concept $(\phi, m(\phi))$ is said to be a sub-concept of another concept $(\phi, m(\phi))$, or $(\phi, m(\phi))$ a super-concept of $(\phi, m(\phi))$, in an information table if $m(\phi) \subseteq m(\varphi)$. A concept $(\phi, m(\phi))$ is said to be a smallest non-empty concept in M if there does not exist another non-empty proper subconcept of $(\phi, m(\phi))$.

Concept learning, to a large extent, depends on the structures of concept space and the target concepts. In general, one may not be able to obtain an effective and efficient learning algorithm, if no restrictions are imposed on the concept space. For this reason, each learning algorithm typically focuses on a specific type of concept.

AN ADAPTIVE LEARNING ALGORITHM FOR CLASSIFICATION

For classification problems, ID3 (Quinlan, 1983) is a well-known algorithm used to generate a decision tree by sequentially choosing the attribute that gives the most information about the class label, the leaf node is added to a branch if the subset of examples for that branch have the same classification labels (i.e., $\Pr(C|[x]) = 1$ or $\Pr(C|[x]) = 0$). In this paper, we introduce an adaptive learning method to construct a decision tree for classification based on three-way decisions with DTRS (Yao, 2010). A subtree is added recursively if the conditional probability lies in between of the two

threshold values α and β. Otherwise, a leaf node is added if the conditional probability $\Pr(C|[x]) \geq \alpha$, or $\Pr(C|[x]) \leq \beta$.

In Search of Effective Granulization

Instead of generating the decision tree based on the information gain, in our approach, the attributes of each inner node of the decision tree are sequentially selected by searching for the most suitable granulization at each level. More specifically, we start from the bigger granule at the top level, if the classification decisions cannot be made based on this granulization, we then search for the smaller granules by adding more attribute as inner nodes, until all the examples are correctly classified or certain condition is met. If the granulization at the current level is sufficient for classification, a finner granulization may not be needed at all; this ensures the generated decision tree to be "almost minimal." Based on these principles, the general scheme of the classification process is shown in Figure 1.

An Adaptive Learning Algorithm

The actual learning process of building the decision tree is a little bit more complicated. There are a few important steps involved in the learning method, which can be described as follows.

Step 1: At the top most level, the attribute has the least attribute values are selected as the root node A, which will give us the largest granulation. If there is more than one attribute satisfies this condition, we choose the attribute that has the minimum number of objects in its deferment area.

Step 2: A new branch is added for each possible value v_i of A. Estimate the conditional probability of each branch with respect to A, where $\Pr(C|[x])$ can be estimated from the frequencies of the training data by putting:

Figure 1. General schema of classification process in search of effective granulization

Classify *(examples)*

```
Use the entire set U as the unlabeled root node of a decision tree;
While there is an unlabeled leaf node in the tree
Choose an unlabeled leaf node;
If the conditional probability is above or equal to α
Then change the node to a labeled leaf node with label = "accept";
Else if the conditional probability is below or equal to β
Then change the node to a labeled leaf node with label = "reject";
Else if the granule meets certain conditions
Then change the node to a labeled leaf node with label = "deferment";
Else replace the unlabeled node with an attribute with each branch corresponds to an
attribute value, divide the granule into unlabeled nonempty sub-granules based on the
attribute value
End
Return a ternary classification tree.
```

$$\Pr(C \mid [x]) = \frac{\mid C \cap [x] \mid}{\mid [x] \mid},$$

where $[x] = m(v_i)$. If $\Pr(C|[x]) \geq \alpha$, objects of this branch belong to the positive region, add a leaf node labeled as C; if $\Pr(C|[x]) \leq \beta$, objects of this branch belong to the negative region, add a leaf node labeled as C^C. Otherwise, the classification of this branch cannot be determined, we then search for the next suitable granularity.

Step 3: At the next level, choose an attribute A_j that has the least attribute values from (Attributes – $\{A\}$) as the child node for the branch. This time, $[x] = m(v_i \wedge v_j)$, where v_j represents the possible values of A_j. Similarly, if there is more than one attribute satisfies this condition, we choose the attribute that has the minimum number of objects in its deferment area. Repeat Step 2. Until we can find a leaf node for each branch. Figure 2 shows such an algorithm.

AN ILLUSTRATIVE EXAMPLE

We illustrate the classification process introduced in the previous section by using an information table as Table 1.

At the top level, attribute Eyes is chosen as the root node since it has the least attribute values. Two branches are added corresponding to two possible value Eyes=blue and Eyes=brown, which divides the data set into two granules:

$$m(Eyes = blue) = \{o_1, o_3, o_4, o_6, o_8\},$$
$$m(Eyes = brown) = \{o_2, o_5, o_7, o_9, o_{10}, o_{11}\}.$$

The conditional probability of each equivalence class can be calculated as follows:

$$\Pr(C \mid [o_1]_{\{Eyes\}}) = \frac{\mid \{o_{1,3,4,6,8}\} \cap \{o_{1,2,3,4,7,8,10}\} \mid}{\mid \{o_{1,3,4,6,8}\} \mid} = \frac{4}{5},$$
$$\Pr(C \mid [o_2]_{\{Eyes\}}) = \frac{\mid \{o_{2,5,7,9,10,11}\} \cap \{o_{1,2,3,4,7,8,10}\} \mid}{\mid \{o_{2,5,7,9,10,11}\} \mid} = \frac{1}{2}.$$

Figure 2. An adaptive learning algorithm in search for effective granulization

```
ALM (S, Decision Attribute, Attributes)
Input:          a training set of examples S,
Output:         a decision tree that correctly classifies all examples in S.
Procedure: Create a root node for the tree;
If all examples are positive, Return the single-node tree Root, with label = +;
If all examples are negative, Return the single-node tree Root, with label = -;
If number of condition attributes is empty, then Return the single node tree Root,
with label = most common value of the decision attribute in the examples;
Otherwise Begin
A =The Attribute that has the least number of attribute values;
Decision Tree attribute for Root = A;
For each possible value, v_i of A
Add a new tree branch below Root, corresponding to the test A = v_i;
```

$$\text{Estimate } Pr(C|[x]) \text{ using } Pr(C|[x]) = Pr\big(C\,|\,[x]\big) = \frac{|C \cap [x]|}{|[x]|};$$

```
Let S(v_i), be the subset of examples that have the value v_i for A;
If Pr(C|[x]) ≥ α
Then below this new branch add a leaf node with label = +;
Else if Pr(C|[x]) ≤ β
Then below this new branch add a leaf node with label = -;
Else if S(v_i) is empty
Then below this new branch add a leaf node with label = most common class value in the
examples;
Else below this new branch add the subtree ALM (S(v_i), Decision Attribute, Attributes
-{A};
End
Return Root
```

Assume the two threshold parameters calculated from the loss functions are $\alpha = 0.6$ and $\beta = 0.4$. We have $Pr(C \mid [o_1]_{\{Eyes\}}) \geq \alpha$, objects $\{o_1, o_3, o_4, o_6, o_8\}$ belong to the positive region, a leaf node can be added to this branch with the class label=+. We also have $\beta \leq Pr(C \mid [o_2]_{\{Eyes\}}) \leq \alpha$, objects $\{o_2, o_5, o_7, o_9, o_{10}, o_{11}\}$ belong to the boundary region and need to be further analyzed.

At the second level, attribute Weight is chosen since it has less attribute values than attribute Hair. A subtree is added to the Eyes=brown branch, with three new branches corresponding to three possible values, that is, Weight=normal, Weight=high, and Weight=low, which divides the data set into three granules:

$$m(Eyes = brown \wedge Weight = normal) = \{o_{10}\},$$
$$m(Eyes = brown \wedge Weight = high) = \{o_{11}\},$$
$$m(Eyes = brown \wedge Weight = low) = \{o_2, o_5, o_7, o_9\}.$$

The conditional probability of each equivalence class can be calculated as follows:

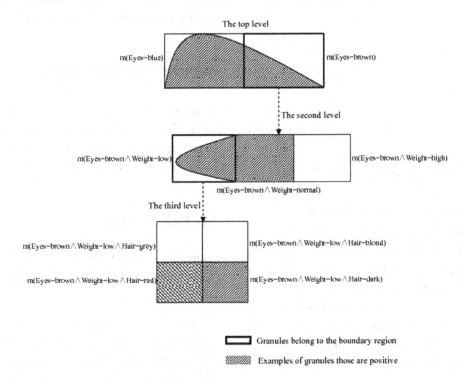

Figure 3. The classification process in search of effective granulization based on three-way decisions

$$\Pr(C \mid [o_{10}]_{\{Eyes, Weight\}}) = \frac{\mid \{o_{10}\} \cap \{o_{1,2,3,4,7,8,10}\} \mid}{\mid \{o_{10}\} \mid} = 1,$$

$$\Pr(C \mid [o_{11}]_{\{Eyes, Weight\}}) = \frac{\mid \{o_{11}\} \cap \{o_{1,2,3,4,7,8,10}\} \mid}{\mid \{o_{11}\} \mid} = 0,$$

$$\Pr(C \mid [o_{2}]_{\{Eyes, Weight\}}) = \frac{\mid \{o_{2,5,7,9}\} \cap \{o_{1,2,3,4,7,8,10}\} \mid}{\mid \{o_{2,5,7,9}\} \mid} = \frac{1}{2}.$$

Assume that we use the same threshold parameters α and β We have $\Pr(C \mid [o_{10}]_{\{Eyes, Weight\}}) \geq \alpha$, objects $\{o_{10}\}$ belong to the positive region, a leaf node can be added to this branch with the class label = +. We also have $\Pr(C \mid [o_{11}]_{\{Eyes, Weight\}}) \leq \beta$, objects $\{o_{11}\}$ belong to the negative region, a leaf node can be added to this branch with the class label=-. Finally, $\beta \leq \Pr(C \mid [o_{2}]_{\{Eyes, Weight\}}) \leq \alpha$, objects $\{o_{2}, o_{5}, o_{7}, o_{9}\}$ belong to the boundary region and need to be further analyzed.

At the third level, attribute Hair is chosen. A subtree is added to the Eyes=brown $\land$ Weight=low branch, with four new branches corresponding to four possible values, that is, Hair=red, Hair=blond, Hair=grey and Hair=dark, which divides the data set into four granules:

$$m(Eyes = brown \land Weight = low \land Hair = red) = \{o_{7}\},$$
$$m(Eyes = brown \land Weight = low \land Hair = blond) = \{o_{5}\},$$
$$m(Eyes = brown \land Weight = low \land Hair = grey) = \{o_{9}\},$$
$$m(Eyes = brown \land Weight = low \land Hair = dark) = \{o_{2}\}.$$

The conditional probability of each equivalence class can be calculated as follows:

$$\Pr(C \mid [o_7]_{\{Eyes,Weight,Hair\}}) = \frac{\mid \{o_7\} \cap \{o_{1,2,3,4,7,8,10}\} \mid}{\mid \{o_7\} \mid} = 1,$$

$$\Pr(C \mid [o_5]_{\{Eyes,Weight,Hair\}}) = \frac{\mid \{o_5\} \cap \{o_{1,2,3,4,7,8,10}\} \mid}{\mid \{o_5\} \mid} = 0,$$

$$\Pr(C \mid [o_9]_{\{Eyes,Weight,Hair\}}) = \frac{\mid \{o_9\} \cap \{o_{1,2,3,4,7,8,10}\} \mid}{\mid \{o_9\} \mid} = 0,$$

$$\Pr(C \mid [o_2]_{\{Eyes,Weight,Hair\}}) = \frac{\mid \{o_2\} \cap \{o_{1,2,3,4,7,8,10}\} \mid}{\mid \{o_2\} \mid} = 1.$$

Assume that we use the same threshold parameters α and β. We have $\{o_7\}$ and $\{o_2\}$ in the positive region, a leaf node can be added to both of these branches with class label = +. We also have $\{o_9\}$ and $\{o_5\}$ in the negative region, a leaf node can be added to both of these branches with class label= -. Up to this point, all the objects in the data set have been classified, the decision tree is complete. This classification process in search for effective granulization based on three-way decisions is illustrated in Figure 3.

CONCLUSION AND FUTURE WORK

Decision-theoretic rough set model provides a ternary classification method for classification problems. The deferred examples must be reexamined by collecting further information. We argue that this process can be done automatically by searching for effective granulization. An adaptive learning algorithm is proposed for this purpose, which generates a decision tree by sequentially choosing the attributes that give the most appropriated granulization. We start from the bigger granule at the top level of the tree, if the classification decisions can be made based on this granularity, a finer granularity may not be needed at all, and this ensures the generated tree to be "almost minimal." An illustrative example is given at the end of the paper to demonstrate this process. For future work, we will test our proposed method in larger scale real world data sets, and compare our results with existing classification algorithms to verify the effectiveness of our method.

ACKNOWLEDGMENT

The first author is supported by an NSERC Alexander Graham Bell Canada Graduate Scholarship. The second author is partially supported by an NSERC Canada Discovery grant.

REFERENCES

Bargiela, A., & Pedrycz, W. (2002). *Granular computing: An introduction.* Boston, MA: Kluwer Academic.

Demri, S., & Orlowska, E. (1997). Logical analysis of indiscernibility. In Orlowska, E. (Ed.), *Incomplete information: Rough set analysis* (pp. 347–380). Heidelberg, Germany: Physica Verlag.

Duda, R. O., & Hart, P. E. (1973). *Pattern classification and scene analysis.* New York, NY: John Wiley & Sons.

Greco, S., Matarazzo, B., & Słowínski, R. (2009). Parameterized rough set model using rough membership and Bayesian confirmation measures. *International Journal of Approximate Reasoning, 49,* 285–300. doi:10.1016/j.ijar.2007.05.018

Herbert, J. P., & Yao, J. T. (2009). Game-theoretic rough sets. *Fundamenta Informaticae, 108*(3-4), 267–286.

Liang, J. Y., & Qian, Y. H. (2008). Information granules and entropy theory in information systems. *Science in China F, 51*(9).

Michalski, R. S., Carbonell, J. G., & Mitchell, T. M. (Eds.). (1983). *Machine learning, an artificial intelligence approach.* San Francisco, CA: Morgan Kaufmann.

Mitchell, T. M. (1982). Generalization as search. *Artificial Intelligence*, *18*, 203–226. doi:10.1016/0004-3702(82)90040-6

Mitchell, T. M. (1997). *Machine learning*. New York, NY: McGraw-Hill.

Pawlak, Z. (1991). *Rough sets: Theoretical aspects of reasoning about data*. Boston, MA: Kluwer Academic.

Pawlak, Z., Wong, S. K. M., & Ziarko, W. (1988). Rough sets: probabilistic versus deterministic approach. *International Journal of Man-Machine Studies*, *29*, 81–95. doi:10.1016/S0020-7373(88)80032-4

Qian, Y. H., Liang, J. Y., & Dang, C. Y. (2009). Knowledge structure, knowledge granulation and knowledge distance in a knowledge base. *International Journal of Approximate Reasoning*, *50*(1), 174–188. doi:10.1016/j.ijar.2008.08.004

Quinlan, J. R. (1983). Learning efficient classification procedures and their application to chess endgames. In Michalski, J. S., Carbonell, J. G., & Michell, T. M. (Eds.), *Machine learning: An artificial intelligence approach* (*Vol. 1*, pp. 463–482). San Francisco, CA: Morgan Kaufmann.

Slezak, D., & Ziarko, W. (2002, December 9). Bayesian rough set model. In *Proceedings of the Conference on the Foundation of Data Mining*, Maebashi, Japan (pp. 131-135).

Slezak, D., & Ziarko, W. (2005). The investigation of the Bayesian rough set model. *International Journal of Approximate Reasoning*, *40*, 81–91. doi:10.1016/j.ijar.2004.11.004

Smith, E. E. (1989). Concepts and induction. In Posner, M. I. (Ed.), *Foundations of cognitive science* (pp. 501–526). Cambridge, MA: MIT Press.

Sowa, J. F. (1984). *Conceptual structures, information processing in mind and machine*. Reading, MA: Addison-Wesley.

van Mechelen, I., Hampton, J., Michalski, R. S., & Theuns, P. (Eds.). (1993). *Categories and concepts, theoretical views and inductive data analysis*. New York, NY: Academic Press.

Wang, Y. (2007). Cognitive informatics: exploring theoretical foundations for natural intelligence, neural Informatics, autonomic computing, and agent systems. *International Journal of Cognitive Informatics and Natural Intelligence*, *1*, 1–10. doi:10.4018/jcini.2007040101

Wang, Y., Widrow, B. C., Zhang, B., Kinsner, W., Sugawara, K., & Sun, F. C. (2011). Perspectives on the field of cognitive informatics and its future development. *International Journal of Cognitive Informatics and Natural Intelligence*, *5*(1), 1–17.

Wang, Y., Zadeh, L. A., & Yao, Y. (2009). On the system algebra foundations for granular computing. *International Journal of Software Science and Computational Intelligence*, *1*(1), 64–86. doi:10.4018/jssci.2009010105

Wille, R. (1992). Concept lattices and conceptual knowledge systems. *Computers & Mathematics with Applications (Oxford, England)*, *23*, 493–515. doi:10.1016/0898-1221(92)90120-7

Yao, Y. Y. (2004a). A partition model of granular computing. In J. F. Peters, A. Skowron, J. W. Grzymala-Busse, B. Kostek, R. W. Swiniarski, & M. S. Szczuka (Eds.), *Transactions on Rough Sets I* (LNCS 3100, pp. 232-253).

Yao, Y. Y. (2004b). Granular computing. *Computer Science*, *31*, 1–5.

Yao, Y. Y. (2007a). Decision-theoretic rough set models. In J. T. Yao, P. Lingras, W.-Z. Wu, M. Szczuka, N. J. Cercone, & D. Slezak (Eds.), *Proceedings of the Second International Conference on Rough Sets and Knowledge Technology* (LNCS 4481, pp. 1-12).

Yao, Y. Y. (2007b). The art of granular computing. In M. Kryszkiewicz, J. F. Peters, H. Rybinski, & A. Skowron (Eds.), *Proceeding of the International Conference on Rough Sets and Emerging Intelligent Systems Paradigms* (LNCS 4585, pp. 101-112).

Yao, Y. Y. (2009). Interpreting concept learning in cognitive informatics and granular computing. *IEEE Transactions on Systems, Man, and Cybernetics. Part B, Cybernetics, 39*(4), 855–866. doi:10.1109/TSMCB.2009.2013334

Yao, Y. Y. (2010). Three-way decisions with probabilistic rough sets. *Information Sciences, 180*(3), 341–353. doi:10.1016/j.ins.2009.09.021

Yao, Y. Y., & Wong, S. K. M. (1992). A decision theoretic framework for approximating concepts. *International Journal of Man-Machine Studies, 37*, 793–809. doi:10.1016/0020-7373(92)90069-W

Yao, Y. Y., Wong, S. K. M., & Lingras, P. (1990). A decisiontheoretic rough set model. In Ras, Z. W., Zemankova, M., & Emrich, M. L. (Eds.), *Methodologies for intelligent systems* (*Vol. 5*, pp. 17–24). New York, NY: North-Holland.

Zadeh, L. A. (1997). Towards a theory of fuzzy information granulation and its centrality in human reasoning and fuzzy logic. *Fuzzy Sets and Systems, 19*, 111–127. doi:10.1016/S0165-0114(97)00077-8

Zhou, B., & Yao, Y. Y. (2008). A logic approach to granular computing. *International Journal of Cognitive Informatics and Natural Intelligence, 2*(2), 63–79. doi:10.4018/jcini.2008040104

Ziarko, W. (1993). Variable precision rough sets model. *Journal of Computer and System Sciences, 46*, 39–59. doi:10.1016/0022-0000(93)90048-2

This work was previously published in the International Journal of Cognitive Informatics and Natural Intelligence, Volume 5, Issue 3, edited by Yingxu Wang, pp. 47-60, copyright 2011 by IGI Publishing (an imprint of IGI Global).

Section 4
Computational Intelligence

Chapter 16
Cognitive Dynamic Systems

Simon Haykin
McMaster University, Canada

ABSTRACT

The main topics covered in this paper address the following four issues: (1) Distinction between how adaptation and cognition are viewed with respect to each other, (2) With human cognition viewed as the framework for cognition, the following cognitive processes are identified: the perception-action cycle, memory, attention, intelligence, and language. With language being outside the scope of the paper, detailed accounts of the other four cognitive processes are discussed, (3) Cognitive radar is singled out as an example application of cognitive dynamic systems that "mimics" the visual brain; experimental results on tracking are presented using simulations, which clearly demonstrate the information-processing power of cognition, and (4) Two other example applications of cognitive dynamic systems, namely, cognitive radio and cognitive control, are briefly described.

1. INTRODUCTION

The first seminal journal paper on Cognitive Radio appeared in February 2005 (Haykin, 2005), which was subsequently followed by the first seminal journal paper on Cognitive Radar that was published in January 2006 (Haykin, 2006a). Emboldened by the publication of those two papers and recognizing that, in reality, Cognitive Radio and Cognitive Radar are two important members

of a broadly defined new integrative field, named Cognitive Dynamic Systems, a predictive point-of-view article with this very title was published in November 2006 (Haykin, 2006b).

In this paper, an overview of Cognitive Dynamic Systems is presented. In a way, this overview provides some highlights of a new book with this very title, due to be published in late 2011 (Haykin, 2011).

The paper is organized as follows. Section 2 discusses the distinction between cognition and adaptation that should be carefully noted. Sec-

DOI: 10.4018/978-1-4666-2476-4.ch016

tion 3 discusses the five underlying principles of Cognitive Dynamic Systems, viewed with human cognition as the frame of reference. The five principles are: the perception-action cycle, multi-scale (layer) memory, attention, intelligence, and language.

With these principles in mind, Section 4 discusses Cognitive Radar with target tracking as the application of interest; the discussion also includes experimental results obtained using simulations. Section 5 briefly describes two other applications: Cognitive Radio and Cognitive Control.

2. DISTINCTION BETWEEN ADAPTATION AND COGNITION

The idea of adaptation may be traced back to the pioneering work of Widrow and associates at Stanford University, California. In particular, Widrow and Hoff (19xx) described an *adaptive filter*, consisting of the following components:

- **Linear Combiner:** Equipped with a set of free parameters.
- **Comparator:** Measures the difference between an externally supplied desired response and the actual output of the linear combiner produced in response to an input signal.
- **Control Mechanism:** Adjusts the parameters of the linear combiner so as to minimize the error signal in some statistical sense.

We may therefore offer the following definition:

Adaptation is a signal-processing paradigm, with a built-in mechanism, which adjusts the free parameters of a typically linear filter of finite-duration impulse response in accordance with statistical variations of the environment.

In direct contrast, we define cognition as follows:

Cognition is an information-processing paradigm with a built-in mechanism, which enables a dynamic system to learn from the experience gained through continued interactions with its environment.

Note that in cognition we speak of information processing rather than signal processing as in the case of adaptation.

3. THE FIVE PRINCIPLES OF COGNITIVE DYNAMIC SYSTEMS

For a dynamic system to be cognitive in the true sense of the word, it has to satisfy the five basic principles of human cognition (Haykin, 2011). The five principles are:

1. The perception-action cycle;
2. Multi-scale memory;
3. Attention;
4. Intelligence;
5. Language.

The fifth principle, language, provides the means for effective and efficient communications between the different parts constituting the cognitive dynamic system. With the language being outside the scope of this paper, the discussion will be focused entirely on principles (1, 2, 3, 4, and 5).

3.1 The Perception-Action Cycle

The human brain has two main parts, left and right. Correspondingly, a cognitive dynamic system also consists of two main parts, one part being responsible for *perception* of the world and the other part being responsible for *action* on or in the world; herewith, the two terms, world and environment, are used interchangeably. For example, in both radar and radio, we have two parts, commonly referred to as the receiver and transmitter. The receiver is responsible for perceiving the environment, while the transmitter is responsible for taking action in the environment.

It follows therefore that if we are to realize the perception-action cycle, that is, principle (i) of cognition, there has to be a *feedback link* that connects the receiver to the transmitter. Typically, such a link is missing from both radar and radio. Hence, the first characteristic that distinguishes both cognitive radar and cognitive radio from their traditional counterparts is the inclusion of this feedback link. From control theory, we know that feedback is a "double-edged sword." In designing a cognitive dynamic system, care must be taken to ensure that *stability* of the system is assured at all times.

In radar, the transmitter and receiver are collocated. Accordingly, inclusion of the feedback link in cognitive radar is relatively straightforward. On the other, in radio, the transmitter and receiver are located in different places; in this second application of cognition, inclusion of the feedback link in cognitive radio is a more demanding proposition.

In conceptual terms, what really matters is the fact that the first physical characteristic of a cognitive dynamic system is the *global feedback loop* that embodies four components:

- The receiver;
- The feedback link from the receiver to the transmitter;
- The transmitter itself;
- The environment.

In other words, we may view a cognitive dynamic system as a *closed-loop feedback control system*. The feedback loop is said to be *global* because the environment lies *inside* the loop. This form of feedback is to be distinguished from *local feedback*, where the feedback loop is localized within the receiver or the transmitter. (Local feedback is a characteristic of adaptive filters.)

Now that we understand that every cognitive dynamic system has global feedback loop, we may describe the perception-action cycle as follows:

- First, the receiver of the system *perceives* the world by extracting the *relevant* information content in the incoming measurements (i.e., received signal). For example, in cognitive radar aimed at the tracking of an unknown target, the relevant information pertains to estimating the *state* of the target. For another example, in cognitive radio, the relevant information refers to underutilized subbands of the radio spectrum.
- Second, the feedback information from the receiver to the transmitter is naturally application-dependent. In the radar example, the feedback information from the receiver to the transmitter is based on the state-estimation error vector. On the other hand, in the radio example, the feedback information refers to where the underutilized subbands are located inside the radio spectrum at a particular point in time and location in space, and similarly for interferers.
- Three, the transmitter exploits the feedback information passed onto it by the receiver so as to *act* on the receiver via the environment for a specific objective in mind. In the tracking radar example, the objective is to improve the tracking accuracy of the state estimated in the receiver. In cognitive radio, the objective is to distribute the underutilized subbands of the radio spectrum among unserviced users as fast as possible and in a fair-minded way so as to provide reliable communication whenever and wherever required.

The measure by which the perception-action cycle is judged is described in conceptual terms as follows (Haykin, 2011):

Information obtained about the environment by the Cognitive Dynamic System is continually defined from one cycle of the perception-action cycle to the next.

In other words, a Cognitive Dynamic System has the built-in capability through feedback from the receiver to the transmitter to continually improve its information about the environment on a cycle-by-cycle basis.

3.2 Memory

Before we proceed to discuss the next principle of cognition, a clear distinction should be made between knowledge and memory:

- Knowledge is static and therefore timeless.
- On the other hand, memory is dynamic and therefore time-varying.

Just as the perception-action cycle in a cognitive system is dynamic, it is only natural for memory to be dynamic too.

In compositional terms, memory consists of a multi-layer neural network (such as the *multi-layer perception*) with one or more layers; each layer consists of computational units, called *neurons*. The term "hidden" is commonly used to signify the fact that the intermediate layers of the network are hidden from the input as well as the output (Haykin, 2009). In addition to the hidden layers, the multi-layer perception also has two other layers:

- **Input Layer:** Provides the interface between the neural network and the outside environment.
- **Output Layer:** Provides one last layer of computational units, neurons, where the actual output of the network is compared against the externally desired response.

The error signal, defined as the difference between these two response provides the mathematical element for adjusting the free parameters of the neural network, starting from the output layer and proceeding backward layer by layer, until all the layers of the network are accounted for. The "supervised-learning" algorithm used to do these adjustments is called the *back-propogation algorithm*[1] (Haykin, 2009); this algorithm includes the celebrated least-mean-square (LMS) algorithm used for the supervised training of a linear adaptive filter consisting of a single computational unit as a special case. To account for statistical variations of the environment, it is advisable to add an unsupervised layer, which could be achieved by using the generalized Hebbian algorithm (Haykin, 2009).

Now that we understand how the memory may be designed using learning machines, the next issue of interest is to address the function of memory in a cognitive dynamic system. To this end, we offer the following (Haykin, 2011):

The function of memory is to predict the consequences of action taken by the system. Naturally, the action may be taken by the transmitter, the receiver, or both.

3.3 Attention

In a cognitive dynamic system, attention is based on the perception-action cycle and memory. However, there is a difference to be noted here: whereas both the perception-action cycle and memory occupy specific physical locations of their own in a cognitive dynamic system, attention manifests itself in the system through *attentional algorithms* that embody the memory and perception-action cycle.

As for the function of attention, we say the following (Haykin, 2011):

Attention in a cognitive dynamic system provides the mechanism needed to prioritize the allocation of computational resources so as to mitigate the information overload problem.

Attention is carried out in the transmitter, the receiver, or both.

3.4 Intelligence

The fourth principle of cognition is intelligence, which is based on the other three principles: the perception-action cycle, memory, and attention. As such, intelligence is the most powerful of all the principles of cognition.

The function of intelligence may be described as follows (Haykin, 2011):

Intelligence provides the algorithmic mechanism, whereby a cognitive dynamic system is enabled to select a decision-making strategy that is optimal in some statistical sense in the face of unpredictable uncertainties.

With decision-making being the last task required of a cognitive dynamic system and the strategy so chosen by the system, we can now see why intelligence is said to be the most powerful of all the principles of cognition.

Another important thing to note here is the fact that with a feedback link connecting the receiver to the transmitter and the deployment of a multi-scale (layer) memory, there will be an abundance of local as well as global feedback loops distributed throughout a cognitive dynamic system. Accordingly, we may go on to say:

Intelligence is facilitated through the local and global feedback loops distributed in the system.

Having covered the principles of cognition, we are now prepared to discuss cognitive radar followed by cognitive radio, which represented two important applications of cognition.

4. COGNITIVE RADAR

As mentioned previously, the transmitter and receiver of a radar system are traditionally collocated, which eases the burden of linking the receiver to the transmitter. Moreover, radar is a remote-sensing system. Putting these two radar features together, we may exploit ideas drawn from the "visual brain" to build a cognitive radar. Making this point even more emphatic; the best way to build a cognitive radio is to *mimic* the visual brain. Needless to say, such an ambitious goal requires not only knowledge of radar, signal processing, information theory and control but also cognition; so, the task involved can be quite demanding.

4.1 Descriptive Details of Cognitive Radar

Figure 1 depicts the block diagram[2] of a cognitive radar that satisfies all the four principles of cognition addressed in Section 3.

Starting with the receiver, it consists of two components:

1. **Environmental Scene Analyzer:** Tracking as the application of interest, the logical choice for implementing this first component of the receiver is to use a *state estimator*. In the simulations to be presented, the target state-space model was highly nonlinear, hence the adoption of the *Cubature Kalman Filter*. This relatively new filter, discussed in detail in Arasaratnan and Haykin (2009), is a method of choice for approximating the optimal Bayesian filter for a nonlinear state-space model, beyond capability of the commonly used extended Kalman filter and under the Gaussian assumption.

2. **Perceptual Memory:** This second component of the receiver consists of a multi-layer perceptor with two hidden unsupervised layers followed by a supervised output layer. The memory is designed using an encoding-decoding principle which is well-suited for the construction of an "optimal" replicator (Hecht-Nielssen, 1995). For the supervised training of the output layer, a *system-model library* is employed. This li-

Figure 1. Cognitive radar

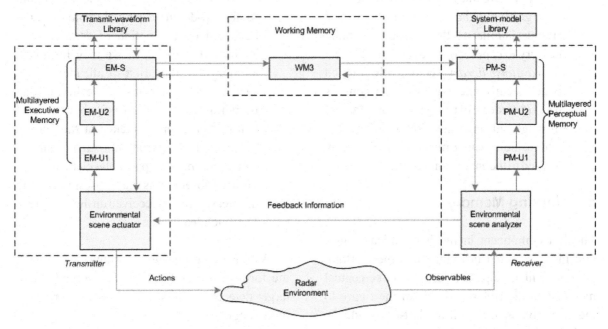

brary is included to account for the fact that the actual nonlinear function and statistical characterization of the system noise for state estimation in a particular cycle may deviate from the corresponding values assigned to the system equation in the state-space model to begin with. Since the perceptual memory is reciprocally coupled to the environmental scene analyzer at the bottom end and it is reciprocally coupled to the system-model library at the top end, as shown in Figure 1, the perceptual memory is enabled to find that particular element stored in the system-model library that is the closest match (i.e., the most highly correlated) to the actual measurements received at a particular cycle. This matching processing is continued from one cycle to the next.

Turning next to the transmitter, we see from Figure 1 that it also consists of two components:

1. **Environmental Scene Actuator:** Here again, with tracking as the application of interest, the logical choice for the environmental scene actuator is approximate dynamic programming (Bertsekas, 2005). We say "approximate" because of the imperfect state estimation problem, which arises because the transmitter does not have direct access to the radar environment; rather, it sees the environment indirectly through the "eyes" of the receiver. To mitigate this problem, the cost-to-go-function (e.g., the mean-squared error of the state-estimation error vector computed from the feedback information) would have to be ensemble-averaged over all possible states (Haykin, 2011; Hecht-Nielssen, 1995). Doing this expectation, we are faced with another problem: the curse of dimensionality. To mitigate this new problem, we resort to the use of the *cubature rule*, the very rule that is used in deriving the Cubature Kalman Filter.

2. **Executive Memory:** This second memory of cognitive radar has a multi-layered structure similar to that of the perceptual memory except for one obvious difference: the *waveform library*, reciprocally coupled to the executive memory at the top, consists of a grid made up of two parameters, namely the range and range rate. Note also that at the bottom, the executive memory is reciprocally coupled to the environmental scene actuator.

4.2 Working Memory

The last component in the hierarchical cognitive radar depicted in Figure 1 is the working memory, which reciprocally couples the perceptual memory on the right-hand side of the figure to the executive memory on the left-hand side. In so doing, the cognitive radar assumes the form of a *synchronized computing machine* that operates in a self-organized manner from one cycle of the perception-action cycle to the next.

4.3 Algorithmic Considerations

Examining Figure 1, we clearly see that both the perception-action cycle and memory occupy distinct space of their own within cognitive radar. When it comes to attention and intelligence, however, they exercise their respective roles in cognitive radar through algorithmic mechanisms that build on the perception-action cycle and memory, as explained next.

To be more specific, we speak of perceptual attention and executive attention:

1. **Perceptual Attention:** Manifests itself in top-down attention and bottom-up attention. To explain top-down attention, suppose that the perceptual memory has picked a certain element in the system-model library as the best match for the environmental measurements received at a particular cycle in the perception-action cycle. A logical algorithm for the top-down perceptual attention would be based on the notion of "explore and exploit." For example, we may formulate an algorithm that, first of all, identifies a subset of the elements in the library that lie within the *localized* neighbourhood of that certain element picked at the current cycle. The top-down attentional mechanism would then pick the particular element in the library subset that is the "closest match" to the measurements received on the next cycle in a Euclidean sense.

As for the bottom-up algorithm, it may take the form of a *hypothesis-testing procedure* for target detection, which sets the stage for tracking to begin.

2. **Executive Attention:** Takes the form of top-down and bottom-up attention algorithms. For the top-down attentional algorithm, it may also proceed based on the explore-exploit strategy in a manner similar to the perceptual attention with a pair of differences:
   a. The waveform library replaces the system model library;
   b. The feedback information (from the receiver to the transmitter) replaces the measurements at the receiver input.

As for the bottom-up attentional algorithm, its role could be simply that of initiating the decision-making.

Finally, the cognitive radar looks to intelligence for the optimal decision-making process picked in the face of environmental uncertainties given the feedback information from the receiver to the transmitter. This process may proceed as follows: Recognizing that desirably the feedback information should involve a one-step prediction of the state into the future, given the current set of measurements; two operations are performed:

- First, the available feedback information is used to formulate its "filtered" version, in which the only term that depends on the next vector of waveform parameters resides in the covariance matrix of the measurement noise vector.
- Second, a cost function is formulated using the filtered feedback information.

Then, the only issue that remains is to find the unknown waveform-parameter vector, for which the cost function based on the filtered feedback information is minimized. Given knowledge of this updated vector, the stage is set for computing the new waveform parameter vector; hence the next cycle of the perception-action cycle may proceed.

Earlier, we said that intelligence builds on the perception-action cycle, memory, and attention, and it is the most powerful of them all. So, it is with cognitive radar.

5. OTHER DYNAMIC SYSTEM APPLICATIONS OF COGNITION

5.1 Cognitive Radio

Another well-studied application of cognition is in radio. Indeed, it is in *cognitive radio* that we see an exponential growth of research interests, when it comes to applied cognition, particularly in the past eight years or so. The motivation for this growth in cognitive radar is to overcome the way in which the radio spectrum is presently underutilized.

In its most basic form, cognitive radio consists of four functional blocks:

1. **Environmental Scene Analyzer:** The purpose of which is to sense the radio environment for the purpose of discovering where underutilized subbands of the radio spectrum are located in time as well as space.

The underutilized subbands are commonly referred to as spectrum holes. One other important function of the environmental scene analyzer is to identify where the interferers are located, also in time as well as space. In effect, the spectrum holes and interferences may be viewed as the spatio-temporal state of the radio environment as seen by cognitive radio (Haykin, Thomson, & Reed, 2009).

2. **Feedback Link:** The function of which is to convey the information on a spatio-temporal state of the radio environment to the transmitter.

3. **Radio Scene Actuator:** Which is made of two functional blocks of its own:

   a. **Dynamic Spectrum Manager (DSM):** Given the feedback information from the receiver, the DSM in responsible for distributing the available spectrum among the competing users in a satisfactory manner, efficiently and effectively, and as fast as possible. In Khozemeih and Haykin (2010), a brain-inspired DSM scheme is described; specifically, the DSM exploits the built-in correlative property of *Hebb's postulate of learning*. This postulate was first described in the classic book on self-organization of behavior (Hebb, 1949).

   b. **Transmit-Power Controller (TPC):** This second functional block of the DSM has the task of allocating the limited battery power among the competing cognitive radio users by treating the network of such users as a game-theoretic problem (Setoodeh & Haykin, 2009).

   c. **Radio Environment:** The natural presence of which closes the global feedback, and with it the perception-action cycle is enabled to operate.

Figure 2. Simulations of cognitive radar: Ground-breaking results

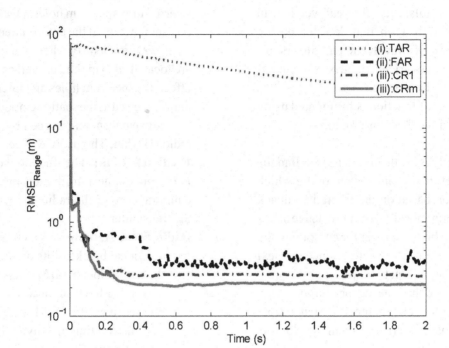

Figure 3. SIMULATIONS of cognitive radar: Ground-breaking results

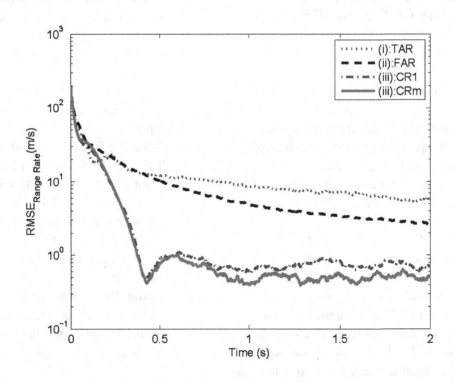

It turns out that the brain-inspired DSM has built-in memory; hence, this necessary process in cognition is satisfied. Moreover, the process of attention is accounted for by the DSM itself as well as attentional information about interferers in the environment sent to the DSM by the environmental scene analyzer. Finally, the combined presence of local loops in the environmental scene analyzer and actuator, together with the global feedback loop responsible for the perception-action cycle, and putting them altogether, the last cognitive process, namely, intelligence, is accounted for.

5.2 Cognitive Control

Work is currently going on how to exploit the ideas gained from the study of cognitive radar, with the objective of formulating the framework for a cognitive controller. Building on this line of thinking, we have successfully demonstrated some of the main principles of cognitive control, using simulations (Figures 2 and 3) based on the following functional blocks:

- **Perceptor:** Plays the role of analyzing the plant output.
- **Controller:** Implemented using the *Q-learning algorithm*, well known in the reinforcement learning literature (Sutton & Barto, 1998).

Currently, we are working on a paper that will demonstrate the practical benefits of cognition control, which, to the best of our knowledge, is the first paper published on cognitive control from an engineering perspective

REFERENCES

Arasaratnan, I., & Haykin, S. (2009). Cubature Kalman Filters. *IEEE Transactions on Automatic Control, 54*(6), 1254–1269. doi:10.1109/TAC.2009.2019800

Bertsekas, D. P. (2005). *Dynamic programming and optimal control* (3rd ed., *Vol. 1*). Hilliard, OH: Athena.

Haykin, S. (2005). Cognitive radio: Brain-empowered wireless communication. *IEEE Journal on Selected Areas in Communications, 23*(2), 201–220. doi:10.1109/JSAC.2004.839380

Haykin, S. (2006a). Cognitive radar: A way of the future. *IEEE Signal Processing Magazine, 23*(1), 30–40. doi:10.1109/MSP.2006.1593335

Haykin, S. (2006b, November). Cognitive dynamic systems. In *Proceedings of the IEEE International Conference on Acoustics, Speech and Signal Processing* (pp. 1369-1372).

Haykin, S. (2009). *Neural networks and learning machines* (3rd ed.). Upper Saddle River, NJ: Prentice Hall.

Haykin, S. (2011). *Cognitive dynamic systems: Perception-action cycle, radar, and radio*. Cambridge, UK: Cambridge University Press.

Haykin, S., Thomson, D., & Reed, J. (2009). Spectrum sensing for cognitive radio. *Proceedings of the IEEE, 97*(5), 849–877. doi:10.1109/JPROC.2009.2015711

Haykin, S., & Xue, Y. (n. d.). *Cognitive radar*. Manuscript submitted for publication.

Haykin, S., Zia, A., Xue, Y., & Arasaratnan, I. (2011). Control theoretic approach to tracking radar: First step towards cognition. *Digital Signal Processing, 21*(5). doi:10.1016/j.dsp.2011.01.004

Hebb, D. (1949). *Self-organization of behavior*. New York, NY: John Wiley & Sons.

Hecht-Nielssen, R. (1995). Replicator neural networks for universal optimal source coding. *Science, 269*(5232), 1860–1863. doi:10.1126/science.269.5232.1860

Khozemeih, F., & Haykin, S. (2010). Self-organizing dynamic spectrum management for cognitive radio networks. In *Proceedings of the 8th Annual Communication Networks and Services Research Conference*, Montreal, QC, Canada.

Setoodeh, P., & Haykin, S. (2009). Robust transmit power control for cognitive radio. *Proceedings of the IEEE*, *97*(5), 915–939. doi:10.1109/JPROC.2009.2015718

Sutton, R. S., & Barto, A. (1998). *Reinforcement learning*. Cambridge, MA: MIT Press.

Widrow, B., & Hoff, M. E. (1960). Adaptive switching circuits. *IRE WESCON Convention Record*, *4*(1), 96–104.

ENDNOTES

[1] The multi-layer perception is simple to train using the back-propogation algorithm. However, there are more refined algorithms for the design of memory, which are based on sparse-coding that is inspired by the human brain; for more details, see Chapter 6 of the CDS book (Haykin, 2006b).

[2] The cognitive radar described in Figure 1 follows from two papers:
a. Paper 7, where the underlying theory of the first step towards cognitizing radar (i.e., implementing the perception-action cycle) for tracking was described for the first time in the literature.
b. Paper 8, building on paper 7, presents the underlying details involved in Figure 1.

This work was previously published in the International Journal of Cognitive Informatics and Natural Intelligence, Volume 5, Issue 4, edited by Yingxu Wang, pp. 33-43, copyright 2011 by IGI Publishing (an imprint of IGI Global).

Chapter 17
A Modular Dynamical Cryptosystem Based on Continuous-Interval Cellular Automata

Jesus D. Terrazas Gonzalez
University of Manitoba, Canada

Witold Kinsner
University of Manitoba, Canada

ABSTRACT

This paper presents a new cryptosystem based on chaotic continuous-interval cellular automata (CCA) to increase data protection as demonstrated by their flexibility to encrypt and decrypt information from distinct sources. Enhancements to cryptosystems are also presented including (i) a model based on a new chaotic CCA attractor, (ii) the dynamical integration of modules containing dynamical systems to generate complex sequences, and (iii) an enhancement for symmetric cryptosystems by allowing them to generate an unlimited number of keys. This paper also presents a process of mixing chaotic sequences obtained from cellular automata, instead of using differential equations, as a basis to achieve higher security and higher speed for the encryption and decryption processes, as compared to other recent approaches. The complexity of the mixed sequences is measured using the variance fractal dimension trajectory to compare them to the unmixed chaotic sequences to verify that the former are more complex. This type of polyscale measure and evaluation has never been done in the past outside this research group.

DOI: 10.4018/978-1-4666-2476-4.ch017

1. INTRODUCTION

Cryptosystems have been developed to handle the challenging task of data protection in the modern information era. The purpose of cryptography is to hide the contents of messages to make them unrecognizable except by someone who has the decryption method available (Anghelescu, Ionita, & Sofron, 2008). Different cryptosystems have been proposed and implemented in either hardware (Anghelescu *et al.*, 2007, 2008), or software (Anghelescu, Sofron, Rîncu, & Iana, 2008), or mixtures of both. Cryptosystems based on *cellular automata* (CA) (Anghelescu *et al.*, 2008) are preferred over continuous chaotic systems (Moulin & Sbodio, 2010; Yifang, Rong, & Yi, 2009). Because of the simplicity and speed of CA-based computations, in contrast to the more costly equivalent models based on differential equations. Systems similar to CA were studied in the late 1950s to generate random sequences in cryptography (Wolfram, 2002).

The cryptosystem proposed in this paper is based on *continuous-interval cellular automata* (CCA) that are generalized CA. The specific interval considered is. It is shown that this cryptosystem is very fast, highly secure, and applicable to many classes of data, including text, sound, and images.

The degree of complexity of a dynamical-system or a cryptosystem based on CA has not been measured in the past. This paper presents such a complexity measure based on the *variance fractal dimension trajectory* (VFDT) (Kinsner, 2007b, 2011b; Kinsner & Grieder, 2008) to compare an unmixed CCA chaotic sequence and a mixed CCA chaotic sequence.

Cryptography is successful if the encoded information cannot be broken, and if it is computationally efficient (Stinson, 2006). Security is an important, challenging, and multi-dimensional research field in networked computing and communication systems (Alpcan & Başar, 2011). Cryptography is one of the many aspects of network security, including: access control, security protocols, information and hardware security,

privacy, risk management, resource allocation among the most important ones (Alpcan & Başar, 2011). It is a good practice to identify potential non secure points, new tools, and the correct time to implement changes (Panayiotou & Bennett, 2009).

Web services bring about many new security problems. Some approaches to manage the access control rely on poor ways to enforce authentication-like feedback (Jin & Peng, 2010) from honest and dishonest accesses. This implies that the system is not able to block undesired accesses making it weak. The need of web-based systems that reduce the dishonest accesses is of vital importance for their users.

Given the dynamic nature of network security (Alpcan & Başar, 2011) one should not rely on static measures, or computationally costly algorithms. Dynamical problems require dynamical solutions (Kinsner, 2007a). A design approach for dynamical cryptosystems is provided considering the important contributions of Shannon: *"Good mixing transformations are often formed by repeated products of two simple non-commuting operations. Hopf has shown, for example, that pastry dough can be mixed by such a sequence of operations. The dough is first rolled out into a thin slab, then folded over, then rolled, and then folded again, etc."* (Shannon, 1949). This idea is the core in the interaction among modules containing a CA to obtain a mixture of different complex behaviours where simple operations are of paramount importance. Implementations based either on high speed hardware, or software, or hybrid are easier to deploy.

Past computing and communication systems have not been designed with security as a priority (Sung, Hsu, & Chen, 2010; Zhao, 2010). Since it is practically impossible for a security expert to oversee all systems all the time (Alpcan & Başar, 2011), the development of more robust protection tools is required. Data from sensors or commands sent through insecure channels to actuators require protection and authenticity verification of the source requesting the execution of a task in high importance applications.

Secure communications for sensitive information is not only compelling for military and government institutions, but also for non-military industries, businesses, and private individuals (Anghelescu, *et al.*, 2008). Today, we can obtain many e-Services from sources such as e-Business (Tang & Zhang, 2005), e-Commerce (Qu, Ma, & Zhang, 2008), e-Health (Ding & Klein, 2010; Boonyarattaphan, Bai, & Chung, 2009), e-Government (Moulin & Sbodio, 2010), and e-Goods (Nenadic, Zhang, Shi, & Goble, 2005). Such e-Services elevate accessibility and information confidence to a high level of importance, and require robust methods capable of offering high reliability of data security. Cryptography is an indispensable part in Internet communications (Yifang *et al.*, 2009) (*e.g.*, cloud computing services (Hu & Klein, 2009). Services based in cloud computing rely highly in the most advanced encryption techniques to ensure data safety (Hu & Klein, 2009). In addition, command encryption used in the military fields is migrating to different civilian industries.

Cellular automata are capable of developing chaotic behaviour using simple operations or rules (Wolfram, 2002), thus offering the benefit of high speed computation. This is the basis in the development of the modular dynamical cryptosystem. "Modular" is used in the sense that different modules, each containing a CCA-based dynamical system, are capable of interacting to generate a complex sequence. It is important that a cryptosystem has the flexibility to encrypt different datasets, rather to focus on specific ones (Moulin & Sbodio, 2010). Fast computation helps in achieving this capability. This research provides a substantial contribution to high-speed implementation and information security of cryptosystems.

The next sections provide background on cryptosystems and their measures, followed by modular dynamical cryptosystems, experimental results and discussion.

2. BACKGROUND

2.1 Cryptography Related Definitions

Cryptography is concerned with changing an open message in *plaintext* (PT) to a *ciphertext* (CT) that is not understood according to a key (Bosworth, 1982; Ferguson, 2010; Katz & Lindell, 2007). The scrambled message (a *cryptogram*) can be either a code or a cipher. A *code* is a scheme of translating words or symbols from the original message into a new collection of corresponding words or symbols. A *cipher* is a scheme of translating individual symbols of the original message into another sequence of symbols. The formation of a code or a cipher requires a key. The *key* is a formula or device by which the code or cipher is created. *Encrypting* (encoding) is the process of changing a PT to produce a cryptogram. The inverse process of deciphering (decoding) a cryptogram is called *decrypting*. A set of algorithms for key generation, encryption, and decryption is called a *cryptosystem*.

A cryptosystem using a single secret key for enciphering and deciphering is *symmetric* (Anghelescu *et al.*, 2008). If there are many receivers, the risk increases that the key is intercepted and decoded. An alternative is to use two keys: one public key for encryption (known to all receivers), and one secret key for decryption (known to only one receiver) as proposed earlier by Diffie and Hellman (1976) and implemented by Rivest, Shamir, and Adelman (1978). The strength of this public-key cryptography is that (1) the deciphering function cannot be derived from the enciphering function (due to the trap-door one-way functions), and (2) the associated message signature can verify the sender to the receiver uniquely.

Cryptosystems should be distinguished from coding and decoding systems for error detection and correction of data for their storage and transmission (Glover & Dudley, 1991; Peterson, 1962; Rhee, 1989; Vanstone & van Oorshot, 1989; Wakerley, 1978).

Cryptoanalysis is concerned with breaking codes and ciphers without any prior knowledge about the key or algorithms (Swenson, 2008, Sinkov, 1968; Gaines, 1956). Both cryptography and cryptoanalysis are called *cryptology*. Cryptology also includes *signal security* (*i.e.*, the methods of protecting messages and communications from interception so that they could not be modified, destroyed, disclosed, or compromised in any way) and *signal intelligence* (the process of obtaining information by intercepting and solving cryptoanalysis).

2.1.1 *Types of Codes and Ciphers*

Secure communications have been used throughout history. For example, Julius Caesar used an alphabet cipher. During the American Revolution, George Washington used a codebook to collect intelligence about the British forces and their movements. Today, information is being gathered by ships, satellites, and electronic eavesdropping.

Simple ciphers include transposition codes such as message reversal, geometrical patterns, route transposition with many of its variations, columnar and double-columnar transposition (both mono and polyliteral), reciprocal ciphers, mono and polyalphabetic substitutions, decimated alphabet ciphers, digraphic substitutions, random table substitutions, as well as linear and nonlinear scalar and matrix scrambling. There are many practical implementations of such cryptosystems (Schneier, 1995).

2.1.2 Data Encryption Standards (DES) & Advanced Encryption Standard (AES)

In 1973, the National Bureau of Standards (NBS) requested the development of a federal standard for computer data protection. In response, IBM developed the *Data Encryption Standard* (DES) (Bosworth, 1982; Gait, 1978; National Bureau

of Standards, 1977). The DES system involves nonlinear ciphering algorithms. It converts 64-bit blocks of PT into 64-bit blocks of CT, using a 56-bit keying parameter whose individual bits are random, and are error protected by 8 odd parity bits. A block of data is subjected to an initial permutation, then to a 16-round computation sequence, and finally to a permutation that is the inverse of the initial one. This DES cryptosystem is symmetric.

Since the original 56-bit DES can be broken quickly, it is now considered obsolete, and a new symmetric advanced *encryption standard* (AES) was developed in 2001, whose keys offer either 128, or 192, or 256 bits.

2.1.3 Public-Key Cryptography

The DES algorithm represents a turning point from classical to modern cryptography, but it is relatively weak, considering today's easy access to extraordinary computing power to break its symmetric key. Public-key cryptography (PKC) was proposed as an attempt to have privacy and an efficient key distribution scheme that eliminates the need for a secure channel to exchange keys (Diffie & Hellman, 1976). Asymmetric key cryptography has become a standard for most of the e-Commerce secure transactions providing identification, authentication, authorization, signature, and verification services. An example of a PKC is the *Rivest-Shamir-Adelman* (RSA) algorithm that uses prime numbers and modular arithmetic for the public and private keys (Gait, 1978; National Bureau of Standards, 1977; Rivest *et al.*, 1978). The security of the PKC comes from the difficulty in factoring large composite prime numbers, while it is easy to compute the product of two large prime numbers. The public and private keys are functions of the product and the two primes.

2.1.4 Elliptic-Curve Cryptography (EEC)

Elliptic-curve cryptography (ECC), proposed by Victor Miller (IBM) and Neal Koblitz (University of Washington) in 1985 (Koblitz, 1995), is an extension of PKC. The security of EEC relies on the *elliptic-curve discrete logarithm problem* (ECDLP) in which exponentiation over the discrete prime field is easy, but the inverse (computing the logarithm) is very difficult. For a given key size, ECC offers considerably greater security than its alternatives. For example, for 156-bit key size in ECC requires an equivalent 1024-bit RSA key. Other equivalent ECC to RSA ratios are 256/3072, 384/7680, 512/15360.

An *elliptic curve* in Cartesian coordinates is the set of solutions (x , y) to an equation of the form (Koblitz, 1994; Menezes, 1993; Rosing, 1998; Washington, 2008)

$$x_2{}^2 = x_1{}^3 + a_1 x_1 + a_2 \qquad (1)$$

where together with an extra point which is called the *point at infinity*. For applications in cryptography, only a Galois *finite field* of elements, , is considered. When is a prime, one can think of as the integers modulo. For a given pair of numbers, the forward computation is simple, but its inverse is practically intractable at this time.

Public key cryptosystems for ECC are analogues of cryptosystems available for other discrete logarithm based systems (such as the multiplicative group of a finite field), including Diffie-Hellman key exchange, ElGamal public key encryption, and ECDSA (an analogue of the US government's digital signature standard). In 2005, the National Security Agency (NSA) decided to adopt elliptic curve based public key cryptography. Research has recently been shifting from elliptic curves to lattices (Gentry, 2009; Micciancio & Regev, 2008; Regev, 2006).

2.1.5 Quantum Cryptography (QC)

In 1984, *quantum cryptography* (QC) was introduced by Bennett and Brassard when they proposed the protocol for secret key distribution known as BB84 (Bennett & Brassard, 1984). This protocol is the most commonly analyzed and implemented. Quantum cryptography ensures the confidentiality of information transmitted between two parties exploiting elementary particles behaviour (*i.e.*, photons). The most mature application of quantum information science is the distribution of secret keys. The Heisenberg uncertainty principle is the basis for QC. It is not possible to know the location and speed of a particle at the same time. The information storage medium is at the atomic scale in the form of *qubits* analogous to traditional bits. A qubit can be either 0 or 1 at the same time by laws of quantum mechanics. By the uncertainty principle, an eavesdropper cannot know everything about a photon carrying a key bit, and part of the information can be destroyed because measurement is destructive in quantum mechanics. Detection of eavesdroppers is possible because any measurement performed on the particle carrying information disturbs it. To accomplish it in QC, a source of truly random numbers (*i.e.*, noise in a resistor) and a classical authenticated channel are needed in the creation of the secret key.

Quantum key distribution is achievable using current technologies (*i.e.*, lasers and fiber optics, and electronic noise sources). Single photon sources (*i.e.*, trapped atoms or ions as nitrogen vacancy colour center in diamonds) are becoming within reach of current technologies. Quantum states and quantum pieces of information are written using Dirac's notation $|0\rangle$ and $|1\rangle$ or $|+\rangle$ and $|-\rangle$ (Van Assche, 2006).

2.2 Continuous-Interval Cellular Automata (CCA) and Chaos Phenomena

Resent developments in cellular automata (CA) are rooted in the work of Konrad Zuse, Stanislaw Ulam, and John von Neumann. It is closely related to the first computing machines (Peitgen, Jürgens, & Saupe, 2004). Wolfram (2002) has prompted another revival of CA. A cellular automaton is a finite state machine (FSM) whose state changes in discrete steps (Peitgen *et al.*, 2004). Given a set of initial conditions and a set of rules, a CA produces a sequence by its inherently growing nature. According to Hachtel and Somenzi (1996), any sequential circuit can be modeled as a FSM, consisting of combinational logic and memory.

Although a CA can be defined by a very simple program, it is capable of producing behaviour of great complexity, and have sensitivity to small changes in initial conditions (Wolfram, 2002). As mentioned already, continuous-interval cellular automata (CCA) are generalization of CA in which a cell can have any value from the interval (Wolfram, 2002).

The behaviour of a CA can be separated into four classes: (1) simple behavior in which the CA arrives at the same final state; (2) distinct possible final states, but all consisting of simple periodic structures; (3) complex behaviour, often random; and (4) interaction between order and randomness.

This paper considers CCA capable of exhibiting behaviour of Class 3 which can yield random sequences or patterns that never repeat (*i.e.*, chaotic). This makes them highly amenable for cryptography.

Rounding flooring errors in computing are due to finite resolution in computers. Certain simple rules in CCA defined by the computation of basic arithmetic operations make them capable of exhibiting chaos. The route to chaos in dynamical systems exhibits cycle stability and period doubling (*i.e.*, bifurcation). The number of

Figure 1. One-point stability prior to the chaotic sequence. Initial conditions at $u[0] = 0.9$ and parameter $g = 1$.

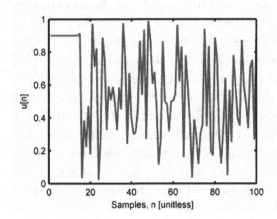

Figure 2. Two-point stability prior to the chaotic sequence. Initial conditions at $u[0] = 0.6$ and the parameter. $g = 19$

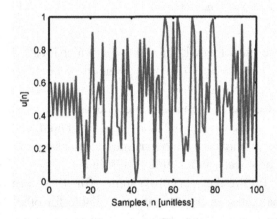

points present on a cycle is composed of powers of two (Kinsner, 2011). A graphical representation of this chaotic nature is provided in Figures 1, 2, and 3 by a unicellular CCA, as described later.

According to the canonical definition of chaos by Devaney (1992) and Kulczycki (2008), the main characteristics of chaotic behaviour are (1) topological transitivity, (2) density of periodic points, and (3) sensitive dependence on initial conditions. It has been shown that the last

Figure 3. Four-point stability prior to the chaotic sequence. The circles are included as an aid to identify the points in the 4 cycle. Initial conditions at $u[0] = 0.6$ and the parameter $g = 7$

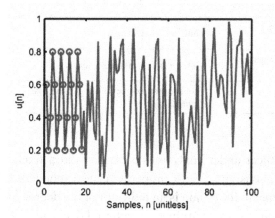

Figure 4. Multi needle strange attractor in the pseudo-phase space. Parameter $g = 5$

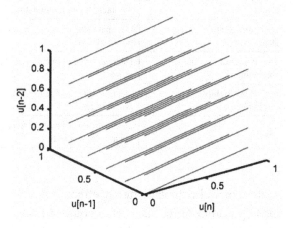

two properties are implied by the first one, thus making it the sufficient condition for a dynamical system to develop chaos, and the third one ensures that the trajectories in a chaotic signal vary greatly for different small values on its initial conditions (Kinsner, 2011). By varying the initial conditions, a totally uncorrelated sequence can be obtained, and as a consequence the number of available waveforms could be infinite (Wu, Liu, Zhao, & Fu, 2001).

A rule (or law) that makes a one-dimensional CCA exhibit chaos is defined by

$$u[n] = \quad g\big(u[n1]\big) \quad g\big(u[n1]\big), \text{ for } n = 1, 2, \dots \quad (2)$$

where is a real parameter, $\in R$, the notation $u[\bullet]$ signifies a discrete map, and the symbol $\bullet$ denotes the floor operator. The initial condition is given at $n = 0$, and just the decimal part is kept at each step. This rule is optimized for one cell CCA to achieve one of the fastest possible computations to produce a chaotic sequence since it only takes one multiplication and one subtraction. The dynamical system described by (2) produces a strange attractor whose points appear in an array of

"needles" in a three-dimensional pseudo-phase space. When the parameter g takes integer odd values a tilted cube of $g \times g$ needles is visible in the pseudo-phase space, and all the needles are of the same length, as illustrated in Figure 4.

The parameter g can have fractional values but when this happens some of the needles can vary in length. As the parameter g increases, the array of needles becomes denser in the pseudo-phase space. For some g the multi-needle attractor exhibits features like point stability and period doubling, characteristic of chaotic phenomena, and then a transition into a chaotic regime, as illustrated in Figures 1, 2, and 3. As seen in the figures, the periods of stability just prior to the chaotic sequences occur when the initial conditions are composed of one digit in the decimal part; if the decimal part includes more digits and is not a negative power of two, then it vanishes. Table 1 lists the peculiarities for the multi-needle attractor which is one of the fastest chaotic as measured by its number of operations. Notice that continuous models (*i.e.*, flows) of chaotic systems are based on differential equations (DE) whose solutions involve more computations. For example, the three-dimensional chaotic

Table 1. Multi needle strange attractor parameters

Regime	Initial Condition Value	Parameter Value
Extinction	All integers and fractional values	Even integers or values less than one
One Point Stability at the Beginning	Decimal value of one digit that is not a negative power of two	Integer ending in one except one
Two Point Stability at the Beginning	Decimal value of one digit that is not a negative power of two	Integer ending in nine
Four Point Stability at the Beginning	Decimal value of one digit that is not a negative power of two	Integer ending in three or seven
Chaos	Decimal value of more than one digit that is not a negative power of two	not being an exact even integer. can take rational values.

weather model with its butterfly effect as developed by Lorenz (1963) is a set of coupled DEs.

It is possible to extend the idea from the unicellular multi-needle attractor to CCA with more cells in n-dimensional systems averaging the summation of the neighbor cells and the current cell. In this model,

$$u[n]$$
$$= \frac{g\sum_{i=-1}^{1}(u[i,n1])}{3} - \frac{g\sum_{i=-1}^{1}(u[i,n1])}{3} \tag{3}$$

where the right and left neighbors of a central cell are added and averaged, and just the decimal part is kept. Since this option has more computational operations, it is not considered here.

2.3 Variance Fractal Dimension Analysis of CCA Chaotic Sequences

This subsection describes a polyscale methodology in order to obtain a measure of a cryptographic sequence complexity (Kinsner, 2011b). Polyscale analysis requires a measurement process in which the scale used is multiplied or divided by a constant factor at each stage. This allows access to the object properties for analysis based on the size of the scale used at a given stage. If the properties vary at different scales as a power law, the

object under analysis is a fractal, and a fractal dimension for those properties can be obtained. A fractal dimension is an indicator of the degree of complexity of an object or pattern. The calculation of a fractal dimension in terms of variance known as variance fractal dimension (VFD) (Kinsner, 2007b, 2011; Kinsner, Cheung, Cannons, Pear, & Martin, 2006; Kinsner & Grieder, 2008, 2010) is used as a tool to determine the complexity of the chaotic sequences produced by CCA.

A time series of a chaotic process can be analyzed directly in time by computing the spread of the increments in the signal amplitude (*i.e.*, through its multiscale variance, σ^2. The variance fractal dimension can be computed in real-time (Kinsner, 2011). An important characteristic of the VFD is that it does not require a window in the Fourier sense, and therefore does not introduce corresponding artifacts (Kinsner, 2011).

The variance fractal dimension, D_σ, is determined by the Hurst exponent, H. The variance, σ^2, of the amplitude increments of a signal $B(t)$ over a time increment Δt is related to the time increment according to the following power law (Kinsner, 2011)

$$\mathrm{Var}\left[B\left(t_2\right) - B\left(t_1\right)\right] \sim \left|t_2 - t_1\right|^{2H} \tag{4}$$

where Var denotes variance, and the symbol $\sim$ reads "is proportional to."

For $\Delta t = \left| t_2 - t_1 \right|$ and $(\Delta B)_{\Delta t} = B(t_2) - B(t_1)$ the exponent H can be calculated from a log-log plot by Shannon (1949)

$$H = \lim_{\Delta t \to 0} \frac{1}{2} \frac{\log_b \left[\mathrm{Var}(\Delta B)_{\Delta t} \right]}{\log_b \Delta t} \qquad (5)$$

where in the analysis performed here, the base b is 2. The embedding Euclidean dimension E (*i.e.*, the number of independent variables in the observed signal), the variance dimension can be computed from

$$D_\sigma = E + 1 - H, 1 \le D_\sigma \le 2 \text{ and } 0 \le H \le 1 \qquad (6)$$

The practical implementation of the technique to calculate the VFD in a digital signal considers the following steps (Kinsner, 2011): First, a sample space of N_T points from the signal is chosen. The range of sizes of Δt at which the spread of ΔB should be computed is obtained by $\Delta t_{K\max} = n_{K\max}$, where $\delta t \le T$, which means that the time interval δt should not exceed the total time T over which the sample space was obtained. Preparation of the parameters for the loop computation of the variance:

$$K_{\max} = \mathrm{int}\left(\frac{\log_b N_T}{\log b} \right) \qquad (7)$$

where b (2 in this case) is the base forming a b-adic sequence for time intervals n_k, $K_{\mathrm{buf}} = \left\lceil \frac{\log_b(8192)}{\log b} \right\rceil$ where the number 8192 represents the number of divisions in the first computation in the loop and is desirable to be greater than 30 for statistical significance, $K_{\mathrm{hi}} = K_{\max} - K_{\mathrm{buf}}$, and $K_{\mathrm{low}} \ge 1$. The main loop to obtain the VFD performs k cycles from K_{hi} to $k = 1$ in which the number of samples is

$n_k = b^k$, the number of windows in the signal is $N_k = \mathrm{int}\left(\frac{N_T}{n_k} \right)$, and the variance for each stage is

$$\mathrm{Var}(\Delta B)_k = \left[\frac{1}{N_k - 1} \right] \left[\sum_{j=1}^{N_k} (\Delta B)_j - \frac{1}{N_k} \left[\sum_{j=1}^{N_k} (\Delta B)_j \right] \right]^2 \qquad (8)$$

The amplitude increment is given by

$$(\Delta B)_j = B(jn_k) - B((j-1)n_k) \text{ for } j = 1, \ldots, N_k \qquad (9)$$

The log values $X_k = \log[n_k]$ and $Y_k = \log\left[\mathrm{Var}(\Delta B)_k \right]$ are stored for the log-log plot and the least-squares fit to obtain the slope s, of the line using

$$s = \frac{K \sum_{i=1}^K X_i Y_i - \sum_{i=1}^K X_i \sum_{i=1}^K Y_i}{K \sum_{i=1}^K X_i^2 - \left(\sum_{i=1}^K X_i^2 \right)^2} \qquad (10)$$

The Hurst exponent is computed by $H = \frac{1}{2} s$, and the VFD is obtained applying (6).

For a non-stationary sequence, this process is repeated on successive windows (either non-overlapping or overlapping) to obtain a VFD trajectory (VFDT) (Kinsner, 2011). If the VFDT is constant, the sequence is a monofractal in time, and if it is not constant, the sequence is multifractal in time.

3. MODULAR DYNAMICAL CRYPTOSYSTEM (MDC)

Continuous-interval cellular automata capable of exhibiting chaos are attractive in cryptography because it is possible to have a great number of possible keys in the keyspace. As defined in Sec-

Figure 5. Modular dynamical cryptosystem structured chart

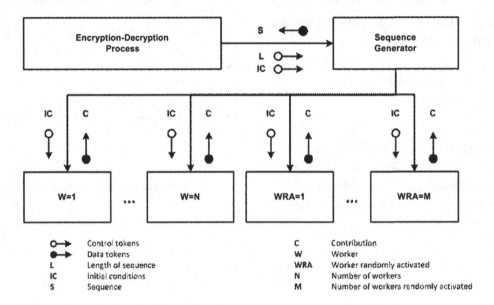

tion 2A, cryptographic techniques are divided into symmetric-key (when the same secret key is used for the encryption and decryption processes), and asymmetric-key (when one key is kept secret and one or more keys are released publicly) (Anghelescu *et al.*, 2008). The sensitive dependence for slight changes in the initial conditions of dynamical systems allows the design of symmetric cryptosystems capable of generating a great number of keys (theoretically infinite) for the encryption and decryption processes.

A proposed cryptosystem model based on chaotic CCA is illustrated in Figure 5. "Modular" denotes the interaction between the different modules each of them containing a dynamical system based on CCA to obtain a final sequence useful in encryption-decryption. The encryption and decryption processes request a sequence of length L from the sequence generator (SG) which acts as a master in the cryptosystem. The sequence S is created chaotically mixing the small sequences C. Each sequence C is provided not only by the workers (W), but also by the workers randomly activated (WRA), with their initial conditions provided by the SG. The initial condi-

tions used by the W and WRA never are equal given the chaotic regime present in the SG. An instance of the chaotic CCA defined by (2) is contained in the SG, the W, and the WRA, with different initial conditions and parameters. The sequences C vary in length chaotically in the same way that the WRA are allowed to operate. Once the sequence S has reached the required length the sequence generation process stops.

As a characteristic of chaotic processes, the variation of the initial conditions in the W and WRA by the SG never exhibits periodicity. A process of chaotically mixing chaotic sequences takes place in the SG, with the benefit of achieving higher secure characteristics in encryption and decryption processes. Other recent approaches (Anghelescu *et al.*, 2007, 2008; Peng, Zhang, & Liao, 2009; Yifang *et al.*, 2009) have limited considerations either for the manner of how CA are used, or the type of information that they are capable to protect. It is possible to encapsulate any dynamical system like the well known Lorenz (1963), Rössler (1977), and Hénon (1976) attractors in a given module in Figure 5. Since those dynamical systems require more computational steps in their implementations than CCA, they are not considered here.

The crytosystem explained is described by the following six-tuple.

$$\left(p,a,c,k,e,d \right) \tag{11}$$

where p represents general plaintexts from different information sources, a restricts all the CCA used in the cryptosystem modules to exhibit chaos, c is the set of ciphertexts, k is a set of possible keys, e is the space of possible encryption rules, and d is the space of possible decryption rules. According to Stinson (2006), a cryptographic system is composed of the quintuple $\left(p,c,k,e,d \right)$, but since CA are of paramount importance here, the a is included to indicate that CCA are capable of exhibiting chaos, and it has to be used to obtain a reliable cryptosystem.

The encryption and decryption keys used are defined as

$$e_K \left[n \right] = \left(p \left[n \right] + u \left[n \right] \right) \mathrm{mod}\, b \tag{12}$$

$$d_K \left[n \right] = \left(p \left[n \right] - u \left[n \right] \right) \mathrm{mod}\, b \tag{13}$$

where $p[n]$ represents the original information sequence, $u[n]$ is the encryption-decryption sequence generated by the SG, and b is the dynamical range that lets a representation of $p[n]$ where no information is lost to achieve perfect encryption and decryption.

4. EXPERIMENTAL RESULTS

The algorithm described in Figure 5 is implemented in Matlab on a laptop (Fujitsu LifeBook A Series, with Windows Vista, Intel Centrino Duo microprocessor, and 1 GB in RAM) to illustrate the different encryption and decryption schemes considered here. It is important to notice that if

dynamical systems defined by differential equations were used, instead of CCA in this limited computer, it would require weeks of processing time to obtain sequences for identical cryptographic applications. Execution time is one feature where CA demonstrate some of their powerful advantages.

The parameter $g = 3$ in the case of the SG, $g_{W1} = 5, g_{W2} = 7,\dots, g_{W10} = 23$ for the W and $g_{WRA1} = 25, g_{WRA2} = 27,\dots, g_{WRA10} = 43$ for the WRA. The initial condition value for the SG is $IC = 0.0802$ in all cases. The blocks constituting the W and the WRA consist of ten CCA in the following experiments. These blocks can have any arbitrary number making the cryptosystem even more powerful.

4.1 Example 1: Text Encryption

A plaintext formed by the repetition of the arbitrary characters YAGGHGTLVJDTGH to form a periodic sequence approximately 19 millions in length. An encryption sequence or key of the same length is provided by the SG applying (2) and then using Equation (12) the plaintext is ciphered. The first 252 ASCII codes of the ciphertext are shown in Figre 6 where no obvious repetitive patterns exist even though a periodic pattern exists in the plaintext. The characters in the ciphertext vary through all the possible ASCII code values.

4.2 Example 2: Sound Encryption

An arbitrary song (with approximately 13.5 million samples in the right and left channel) is chosen as plaintext. A key of 27 million samples in length is generated by the SG to encrypt the information in both channels. The original information amplitude varies between -1 and 1. The information values are changed to a range from 0 to 2 as minimum and maximum respectively, then (12) is applied, and finally restoring the values to -1 and 1 again the information is ciphered.

Figure 6. Sequence of encrypted ASCII code. Just the first 252 ASCII codes are shown. Approximately 19 million characters where ciphered.

229	014	078	243	015	203	002	089	210	142	156	111
066	020	131	202	110	004	093	082	208	116	089	092
012	114	171	182	167	201	113	233	165	233	152	184
165	156	215	019	192	014	049	126	161	059	120	159
147	159	023	083	027	119	188	136	244	109	119	211
079	040	001	200	028	101	208	155	104	244	143	195
070	084	005	141	155	187	001	116	221	204	211	239
013	237	151	240	020	084	095	079	133	073	036	041
151	146	128	183	141	003	196	163	166	231	184	051
169	194	050	049	018	227	102	233	140	137	000	145
049	026	254	224	200	053	006	038	224	025	212	009
078	248	244	000	008	097	068	048	125	231	119	149
057	142	123	152	162	024	192	050	205	023	125	039
200	211	163	121	001	197	018	185	081	218	103	106
079	223	094	068	133	110	054	151	014	198	226	190
090	020	026	116	211	067	248	243	161	238	055	087
084	051	230	058	000	237	188	124	111	190	043	218
210	162	063	126	001	183	113	205	220	195	244	111
169	158	099	220	104	152	181	144	111	162	081	110
085	054	115	200	104	130	197	219	043	127	229	119
213	243	075	048	075	089	114	051	223	133	202	141

Figure 7 shows 5,000 samples, whit both encrypted channels showing blocks filled in colour. This means that a sample takes any value in the corresponding dynamic range producing a sequence that provides incomprehensible meaning when it is played back. In fact, when one listens to this encrypted sequence it seems to be a strong rainy storm no matter what was the original information contained in the sound sequence before.

4.3 Example 3: Image Encryption

Pictures in grayscale and colour are also included in the scope of cryptographic examples presented in this paper. For the grayscale case, the picture of Figure 8 acts as plaintext. An encryption key approximately 4 million in length is generated to cipher the information contained in the picture of 1,728×2,304 pixels. The encryption sequence is reshaped to get a matrix the same size of the original picture and then by applying (12) the ciphered picture presented in Figure 9 is obtained.

Figure 10 presents a similar process done to the colour picture with the slight difference that it is a three-dimensional array 1,728×2,304×3 that

Figure 7. Segment from a sound sequence approximately 13.5 million samples in total length. Just five thousand samples are shown for both right and left channels to let the reader appreciate how the transformation of the wave in both audio channels.

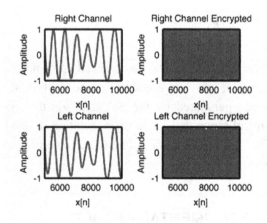

contains the values for the red, green and blue components. A sequence approximately 12 million in length is generated and reshaped into a three-dimensional array the same size mentioned before, and by the use of (12) the colour picture is ciphered

Figure 8. Grayscale picture 1,728×2,304 pixels of some buildings at downtown in Los Angeles, CA, USA

Figure 10. Colour picture 1,728×2,304×3 pixels of Houdini's star in the Hollywood Walk of Fame at Los Angeles, CA, USA

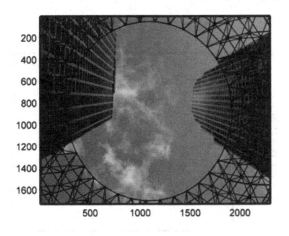

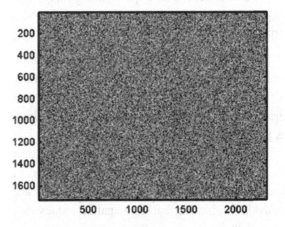

Figure 9. Grayscale encrypted picture 1,728×2,304 pixels of some buildings at downtown in Los Angeles, CA, USA

Figure 11. Encrypted colour picture 1,728×2,304×3 pixels of Houdini's star in the Hollywood Walk of Fame at Los Angeles, CA, USA. A sequence approximately 12 million in length was used.

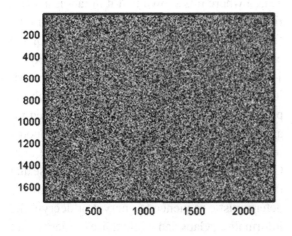

as illustrated in Figure 11. For both cases, no trace of the original information in the original grayscale and colour pictures was left after the encryption process.

4.4 Variance Fractal Dimension Analysis

Figure 12 shows the log values X_k and Y_k and the slope fitting following the procedure just described for the realization of a Gaussian white noise

Figure 12. Gaussian white noise VFD. This is based on a sequence realization of 10 million. The last ten binary orders of magnitude in the computation are displayed. As expected the variance is 1, and the variance fractal dimension is almost 2.

Figure 13. Uniformly distributed on the interval pseudorandom numbers VFD. This is based on a sequence realization of 10 million. The last ten binary orders of magnitude in the computation are displayed.

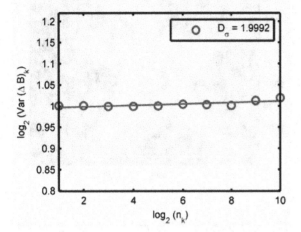

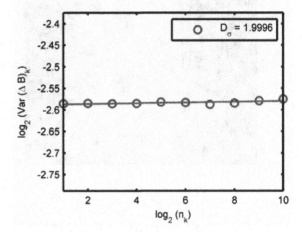

(GWN) signal with 10 million samples. The variance fractal dimension for the GWN signal is $D_\sigma \cong 2$ as expected for a space filling curve. The parameters used for the loop computation of the VFD are $K_{max} = 23$, $K_{buf} = 13$, $K_{hi} = 10$, and $K_{low} = 1$. In general the first possible ten points in the loglog plot are not considered in any of the VFD determined due to some artifacts that can arise and just the strongest part where the slope is preserved is computed and displayed.

Similarly, Figure 13 displays the realization of a uniformly distributed random sequence in the interval $[0,1]$ where the same conditions are considered. Notice that the negative values displayed in the y axis in the figure is due to the logarithmic base $b = 2$ representation of the variance value and it should not be confused with an attempt to represent negative variances (which would be against to the mathematical formulations and concepts of probability). The variance in this case is very small compared with the GWN basis presented before.

Figure 14 depicts the VFD of an unmixed realization of the attractor used through this research where it is shown that it in fact is a space filling curve by the value of two obtained in the fractal dimension considered. The VFD of the chaotically mixed sequence is presented in Figure 15 where it is seen that the same level of fractality is preserved, but the exponent for which the power law holds is higher than the unmixed realization of the CCA presented before. Continuous-interval cellular automata have advantage over GWN or pseudorandom numbers in the sense that there is no need to store the realization of the deterministic sequence to encrypt or decrypt the information. Gaussian white noise sequences are not deterministic, and true random number generators that are able to generate non periodic sequences of numbers are not easy to design.

The histograms of the plaintexts and ciphertexts are presented in Figures 16 through 27, respectively. The plaintexts histograms for the sound wave, the grayscale picture, and the different RGB (red, green, and blue) channels in the colour pic-

Figure 14. Unmixed CCA sequence VFD. This is based on a sequence realization of 10 million. The last ten binary orders of magnitude in the computation are displayed.

Figure 15. Mixed CCA sequence VFD. This is based on a sequence realization of 10 million. The last ten binary orders of magnitude in the computation are displayed.

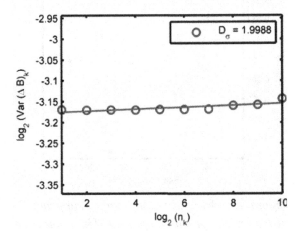

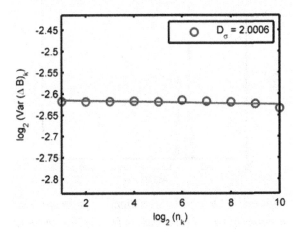

ture show diverse shapes. It is seen that the information after encryption has a uniform distribution even though different sources as plaintext are considered. The figures showing the ciphertexts histograms are configured to display signal amplification. This allows looking for specific signal characteristics as hooks. All the ciphertexts histograms have a variability range that changes in a scale of thousandths as observed in Figures 22, 23, 24, 25, 26, and 27. It is well known that a cryptosystem that exhibit good uniform distribution characteristics after ciphering the elements of the plaintext provides resistance to statistical analysis (Yifang *et al.*, 2009).

5. DISCUSSION OF EXPERIMENTAL RESULTS

Four different sources of information (data) are considered in this paper to show the wide areas of applications that MDC can reach and the flexibility for its implementation. In all the cases presented, no information loss took place during the encryption and decryption processes. The main

goal of cryptography (*i.e.*, to keep the information intact and hidden) has been demonstrated. Similar protection can be achieved in all the four cases. This provides an intuitive understanding of how versatile the MDC is for robust cryptographic applications. Thus, any institution that needs to secure or protect information could benefit from the advantages of the MDC.

The average processing time was 23, 81, 6, and 13 seconds for the text, music, grayscale picture, and colour picture respectively. The text, grayscale picture, and colour picture used vectors capable of storing 8 bits data. The audio case used a vector with data storage capacity of 64 bits. These processing times appear to be highly competitive, considering that a very limited hardware was used to run the different experiments.

A desirable property of any cryptosystem is that a small change of one bit of the key should produce a change in many bits of the CT (Nichols, 1999). This is known as "avalanche effect" and is present on most of the modern cryptosystems in different degrees according to how they mix or "scramble" the bits. The proposed cryptosystem inherently achieves this objective because

Figure 16. Original text histogram. The number of characters considered are approximately 19 million.

Figure 19. Original color image histogram illustrating the red component. The number of pixels are approximately 12 million.

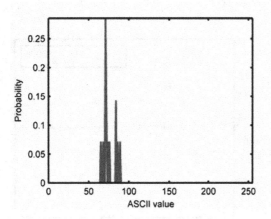

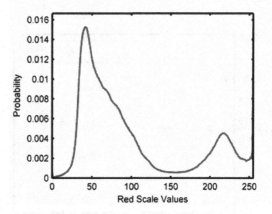

Figure 17. Original sound signal histogram. The number of samples considered are approximately 27 million.

Figure 20. Original color image histogram illustrating the green component. The number of pixels are approximately 12 million.

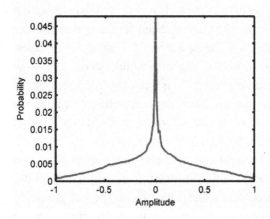

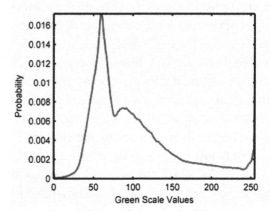

Figure 18. Original gray scale image histogram. The number of pixels considered are approximately 4 million.

Figure 21. Original color image histogram illustrating the blue component. The number of pixels are approximately 12 million.

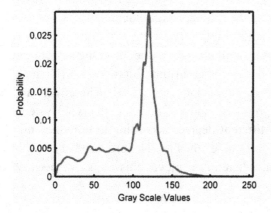

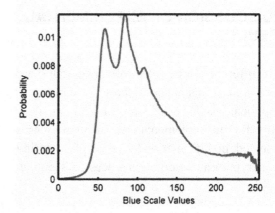

Figure 22. Histogram of the text after encryption. The number of characters are approximately 19 million

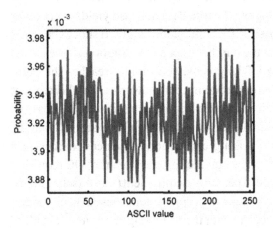

Figure 23. Histograms of the sound signal after encryption. The number of samples considered are approximately 27 million.

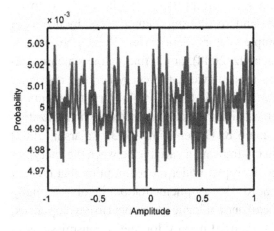

Figure 24. Histograms of the gray scale image after encryption. The number of pixels considered are approximately 4 million.

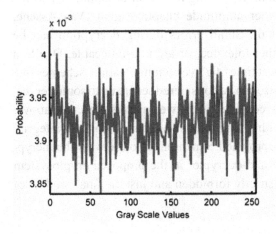

Figure 25. Histogram of the red component in the color image after encryption. The number of pixels are approximately 12 million.

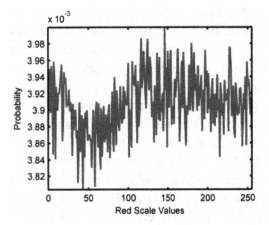

Figure 26. Histogram of the green component in the color image after encryption. The number of pixels are approximately 12 million.

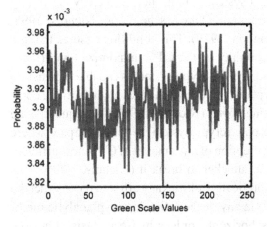

Figure 27. Histogram of the blue component in the color image after encryption. The number of pixels are approximately 12 million.

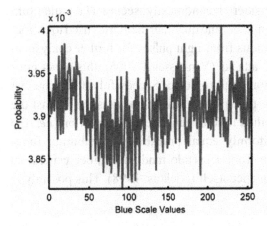

a small change introduced in a chaotic system initial conditions produces a complete different sequence (Kinsner, 2011; Peitgen *et al.*, 2004; Strogatz, 2000).

Applications in both symmetric and asymmetric operation schemes are possible with the presented cryptosystem if appropriate protocols are used for key exchange. The symmetric case can follow the classical protocol where the key is distributed through a secure channel. The Menezes-Qu-Vanstone (MQV) protocol widely used in ECC can be implemented for the asymmetric case (Rosing, 1998). This protocol avoids the "man in the middle" (MIM) attack by "perfect forward secrecy" and performs authentication of the key holders (Rosing, 1998).

Since the length of the encryption sequences used is the same as the different PTs used, this cryptosystem exhibits "perfect secrecy" according to Shannon (Van Assche, 2006; Nichols, 1998; Shannon, 1949). This condition ensures that the uncertainty of the CT is maximized and (theoretically) the cryptosystem is capable of resisting cryptanalysis even in the presence of infinite computing power (Nichols, 1998). Consequently, the presented cryptosystem is a "one-time pad" where interception of any amount of CT is not sufficient to an attacker to break it (Nichols, 1998). This theoretically nullifies any effort to reconstruct the PT via any or the following cryptanalytic methods: horizontal or lengthwise analysis, cohesion, reassembly via Kasiski or Kerckhoff's columns, repeats, or internal framework erection (Nichols, 1998). However, since no cryptosystem can be considered canonically secure (*i.e.*, quantum computers and quantum memories used to extract photons from light pulses for further analysis in QC attacks (Van Assche, 2006), this paper does not attempt to do so. Successful attacks against one-time pad cryptosystems must be against the method used to generate the encryption (*i.e.*, the uniformly equal probability distribution function from a pseudo random number generator) sequence itself (Nichols, 1998). This possibly is

one of the most difficult challenges posed for an attacker. Even if the attacker could succeed in characterizing the system, then brute force might bring out the true PT, but it also yields every other possible PT of the same length. It should be noted that the number of possible solutions increases as the PT lengthens, and rapidly reaches the point where more computing power and a higher amount of time are required to come up with all possible solutions of an attack of this nature (Nichols, 1998). Long encryption sequences to encrypt a PT can be obtained also by means of QC requiring highly specialized equipment (Van Assche, 2006).

Attacks to cryptosystems based on chaos can succeed under the following conditions: (1) identification of the chaotic system in the time domain by one of its state variables, (2) recovering the message signal by using an approximate model with some errors which can be easily removed by standard filtering methods (*i.e.*, by means of sample vectors dictionaries (Rohde, Nichols, & Bucholtz, 2008), or (3) finding the synchronization used between the encrypter and decrypter (Yang, Wu, & Chua, 1997). Cryptosystems or communications schemes that use synchronization (Cuomo & Oppenheim, 1993; Rajarshi, 2007) usually transmit a variable between the encrypter and decrypter under the assumption that it is not noticed. These implementations involve normally at least one analogue component to develop chaos, and are used mainly for audio transmissions to "mask" a PT. In the traditional cryptography based on synchronized chaos, the encrypted signal is obtain by adding the lower amplitude PT to the higher amplitude chaotic signal (Van Assche, 2006). Small errors during decryption can be either tolerated, or are not noticeable. Due to a short time of a live communication, schemes like these can be considered secure up to some degree, but considering the weaknesses presented above, neither long term data security, nor data integrity are provided. Synchronization between the encrypter and decrypter in the proposed cryptosystem is strictly forbidden and just the final product of encryption is considered secure for storage or transmission. This avoids exposing state variables and, as a consequence, the door to reconstruct a model of the system based on monitoring one of its state variables is closed.

6. CONCLUSION

This paper presents an efficient new cryptosystem based on modular continuous-interval cellular automata. The efficiency of the continuous-interval cellular automata (CCA) is the basis to generate high-speed chaotic sequences used in the new modular dynamic cryptosystem. Distinct modules containing CCA have been integrated to generate short chaotic sequences that can be combined to form a chaotically mixed sequence with higher variance. Since the proposed cryptosystem performs equally well on different classes of data (text, sound, images), it can be used in different applications that require storing and/or transmitting information in a secure way. The introduction of a CCA capable of exhibiting a chaotic regime is the core of the proposed cryptosystem, with the advantage of being fast and flexible for implementation. This overcomes the difficulty of implementing quantum cryptography (QC), often considered as a technological challenge. Shannon's observation mentioned in the introduction about good mixing transformations made by non commuting operations is transcendental and extremely important through the whole process of the cryptosystem operation.

The perfectly reversible encryption and decryption processes provided by the new cryptosystem could be of value to both civilian and military industries because of the very large number of keys provided by the variation of initial conditions in the chaotic sequences. Most of the approaches presented in the available literature consider CA with limited capabilities, limited number of keys, and a unique module that describes the whole cryptosystem, making it vulnerable or predict-

able. The multiscale complexity measure of the amount of chaos that a sequence in a CA contains is done here for the first time. This provides an important contribution as an alternative way of how CA sequences should be studied, analyzed, and compared.

The inherent strong "avalanche effect" in the presented cryptosystem is one of its most important characteristics. This is obtained by the nature of a chaotic sequence based on dependence on initial conditions and is amplified by means of the chaotic mixing. Also the "one-time pad" profile used on its development is remarkable to obtain "perfect secrecy" on any of the PTs enciphered. This has several advantages: (1) it eliminates known classical attacks, (2) establishes a fence for the known attacks (*i.e.*, for the method used to obtain the encryption sequence in "one-time pad" cryptosystems, identification of a chaotic system in the time domain, model reconstruction, or synchronization search), provided that the encryption sequence is at least the same length as the PT, and that synchronization is not used; (3) provides an opportunity to resist attacks not reported so far; and (4) attempts to establish a possible new research direction on how chaotic systems should be used and implemented for cryptographic applications.

It is also important to note that the proposed cryptosystem can be used for symmetric and asymmetric applications.

ACKNOWLEDGMENT

This work was supported in part by the Natural Sciences and Engineering Research Council (NSERC) of Canada.

REFERENCES

Alpcan, T., & Başar, T. (2011). *Network security: A decision and game-theoretic approach*. Cambridge, UK: Cambridge University Press.

Anghelescu, P., Ionita, S., & Sofron, E. (2008). FPGA implementation of hybrid additive programmable cellular automata encryption algorithm. In *Proceedings of the International Conference on Hybrid Intelligent Systems* (pp. 96-101).

Anghelescu, P., Sofron, E., & Ionita, S. (2007). VLSI implementation of high-speed cellular automata encryption algorithm. In *Proceedings of the International Conference on Semiconductor* (Vol. 2, pp. 509-512).

Anghelescu, P., Sofron, E., Rîncu, C. I., & Iana, V. G. (2008). Programmable cellular automata based encryption algorithm. In Proceedings of the International Conference on Semiconductor (Vol. 2, pp. 351-354).

Bennett, C. H., & Brassard, G. (1984). Quantum cryptography: Public key distribution and coin tossing. In *Proceedings of the IEEE International Conference on Computers, Systems and Signal Processing*, Bangalore, India (pp. 175-179).

Boonyarattaphan, A., Bai, Y., & Chung, S. (2009). A security framework for e-Health service authentication and e-Health data transmission. In *Proceedings of the 9th International Symposium on Communications and Information Technology* (pp. 1213-1218).

Bosworth, B. (1982). *Codes, ciphers, and computers: An introduction to information security.* Rochelle Park, NJ: Hayden.

Cuomo, K. M., & Oppenheim, A. V. (1993). Circuit implementation of synchronized chaos with applications to communications. *Transactions of the American Physical Society Journal*, 65-68.

Devaney, R. L. (1992). *A first course in chaotic dynamical systems: Theory and experiment.* Reading, MA: Addison-Wesley.

Diffie, W., & Hellman, M. E. (1976). New directions in cryptography. *IEEE Transactions on Information Theory*, *22*(6), 644–654. doi:10.1109/TIT.1976.1055638

Ding, Y., & Klein, K. (2010). Model-driven application-level encryption for the privacy of e-Health data. In *Proceedings of the International Conference on Availability, Reliability, and Security* (pp. 341-346).

Ferguson, N. (2010). *Cryptography engineering: Design principles and practical applications.* New York, NY: Wiley.

Gaines, H. F. (1956). *Cryptanalysis: a study of ciphers and their solution.* New York, NY: Dover.

Gait, J. (1978). Encryption standard: Validating hardware techniques. *Dimensions/NBS*, *62*(7-8), 22–24.

Gentry, C. (2009). Fully homomorphic encryption using ideal lattices. In *Proceedings of the 41st Annual ACM Symposium on Theory of Computing*, Bethesda, MD.

Glover, N., & Dudley, T. (1991). *Practical error correction design for engineers* (2nd ed.). Broomfield, CO: Cirrus Logic-Colorado.

Hachtel, G. D., & Somenzi, F. (1996). *Logic synthesis and verification algorithms.* New York, NY: Springer Science & Business Media.

Hénon, M. (1976). A two-dimensional mapping with a strange attractor. *Communications in Mathematical Physics*, *50*, 69–77. doi:10.1007/BF01608556

Hu, J., & Klein, A. (2009). A benchmark of transparent data encryption for migration of web applications in the cloud. In *Proceedings of the IEEE 8th International Conference on Dependable, Autonomic and Secure Computing* (pp. 735-740).

Jin, S., & Peng, J. (2010). Access control for web services based on feedback and decay. In *Proceedings of the IEEE 9th International Conference on Cognitive Informatics* (pp. 501-505).

Katz, J., & Lindell, Y. (2007). *Introduction to modern cryptography: Principles and protocols*. Virginia Beach, VA: Chapman & Hall/CRC.

Kinsner, W. (2007). A unified approach to fractal dimensions. *International Journal of Cognitive Informatics and Natural Intelligence, 1*(4), 26–46. doi:10.4018/jcini.2007100103

Kinsner, W. (2007). Towards cognitive machines: Multiscale measures and analysis. *International Journal of Cognitive Informatics and Natural Intelligence, 1*(1), 28–38. doi:10.4018/jcini.2007010102

Kinsner, W. (2011a). *Fractal and chaos engineering*. Winnipeg, MB, Canada: University of Manitoba.

Kinsner, W. (2011b). It's time for polyscale analysis and synthesis in cognitive systems. In *Proceedings of the IEEE 10th International Conference on Cognitive Informatics and Cognitive Computing*.

Kinsner, W., Cheung, V., Cannons, K., Pear, J., & Martin, T. (2006). Signal classification through multifractal analysis and complex domain neural networks. *IEEE Transactions on Systems, Man, and Cybernetics, 36*(2), 196–203. doi:10.1109/TSMCC.2006.871148

Kinsner, W., & Grieder, W. (2008). Speech segmentation using multifractal measures and amplification of signal features. In *Proceedings of the IEEE 7th International Conference on Cognitive Informatics* (pp. 351-357).

Kinsner, W., & Grieder, W. (2010). Amplification of signal features using variance fractal dimension trajectory. *International Journal of Cognitive Informatics and Natural Intelligence, 4*(4), 1–17. doi:10.4018/jcini.2010100101

Koblitz, N. (1994). *A course in number theory and cryptography* (2nd ed.). New York, NY: Springer.

Kulczycki, M. (2008). Noncontinuous maps and Devaney's chaos. *Regular -and Chaotic Dynamics, 13*(2), 81-84.

Lorenz, E. N. (1963). Deterministic nonperiodic flow. *Journal of the Atmospheric Sciences, 20,* 130–141. doi:10.1175/1520-0469(1963)020<0130:DNF>2.0.CO;2

Menezes, A. J. (1993). *Elliptic curve public key cryptosystems*. New York, NY: Springer.

Micciancio, D., & Regev, O. (2008). Lattice-based cryptography. In Bernstein, D. J., Buchmann, J., & Dahmen, E. (Eds.), *Post quantum cryptography* (pp. 147–191). Berlin, Germany: Springer-Verlag.

Moulin, C., & Sbodio, M. L. (2010). Improving the accessibility and efficiency of e-Government processes. In *Proceedings of the IEEE 9th International Conference on Cognitive Informatics* (pp. 603-610).

National Bureau of Standards. (1977). *Data encryption standard*. Washington, DC: NTIS.

Nenadic, A., Zhang, N., Shi, Q., & Goble, C. (2005). DSA-based verifiable and recoverable encryption of signatures and its application in certified e-Goods delivery. In *Proceedings of the IEEE International Conference e-Technology, e-Commerce and e-Service* (pp. 94-99).

Nichols, R. N. (1998). *ICSA guide to cryptography*. New York, NY: McGraw-Hill.

Panayiotou, C., & Bennett, B. (2009). Critical thinking attitudes for reasoning with points of view. In *Proceedings of the IEEE 8th International Conference on Cognitive Informatics* (pp. 371-377).

Peitgen, H. O., Jürgens, H., & Saupe, D. (2004). *Chaos and fractals: New frontiers of science*. New York, NY: Springer Science & Business Media.

Peng, J., Zhang, D., & Liao, X. (2009). Design of a novel image block encryption algorithm based on chaotic systems. In *Proceedings of the IEEE 8th International Conference on Cognitive Informatics* (pp. 215-221).

Peterson, W. W. (1962). *Error-correcting codes*. Cambridge, MA: MIT Press.

Qu, Z., Ma, T., & Zhang, Y. (2008). Application of parameter modulation in e-Commerce security based on chaotic encryption. In *Proceedings of the International Symposium in Electronic Commerce and Security* (pp. 390-393).

Rajarshi, R. (2007). Synchronization, chaos and consistency. In *Proceedings of the Quantum Electronics and Laser Science Conference* (pp. 1-2).

Regev, O. (2006). Lattice-based cryptography. In C. Dwork (Ed.), *Proceedings of the 26th Annual International Cryptology Conference on Advances in Cryptography* (LNCS 4117, pp. 131-141).

Rhee, M. Y. (1989). *Error correcting coding theory*. New York, NY: McGraw-Hill.

Rivest, R., Shamir, A., & Adelman, L. (1978). A method of obtaining digital signatures and public-key cryptosystems. *Communications of the ACM, 21*(2), 120–126. doi:10.1145/359340.359342

Rohde, G. K., Nichols, J. M., & Bucholtz, F. (2008). Chaotic signal detection and estimation based on attractor sets: application to secure communications. *Chaos (Woodbury, N.Y.), 18*(1). doi:10.1063/1.2838853

Rosing, M. (1998). *Implementing elliptic curve cryptography*. Greenwich, CT: Manning.

Rössler, O. (1977). Continuous chaos. In *Proceedings of the International Workshop on Synergetics at Schloss Elmau*.

Schneier, B. (1995). *Applied cryptography: Protocols, algorithms and source code in C* (2nd ed.). New York, NY: John Wiley & Sons.

Shannon, C. E. (1949). Communication theory of secrecy systems. *The Bell System Technical Journal, 28*, 656–715.

Sinkov, A. (1968). *Elementary cryptoanalysis: A mathematical approach*. New York, NY: Random House.

Stinson, D. R. (2006). *Cryptography theory and practice* (3rd ed.). Boca Raton, FL: Chapman & Hall/CRC.

Strogatz, S. H. (2000). *Nonlinear dynamics and chaos*. Cambridge, MA: Westview Press, Perseus Books.

Sung, W. T., Hsu, Y. C., & Chen, K. Y. (2010). Enhance information acquired efficiency for wireless sensors networks via multi-bit decision fusion. In *Proceedings of the IEEE 9th International Conference on Cognitive Informatics* (pp. 154-159).

Swenson, C. (2008). *Modern cryptanalysis: Techniques for advanced code breaking*. New York, NY: John Wiley & Sons.

Tang, Y., & Zhang, L. (2005). Adaptive bucket formation in encrypted databases. In *Proceedings of the IEEE International Conference in e-Technology, e-Commerce and e-Service* (pp. 116-119).

Van Assche, G. (2006). *Quantum cryptography and secret-key distilation*. Cambridge, UK: Cambridge University Press. doi:10.1017/CBO9780511617744

Vanstone, S. A., & van Oorshot, P. C. (1989). *An introduction to error correcting codes with applications*. Boston, MA: Kluwer Academic.

Wakerley, J. (1978). *Error correcting codes, self-checking circuits and applications*. New York, NY: North-Holland.

Washington, L. C. (2008). *Elliptic curves: Number theory and cryptography* (2nd ed.). Boca Raton, FL: Chapman and Hall/CRC.

Wolfram, S. (2002). *A new kind of science.* Champaign, IL: Wolfram Media.

Wu, X., Liu, W., Zhao, L., & Fu, J. S. (2001). Chaotic phase code for radar pulse compression. In *Proceedings of the IEEE National Radar Conference* (pp. 279-283).

Yang, T., Wu, C. W., & Chua, L. O. (1997). Cryptography based on chaotic systems. *IEEE Transactions on Circuits and Systems. I, Fundamental Theory and Applications, 44*(5), 469–472. doi:10.1109/81.572346

Yifang, W., Rong, Z., & Yi, C. (2009). A self-synchronous stream cipher based on composite discrete chaos. In *Proceedings of the IEEE 8th International Conference on Cognitive Informatics* (pp. 210-214).

Zhao, F. (2010). Sensors meet the cloud: Planetary-scale distributed sensing and decision making. In *Proceedings of the IEEE 9th International Conference on Cognitive Informatics* (p. 998).

This work was previously published in the International Journal of Cognitive Informatics and Natural Intelligence, Volume 5, Issue 4, edited by Yingxu Wang, pp. 83-109, copyright 2011 by IGI Publishing (an imprint of IGI Global).

Chapter 18
Time and Frequency Analysis of Particle Swarm Trajectories for Cognitive Machines

Dario Schor
University of Manitoba, Canada

Witold Kinsner
University of Manitoba, Canada

ABSTRACT

This paper examines the inherited persistent behavior of particle swarm optimization and its implications to cognitive machines. The performance of the algorithm is studied through an average particle's trajectory through the parameter space of the Sphere and Rastrigin function. The trajectories are decomposed into position and velocity along each dimension optimized. A threshold is defined to separate the transient period, where the particle is moving towards a solution using information about the position of its best neighbors, from the steady state reached when the particles explore the local area surrounding the solution to the system. Using a combination of time and frequency domain techniques, the inherited long-term dependencies that drive the algorithm are discerned. Experimental results show the particles balance exploration of the parameter space with the correlated goal oriented trajectory driven by their social interactions. The information learned from this analysis can be used to extract complexity measures to classify the behavior and control of particle swarm optimization, and make proper decisions on what to do next. This novel analysis of a particle trajectory in the time and frequency domains presents clear advantages of particle swarm optimization and inherent properties that make this optimization algorithm a suitable choice for use in cognitive machines.

DOI: 10.4018/978-1-4666-2476-4.ch018

1. INTRODUCTION

One of the goals of a cognitive system is to emulate the behavior of human cognition (Wang, 2002; Wang, Zhang, & Kinsner, 2010). There are several cognitive systems that are being implemented at different levels of complexity, including cognitive radio (Haykin, 2005; Haykin, Reed, Li, & Shafi, 2009a, 2009b), cognitive radar (Haykin, 2006), and cognitive networks (Hossain & Bhargava, 2007). All these examples of cognitive systems share four phases used to (1) acquire, (2) process, (3) interpret, and (4) express information for actions (Wang, 2002). The acquisition phase consists of sensors gathering data from both the environment and other interacting systems or living creatures, as well as the system itself. This phase is also responsible for representing the data to maximize the efficiency of the subsequent phases. Failure to pick a suitable representation can have a significant effect on the performance of the cognitive machine (Gavrilova, 2009). The second phase of the system is to process the raw data by extracting the useful information from the system, and to perform pertinent computations to determine how the cognitive machine is to act in the future. As described in (Kinsner, 2004), the processing and interpreting of information must be done at multiple scales in order to extract the pertinent features for very general and very specific decision making processes. The main objective of the multi-scale analysis is to reveal any long-term correlations in the behavior or the underlying processes that can be used to classify and/or improve the performance of the system (Kinsner, 2004). Cognitive processes are considered to be inherently long-range dependent since the environment in which systems is placed also experiences a long-range correlated behavior. The interpretation phase determines which actions to execute in which order and finally, the last stage expresses the decisions in ways that can be conveyed to other parts of the system (actuators) and, in some cases, to a user.

As part of the processing and interpretation stages, a cognitive system must be able to look at large amounts of data and make quick (often faster than real time, or hyper-real time) decisions on what to do next. The decision process often requires more alternatives to be considered in a short window of time than it is physically possible for a real-time system (Kinsner, 2004). Thus, in order to make good decisions without exploring all possible paths, a cognitive system requires optimization techniques that can survey the possible options, and quickly select the best or most suitable option possible. Furthermore, given the long-term dependence found in cognitive machines, an optimization algorithm that can reveal correlated behaviors can help better predict future behaviors.

There are many applications of this research. For example, scheduling tasks on multi-core systems for space applications requires an intelligent system capable of autonomously reading status information on the system to select which routines to execute in order to keep the satellite operational. The status of the components can be used as drivers for the evolutionary algorithm to schedule tasks even when unpredictable situations arise (i.e., one processor fails). Furthermore, in these real-time applications, one cannot allocate tasks dynamically using near-optimal schedules. Therefore, long-term correlations about the system and an intricate knowledge of the behavior of the satellite can be used in predictive scheduling to maximize the use of available resources.

This paper first reviews the requirements for an ideal optimization technique for use in cognitive systems, and uses a novel analysis of a *particle swarm optimization* trajectories in the time and frequency domains to show how the algorithm is inherently designed to satisfy these requirements.

2. EVOLUTIONARY OPTIMIZATION

This section presents an overview of optimization techniques and a description of the *particle swarm optimization* (PSO) algorithm originally developed by Kennedy and Eberhart (1995). In this algorithm, the particles act like entities with cognitive elements in their behavior because they are aware of the environment through social interactions.

2.1 Optimization Techniques

The objective of optimization problems is to modify parameters in $\mathbb{R}$ to find a minimum (or maximum) value for a given function, f (Gerald & Wheatley, 2004; Berg, 2001).

Find $x' \in \mathbb{R}^D$ for which $f(x') \le f(x), \forall x \in \mathbb{R}^D$

In order to find an acceptable solution without exploring the entire parameter space, optimization problems require: (1) knowledge of the properties of parameter space, and (2) a cost function to evaluate the problem (Kennedy & Eberhart, 2001).

The properties of the parameter space and the cost function can be used to select an appropriate technique to solve a problem. For example, linear functions can be solved using *least squares* (LS) and unimodal functions can use *convex optimization* (CO) to find a good solution based on the directions of the function gradients with respect to each dimension (Boyd & Vandenberghe, 2004). However, for non-linear problems, gradients can point to a local solution that may be far from optimal. To avoid local minima, iterative optimization algorithms are designed to encourage exploration of the parameter space as the algorithm converges on a solution (Shaw & Kinsner, 1996).

Furthermore, optimization problems can be classified as constrained or unconstrained. In constrained optimization problems, the solution

is guaranteed to be within a certain area of the parameter space. Thus, depending on the size of the area different techniques can be used to first narrow down the search area and then find a solution. Unconstrained problems start with an educated guess of where the solution will reside, but do not limit the algorithms from exploring parameters outside that range if the original guess was incorrect. The latter type of problems is more difficult to solve and generally requires special care to avoid local solutions while quickly converging on a solution.

For cognitive machines, the additional underlying assumption of any optimization problem is that the cost function, that defines the shape of the parameter space, does not experience significant changes during the period required to find a solution. If in fact the cost function is changing, an optimization algorithm may get stuck in a local solution or diverge to never find a solution. Thus, the final answer obtained may be meaningless like to those obtained in analog-to-digital conversions when the aperture time of a signal is shorter than the conversion time (Kinsner, 2010). However, unlike the well-established knowledge of quantization steps in analog-to-digital conversions that quantify the maximum rate of change before a converter ceases to work, the significance of the change in optimization problems remains an open research topic of great interest for real-time and cognitive systems. Therefore, the remaining of this paper assumes that the parameter space is not changing during one run of an optimization algorithm.

2.2 Evolutionary Optimization Algorithms

Evolutionary algorithms form one approach to solve unconstrained, non-linear optimization problems. Figure 1 can describe most evolutionary algorithms. Such an algorithm (1) starts with a random solution, then (2) iteratively generates a new solution(s) through some evolutionary means (i.e., mutation, crossover, social interac-

Figure 1. Structure of a typical evolutionary optimization algorithm

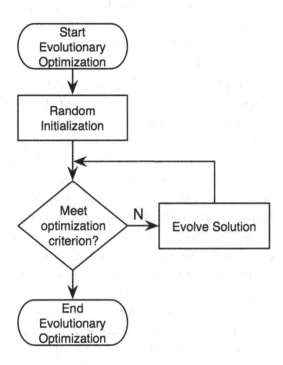

tions), and (3) checks a terminating condition based on the quality of the solution and the number of iterations it took to reach that point (Floreano & Mattiussi, 2008). Some examples of evolutionary algorithms include (1) simulated annealing (SA), (2) genetic algorithms (GA), (3) continuous ant colony optimization[1] (CACO), and (4) particle swarm optimization (PSO). SA works by making random perturbations of a candidate solution and using a temperature gradient to control the number of candidate solutions producing worst results that are accepted over time (Laarhoven & Aarts, 1987). In GA, a collection of genes representing the parameters being optimized are grouped into a chromosome that evolves through mating and mutation (Floreano & Mattiussi, 2008). While in CACO, ants travel between the nest and a source of food trying to find the shortest path connecting the two sites. The optimization and

evolution occurs through different amounts of pheromones left by the ants along the path that correlate to the fitness level of that path (Socha, 2009). Lastly, in the PSO algorithm a set of particles search for a solution in a D-dimensional parameter space based on the particle's best position and the position of the neighborhood's best particle. After a number of iterations, the particles converge on a good solution.

In the small subset of optimization algorithms described, SA considers a single coordinate in the search space at a time, while GA, CACO and PSO are population-based algorithms that consider multiple coordinates in the parameter space at the time. Intuitively, population-based algorithms require more resources but are quicker at solving high-dimensional problems with a large parameter space. Angeline (1998) further fragments the classes of population-based evolutionary algorithms based on the used of an explicit selection function capable of reallocating the resources of the poorly performing members of the population. In GA, the selection process eliminates the genes producing relatively bad solutions are replaced with the children of the better solutions found in the set. This in essence guides the search towards the better solutions by jumping to new areas of the parameter space that are closer to the optimal solution. The jumps depend solely on the crossover operator and thus it is hard to establish a path taken by genes on a way to a solution. CACO experiences similar behaviors as the presence of more or less pheromones can act as the natural selection mechanism. As the pheromones disappear over time, some of the less popular paths taken by the ants are eliminated. While, in PSO the solutions are guided by the information from previous iterations is preserved through the particle's personal best position. As a consequence, rather than generating a brand new solution like in GA or CACO, PSO manipulates the direction of the poorly performing members of the population through information collected about the best solutions found up to that point in time. Thus, each particle exhibits a trajectory that can be studied to understand the behavior of the algorithm.

2.3 Particle Swarm Optimization

Kennedy and Eberhart's PSO algorithm is based on the interactions between organisms as seen in flocks of birds and schools of fish (Kennedy & Eberhart, 1995). The original algorithm is selected as a suitable starting point for a conceptual study of particle dynamics. In the algorithm, K organisms are represented by particles that fly in discrete steps through a virtual D-dimensional parameter space looking for the optimal solution (or at least one that satisfies some criterion). In each iteration, the particles move to a new position defined by a weighted sum of the particle's previous best position and the position of the neighborhood's current best particle (Kennedy & Eberhart, 2001). Thus, one can view each particle as a small cognitive entity that is aware of the environment through social interactions, where the weights mimic the struggles between independence and peer-pressure.

Algorithm 1 shows the pseudo-code for the original PSO algorithm (Kennedy & Eberhart, 1995). Vector $\mathbf{x}_k(n)$ has length D, and represents the current state of the kth particle for each iteration n, $\mathbf{p}_k$ is the particle's previous best position, and $\mathbf{p}_g$ is the neighborhood's current best position. For each iteration, n, the position of the particle is updated by velocity, $\mathbf{v}_k(n)$, computed as a weighted sum of the personal and social components. The weights φ_1 and φ_2 are selected from a uniform distribution between $[0,2]$ as this assumes no knowledge of the parameter space and thus provides an unbias balanced between the influences emerging from $\mathbf{p}_k$ and $\mathbf{p}_g$ respectively. The calculated velocity at each iteration is constrained to the range shown on line 9 to prevent particles from diverging away from a solution by taking larger steps at each iteration.

There are many variations to improve the performance of the PSO algorithm by (1) experimenting with different number of particles (Angeline, 1998; Trelea, 2003), (2) introducing an inertia weight to decrease exploration as t increases (Shi & Eberhart, 1999; Berg 2001; Clerc & Kennedy, 2002; Kennedy, 2007), (3) selecting fixed weights that guarantee convergence (Clerc

Algorithm 1. Particle Swarm Optimization

```
1: repeat
2:    for k=1 to K do
3:        if G(x_k) < G(x_k)
4:           p_k  ←x_k
5:        endif
6:        g ← index of best neighbor
7:        for d=1 to D do
8:               v_{k,d}(n) ← v_{k,d}(n-1) + φ_1(p_{k,d} -x_{k,d}(n-1)) + φ_2(p_{g,d} - x_{k,d}(n-1))
9:               v_{k,d}(n) ∈ (-v_{max},v_{max})
10:              x_{k,d}(n) ← x_{k,d}(n-1) + v_{k,d}(n)
11:          end for
12:   end for
13: until termination criterion
14: return p_{best}
```

& Kennedy, 2002; Chuan & Quanyuan, 2007; Berg, 2001, Bergh & Engelbrecht, 2006; Eberhart & Shi, 2000; Trelea, 2003), (4) adjusting the termination criterion (Zielinski, Peters, & Laur, 2005), (5) using different neighborhood topologies (Kennedy, 1999; Mendes, 2004; Kennedy & Mendes, 2002; Mendes, Kennedy & Neves, 2004; Schor, Kinsner & Anderson, 2010), and (6) incorporating chaos into the parameter space exploration to prevent repeating the path taken by the particles similar to the use of chaos in simulated annealing (Shaw & Kinsner, 1996; Liu et al., 2005).

Given the large number of improvements provided in the literature, new suggestions for baselines have been proposed (Bratton & Kennedy, 2007), but, despite all the modifications, the basic principle in the algorithm has remained the interaction between particles. Each particle relies on the exchange of information from the neighborhood in order to determine where to move to next. Additional discussions on the exchange of information are provided in (Schor, Kinsner & Anderson, 2010).

Many of the improvements highlighted above can reduce the number of iterations required to find a solution under benchmarked tests. However, some of these upgrades can also alter the inherited nature of the algorithm and remove one of the basic characteristics that make it suitable for cognitive systems. For example, using fixed weights as proposed by Eberhart and Shi (2000) can diminish the amount of exploration as compared to random weights that vary the distribution of social and personal weights (hence limiting the free will of the particles). Therefore, the remaining of this paper deals with the original PSO algorithm proposed by Kennedy and Eberhart in 1995 in order to find improvements required by cognitive systems. Section 5.1 provides some additional comments on the differences in the behavior of the original algorithm compared to some of the popular variations found in the literature.

3. REQUIREMENTS FOR OPTIMIZATION IN COGNITIVE PROCESSES

As mentioned in Section 1, a cognitive system requires optimization algorithms for acquiring, processing and interpreting information from the environment in real-time. The system also needs to be able to adapt to the changing environment and predict most likely future behaviors based on optimally selected past and current knowledge. Based on this brief description of a cognitive system, one can identify several requirements for optimization algorithms including (1) convergence rate, (2) degree of exploration of the parameter space, (3) storage and system size, (4) adaptability, and (5) multi-scale capabilities.

3.1 Convergence Rate

The *convergence rate* is a common requirement and a means of comparing optimization algorithms. It is expressed as the number of iterations until a termination criterion is reached (Floreano & Mattiussi, 2008). This measure implicitly requires that all iterations last approximately the same amount of time such that the number of iterations can be compared using a linear scale. In a cognitive system, the speed is crucial for maintaining real-time performance and avoiding overruns caused by slower iterative algorithms.

Although much of the literature on evolutionary optimization algorithms uses this metric to compare different algorithms, the convergence rate alone does not provide enough information to judge the an algorithm effectively on a large set of problems. The convergence rate is only sufficient when comparing two similar algorithms with small variations to a small subset of parameter that do not change the overall nature of the algorithm. For example, two implementations of a GA with varying mutation rates can be compared to find which rate is most beneficial for a particular application. However, larger changes, and, for that matter, different optimization algorithms cannot be compared using the convergence rate alone, as other factors may play a significant role in the performance of the algorithm. For example, increasing the number of particles in PSO can lead to a faster convergence for the algorithm, but one must also consider the increased memory requirements and cost function evaluations that emerge from the change. The following sections describe other critical requirements and metrics that must be evaluated in an evolutionary algorithm.

3.2 Degree of Exploration

The *degree of exploration of the parameter space* gauges how much of the parameter space is explored by an optimization algorithm to find a solution. This allows us to compare two algorithms that take approximately the same number of iterations to find a solution. An argument could be made that if two algorithms have the same convergence rate, then they both explore the same number of solutions, and thus, the degree of exploration of the parameter space is the same. The problem with this argument is that it assumes that the parameter space is uniformly distributed such that the optimal solution is equally likely to be found at any point. This is obviously not a valid assumption, since the parameter space can have valleys of local solutions where algorithms can be trapped if they do not sample many regions of the parameter space. For example, local topologies in PSO can lead to a higher degree of exploration because particles can explore more of the space while the information about a global optimum is transmitted to the entire set (Schor, Kinsner & Anderson, 2010).

The population-based evolutionary algorithms are inherently more likely to explore more of the parameter space if the initial solutions are not randomly initialized to the same small region (Angeline, 1998). However, exploring more paths is not advantageous if the same combinations of solutions are repeatedly being tested, so chaotic trajectories are desirable (Shaw & Kinsner, 1996).

Furthermore, in cognitive machines, where the real-time performance is more important than the use of an optimal solution, a fixed number of iterations can be used to optimize the system. In this format, the same amount of work and resources are used each time the optimization algorithm is used and the best solution achieved during that time is returned to the system regardless of how close that is to the optimal solution (Zielinski et al., 2005). In these types of problems, the *degree of exploration of the parameter space* is of critical importance.

Despite its importance, there are no metrics capable of quantifying the degree of exploration in evolutionary optimization algorithms. This is specially difficult in unconstrained problems where solutions are only bounded during initialization, but are then allowed to take on any value. Furtheremore, since the structure of the parameter space is not known a priori, a measurement about one point, $\mathbf{x}_i$ does not guarantee that a point $\mathbf{x}_j$ in the vicinity ($\| \mathbf{x}_i - \mathbf{x}_j \| < \epsilon$ for small ϵ) has a similar fitness. This problem is even more pronounced when the cost function has many discontinuities.

3.3 Storage and System Size

The *storage and system size* refer to the amounts of resources used by an optimization algorithm during the search process. The resources consists of a combination of both memory used to store the state of the algorithm at each iteration and the computaitonal complexity of the algorithm. For example, solving a 2-dimensional problem in particle swarm with 20 particles using floating point arithmetic requires 480 bytes (estimated as 20 particles * 2 dimensions * [1 position + 1 velocity + 1 personal best] * 4 bytes). Thus, if one increases the number of particles to 30 to explore more of the parameter space, the minimum number of bytes require to store the state of the problem grows to 720 bytes. And, the problem is even more complicated when dealing with high dimensional optimization problems. Thus, there is a clear trade off between increased performance by exploring more of the parameter space and the memory requirements for the system. One can make a similar argument for the processing power used. The computational complexity of the cost function has a significant effect on the performance of the algorithm due to the number of function evaluations required. Following the logic from the previous example, increasing the number of particles would require a much larger number of cost function evaluations to compare the state of each particle and find where the global best is stored. Which, depending on the complexity of the function could have a very significant effect on the performance of the algorithm. And, even if a solution is found in fewer iterations, each iteration could require many clock cycles to be processed (i.e., complexity of Sphere function vs. the Griewank function).

Therefore, there is a serious constraint that is applied to the speed and degree of exploration requirement that lies in the balancing act required to solve the optimization problem using limited resources. In order to define an optimization algorithm suitable for use in cognitive systems, one must select those algorithms that use fewer resources such as memory and processing power.

3.4 Adaptability

Another important requirement for optimization of cognitive systems is the ability of the algorithm to *adapt* to changes in the environment to ensure a good solution is found in most scenarios. This requirement is particularly difficult in cognitive machines because of the unknowns related to the environment measurements that drive cognition. The work of Eberhart and Shi (2007) further describes some of the adaptation difficulties related to optimization algorithms as: (1) the large problem spaces with many candidate solutions to evaluate, (2) asymmetrical high-dimensional problems where it is difficult to determine which variables, and which combinations of variables, to

modify, (3) complex nonlinear fitness functions with many local optima and/or discontinuities, and (4) the changes experienced by the parameter space over time (even when assuming the changes do not affect a single optimization run). Consequently, one must find a tangible way to optimize a function in a cognitive system in spite of the unknowns in the environment readings that make up the parameter space.

Intuitively, one can use a large number of test functions (i.e., common functions used to evaluate evolutionary algorithms that include the Sphere, Rastrigin, Rosenbrock, and Griewank function). These types of functions are strategically selected because of the different characteristics of each function, but doing so leads to another problem described by Wolpert and Macready (1997) as the no-free-lunch (NFL) theorem. The NFL theorem states that even if one shows than an algorithm works well on one class of problems, another class of problems that was not fully considered offsets its performance for the algorithm in question. This means that one cannot define an overall general optimizer that will work on any application. As a way to combat the problem, we suggest that in order to improve the performance of an optimization algorithm, one must incorporate problem-specific (parameter space) knowledge into the behavior of the algorithm (Wolpert & Macready, 1997). This approach is very effective in problems where the variations in the parameter space are characterized by changes in the location of local and optimal solutions and not by changes in the overall structure of the parameter space (i.e., no sudden discontinuities added to the problem). Since this paper does not attempt to solve a particular problem, no parameter specific information is incorporated into the algorithm. However, these types of modifications (i.e., putting bounds on some parameters being optimized that match physical elements of the system) can improve the adaptability of optimization algorithms for a particular cognitive machine.

3.5 Multi-Scale Capabilities

Furthermore, one characteristic that distinguishes cognitive machines from other systems is the ability to perform tasks at multiple levels that mimic the instinctive and rational decisions of humans. That means, a machine must have fast reflexes in the form of fast algorithms to deal with simple tasks and complement that with a level of consciousness representative of the brain functions. More specifically, cognitive systems operate on temporal, spatial, and spatio-temporal signals that carry large amounts of information that must be analyzed at different scales to keep up the real-time aspects of the system (Kinsner, 2004). At the reflex level, the features and overall characteristics of a signal can be sufficient for optimization, however, for more complex decisions, additional information such as entropy based metrics are required to make decisions. An in-depth review of multi-scale metrics for use in cognitive systems is provided in (Kinsner, 2004, 2005).

But, the use of multi-scale metrics to define the cost function of an optimization algorithm is not sufficient for cognitive systems. The real advantage of using multi-scale features are in the study of the behavior of an optimization algorithm as it approaches a solution and in determining whether there are long term dependencies that can be exploited to improve the performance of the algorithm. Moreover, multi-scale feature extraction can be used as a good-bad indicator in the optimization process to guide the search towards the optimal solution.

4. DYNAMICS OF EVOLUTIONARY ALGORITHMS

Since, as described in Section 2, there are many different types of evolutionary optimization algorithms, a comparison of the performance of differ-

ent types requires a generic class of tests that can be applied to any algorithm described by Figure 1. The convergence rate and system size are the only elements that can be easily measured for a particular algorithm. Furthermore, the degree of exploration is hard to quantify given the lack of knowledge about the parameter space. However, as suggested by Angeline (1998), population-based algorithms are by definition exploring more of the parameter space as shown through a time series analysis by Schor, Kinsner, and Anderson (2010). That leaves two difficult questions of gauging the adaptability and multi-scale capabilities for an algorithm.

Immediately, one can deduce that the techniques used in the assessment must not in any way depend on how the new solutions are created as part of the evolutionary process but rather on the trajectory towards a solution. Consequently, one needs to consider both the theoretical and practical aspects of the algorithm in question as if it were a generic evolutionary algorithm. Furthermore, to ensure that any results obtain are of value to other researches, the all metrics must be implementation independent and must focus on the underlying principles that drive the evolutionary optimization.

One way to study the dynamics of algorithms described by Figure 1, is to investigate the trajectory as the solution evolves from a random guess to the final solution. This trajectory can be expressed as a set of K time series along each dimension being optimized against the iteration number. In the case of population-based algorithms, one can trace either the best of the set or a single individual in the population. But, as briefly discussed in Section 2.2, some algorithms do not show a continuous trajectory towards a solution, but rather generate a new set of solutions at each iteration. Therefore, in order to show the benefit of a time series analysis, the remaining of this paper tests whether PSO meets the requirements for optimization algorithms for cognitive machines.

5. ANALYSIS OF DYNAMICS OF PARTICLE SWARM

The dynamics of particles in the PSO algorithm can be studied through the trajectory of a single particle as it travels from its random initial location to an optimal place in the parameter space (Trelea, 2003; Schor, Kinsner, & Anderson, 2010; Schor & Kinsner, 2010). From Algorithm 1, it is evident that the path taken by each particle depends on its initial placement, the relative position of its neighbors, the function being optimized, and the randomly generated weights. To address this issue, Sec. 5.1 describes how an ensemble can be used to identify an average particle to be used in this study. Given an acceptable model, Section 5.2 extracts the time series of interests to be analyzed in the time domain (Section 5.3) and frequency domain (Section 5.4). Other domains, such as the Wavelet domain, are not addressed in this paper.

5.1 Selecting a Particle to Study

In order to select a particle for this study, one must first identify what function is being optimized. In this paper, focus has been placed on two distinct functions (Sphere and Rastrigin) that are widely used to benchmark the performance of various continuous optimization algorithms like PSO (Kennedy & Eberhart, 2001). Since this is a generic analysis, no problem-specific information is incorporated into the algorithms

as suggested by Wolpert and Macready (1997). The functions, their effective ranges, and global minima are summarized in Table 1. The Sphere is a unimodal function that has an increasing slope the further away one is from the optimal solution, thus quickly guiding all particles in the correct direction. This function does not have any local solutions where particles can get stuck, and therefore, can be used as the simplest model to explore the behavior of particles in PSO. Rastrigin is a multimodal function with a parabolic mesh-like that leads to the optimal solution. The mesh is made up of cones that form local minima and can act as traps for particles. Many algorithms that depend on an inertial weight or temperature gradient (i.e., simulated annealing) must be fine tuned to avoid local solutions.

In addition, this study uses 20 particles to test the Sphere and Rastrigin functions. The number of particles purposely approaches the lower bounds described in the literature (Kennedy & Eberhart, 2001) such that the resources are minimized as required for real-time computation. The particles are connected in a global topology, thus at each iteration, the best particle in the population affects the trajectory of all the other particles in the set. Although there is evidence in favor of using different topologies to speed up convergence (Schor, Kinsner, & Anderson, 2010), the global topology makes all particles the same, thus there is no need to study the trajectory of special pieces of the set (i.e., nodes and root of a star topology). The re-

Table 1. Test functions used for the analysis of particle trajectories. The last column indicates the threshold defined in Sec. 5.2 that separates the transient from the steady state portion of the particle's trajectory.

Name	Function	Symmetric Range (global min for D=2)	Threshold
Sphere	$f_1(\mathbf{x}_j) = \sum_{d=1}^{D} x_{j,d}^2$	$[-100,100]^D$ (0,0)	300
Rastrigin	$f_2(\mathbf{x}_j) = \sum_{d=1}^{D} \left(x_{j,d}^2 - 10\cos\left(2\pi x_{j,d}\right) + 10 \right)$	$[-5.12,5.12]^D$ (0,0)	300

maining parameters are set to $V_{max} = 0.2$, $D = 2$, $K = 20$, $\varphi_1 = 0.1 \times U(0.1)$, and $\varphi_2 = 0.1 \times U(0.1)$, where $U(a, b)$ represents a uniformly distributed random number in $[a, b]$. The small V_{max} and φ variables are purposely selected to slow down convergence and thus extend the length of the transient to facilitate the analysis. Lastly, for this experiment, the termination criterion is when the number of iterations reaches 4000. Note, for clarity, some graphs show fewer than the full 4000 iterations; however the analysis is conducted over the full range. The function converges very quickly, so the extra iterations simply further optimize it to a higher resolution. The implementation of the PSO algorithm used in this paper does not have an inertia weight to control the amount of exploration over time, thus, any trajectories obtained would demonstrate the level of adaptability required from an optimization algorithm for use in cognitive machines.

Figures 2 and 3 show the trajectory of a typical particle through the parameter space of the Sphere and Rastrigin functions respectively. The obvious question is then, are these behaviors typical for the particles? This is a rather difficult question to answer because of the stochastic nature of the algorithm. The random initialization and varying φ_1 and φ_2 can generate a butterfly effect from small differences (Kinsner, 2009).

There are many ways to reduce the number of unknowns in order to answer this question. One option is to fix the weights such that once initialized, the behavior of the particles becomes deterministic. However, as previously discussed, this reduces the effective region of the parameter space being explored and it assumes that there environment does not experience significant changes over time. Any option that fixes the behavior of some particles to remove those unknowns from the set are not considered as they remove the inherited social interactions that drive the algorithm. Thus,

Figure 2. 3D representation of the trajectory of a particle through the parameter space of the Sphere function

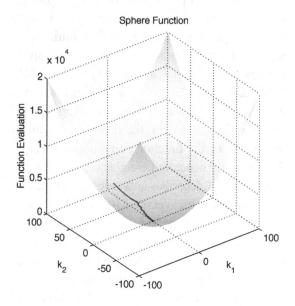

Figure 3. 3D representation of the trajectory of a particle through the parameter space of the Rastrigin function

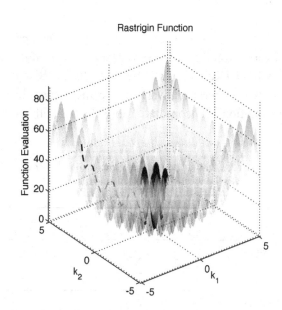

the approach taken in this paper is to fix the initial conditions for all particles, and then run the same optimization problem many times to produce an ensemble. Then, one can study how the distribution of a single particle's position varies over time.

The histograms over time for the two functions being studied are shown in Figures 4 and 5 respectively. These images are generated by performing 200 runs where the particles are initiated with the same position and velocity vectors before the seeds for the random number generator that drives the social and personal weights are initialized. Looking at these histograms, one must remember that the algorithm was purposely slowed down through V_{max} and φ, therefore although it appears as the vast majority of the particles follow the exact same path, there are small variations hidden in the histogram ($V_{max} = 0.2$). The features become more prominent in the Rastrigin histogram due to the smaller range being optimized. Increasing V_{max} and φ causes the particles to take larger steps and therefore have a larger spread in the histogram at every step. In these trajectories, it is apparent that the particles are first pulled in a diagonal along all dimensions being optimized. However, once the particle reaches an optimal value along one dimension it begins to oscillate around the minimum found. During this stage, the particle explores the local area along that dimension while the other dimension continues to make longer strides towards the solution. This behavior is seen in all symmetric functions as the underlying gradient pushes towards the solution.

5.2 Extracting a Time Series

There are three important pieces of information embedded in the trajectories shown in Figures 4 and 5. Those are the particle's position vector, $\mathbf{x}_k(n)$, its velocity vector (or relative change in position) $\mathbf{v}_k(n) = \mathbf{x}_k(n) - \mathbf{x}_k(n-1)$, and its fitness over time. And, one can go further by splitting

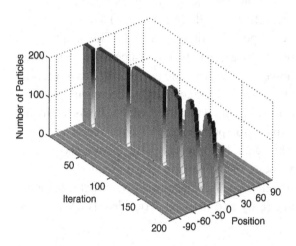

Figure 4. Histogram showing the trajectory an ensemble of 200 particles initialized to the same conditions in the Sphere function

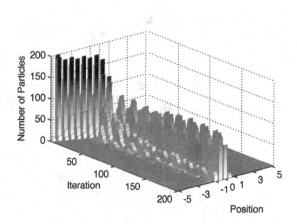

Figure 5. Histogram showing the trajectory an ensemble of 200 particles initialized to the same conditions in the Rastrigin function

the position and velocity vectors into separate time series for each dimension being optimized. This information forms the time series that describes a single particle in PSO.

Figures 6 and 7 show the time series of interest for the particles appearing in Figures 2 and 3 respectively. In these graphs, (a) shows the particle's position along d_1 and d_2. This graph does not include the particle's best position over time; however that curve can be described as a smooth curve that is monotonically decreasing to the optimal point, while the position shows the local exploration performed by the particle over time. The second plot (b) in these figures shows the velocity of the particle over time along the two dimensions. This diagram shows a weakly stationary (only two moments considered) time series

that is bounded by V_{max} as shown in Algorithm 1. The last plot (c) shows the fitness of the particle's current and best position over time. This information indicates that the particle is converging towards zero for all functions, which corresponds to the global minimum.

The trajectories shown in Figure 2 can be broken down into two components: the transient and the steady state. This separation is important because the properties during the transient are completely different from those of the remaining of the time series (Kantz & Schreiber, 2004). The transient is defined as the time from the initial state until the time when the position of all particles along all dimensions is below a certain threshold. In this case, the threshold for both functions was defined as 300 iterations. Nor-

Figure 6. Time series showing the particle's position, velocity, and fitness over time in the Sphere function. (a) The particle's position over time along the 2 dimensions being optimized. (b) The velocity of the particle over time along the 2 dimensions being optimized. (c) The fitness of the particle as it converges to the global optimum when the fitness equals zero. The fitness for the best position and the current position are almost indistinguishable for the Sphere function.

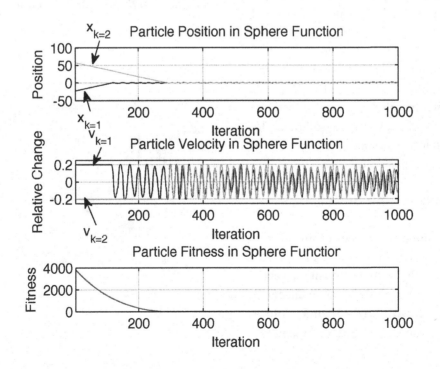

Figure 7. Time series showing the particle's position, velocity, and fitness over time in the Rastrigin function. (a) The particle's position over time along the 2 dimensions being optimized. (b) The velocity of the particle over time along the 2 dimensions being optimized. (c) The fitness of the particle as it converges to the global optimum when the fitness equals zero. This plot highlights the monotonically decreasing fitness of the best position and contrasts the variation in the current position's fitness as the particle explores many of the local solution spaces.

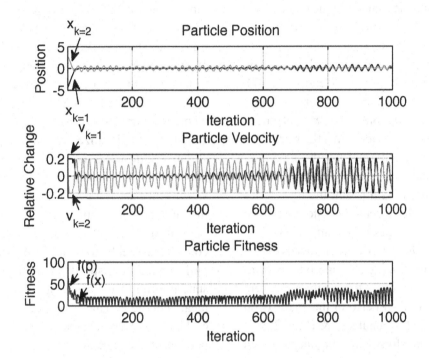

mally, when using larger step sizes, the threshold would be larger for the Rastrigin function than for the Sphere function as it takes longer for particles to converge on a good solution, however, due to the small step sizes, the two can be set to the same value.

Similar graphs to can be constructed for the remaining 19 particles used to optimize each function. The features described above apply to all particles in the set. The only noticeable difference is that, for some graphs, the velocity time series is constant for the first few iterations of the algorithm remains constant at one extremum. This can be explained in the context of the algorithm because the difference between the particle's current position and the best particle in the set is used

to direct the particle through the parameter space. For example, if particle k_1 is located near the optimal value (in this case the origin) at $(d_1, d_2) = (0.5, 2)$ and particle k_2 is initially placed at $(d_1, d_2) = (97, 5)$, and then k_1 is attracted to k_2 with a greater acceleration along the d_1 dimension than along the d_2 dimension because the difference is greater. This means that for many φ_2 the velocity is maximized.

Out of the position, velocity, and fitness time series extracted for each particle, the position depends on the function and particle selected for the study, the velocity appears to behave in a similar fashion for all particles regardless of the function being optimized, and the fitness does not carry enough information other than to classify the performance of different implementations of the algorithm. Thus, the remaining of the paper deals with a multiscale analysis of the velocity trajectories in order to extract the complexity measures required in the classification of the signal behavior and its control.

Before starting the analysis of the trajectories, it is important to note that the position and velocity time series for never versions of the algorithm, such as for example those generated by the new standard proposed by Bratton and Kennedy (2007), have some notably distinct behaviors. In those algorithms, the constriction coefficient that guides the particles to convergence limits the size of the step taken at each iteration and uses the maximum velocity constraints as a safety net that is not intended to be used. As a result, the time series shows a particle whose velocity is slowing down as the particle converges to a solution. Thus, to properly analyze these particles, one needs to first apply a transformation (i.e., Lamperti transform) to the data to remove the temperature gradient like behavior, and then perform the analysis on the resulting signal (Rangarajan & Ding, 2003). The use of such transforms is an ongoing research topic for the authors and thus will not be discussed further in this paper.

5.3 Time Domain Analysis

The first analysis performed on the time domain is to determine whether the signal is stationary. If a stochastic process has constant first moment (mean) and second moment (variance or covariance, if necessary), then the process is called *wide-sense stationary* (WSS) or weakly-stationary for short. And, if all the moments of a stochastic process are unchanging, then the process is called *strict-sense stationary* (SSS) or strongly-stationary. In swarm optimization, although the source of information is known, the random numbers, personal best, and social influences are dynamic. Thus, it is hard to determine whether the signal is stationary from a single time series. However,

weakly-stationarity can be examined by looking at a sliding window of the data and making sure that the mean and variance constant over time (Kinsner, 2009).

Figures 8, 9 10, and 11 show the first two moments of the velocity time series along the d_1 dimension for the same particle mentioned in the previous sections. Figures 8 and 10 depicts the transient's statistical properties and Figures 8 and 10 shows the steady state properties for the two functions being studied. The circles indicate one moment calculated over a window of 32 and 128 elements respectively and the line is simply added to help visualize the constant pattern. The sliding window size is selected such that it lies between the 30 elements needed for statistical significance and the size of main features seen in the envelope of the function. The figures show that during the transients, the particles can be considered weakly-stationary following the initial pull to maximum speed. In contrast, the steady state portion of the trajectory shows a constant mean, but the variance experiences more changes than during the transient. Thus, the steady state portion of the trajectory can be classified as a heteroscedastic signal.

Another important measure from the time series is the correlation. A high correlation means the time series has a long-term dependence that can serve as an indication that the solutions are evolving over time. In cognitive systems, knowing that the optimization trajectory has a long-term dependence can be exploited to speed up convergence by use of predictive techniques (Kinsner, 2009).

Figures 12 and 13 show the correlation for the particle's trajectory along one dimension for the transient for the Sphere and Rastrigin functions. The graphs show a strong correlation for the first few iterations consistent with the portion of the time series that maximizes the velocity. However, the graphs do not show a very noticeable long term correlation. In the case of the Rastrigin function,

Figure 8. First two moments of the particle's trajectory during the transient of the Sphere function. (a) shows the velocity of the particle over the time, and (b) and (c) show the mean and variance respectively calculated using a sliding window of 32 elements.

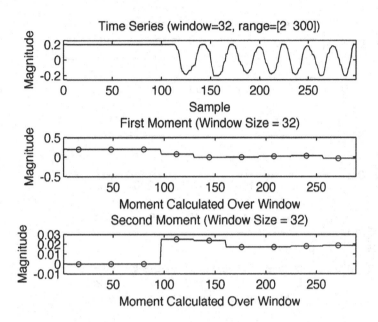

Figure 9. First two moments of the particle's trajectory during the transient of the Rastrigin function. (a) shows the velocity of the particle over the time, and (b) and (c) show the mean and variance respectively calculated using a sliding window of 32 elements.

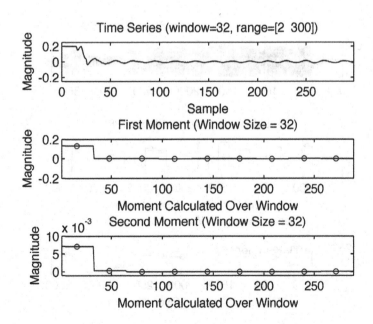

Figure 10. First two moments of the particle's trajectory during the steady state portion of the Sphere function time series. (a) shows the velocity of the particle over the time, and (b) and (c) show the mean and variance respectively calculated using a sliding window of 32 elements.

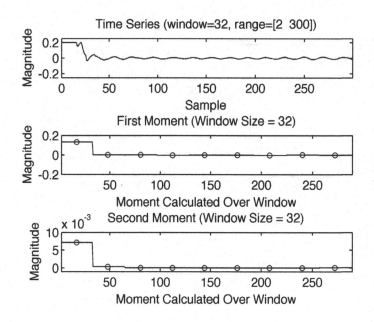

Figure 11. First two moments of the particle's trajectory during the steady state portion of the Rastrigin function time series. (a) shows the velocity of the particle over the time, and (b) and (c) show the mean and variance respectively calculated using a sliding window of 32 elements.

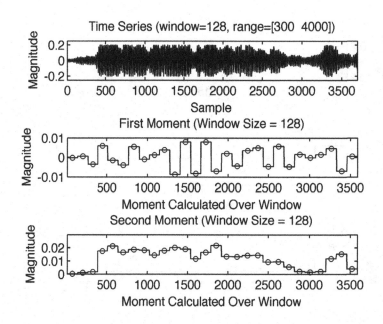

Figure 12. Autocorrelation of the particle's trajectory during the transient in the Sphere function. (a) The particle's velocity. (b) Its autocorrelation.

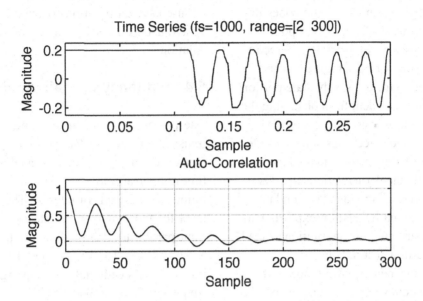

Figure 13. Autocorrelation of the particle's trajectory during the transient in the Rastrigin function. (a) The particle's velocity. (b) Its autocorrelation.

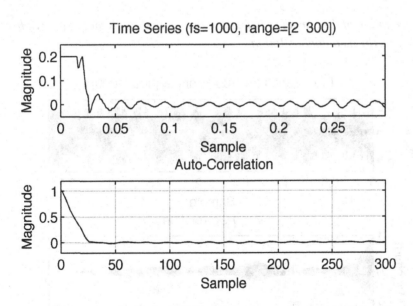

the long term correlation is even less noticeable once the particle reaches the vicinity of the solution and is trapped within one of the cones that make up the parameter space. At this point, the correlation decreases as the particle is forced to switch directions many times. Figures 14 and 15 show the correlation for the steady state portion of the time series. As evident from observing the envelope for the velocities shown, there is a correlation that drives the trajectories of the particles is stronger for short term changes in position, and then weakens for longer term relationships. Overall, the correlation is lower than that found in the transient because the particle is now exploring the parameter space surrounding the solution instead of traveling towards a destination.

In order to gain a better understanding of the algorithm the random weights φ_1 and φ_2 can be fixed as suggested by (Eberhart & Shi, 2000; Trelea, 2003). The results for fixed weights show a very high autocorrelation which is especially high for the steady state. In this case, rather than exploring the particle around the solution, the

particle reaches a quasi-periodic state as described in (Clerc & Kennedy, 2002). If this behavior is not used as a termination criterion, then it should be avoided as it does not help in finding a better solution.

5.4 Frequency Domain Analysis

The power spectrum of the time series for the transient and steady state portions are computed to confirm if the correlation analysis from the time domain and also gauge the amount of exploration in the signals of interest. The power spectrum for the transient and steady states for the same particle used throughout this paper are plotted in Figures 16, 17, 18, and 19 for both the transient and steady state. A sampling or reference frequency of $f_s = 1000\ Hz$ is used throughout. The spectrums for the steady state show a peak at 500 Hz that is a side effect of the sampling frequency selected.

Intuitively, the plots agree with the observations from previous sections. During the transient,

Figure 14. Autocorrelation of the particle's trajectory during the steady state in the Sphere function. (a) The particle's velocity. (b) Its autocorrelation.

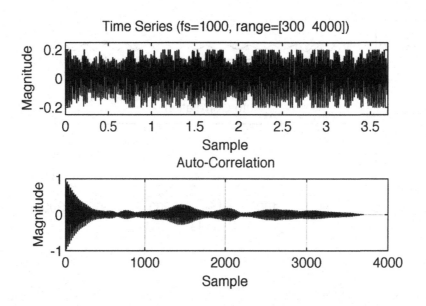

Figure 15. Autocorrelation of the particle's trajectory during the steady state in the Rastrigin function. (a) The particle's velocity. (b) Its autocorrelation.

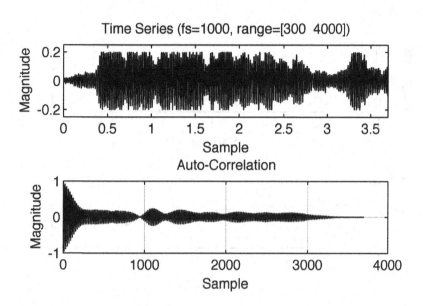

Figure 16. Power spectrum of particle's trajectory during the transient of the Sphere function. (a) The particle's velocity. (b) Its power spectrum.

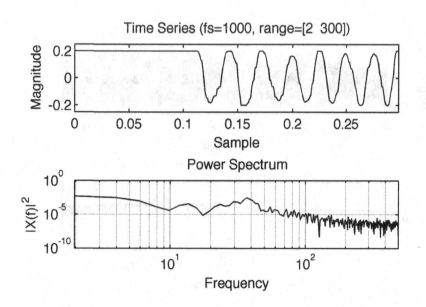

Figure 17. Power spectrum of particle's trajectory during the transient of the Rastrigin function. (a) The particle's velocity. (b) Its power spectrum.

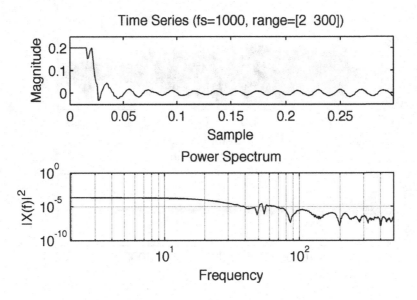

Figure 18. Power spectrum of particle's trajectory during the steady state of the Sphere function. (a) The particle's velocity. (b) Its power spectrum.

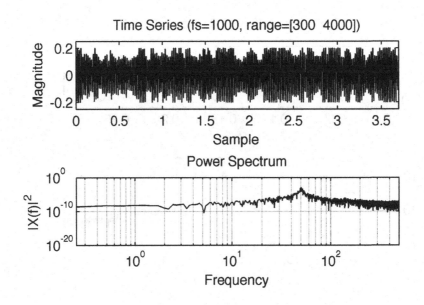

Figure 19. Power spectrum of particle's trajectory during the steady state of the Rastrigin function. (a) The particle's velocity. (b) Its power spectrum.

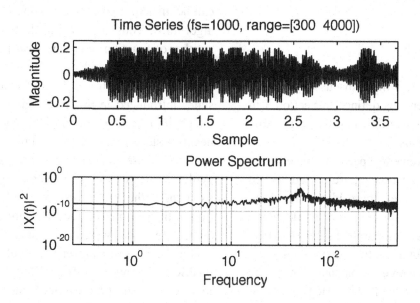

the overall shape of the spectrum appears to have a negative slope with some variation around certain frequencies that correspond to the amplitude modulation of the overall signal. The steepness of the slope is determined with the path taken by the particles. When a particle is the best in the neighborhood in the Rastrigin function, the particle begins exploring the parameter space locally and when some particles are optimizing the Rastrigin function and get trapped in a local solution during the transient, then the steepness of the slopes is reduced. In the first case this is because the particle is exploring the parameter space and without an external force telling it in which general direction to travel. The latter case exhibits a higher slope because the particle requires more energy to escape these local solutions. That is, the weights during some iterations may not be sufficient to escape the local minimum and therefore, small steps are taken to escape that space.

The negative slope shown in Figures 16 and 17 are characteristic of self-affinity and long-term dependences. These show that the personal best solutions stored by each particle act as a

memory for the system that enables particles to constantly move towards a better solution. For comparison, if the optimization algorithm selected new random solutions at each iteration with no influence from the past, then the slope would be equal to zero showing that there is no correlation from one iteration to the next (as seen in white noise). At the other extreme, a very steep slope is indicative of very long-term correlated trajectories typical of a steepest decent algorithm with added randomness to explore the path on a way towards a solution (as seen in deep black noise). Thus, the balanced mixture of random behavior with long memory is the specific behavior that makes PSO suitable for cognitive machines. Additional research is required to measure the fractal dimension and determine how that can be useful in improving the algorithm or its use, however, this kind of behavior is generally a good indication that the particles are exploring the terrain as they approach the solution. Preliminary attempts suggest that adjusting the strength of personal and global best based on the age of the values (how long they have remained the same) can be used to fine tune the slope.

6. CONCLUSION

In this section, PSO will be discussed in terms of the five requirements defined in Section 3 in order to assess whether the algorithm is suitable for cognitive systems.

The two requirements speed of convergence and storage and system size, are not addressed in this paper. However, these ideas have been addressed by other authors. For example, some literature provides a comparative study of evolutionary optimization algorithms that suggests that PSO has a fast convergence rate and that can be resilient to local optima (Elbeltagi, Hegazy, & Grierson, 2005). The storage and system size is very negligible for PSO. Minimum implementations require that only the particle's position, velocity, and best position be stored, which for 20-100 particles (Bratton & Kennedy, 2007) does not impose a large demand on the system. Furthermore, the amount of processing required depends on the function being optimized and to a lesser extent on the equation to update the particle's velocity. Thus, for all intents and purposes, PSO's storage and system size is adequate for cognitive systems.

Suggestions for gauging the degree of exploration are described by Schor, Kinsner, and Anderson (2010) through a time series analysis (Figure 3). The results from that paper show that the degree of exploration depends on the local topology selected. Topologies exhibiting multiple paths for information flow, small clusters, and root/filtering nodes can increase the degree of exploration without sacrificing the convergence speed. This is further reinforced by the two stage search approach defined in this paper, where the particles use a global search during a transient phase and then focus the search over a smaller area as the particles enter a steady state.

The adaptability of PSO to changing environments is the key for all cognitive systems and can be difficult to assess in comparison to other algorithms. Thus, all one can do is test a large number of functions and whenever possible add domain-specific knowledge to help the search. In addition, the novel division of transient and steady states introduced in this paper gives additional insight into the adaptability of the algorithm. The negative slope in the power spectrum during the transient phase indicates that the particles are evolving towards a solution. The steepness of the slope shows the balance between random exploration of the parameter space and quickly moving and converging towards a solution.

Lastly, this paper presents a novel perspective on PSO and its multi-scale capabilities. As shown in Figures 16 and 17, the trajectories for particles during the transient have exhibited some correlation and self-affine characteristics typical of fractals. This indicates that the trajectories taken by the particles have a long-term dependence (Kantz & Schreiber, 2004; Rangarajan & Ding, 2003). Further studies into the properties of transients are required to determine how these characteristics can be exploited to improve the performance of the algorithm.

Based on the preliminary work presented in this paper, it appears that particle swarm optimization is a suitable algorithm for cognitive systems. The long-term dependence experienced during the transient as the particle approaches the solutions gives an indication on the rate of convergence. Once the particle has reached the vicinity of the solution, the behavior changes forcing a greater amount of exploration and finds the optimal solution. Using this information and the characteristics and metrics described in this paper, the behavior of the particle swarm can be further controlled for use in cognitive machines.

ACKNOWLEDGMENT

This work was partially supported by MITACS Accelerate Program, Bristol Aerospace Ltd. in Winnipeg, Manitoba, and the Natural Sciences and Engineering Research Council (NSERC) of Canada.

REFERENCES

Angeline, P. (1998). Evolutionary optimization versus particle swarm optimization: Philosophy and performance differences. In V. W. Porto, N. Saravanan, D. Waagen, & A. E. Eiben (Eds.), *Proceedings of the 7th International Conference on Evolutionary Programming 4* (LNCS 1447, pp. 601-610).

Bergh, F. (2001). *An analysis of particle swarm optimizers*. Unpublished doctoral dissertation, University of Pretoria, South Africa.

Bergh, F., & Engelbrecht, A. (2006). A study of particle swarm optimization particle trajectories. *Information Sciences*, *176*(8), 937–971. doi:10.1016/j.ins.2005.02.003

Boyd, S., & Vandenberghe, L. (2004). *Convex optimization*. Cambridge, UK: Cambridge University Press.

Bratton, D., & Kennedy, J. (2007, April). Defining a standard for particle swarm optimization. In *Proceedings of the IEEE Symposium on Swarm Intelligence*, Honolulu, HI (pp. 120-127). Washington, DC: IEEE Computer Society.

Chuan, L., & Quanyuan, F. (2007, August). The standard particle swarm optimization algorithm convergence analysis and parameter selection. In *Proceedings of the Third International Conference on Natural Computation*, Hainan, China (Vol. 3, pp. 823-826). Washington, DC: IEEE Computer Society.

Clerc, M., & Kennedy, J. (2002). The particle swarm - explosion, stability, and convergence in a multidimensional complex space. *IEEE Transactions on Evolutionary Computation*, *6*(1), 58–73. doi:10.1109/4235.985692

Eberhart, R., & Shi, Y. (2000, August). Comparing inertia weights and constriction factors in particle swarm optimization. In *Proceedings of the Congress on Evolutionary Computation*, La Jolla, CA (Vol. 1, pp. 84-88). Washington, DC: IEEE Computer Society.

Eberhart, R., & Shi, Y. (2007). *Computational intelligence: Concepts to implementations* (p. 512). San Francisco, CA: Morgan Kaufmann.

Elbeltagi, E., Hegazy, T., & Grierson, D. (2005). Comparison among five evolutionary-based optimization algorithms. *Advanced Engineering Informatics*, *19*, 43–53. doi:10.1016/j.aei.2005.01.004

Floreano, D., & Mattiussi, C. (2008). *Bio-inspired artificial intelligence*. Cambridge, MA: MIT Press.

Gavrilova, M. (2009). Adaptive computation paradigm in knowledge representation: traditional and emerging applications. *International Journal of Software Science and Computational Intelligence*, *1*(1), 87–99. doi:10.4018/jssci.2009010106

Gerald, C., & Wheatley, P. (2004). *Applied numerical analysis* (7th ed.). Reading, MA: Addison-Wesley.

Haykin, S. (2005). Cognitive radio: Brain-empowered wireless communications. *IEEE Journal on Selected Areas in Communications*, *23*(2), 201–220. doi:10.1109/JSAC.2004.839380

Haykin, S. (2006). Cognitive Radar: A way of the future. *IEEE Signal Processing Magazine*, *30*, 30–40. doi:10.1109/MSP.2006.1593335

Haykin, S., Reed, J. H., Li, G. Y., & Shafi, M. (Eds.). (2009a). Cognitive radio – Part 1: Practical perspectives. *Proceedings of the IEEE*, *97*(4).

Haykin, S., Reed, J. H., Li, G. Y., & Shafi, M. (Eds.). (2009b). Cognitive radio – Part 2: Fundamental issues. *Proceedings of the IEEE, 97*(5).

Hossain, E., & Bhargava, V. (Eds.). (2007). *Cognitive wireless communication networks* (1st ed.). New York, NY: Springer. doi:10.1007/978-0-387-68832-9

Kantz, H., & Schreiber, T. (2004). *Nonlinear time series analysis* (2nd ed.). Cambridge, UK: Cambridge University Press.

Kennedy, J. (1997, April). The particle swarm: Social adaptation of knowledge. In *Proceedings of the IEEE International Conference on Evolutionary Computation*, Indianapolis, IN, (pp. 303-308). Washington, DC: IEEE Computer Society.

Kennedy, J. (1999, July). Small worlds and mega-minds: effects of neighborhood topology on particle swarm performance. In Washington, DC: IEEE Computer Society. *Proceedings of the Congress on Evolutionary Computation, 3,* 1931–1938.

Kennedy, J. (2007). Some issues and practices for particle swarms. In *Proceedings of the Symposium on Swarm Intelligence*, Honolulu, HI (pp. 162-169). Washington, DC: IEEE Computer Society.

Kennedy, J., & Eberhart, R. (1995, June). Particle swarm optimization. In *Proceedings of the IEEE International Conference on Neural Networks*, Cambridge, UK, (Vol. 4, pp. 1942-1948). Washington, DC: IEEE Computer Society.

Kennedy, J., & Eberhart, R. (2001). *Swarm intelligence*. San Francisco, CA: Morgan Kaufmann.

Kennedy, J., & Mendes, R. (2002, May). Population structure and particle swarm performance. In *Proceedings of the Congress on Evolutionary Computation*, Honolulu, HI (Vol. 2, pp. 1671-1676). Washington, DC: IEEE Computer Society.

Kinsner, W. (2004, August). Is entropy suitable to characterize data and signals for cognitive informatics? In *Proceedings of the 3rd IEEE International Conference on Cognitive Informatics*, British Columbia, Canada (pp. 6-21). Washington, DC: IEEE Computer Society.

Kinsner, W. (2005, August). A unified approach to fractal dimensions. In *Proceedings of the 4th IEEE International Conference on Cognitive Informatics*, Irvine, CA (pp. 58-72). Washington, DC: IEEE Computer Society.

Kinsner, W. (2010). *Fractal and chaos engineering (course notes)*. Winnipeg, MB, Canada: University of Manitoba.

Liu, B., Wang, L., Jin, Y.-H., Tang, F., & Huang, D.-X. (2005). Improved particle swarm optimization combined with chaos. *Chaos, Solitons, and Fractals, 25*(5), 1261–1271. doi:10.1016/j.chaos.2004.11.095

Mendes, R. (2004). *Population topologies and their infuence in particle swarm performance*. Unpublished doctoral dissertation, Universidade do Minho, Portugal.

Mendes, R., Kennedy, J., & Neves, J. (2004). The fully informed particle swarm: Simpler, maybe better. *IEEE Transactions on Evolutionary Computation, 8*(3), 204–210. doi:10.1109/TEVC.2004.826074

Rangarajan, G., & Ding, M. (Eds.). (2003). *Processes with long-range correlations: theory and applications*. Berlin, Germany: Springer-Verlag. doi:10.1007/3-540-44832-2

Schor, D., & Kinsner, W. (2010, July). A study of particle swarm optimization for cognitive machines. In *Proceedings of the 9th International Conference on Cognitive Informatics*, Beijing, China (pp. 26-33). Washington, DC: IEEE Computer Society.

Schor, D., Kinsner, W., & Anderson, J. (2010, May). A study of optimal topologies in swarm intelligence. In *Proceedings of the IEEE Canadian Conference on Electrical and Computer Engineering*, Calgary, AB, Canada (pp. 1-8). Washington, DC: IEEE Computer Society.

Shaw, D., & Kinsner, W. (1996, May). Chaotic simulated annealing in multilayer feedforward networks. In *Proceedings of the IEEE Canadian Conference on Electrical and Computer Engineering*, Calgary, AB, Canada (Vol. 1, pp. 265-269). Washington, DC: IEEE Computer Society.

Shi, Y., & Eberhart, R. C. (1999). Empirical study of particle swarm optimization. In []. Washington, DC: IEEE Computer Society.]. *Proceedings of the Congress on Evolutionary Computation, 3*, 1945–1950.

Socha, K. (2008). *Ant colony optimization for continuous and mixed-variable domains*. Unpublished doctoral dissertation, Universite Libre de Bruxelles, Belgium.

Trelea, I. C. (2003). The particle swarm optimization algorithm: Convergence analysis and parameter selection. *Information Processing Letters, 85*(6), 317–325. doi:10.1016/S0020-0190(02)00447-7

Wang, Y. (2002, August). On cognitive informatics. In *Proceedings of the 1st IEEE International Conference on Cognitive Informatics*, Calgary, AB, Canada (pp. 34-42). Washington, DC: IEEE Computer Society.

Wang, Y., Zhang, D., & Kinsner, W. (Eds.). (2010). *Advances in cognitive informatics and cognitive computing*. Berlin, Germany: Springer-Verlag.

Wolpert, D., & Macready, W. (1997). No free lunch theorems for optimization. *IEEE Transactions on Evolutionary Computation, 1*(1), 67–82. doi:10.1109/4235.585893

Zielinski, K., Peters, D., & Laur, R. (2005, December). Stopping criteria for single-objective optimization. In *Proceedings of the 3rd International Conference on Computational Intelligence, Robotics, and Autonomous Systems* (pp. 1-6).

ENDNOTES

[1]   The standard ant colony optimization (ACO) algorithm is designed to solve discrete problems, thus its continuous counterpart is deliberately described such that one can make a clear comparison between other continuous optimization algorithms.

This work was previously published in the International Journal of Cognitive Informatics and Natural Intelligence, Volume 5, Issue 1, edited by Yingxu Wang, pp. 18-42, copyright 2011 by IGI Publishing (an imprint of IGI Global).

Chapter 19
On Machine Symbol Grounding and Optimization

Oliver Kramer
Bauhaus-University Weimar, Germany

ABSTRACT

From the point of view of an autonomous agent the world consists of high-dimensional dynamic sensorimotor data. Interface algorithms translate this data into symbols that are easier to handle for cognitive processes. Symbol grounding is about whether these systems can, based on this data, construct symbols that serve as a vehicle for higher symbol-oriented cognitive processes. Machine learning and data mining techniques are geared towards finding structures and input-output relations in this data by providing appropriate interface algorithms that translate raw data into symbols. This work formulates the interface design as global optimization problem with the objective to maximize the success of the overlying symbolic algorithm. For its implementation various known algorithms from data mining and machine learning turn out to be adequate methods that do not only exploit the intrinsic structure of the subsymbolic data, but that also allow to flexibly adapt to the objectives of the symbolic process. Furthermore, this work discusses the optimization formulation as a functional perspective on symbol grounding that does not hurt the zero semantical commitment condition. A case study illustrates technical details of the machine symbol grounding approach.

DOI: 10.4018/978-1-4666-2476-4.ch019

1. INTRODUCTION

The literature on artificial intelligence (AI) defines "perception" in cognitive systems as the transduction of subsymbolic data to symbols (e.g., Russell & Norvig, 2003). Auditory, visual or tactile data from various kinds of sense organs is subject to neural pattern recognition processes, which reduce it to neurophysiological signals that our mind interprets as symbols or schemes. The human visual system has often referred to as an example for such a complex transformation. Symbols are thought to be representations of entities in the world, having a syntax of their own. Even more importantly, symbols are supposed to be grounded by their internal semantics. They allow cognitive manipulations such as inference processes and logical operations, which made AI researches come to believe that thinking can be referred to as the manipulation of symbols, and therefore could be considered to be computations (Harnad, 1994). Cognition becomes implementation-independent, systematically interpretable symbol-manipulation.

However, how do we define symbols and their meaning in artificial systems, e.g., for autonomous robots? Which subsymbolic elements belong to the set that defines a symbol, and – with regard to cognitive manipulations – what is the interpretation of a particular symbol? These questions are the focus of the "symbolic grounding problem" (SGP) (Harnad, 1990), and the "Chinese room argument" (Searle, 1980), both of which concentrate on the problem of how the meaning and the interpretation of a symbol is grounded in action. Several strategies have been proposed to meet these challenges. For a thorough review cf. (Taddeo & Floridi, 2005).

To my mind the definition of a symbol and its interpretation is mostly of functional nature. The intention and the success in solving problems to achieve goals must guide the meaning and

thus the definition of symbols. Hence, it seems reasonable to formulate the symbol definition as optimization problem. Optimal symbols and their interpretations yield optimal success of an autonomous agent. In many artificial systems symbols are defined by an interface algorithm that maps sensory or sensorimotor data to symbol tokens, e.g., class labels. Optimizing a symbol with regard to the success of cognitive operations means optimizing the interface design. In many artificial systems the interface design is part of an implicit system modeling process – regrettably often without much effort spent on an optimal architecture.

The paper is structured according to three perspectives it introduces. First, the formal perspective in Section 2 will formulate the interface design as global optimization problem. The concepts of symbols and higher cognitive operations are formalized. The interface between subsymbolic and symbolic representations is introduced in an optimization formulation while potential objectives, free parameters and a two-level optimization process are discussed. An algorithmic perspective is shown in Section 3, where I discuss typical data mining and machine learning tasks like classification, clustering and dimensionality reduction in the context of interface design and symbol grounding. I propose not to restrict to connectionist approaches, but to make use of recent data mining and machine learning techniques – from K-means to kernel methods. The cognitive perspective in Section 4 discusses the consequences of the interface optimization formulation on the symbolic grounding problem. To my mind – as only the agent's objective has to be formulated explicitly, and this is implicit to any biological form of life[1], the optimization formulation is close to fulfilling the zero semantical commitment condition. Last, I present a case-study of interface optimization in Section 5. In Section 6 the most important results are summarized.

2. INTERFACE DESIGN AS OPTIMIZATION PROBLEM

Cognitive operations operate on a symbolic level. After the characterization of symbolic algorithms, I formulate the definition of a symbol via its connection to subsymbolic representations. An interface algorithm maps the subsymbolic data onto symbols. With regard to the objectives of the cognitive system the interface design is formulated as global optimization problem.

2.1 Symbols and Interfaces

The definition of higher cognitive operations of autonomous agents is no easy undertaking and faces similar problems like the definition of intelligence in cognitive sciences and psychology. Most of the higher cognitive operations involve the perception of sensorial information. Spatial reasoning involves visual perception, while the use of language involves auditory perception. Hence, higher cognitive operations include an appropriate interface I, and algorithmic operations on the higher level, so called symbolic algorithms. Because of the difficulties we face with regard to a definition of what intelligent algorithms are, one can characterize symbolic operations by giving examples, e.g., deduction processes in propositional logic, or evolvement, understanding and usage of language, as well as spatial reasoning. This characterization takes into account what one can only loosely define as "more sophisticated" intelligence. In general, most algorithms from classic artificial intelligence – from depth-first search to planning and reasoning – belong to the class of symbolic algorithms. In most cases – and this is frequently claimed to be important – a cognitive system is situated into a real environment, this is denoted as embodied intelligence, see Ziemke (2004).

In the following, I assume that an autonomous agent performs cognitive operations with a symbolic algorithm, i.e., an algorithm that operates on the level of symbols.

Definition 2.1 (Symbolic Algorithm): A symbolic algorithm A performs (cognitive) operations on a set of symbols S.

If possible, we measure the success of the operations by a quality measure fA. But agents are situated in a complex world, and have to translate their perception to the symbolic level. A cognitive symbol is the basis of many approaches in artificial intelligence – although not always explicitly stated. The meaning of a symbol $s \in S$ is based on its interpretation on the symbolic level. On the one hand symbols are only tokens – and may be defined shape-independent (Harnad, 1994). But, the effect they have on the symbolic algorithm A can be referred as meaning or interpretation of the symbol.

I assume that in autonomous agents an interface I exists that maps the subsymbolic data from a high-dimensional set D onto the set of symbols S. This translation is accomplished by an interface algorithm. An interface performs a mapping $I: D \rightarrow S$ from subsymbolic data D to the set of symbols S.

Definition 2.2 (Interface): The interface from subsymbolic to symbolic representations $I: D \rightarrow S$ maps each data sample $d \in D$ to a symbol $s \in S$.

Mapping I defines the set of symbols and may be implemented by any interface algorithm. From the perspective of cognitive economy, it makes sense that $|S| << |D|$. In case of a self-organizing map, see Paragraph 3.2, a cognitive symbol is a Voronoi cell in data space and its corresponding winner neuron n*, i.e., the corresponding symbol s. These interfaces exploit the intrinsic structure of data space, but most are parameterized and thus can be subject to optimization with regard to certain properties. A bias of the search problem is necessary since the learning of representation and training of the learning algorithm is a hopeless undertaking due to an exponential increase of the search space as already Mayo (2003) states.

The exploitation of the intrinsic structure of the subsymbolic data space is such a bias, and to my mind the most adequate. Interface design concerns the choice of a proper interface algorithm, appropriate parameterizations, and also the choice of adequate features. Note, that feature selection – a very successful technique to reduce the solution space – is also a special case of interface optimization.

2.2 Interface Optimization

Now, we formulate the design of interface I as optimization problem: we want to find the optimal mapping $I*: D \rightarrow S$ with regard to the success fA of the symbolic algorithm A.

Definition 2.3 (Interface Optimization): The optimal interface I* maximizes the success fA, i.e., $I \; I_* = \mathrm{argmax}_I \left\{ f_A(I) \big| I \in \psi \right\}$

For this optimization formulation we have to define a quality measure fA with regard to the symbolic algorithm A. The set of interfaces ψ may consists of the same algorithm with various parameterizations, e.g., K-means with different numbers of clusters k. The optimization problem may be solved offline, i.e., the systems runs until a termination condition is reached. Afterwards, the feedback fA from the symbolic algorithm is sent to the optimizer. The optimizer chooses an interface variant or a new parameterization and so forth. If feedback fA allows, an online-adaptation of the interface is another promising possibility. In this scenario the optimizer adapts the interface during runtime of the system – this is only possible if the feedback is available online. In the following, we will discuss typical free parameters and possible feedback for the interface optimization process. Figure 1 illustrates the optimization approach exemplarily for a classification task. The optimization module biases the classification task with regard to an external goal-directed feedback.

2.2.1 Optimization

In practice, the model constructor does not spend effort into the explicit design of the interface between subsymbolic and symbolic representations.

Figure 1. Data flow model for classification tasks (lower part), complemented by the optimization module that biases the classification task with regard to an external goal-directed feedback (upper part)

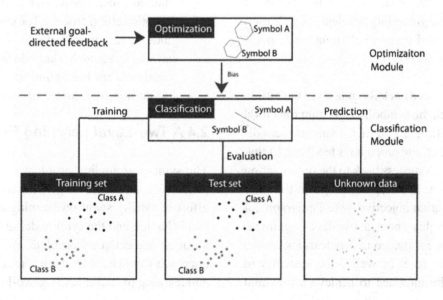

It is frequently an implicit result of the modeling process. Classification algorithms are applied taking into account the learning signal that a separate source delivers as class label. Clustering algorithms find the structure in the data with regard to special properties, e.g., data density. Most system designers rely on the correctness of the class label or on the abilities of the clustering algorithms not concentrating on the requirements of the symbolic algorithm. The definition as optimization problems helps to get aware that the design is important and to make the application of optimization techniques more obvious.

2.3 Optimization Criteria

For the adaptation of an optimal interface In a clear optimization objective has to be specified. The main objective is to map high-dimensional sensory data to a meaningful set of symbols (of arbitrary shape). How can this mapping be measured in terms of a feedback fA from the symbolic algorithm? The feedback depends on the goal of the autonomous agent. If it can explicitly be expressed by a function fA, an optimization algorithm is able evolve the interface. In general, we see the following scenarios to get the feedback of the symbolic algorithm, of which only the first two fulfill the zero semantical commitment condition of the symbolic grounding problem – a definition and discussion of the symbol grounding aspects will follow in Section 4.

1. **Offline-Feedback Response:** In the offline approach the symbolic algorithm runs for a defined time, e.g., until a termination condition is met, and propagates feedback fA that reflects its success back to the optimization algorithm. If interface design is the only optimization objective – see Paragraph 2.4 for thoughts about a two-level optimization process that considers learning on the symbolic level as well – the system will adapt the interface to achieve a maximal

response. This process might be quite slow if the symbolic algorithm is supposed to run for a long time to yield fA.

2. **Online-Feedback Response:** If the symbolic algorithm delivers the feedback fA during runtime, this feedback can be used to define symbols online. For example in a reinforcement learning scenario, where artificial agents have to learn from rewards in uncertain and dynamic environments, the temporal information of the rewards can guide the interface process online. If an agent is in a place of the environment where many varying rewards are available the online-feedback response might lead to a more granular resolution of states in comparison to places where no feedback is available.

3. **Data Driven:** The intrinsic structure of the sensorial data itself usually guides clustering approaches and might also be important as strategy to ground the meaning of symbols, e.g., to differentiate between concepts that do not belong to the same cluster. With regard to its intrinsic structure, clustering yields a reasonable discretization into meaningful symbols in these cases.

4. **User-Driven:** The practitioner should include as much knowledge as possible into the optimization process. The knowledge can guide the response fA manually. But the user can also integrate his knowledge in form of constraints for the optimization problems, e.g., defining the number of symbols a priori.

2.4 A Two-Level Learning Problem

The symbol grounding problem is not the only problem an autonomous agent has to solve. More effort is usually spent on learning of strategies and behavior, and the symbol definition remains an often neglected subproblem. An appropriate approach would be to treat interface optimization and learning of strategies as a two-level learning problem. One approach would be to optimize both levels alternately: With a fixed set of symbols the learning problem can be optimized, with a fixed learning strategy the set of symbols can be optimized. Whether interface optimization and learning can be solved in parallel surely depends on whether the increase of the size of the search space does not make the whole optimization problem unsolvable. The two-level learning problem is related to the two-level mapping from the subsymbolic level to the meanings of symbols that I will discuss in Section 4.

3. MACHINE LEARNING INTERFACES

The translation of high-dimensional subsymbolic data into symbols are tasks that are well known in data mining and machine learning under the terms classification, clustering and dimensionality reduction. Many symbol grounding related work exclusively concentrated on neural networks in the past (Harnad, 1987, 1990; Cangelosi, 2002; Sun & Peterson, 1998), perhaps due to a historical affinity to connectionist models. To overcome the restriction this section shows the relation between symbol grounding and machine learning: assigning unknown objects to known concepts is known as classification, grouping objects is known as clustering, finding low dimensional representations for high-dimensional data is denoted as dimension reduction.

3.1 How are Machine Learning Algorithms Related to Symbol Grounding?

The problem of iconization, discrimination and identification formulated by Harnad (1987) is closely related to the question how to map high-dimensional data to classes or clusters. Classification, clustering and dimensionality reduction are similar in this context. They perform a mapping from a high-dimensional data space D to a low dimensional set of symbols S that may be a class, a cluster, or a low dimensional manifold. The three machine learning tasks implement the nature of dimensionality reduction as follows: Classification algorithms deliver a subsymbolic to symbolic mapping $I: D \rightarrow S$ with regard to explicitly labeled data samples in a supervised way. In a training phase mapping I is learned by reducing the classification error. A learned interface is used to classify unknown data, i.e., to assign symbols to classes of similar high-dimensional input data. Clustering algorithms deliver the subsymbolic to symbolic mapping D to a set of clusters S with regard to the intrinsic structure of the subsymbolic data and the properties of the algorithm in an unsupervised way (see Section 2.3, data-driven feedback) Frequently, the dimensionality of observed data is much higher than the intrinsic dimensionality. A 3D-object for example has got an intrinsic dimensionality of 3, but on a digital image the dimensionality of the data vector is much higher depending on the resolution of the picture. Last, dimension reduction methods have a similar task like classification and clustering. For high-level data low-level representation have to be found, e.g., a mapping from subsymbolic to symbolic data $I: D \rightarrow S$ or the mapping from $Rm \rightarrow Rn$ with $m > n$. I come to the conclusion that classification, clustering and dimensionaity reduction algorithms from machine learning are eligible algorithms for the interface I from subsymbolic to symbolic representations.

3.2. Examples for Related Machine Learning Algorithms

In the last years, kernel methods became quite popular in machine learning and data mining. It is not the scope of this work to review these methods. For a detailed introduction I refer to textbooks like Bishop's (2006) or Hastie et al.'s (2009). Here, I only comment on the properties of three possible interface algorithms with regard to the interface problem.

A simple but successful clustering technique is K-means clustering (Bishop, 2006). K-means needs one essential parameter: the number of clusters k – that we denote as number of symbols. Each cluster Cj ∈ S with 1 ≤ j ≤ k can be described by its cluster center cj, the barycenter of the cluster elements. This concept shows that both clustering and cognition share similar ideas: If the distances between the elements in the data space and the cluster centers are minimal, then clusters of elements should be represented by the same center whilst far-out accumulations of elements belong to different centers. This principle is similar to the idea of semantic distances of mental models. K-means work as follows. At the beginning it randomly generates k initial cluster centers cj. In order to minimize the sum of distances, K-means works iteratively in two steps. In the first step each data element xi is assigned to the cluster Cj with minimal distance. In the next step K-means computes the new cluster centers cj as average of the data elements that belong to Cj. K-means continues with the cluster center computation, and so forth. The algorithm ends if the cluster assignment does not change or if the change falls below a threshold value ε. The process converges, but may get stuck in local optima. K-means allows specifying the number of clusters. If we use K-means as interface algorithm, we can treat k as free parameter that can be optimized with regard to fA. The optimal number of symbols to solve cognitive tasks may frequently not be known in advance. Perspectives are the number of states in reinforcement learning scenarios, or the number of words in language learning scenarios. But also other clustering algorithms may be applied, e.g., distance based approaches like DBSCAN that are based on the distances between the data samples (Ester et al., 1996). In Section 5 we employ a kernel density approach in the context of a simple case study.

In comparison to clustering algorithms, most dimensionality reduction algorithms maintain the structure of the data space, e.g., neighbored data samples in data space are neighbored on a low-dimensional manifold. A recommendable example is the self-organizing map by Kohonen (1990). Its number of neurons and the learning parameters are eligible free parameters for optimization. In each generation the self-organizing map updates the weights w of a winner neuron and its neighborhood with the help of learning parameters η and a neighborhood parameter h, so that they are pulled into the direction of data sample x the algorithms lead to a mapping from the feature space D to the map. The mapping maintains the topology of the neighborhood: Close data samples in the high-dimensional space lie close together on the map. Whether this property is important for the interface depends on the interpretation of the symbols with symbolic algorithm A.

3.3. Optimization Algorithms

When the optimization objectives are clearly specified, and a feedback fA of a given interface I is available, the choice of an adequate optimization algorithm has to be answered. If no more information is available than the feedback, i.e., no explicitly given functions or derivatives, we recommend to apply evolutionary algorithms. A comprehensive survey of evolutionary algorithms is given by Eiben and Smith (2003). Evolutionary computation comprises stochastic methods for global optimization, i.e., optimization problems with multiple local optima. They are biologically inspired and imitate principles that can be observed in natural evolution like mutation, crossover and selection. If the optimization problem is not supposed to suffer from multiple local optima, deterministic direct search methods like Powell's conjugate gradient algorithm (Powell, 1964) or similar optimization algorithms for convex optimization may be applied.

4. PERSPECTIVE OF THE SYMBOL GROUNDING PROBLEM

Now, I discuss the interface optimization problem from the perspective of the symbol grounding problem. After its short introduction, I describe the implicit evolvement of symbol meaning. Guided by seven features of a valid solution to the symbol grounding problem I will discuss the optimization formulation as valid solution of the symbol grounding problem.

4.1 Related Approaches

Harnad (1987) proposes three stages of grounding, i.e., iconization, discrimination and identification – that comply with the tasks that data mining algorithms solve, see Section 3. In the taxonomy of symbol grounding approaches the idea to bind the representation to functional or intentional properties can be found in the representational approaches. Mayo et al. (2003) proposes a functional organization of the representations and introduces task-specific categories where symbols are formed in order to solve task-related problems. They introduce a bias to put sensory data into a category that best contributes to the solution of a particular problem.

Sun's (2000) intentional model is also related to the optimization view. He introduces a two-level approach, whose first level concerns behavior guided by the external world and innate bias. Conceptual representations are learned on the second level. On the first level the autonomous agent explores the world by trial-and-error bound to its objectives. On the second level the first level intentional data is used to evaluate courses of action to achieve objectives. The top down process that guides selection of actions is similar to the top-down interface optimization feedback principle introduced in this work.

4.2 A Valid Solution to the Symbol Grounding Problem?

On page 33 of their work Taddeo et al. (2005) postulate seven features a valid solution of the symbol grounding problem needs. We will shortly discuss the interface optimization problem in the context of these seven properties citing the postulates.

1. The optimization approach is a "bottom-up, sensorimotor approach" as subsymbolic sensorimotor data is mapped from the bottom to the level of symbols, and from there implicitly to their meanings.
2. It is a "top-down feedback approach that allows the harmonization of top level grounded symbols and bottom level, sensorimotor interactions with the environment": The feedback fA from the overlying symbolic algorithm guides the grounding of the symbols explicitly during optimization. Hence, the mapping is performed bottom-up, while the feedback for the optimization process is fed top-down and controls the harmonization between both levels.
3. "The availability of some sort of representational capacities in the autonomous agent" holds true as the interface is defined as general mapping. Any mapping with appropriate representational capacities may be chosen. In particular data mining algorithms map parts of D to S according to intrinsic structures of the subsymbolic data. Their representational capacities are the reason for their success
4. "The availability of some sort of categorical/abstracting capacities in the autonomous agent" is an apparent feature when mapping subsymbolic to symbolic data with an interface algorithm: symbols abstract from sensorimotor patterns and form a category (class or cluster). Decontextualization may be implemented on the symbolic level.

5. "The availability of some sort of communication capacities" to "avoid the Wittgensteinian problem of a private language" is also valid from the optimization point of view. First, the development of language is a task that can itself be seen as objective and for that a feedback fA can be defined. With a 'teacher' the mapping can also be learned in a supervised way, i.e., it can be treated as classification approach. Second, symbols and their meaning may be exchanged exclusively on the symbolic level. Both points of views are consistent with the communication capacity condition.

6. "An evolutionary approach in the development of (1) to (5)." The whole symbol grounding and semantic elaboration process is a process the autonomous agent has to evolve. My formulation of the symbol grounding problem as optimization problem strengthens this assumption. Evolutionary algorithms are an appropriate choice to solve the optimization problem. This issue has already been discussed in Paragraph 3.3.

7. Satisfaction of the zero semantical commitment condition in the development of (1) to (6).

To my mind the optimization formulation fulfills the first six conditions. Due to its importance, the last property is discussed in the next paragraph.

4.3 Zero Semantical Commitment Condition

While it may not be difficult to ground symbols in one way or other, finding an answer to the question of how an autonomous agent is able to solve this task on its own thereby elaborating its own semantics renders much more difficult. In biological systems, genetic preconditions and the interaction with the environment and other autonomous agents seem to be the only sources this elaboration is based on. Therefore, the interpretation of symbols must be an intrinsic process to the symbolic system itself without the need for external influence. This process allows the agent to construct a sort of "mental" representation that increases its chances of survival in its environment. Harnad derived three conditions from this assumption: First, no semantic resources are pre-installed in the autonomous agent (no innatism, or nativism respectively), second, semantic resources are not uploaded from outside (no externalism), and third, the autonomous agent possesses its own means to ground symbols (using sensors, actuators, computational capacities, syntactical and procedural resources, etc.) (Harnad, 1990, 2007). Taddeo and Floridi (2005) called this the "zero semantical commitment condition."

Is the zero semantical commitment condition fulfilled in the interface optimization view? The interface part – as parameterized mapping function – as well as the optimization algorithm are both computational and procedural resources that are allowed in condition 3. Up to here, we can assume that no knowledge about the meaning of symbols is integrated into the autonomous agent. Neither are semantic resources uploaded from outside. The only external knowledge that is used during learning and optimization is a learning signal from the symbolic level that reflects the success. Such a learning signal must exist in every learning scenario. Without any kind of learning signal, learning is not possible at all. If available in form of neurotransmitters in case of nervous systems or in form of survival and reproduction in case of the Darwinian principle of surviving of the fittest, external feedback is the basis of every biological kind of learning. Consequently, to my mind the optimization view does not violate any of the above conditions for symbol grounding with respect to any innatism or externalism, and therefore is a perspective towards the fulfillment of the zero semantical commitment condition.

The formulation of objectives and the appropriate definition of feedback fA is an open, and problem-dependent, question. Similarities to reinforcement approaches are obvious. But it has to be considered that the objective function itself does not violate the zero semantical commitment condition as externalism may be introduced, if not the general objective of the autonomous agent is reflected in the feedback, but symbol grounding information in an explicit form. Hence, the only condition for the feedback is to exclusively reflect the fulfillments of the agent's needs and other general objectives.

5. A CASE STUDY

To concretize the described optimization perspective, I present an artificial toy scenario that is ought to illustrate the working mechanisms of the proposed approach. I start with an overview:

1. **Perception:** Random Gaussian data clouds are produced at different times representing subsymbolic observations d $\in$ D
2. **Interface:** A clustering approach I clusters the data observations depending on two parameters, leading to an assignment of observations to symbols.
3. **Hebbian translation and Interface:** Similar to the Hebbian / STDP learning rule[2] temporal information is used to translate concepts into propositional formulas. Basic inference processes (A) are used to evaluate the interface.
4. **Optimization:** Free parameters are optimized, in particular w.r.t. the interface, i.e., kernel density clustering parameters.

5.1 Perception

Let us assume that a cognitive agent observes data clouds consisting of N-dimensional points x $\in$ RN. These clouds represent subsymbolic sensory input he perceives in an observation space. The temporal context of appearance and disappearance will induce a causal meaning. Observations that

belong to one concept are subsumed to concept di = {x1,...,xN} at time t. Such a data cloud is produced with the Gaussian distribution N (v, σ)^3. Temporal information like appearance and disappearance of data clouds is determined by a fixed scheme, see Section 5.3.

5.2 Interface

The machine learning interface has the task to assign the observations to symbols. We employ a clustering approach that assigns each cluster to a symbol. We employ a simple kernel density clustering scheme described in the following. Cluster centers are placed in regions with high kernel densities, but with a least distance to neighbored high kernel densities. For this sake, iteratively the points with the highest relative kernel density (Parzen, 1962)

$$d(x_j) = \sum_{i=1, i \neq j}^{N} K_h(x_i - x_j) > \varepsilon$$

given a bandwidth h are identified, with a minimum distance $\rho \in$ R+ to previously computed codebook vectors C, d(xj,ck) > ρ for all ck $\in$ C, and a minimum kernel density $\varepsilon \in$ R+. These points are added to set C of cluster-defining codebook vectors. A symbol si corresponds to a codebook vector ci $\in$ C with all closest observations resulting in a Voronoi-tesselation of the observation space D. The clustering result significantly depends on the two free parameters ρ and ε that will be subject to the optimization process.

5.3 Hebbian Translation and Inference

From the temporal information, i.e., appearance, disappearance, and order, logical relations are induced, and translated into propositional logic formulas. This is an important and new step, and probably the most interesting contribution of this toy scenario. We employ two important rules.

1. If two symbols occur at once, e.g., s1 at t1 and s2 at t2 with $|t1-t2| \leq \theta1$, this event induces the formula $s1 \wedge s2$. Two concepts that occur at once are subsumed and believed to belong together from a logical perspective. This rule can be generalized to more than two symbols.

2. If symbol s2 at t2 occurs within time window $[\theta1, \theta2]$ after symbol s1 at t2, i.e., if $\theta1 < t2 - t1 < \theta2$, this induces the implication rule $s1 \rightarrow s2$. This means, s2 follows from the truth of s1 (another interpretation is that s2 may be caused by s1).

Translated into propositional logic formulas, inference processes are possible, which represent higher cognitive processes. For this sake a simple inference engine is employed. The logical formulas induced by the Hebbian process are compared to the original formulas that are basis of the data generation process. They are evaluated testing a set of evaluation formulas of the form $(A, \blacklozenge)$ with formula A over the set of symbols, with $\blacklozenge \in \{true, false\}$ corresponding to the data generating set.

5.4 Optimization

The optimization process has the task to find parameter settings for ρ and ε that allow an optimal inference process w.r.t. a set of evaluation formulas. For this sake we employ a $(\mu + \lambda)$-evolution strategy (Beyer & Schwefel, 2002). To evaluate the quality of the interface, we aggregate two indicators: (1) the number Nf of concepts (clusters) that have been found in relation to the real number of symbols Ns, and (2) the number K of correct logical values when testing the evaluation set formulas for feasibility. Both indicators are subsumed to a minimization problem formulation expressed in the fitness function

$$f_A := \left| N_f - N_s \right| - K$$

The problem of assigning the symbols to correct atoms is solved by trying all possible assignments; the highest K of matching logical values is used for evaluation. From another perspective, f is a measure for the performance of the higher cognitive process A.

5.5 Results

As a first simple test case 15 logical formulas with 10 atoms (symbols) have been generated. As evaluation set 20 formulas, 10 feasible and 10 infeasible, are used. Each symbol is represented by a data cloud with different parameterizations. A $(15 + 100)$-ES optimizes the parameter of the kernel density clustering heuristic. The optimization is stopped when no improvement could have been achieved for $t = 100$ generations. The experiments have shown that the system was able to evolve reasonable parameters for ρ and ε. The clustering process is able to identify and distinguish between concepts. In most runs the correct number symbols have been retrieved from the data cloud, the other runs only differ in at most 3 symbols. In 83 of 100 runs at least 15 formulas of the evaluation set match the observations. With this simple case study, I hope to have demonstrated how the machine symbol grounding perspective can be instantiated.

6. SUMMARY AND CONCLUSION

From the formulation of the interface between subsymbolic and symbolic representations as optimization problem various consequences arise. The optimization process will improve the performance of interfaces and hence the success in solving cognitive tasks. Learning becomes a two-level optimization problem: interface learning and learning on the symbolic level. Many approaches from machine learning for dimensionality reduction, clustering and optimization are adequate methods for the interface problem. The bias of the intrinsic structure in the data that is exploited by data mining algorithms leads to a reasonable reduction of the solution space.

Having this optimization nature in mind, the designer of a cognitive system can invest more time into careful tuning and control of interface properties. Although the flexibility of most current dimensionality reduction and clustering methods is quite high, in the future the designers of artificial intelligent systems may spend more effort in the development of adaptive and evolvable interface algorithms. A purely mathematical and algorithmic formulation only allows a narrow view on the optimization problem. But a fruitful impact for cognitive modeling and the answering to the question how to measure the success of higher cognitive functions from cognitive sciences and psychology becomes important.

Last, the binding of the symbol grounding to the objective of the acting autonomous agent leads to the fulfillment of the zero semantical commitment condition as neither internal nor external knowledge, except the objective of the autonomous agent and its learning algorithms are explicitly integrated into the agent. Interface and optimization algorithm are computational and procedural resources.

REFERENCES

Beyer, H.-G., & Schwefel, H.-P. (2002). Evolution strategies - A comprehensive introduction. *Natural Computing, 1*, 3–52. doi:10.1023/A:1015059928466

Bishop, C. M. (2006). *Pattern recognition and machine learning*. New York, NY: Springer.

Cangelosi, A., & Greco, S. H. (2002). Symbol grounding and the symbolic theft hypothesis. In Cangelosi, A., & Parisi, D. (Eds.), *Simulating the evolution of language* (pp. 91–210). New York, NY: Springer. doi:10.1007/978-1-4471-0663-0_9

Eiben, A., & Smith, J. (2003). *Introduction to evolutionary computing*. Berlin, Germany: Springer-Verlag.

Ester, M., Kriegel, H.-P., Sander, J., & Xu, X. (1996). A density-based algorithm for discovering clusters in large spatial databases with noise. In *Proceedings of the 2nd International Conference on Knowledge Discovery and Data Mining* (pp. 226-231).

Harnad, S. (1987). *Categorical perception: the groundwork of cognition: Applied systems and cybernetics* (pp. 287–300). Cambridge, UK: Cambridge University Press.

Harnad, S. (1990). The symbol grounding problem. *Physica D. Nonlinear Phenomena, 42*, 335–346. doi:10.1016/0167-2789(90)90087-6

Harnad, S. (1994). Computation is just interpretable symbol manipulation: Cognition isn't. *Minds and Machines, 4*, 379–390. doi:10.1007/BF00974165

Harnad, S. (2007). Symbol grounding problem. *Scholarpedia, 2*(7), 2373. doi:10.4249/scholarpedia.2373

Hastie, T., Tibshirani, R., & Friedman, J. (2009). *The elements of statistical learning*. Berlin, Germany: Springer-Verlag.

Kohonen, T. (1990). The self-organizing map. *Proceedings of the IEEE, 78*(9), 1464–1480. doi:10.1109/5.58325

Mayo, M. J. (2003). Symbol grounding and its implications for artificial intelligence. In *Proceedings of the Twenty-Sixth Australian Computer Science Conference* (pp. 55-60).

Parzen, E. (1962). On the estimation of a probability density function and mode. *Annals of Mathematical Statistics, 33*(3). doi:10.1214/aoms/1177704472

Powell, M. (1964). An efficient method for finding the minimum of a function of several variables without calculating derivatives. *The Computer Journal*, *7*(2), 155–162. doi:10.1093/comjnl/7.2.155

Russell, S. J., & Norvig, P. (2003). *Artificial intelligence: A modern approach* (2nd ed.). Upper Saddle River, NJ: Pearson Education.

Searle, J. R. (1980). Minds, brains, and programs. *The Behavioral and Brain Sciences*, *3*, 417–457. doi:10.1017/S0140525X00005756

Sun, R. (2000). Symbol grounding: A new look at an old idea. *Philosophical Psychology*, *13*, 149–172. doi:10.1080/09515080050075663

Sun, R., & Peterson, T. (1998). Some experiments with a hybrid model for learning sequential decision making. *Information Sciences*, *111*, 83–107. doi:10.1016/S0020-0255(98)00007-3

Taddeo, M., & Floridi, L. (2005). Solving the symbol grounding problem: a critical review of fifteen years of research. *Journal of Experimental & Theoretical Artificial Intelligence*, *17*(4), 419–445. doi:10.1080/09528130500284053

Ziemke, T. (2004). Embodied AI as science: Models of embodied cognition, embodied models of cognition, or both? In F. Iida, R. Pfeifer, L. Steels, & Y. Kuniyoshi (Eds.), *Proceedings of the International Seminar on Embodied Artificial Intelligence* (LNCS 3139, pp. 27-36).

ENDNOTES

[1] Survival, reproduction, and each objective that is connected to the first two objectives

[2] STDP means spike-timing-dependent plasticity and is known to be one of the most important learning rules in the human brain.

[3] $N(\nu, \sigma)$ represents Gaussian distributed numbers with expectation value ν, and standard deviation σ.

This work was previously published in the International Journal of Cognitive Informatics and Natural Intelligence, Volume 5, Issue 3, edited by Yingxu Wang, pp. 73-85, copyright 2011 by IGI Publishing (an imprint of IGI Global).

Chapter 20
Image Dimensionality Reduction Based on the Intrinsic Dimension and Parallel Genetic Algorithm

Liang Lei
Chongqing University, China

TongQing Wang
Chongqing University, China

Jun Peng
Chongqing University, China

Bo Yang
Chongqing University, China

ABSTRACT

In the research of Web content-based image retrieval, how to reduce more of the image dimensions without losing the main features of the image is highlighted. Many features of dimensional reduction schemes are determined by the breaking of higher dimensional general covariance associated with the selection of a particular subset of coordinates. This paper starts with analysis of commonly used methods for the dimension reduction of Web images, followed by a new algorithm for nonlinear dimensionality reduction based on the HSV image features. The approach obtains intrinsic dimension estimation by similarity calculation of two images. Finally, some improvements were made on the Parallel Genetic Algorithm (APGA) by use of the image similarity function as the self-adaptive judgment function to improve the genetic operators, thus achieving a Web image dimensionality reduction and similarity retrieval. Experimental results illustrate the validity of the algorithm.

DOI: 10.4018/978-1-4666-2476-4.ch020

1. INTRODUCTION

The magnitude of the current Internet has already become huge, and it is still growing rapidly. The image set based on the Internet will be shown as a distributed image database in a striking size. Faced Web image retrieval, Internet users are usually reluctant to spend too much time to wait for search results. Rather, they are more in pursuit of image retrieval speed and accuracy. Therefore, Web image retrieval technology is focused on finding out a particularly efficient search algorithm in the premise of a considerable accuracy. Since image retrieval technology, as an application of artificial intelligence, play a pivotal role in a very important engineering application of computational Intelligence, i.e., pattern recognition and computer vision, the subject topic of this paper is within the broad scope of computational Intelligence.

In the content-based Web image retrieval, image dimension reduction is the key technology to improve retrieval efficiency. Under normal circumstances, the image feature vector has its dimensions in order of magnitude of 10^2, while the performance of the index structure may rapidly decline as the number of dimensions increases. Especially in the higher dimension (>10), it is even less efficient than sequential scan. How to reduce more of the image dimensions without losing the main features of the image has become a hot spot in the research of Web content-based image retrieval. In this paper we focus on Similarity Calculation, between two images, by Adaptive Parallel Genetic Algorithm based on Random Operator (APGARO).

The paper is organized as follows. In Section 2, we described the commonly used methods for image dimension reduction. The dimensionality reduction method that is based on the HSV features was proposed in Section 3. We described in Section 4 parallel genetic algorithm. An Adaptive Genetic Operator based on Random Operator (AGORO) in Section 5 was used to achieve a Web image dimensionality reduction and similarity retrieval.

In Section 6, the results showed that this method has greatly improved the image retrieval in time and precision rates.

2. COMMONLY USED METHODS FOR IMAGE DIMENSION REDUCTION

Web Image Dimensional Reduction has a basic principle, that is, the sample is mapped to a low-dimensional space from the input space via a linear or nonlinear mode, and thus to obtain a compact low-dimensional expression on the original data sets. Traditional linear dimensionality reduction methods are featured with simplicity, easiness to explain and extendibility, etc., making it a major research direction in high-dimensional data processing. The existing linear dimension reduction methods include Principal Component Analysis (PCA) (Banerjee, 2009; Zhu, 2009; Zhang, 2009; Fan, 2008), Independent Component Analysis (ICA) (Rahman, 2009; Wang, 2009; Müller, 2009), Fisher Discriminated Analysis (FDA) (Zachary, 2000), Principal Curves, Projection Pursuit (PP), Local Linear Projection (LLP), as well as Self-Organizing Map (SOM) that is based on neural networks (Xiao, 2007). These methods are actually ways to find the best linear model under different optimization criteria, and this is also common to linear dimension reduction methods. However, with the advent of the information age, especially in the Web environment, a large number of high-dimensional nonlinear data will inevitably come along. Traditional linear dimension reduction methods are difficult to directly be used to analyze high-dimensional and non-linear data sourced from the real world. This may attribute to the following main reasons: the dimension of expansion leading to a rapid increase in computational complexity; high-dimensional may lead to a relatively small sample size, causing the statistical damage on some of the asymptotic properties; traditional methods in dealing with high dimensional data cannot meet the robustness requirements. Therefore, the study of high-dimensional nonlinear data confronts many difficulties. This is mainly because that high-dimensional factor may bring about the sparse data, and the curse of dimensionality, while the non-linear feature makes the rapid maturing of the existing linear model that no longer applies.

At present there are mainly two types of non-linear dimensionality reduction methods, which are based on the kernel and on manifolds, respectively. The former involves the use of Mercer's kernel and the corresponding Reproduction Kernel Hilbert Space (RKHS), no need to create a complex hypothesis space. By defining the Mercer kernel, feature space can be implicitly defined (Yang, 2009; Lee, 2009). Methods based on kernel have the difficulties in how to select an appropriate kernel function. In addition, the use of kernel function does not require knowing the specific characteristics of space, resulting in the kernel function method in the lack of physical intuition. This is also a drawback to inhibit the rapid development of kernel-based methods. The latter is the non-linear dimensionality reduction method developed in recent years. Based on manifold learning, this method is essentially to have the data in high dimensional space mapped to a low-dimensional non-linear sub-space (that is, low-dimensional manifolds), thus to achieve lower dimension. Manifold learning mainly includes the use of locally linear embedding (LLE), isometric mapping (Isomap), Laplacian eigenmaps (LE) and other algorithms (Banerjee, 2009).

Based on the multidimensional scaling algorithm, Isomap attempt stop reserve globally the geodesic distances between any pair of data points. LLE tries to discover the nonlinear structure by exploiting the local geometry of the data. Later, different manifold learning algorithms have been proposed, such as manifold charting (Brand, 2003), Laplacian eigenmap (LE) (Belkin, 2003), Hessian LLE (HLLE) (Donoho, 2003), local tangent space alignment (LTSA) (Zhang, 2004), maximum variance unfolding (MVU) (Weinberger, 2004, 2006), conformal eigenmap (Sha, 2005), and other extensive work (Lee, 2000, 2004, 2005), etc.

Most NLDR algorithms can be considered into a common framework of thinking globally and fitting locally, in which the locally geometrical information is collected together to obtain a global optimum (Saul, 2004).

Locally geometrical information of data is explored and exploited in different ways. In LLE (Roweis, 2000), each data point is linearly reconstructed with its neighbors and such a linear representation is maintained in a lower-dimensional space. LE calculates the similarity of any pair of neighboring data points to define the graph Laplacian (Belkin, 2003). HLLE, LTSA and LSE explore the local relations between neighboring data points in tangent spaces. Local coordinates are mapped, linearly (Zhang, 2004) or nonlinearly (Xiang, 2006), to the global coordinate system with lower dimensionality. In contrast, MVU (Weinberger, 2004, 2006) and conformal eigenmap (Sha, 2005) utilize the locally geometrical relations in a straightforward way. In MVU, Euclidean distances between neighboring data points are globally preserved in lower-dimensional space. Such an idea is extended in conformal eigenmap by preserving angle information and similar results can be obtained.

Of course, LLE algorithm has its advantages, namely, it can learn any dimension of local linear low-dimensional manifolds, but the manifold that needs to learn can only be unclosed and locally linear. Isomap algorithm is applied to learn the internally flatted low-dimensional manifold, yet not suitable for learning the internal manifold with greater curvature (Zhao, 2005); LE algorithm can maximize the local structure of data to be retained with sharp reduction in the number of dimensions. This advantage makes it an obvious advantage in Web image retrieval. However, when applied in the Web image retrieval, LE algorithm has a significant flaw, that is, it does not have a transformation matrix, for which the new data point cannot be directly mapped to low-dimensional space, in addition to the too large a calculation volume.

3. INTRINSIC DIMENSION ESTIMATION BASED ON THE HSV IMAGE FEATURES

The sample data in high-dimensional (D-dimensional) space is generally not likely to be filled with the R^D space; otherwise the data set will not bring any useful information. These data are in fact, on low-dimensional manifolds in a high-dimensional space, i.e. a lower-dimensional "curve surface" (Zhao, 2005). The dimensions of the manifold should be the data intrinsic dimension, while D merely refers to representation dimension. In other words, the intrinsic dimension of a data set refers to the actual dimension of space objects expressed by the data set, regardless of their dimensions in the space. For example, European-style three-dimensional spatial objects may have a curve with the intrinsic dimension of only 1. Geometrically, the intrinsic dimension is the dimension d of low-dimensional manifolds in which sampling points in D-dimensional space can be embedded approximately, and d<D. Practically, the starting point for dimension reduction is to retain the useful information of the original data, based on which more of the dimensions might be reduced to simplify the high-dimensional data representation. Therefore, the estimation of intrinsic dimension is not always required to be very precise.

This section will give a detailed description of the dimensionality reduction process on the HSV color features, and then to propose a similar search algorithm.

3.1 A Maximum Likelihood Estimator of Intrinsic Dimension

The existing approaches to estimating the intrinsic dimension can be roughly divided into two groups: eigenvalue or projection methods, and geometric methods.

Eigenvalue methods, from the early proposal of Fukunaga (1991) to a recent variant (Bruske, 1998) are based on a global or local PCA, with intrinsic dimension determined by the number of eigenvalues greater than a given threshold. Global PCA methods fail on nonlinear manifolds, and local methods depend heavily on the precise choice of local regions and thresholds (Verveer, 1995). The eigenvalue methods may be a good tool for exploratory data analysis, where one might plot the eigenvalues and look for a clear-cut boundary, but not for providing reliable estimates of intrinsic dimension.

The geometric methods exploit the intrinsic geometry of the dataset and are most often based on fractal dimensions or Nearest Neighbor (NN) distances.

Literature (Levina, 2005) derives the maximum likelihood estimator (MLE) of the dimension m from i.i.d. observations $X_1,...,X_n$ in R^p. However, the estimator is considered for m fixed, n→∞.

3.2 HSV Model

RGB color model oriented for the hardware devices may have a certain distance with the human visual perception, and not so convenient in use. For example, for a given a colored signal, it can be very difficult to determine the R, G and B components that might be found within. HSV color model is a perception-oriented non-linear color model, by which color signals can be expressed as three kinds of attributes (Yang, 2008), namely, Hue, Saturation and Value. Where H refers to the wavelength of the light that is reflected from an object or comes through it, and generally can be distinguished by color name, such as red, orange, green, and measured with an angle 0 ~ 3600. The saturation S refers to the color depth, which is measured in percentage, ranging from 0 to 100% (full saturation). Value V is the color brightness, and also indicated in percentage, ranging from 0 (black) to 100% (white).

This color model is expressed by the Munsell three-dimensional coordinate system. Due to the psychological perception of independence between the coordinates, changes in the color components can be independently perceived. Meanwhile, this color model is featured with linear scalability. The perceivable color difference is proportional to the Euclidean distance in the corresponding values of the color components. This can be better to reflect the human's ability in color perception. CBIR application of this model is more suitable to be determined with the user's eye.

3.3 RGB Model Conversion to HSV Model

A picture is normally to be three RGB color value, and the conversion from RGB to HSV space is as follows:

The given values in RGB color space r, g, b ∈ [0,1, ..., 255] can be then converted into values h, s and v in HSV space, with the calculation as follows:

r ', g', b' are defined as:

$$r' = \frac{\max(r,g,b) - r}{\max(r,g,b) - \min(r,g,b)} \tag{1}$$

$$g' = \frac{\max(r,g,b) - g}{\max(r,g,b) - \min(r,g,b)} \tag{2}$$

$$b' = \frac{\max(r,g,b) - b}{\max(r,g,b) - \min(r,g,b)} \tag{3}$$

then,

$$v = \frac{\max(r,g,b)}{255}$$

$$s = \frac{\max(r,g,b) - \min(r,g,b)}{\max(r,g,b)} \tag{4}$$

$$h' = \begin{cases} 5 + b', & r = \max(r,g,b) \, and \, g = \min(r,g,b) \\ 1 - g', & r = \max(r,g,b) \, and \, g \neq \min(r,g,b) \\ 1 + r', & g = \max(r,g,b) \, and \, b = \min(r,g,b) \\ 3 - b', & g = \max(r,g,b) \, and \, b \neq \min(r,g,b) \\ 3 + g', & b = \max(r,g,b) \, and \, r = \min(r,g,b) \\ 5 - r', & other \end{cases} \tag{5}$$

$$h = 60 \times h' \tag{6}$$

Here r, g, b ∈ [0,255], h ∈ [0°,360°], s ∈ [0,1], v ∈ [0,1].

3.4 Quantization of HSV Color

Based on the corresponding values h, s and v obtained, HSV space should also be given the appropriate quantization before the histogram calculation. This can significantly reduce the required computation.

In general, HSV space should be made in non-uniform quantization, that is, being quantified into (8,3,3), (18,3,3), (12,5,5) and (6,2,2), etc., and the corresponding features of dimensions are 72,162,300 and 24, respectively (Bai, 2009; Rahman, 2009). Experimental results showed that, HSV quantization levels respectively were 8, 3 and 3, and the best effect could be the result of 72-dimensional feature vector. By 72-dimensional quantitative level, the three color components can be synthesized into matrix L, then

$$L = h \bullet Q_s \bullet Q_v + s \bullet Q_v + v \tag{7}$$

where, Q_s and Q_v are the quantitative series of component s, v. Take $Q_s = 3$, $Q_v = 3$, so Formula (7) can be expressed as

$$L = 9H + 3S + V(0 \leq L \leq 71) \tag{8}$$

3.5 Intrinsic Dimension Estimation by Similarity Calculation

As for the 72-dimensional HSV color features, each dimension is the representative of a proportion of a certain color image (Wang, 2009; Müller, 2009). Therefore, the similarity of two images can be simply expressed by histogram intersection. HSV-based features of 2 images can be set as follows:

$$X = \left(Hsv_{x1}, Hsv_{x2}, \cdots, Hsv_{x71} \right) \tag{9}$$

$$Y = \left(Hsv_{y1}, Hsv_{y2}, \cdots, Hsv_{y71} \right) \tag{10}$$

Then using the histogram intersection method, similarity between the image represented by X and the image represented by Y can be expressed as:

$$Sim(X, Y) = \sum_{i=0}^{71} \min(Hsv_{xi}, Hsv_{yi}) \tag{11}$$

In fact, if a dimension has the greatest property value, then the corresponding color may have the largest proportion in the image, that is, the color can be one of the main colors of the corresponding image. Therefore, for an image, if the sum of the feature value on some of the dimensions exceeds 80% or more, that is, at least 80% of the colors of an image are determined by these dimensions, then those only retained dimensions can be sufficient to provide an accurate representation of the original image's color information. Based on this idea, the resulted dimension reduction can retain the fundamental features of the original data to the maximum level, without causing the loss of key messages. Of course, as far as HSV features of a image is concerned, it is not to say that any dimension with property values more than 80% can be considered as its intrinsic dimension. Rather, the 72 attribute values should be sorted from top to bottom, followed by sequential accumulation in descending order. If the top d maximum values are amounted to over 80%, then d is its intrinsic dimension, with the corresponding one-dimensional d as its intrinsic dimension.

4. PARALLEL GENETIC ALGORITHM

Genetic algorithm is a heuristic random search method based on natural evolution which requires considerable amount of CPU time. Since the optimization problem has to be solved in given computing and time constraints, parallel genetic algorithm is an attempt to speed up the program without interfering with other properties of the algorithm.

Existing parallel implementations of genetic algorithm can be classified into following categories:

1. **Distributed GAs (Parallel Island Models):** Such algorithms assume that several subpopulations evolve in parallel. The models include a concept of migration (movement of an individual string from one subpopulation to another).
2. **Parallel GAs:** In that case several parallel processes work over one common population.

The parallel genetic algorithm (PGA) can be implemented using several threads. The main benefits that arise from multithreading are: better program structure (any program in which many activities do not depend upon each other can be redesigned so that each activity is executed as a thread) and efficient use of multiple processors (numerical algorithms and applications with a high degree of parallelism, such as matrix multiplication or, in this case, genetic algorithm, can run much faster when implemented with threads on a multiprocessor).

For every algorithm that we want to execute in multiple threads, first we have to identify independent parts and assign to each a thread. One or more threads can be assigned to each genetic operator (selection, crossover and mutation). Additionally, we can assign a thread for user interface, a thread for parameter control, a thread for results comparison with other methods (e.g., we can implement a completely random search mechanism and compare its results with the genetic algorithm), etc.

4.1 Genetic Operators as Independent Parts of GA

The parallel steady-state genetic algorithm with tournament bad individual selection was implemented. In this implementation, the genetic algorithm consists of two threads: one performs tournament selection and crossover and the other mutation (Budin, 1998).

```
(1) initialize population{
   create thread for tournament se-
lection and
   crossover;
   create thread for mutation;
   wait while termination criterion
is not
   reached;
   delete all threads;
   }
(2) Thread for mutation
   forever{
   choose randomly one individual
and mutate it;
   }
(3) Thread for tournament selection
and crossover
   {
   choose randomly three individu-
als;
   delete the worst of three chosen
individuals;
   new individual =
crossover(survived parents);
   }
```

The major problem of that simple parallel implementation is that it has no control over mutation probability. The consequence is a very bad algorithm behavior. The results are slightly better than random search, but also useless.

4.2 Parallel Genetic Algorithm with Equal Threads

If we want to make a good use of multiprocessor system with more than two processors, the genetic algorithm has to be divided into more than two threads. The idea is in dividing the genetic algorithm into required number of equal and independent parts (Budin, 1998).

```
(1) Extended parallel genetic algo-
rithm 2{
   initialize population;
   create several equal evolution
threads;
   wait while termination criterion
is not
   reached;
   delete all threads;
   }
(2) Evolution thread
   forever{
   perform tournament selection;
   delete selected individual;
   perform crossover;
   replace deleted individual;
   perform mutation;
   }
```

This is the same algorithm like the described extended parallel genetic algorithm (EPGA), but it is divided in a different manner. Each thread performs all genetic operators like the nonparallel genetic algorithm. One thread operates on only a part of the population, because the tournament selection works over only three chromosomes in each iteration. The other thread can work over the same chromosomes (one, two or all three) at the same time without any synchronization. This kind of parallel algorithm works the same with one or more threads.

5. ADAPTIVE GENETIC OPERATORS

In the course of the practical application of the parallel genetic algorithm (PGA), the fixed crossover and mutation probabilities are prone to premature convergence, only to get a local optimal solution (Zachary, 2000; Liu, 2008). Therefore, the algorithm needs to be improved to enable the adaptive regulation of crossover probability and mutation probability. Furthermore, optimization algorithm would get faster and not prone to falling into local optimal solution.

Literature (Chen, 2009; Chitrakala, 2009) proposed an improved self-adaptable genetic algorithm (IAGA), the crossover probability Pc and mutation probability Pm are shown as the Formulas (12) and (13).

$$P_c = \begin{cases} P_{c1} - \dfrac{(P_{c1} - P_{c2})\left(f' - f_{avg}\right)}{f_{\max} - f_{avg}}, f' \geq f_{avg} \\ P_{c1}, f' \leq f_{avg} \end{cases}$$

(12)

$$P_m = \begin{cases} P_{m1} - \dfrac{(P_{m1} - P_{m2})\left(f_{\max} - f\right)}{f_{\max} - f_{avg}}, f \geq f_{avg} \\ P_{m1}, f \leq f_{avg} \end{cases}$$

(13)

Here, f_{max} is the population's largest fitness; f_{avg} for the average fitness of the population in each generation; f' is the larger fitness in the two individuals to get crossover; f for the fitness of an individual to get mutation. Usually, P_{c1} is considered as 0.9, P_{c2} as 0.6, P_{m1} as 0.1, and P_{m2} as 0.01.

When using this algorithm, however, the four parameters, P_{c1}, P_{c2}, P_{m1} and P_{m2}, are considered as fixed values, that is, P_{c1}- P_{c2} in (1) and P_{m1}- P_{m2} in (2) are constant. This may lead to poor capacity of the individual variation and further to get stagnation. And the elite retention policies, while effective in the protection and promotion of outstanding individuals, but too large the number of individuals would cause the population to evolve into a stagnant state, resulting in local convergence.

To this end, this paper proposes an Adaptive Parallel Genetic Algorithm (APGA) is made with regard to the random variation of crossover operator and mutation operator in a certain range.

$$P_c = \begin{cases} rand(P_{c1}) - \dfrac{[rand(P_{c1}) - rand(P_{c2})]\left(f' - f_{avg}\right)}{f_{\max} - f_{avg}}, f' \geq f_{avg} \\ rand(P_{c1}), f' \leq f_{avg} \end{cases}$$

(14)

$$P_m = \begin{cases} rand(P_{m1}) - \dfrac{[rand(P_{m1}) - rand(P_{m2})]\left(f_{\max} - f\right)}{f_{\max} - f_{avg}}, f \geq f_{avg} \\ rand(P_{m1}), f \leq f_{avg} \end{cases}$$

(15)

where, $0.75 < P_{c1} \leq 0.9$, $0.6 \leq P_{c2} \leq 0.75$, $0.05 < P_{m1} \leq 0.1$, $0.01 \leq P_{m2} = 0.05$.

The APGA can not only automatically change with fitness, but also enable the random variation of P_{c1}- P_{c2}, P_{m1}- P_{m2}. This leads to a corresponding increase in the crossover probability and mutation probability of the individuals with excellent performance in population.

6. TEST ANALYSIS

Image data used in the experiments came from three image search engines: Baidu, Google and Sogou. In Baidu, Google and Sogou, enter the "white rabbit", respectively. And from the search results, take the top 200 images, respectively, a total of 600 images with different sizes of pixel.

6.1 Adaptive Parallel Genetic Algorithm (APGA) Design

Image Retrieval has the images as search objects, so we can take 72-dimensional HSV features of each image as a chromosome. Based on the most commonly used binary coding, binary symbols of fixed-length string can be used to express the image in groups. Each chromosome is composed of 72 binaries, with its allele composed of two-valued symbol set $(0, 1)$. Allele as 1 represents the features being selected, 0 is not selected. Since the intrinsic dimension of each image may be different, then the number of 0 or 1 are also different, thus constituting an initial population, on which genetic algorithm can be running. Through selection, crossover and mutation operations, the optimal image of the final subset can be obtained to provide to users. See the follows for detailed search process:

Step 1: The collected images were randomly divided into two groups to form two populations.

Step 2: Take Figure 1 as the target retrieval image, Sim (X, Y) in Formulas (11) was based for the calculation of the 600 images and Sim (X, Y) in Figure 1, with the resulted Sim (X, Y) as the respective fitness.

Step 3: The roulette wheel selection method is employed to select high quality individuals for crossover and mutation.

Step 4: The Formulas (14) and Formulas (15) methods are based for chromosome crossover and mutation.

6.2. Test Results

According to the above mentioned **APGA** method, obtained fifteen images that were most similar to the target image, as shown in Figure 2.

In order to test the performance of the APGA algorithm, a number of experiments were performed on our image retrieval for the 600 images, and the methods to be compared include the tra-

Figure 1. Target image

ditional parallel genetic algorithm (PGA) and the Adaptive Genetic Algorithm based on Random operator (AGAR) as provided in Reference (Lei, 2010), as shown in Figure 3, Figure 4 and Figure 5.

In Figure 4, Convergence Speed of the three algorithms is unstable because genetic operator is uncertain. From the test results, the APGA had a large advantage in retrieval time, convergence speed and accuracy.

7. CONCLUSION

This paper made an analysis of the commonly used methods for Web image dimensionality reduction, and then put forward the image dimensionality reduction method that is based on the HSV color model. By quantifying 72-dimensional HSV color features, the concept of intrinsic dimensions was introduced to reduce the workload on the calculation of the image dimensional reduction. After analysis of the traditional parallel genetic algorithm, the APGA algorithm based on random operator was proposed. The test results proved that this improved genetic algorithm significantly increased the aspects of image retrieval time and precision rates. But when it comes to retrieve the mass images in different pixels, retrieval efficiency is also not satisfactory, and this is the future direction on which our efforts should be made.

Figure 2. Fifteen images most similar to the target image

Figure 3. Retrieval time contrast

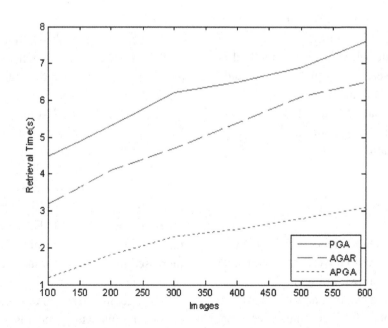

Figure 4. Convergence speed contrast

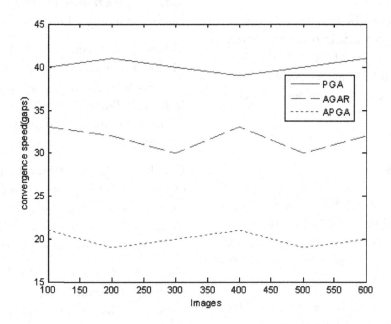

Figure 5. Accuracy contrast

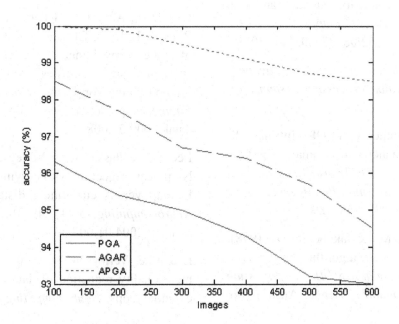

ACKNOWLEDGMENT

This work is supported by the education department of Chongqing of China (No. KJ091409) and Chunhui plan of Ministry of Education of China (No. Z2008-1-63019). The authors are grateful for the anonymous reviewers who made constructive comments.

REFERENCES

Bai, X., & Liu, W. J. (2009). Research of image retrieval based on color. In *Proceedings of the International Forum on Computer Science-Technology and Applications*, Chongqing, China (Vol. 2, pp. 283-286).

Banerjee, M., Kundu, M. K., & Maji, P. (2009). Content-based image retrieval using visually significant point features. *Fuzzy Sets and Systems*, *160*(23), 3323–3341. doi:10.1016/j.fss.2009.02.024

Belkin, M., & Niyogi, P. (2003). Laplacia neigenmaps for dimensionality reduction and data representation. *Neural Computation*, *15*(6), 1373–1396. doi:10.1162/089976603321780317

Brand, M. (2003). Charting a manifold. *Advances in Neural Information Processing Systems*, *15*, 961–968.

Bruske, J., & Sommer, G. (1998). Intrinsic dimensionality estimation with optimally topology preserving maps. *IEEE Transactions on Pattern Analysis and Machine Intelligence*, *20*(5), 572–575. doi:10.1109/34.682189

Budin, L., Golub, M., & Jakobović, D. (1998). Parallel adaptive genetic algorithm. In *Proceedings of the International ICSC/IFAC Symposium on Neural Computation*, Vienna, Austria (pp. 157-163).

Chen, X. Y. (2009). *Image registration and its MATLAB programming process*. Beijing, China: Electronic Industry Press.

Chitrakala, S., Shamini, P., & Manjula, D. (2009). Multi-class enhanced image mining of heterogeneous textual images using multiple image features. In *Proceedings of the IEEE International Advance Computing Conference*, Patiala, India (pp. 496-501).

Donoho, D. L., & Grimes, C. (2003). Hessian eigenmaps: Locally linear embedding techniques for high dimensional data. *Proceedings of the National Academy of Arts and Sciences*, *100*(10), 5591–5596. doi:10.1073/pnas.1031596100

Fan, Z. Z., & Liu, E. G. (2008). New approach on image retrieval based on color information entropy. *Journal of Computer Applications*, *25*(1), 281–282.

Fukunaga, K., & Olsen, D. R. (1971). An algorithm for finding intrinsic dimensionality of data. *IEEE Transactions on Computers*, *20*, 176–183. doi:10.1109/T-C.1971.223208

Lee, D. J., Antani, S., Chang, Y., Gledhill, K., Rodney, L. L., & Christensen, P. (2009). CBIR of spine X-ray images on inter-vertebral disc space and shape profiles using feature ranking and voting consensus. *Data & Knowledge Engineering*, *68*(12), 1359–1369. doi:10.1016/j.datak.2009.07.008

Lee, J., Lendasse, A., & Verleysen, M. (2004). Nonlinear projection wit hcurvilinear distances: Isomap versus curvilinear distance analysis. *Neurocomputing*, *57*(1), 49–76. doi:10.1016/j.neucom.2004.01.007

Lee, J., & Verleysen, M. (2005). Nonlinear dimensionality reduction of data manifolds with essential loops. *Neurocomputing*, *67*(1-4), 2–53.

Lee, J., & Verleysen, M. (2007). *Nonlinear dimensionality reduction*. Berlin, Germany: Springer-Verlag. doi:10.1007/978-0-387-39351-3

Lei, L., Wang, X., Yang, B., & Peng, J. (2010). Image dimensionality reduction based on the HSV feature. In *Proceedings of the 9th IEEE International Conference on Cognitive Informatics*, Beijing, China (pp. 127-131).

Levina, E., & Bickel, P. J. (2005). Maximum likelihood estimation of intrinsic dimension. In *Advances in neural information processing systems* (pp. 777–784). Cambridge, MA: MIT Press.

Liu, J. S., Zhang, D. Y., Liu, S. W., Fang, Y., & Zhang, M. (2008). The evaluation of wavelet and data driven feature selection for image understanding. In *Proceedings of the 1st International Conference on BioMedical Engineering and Informatics*, Sanya, China (Vol. 2, pp. 268-271).

Müller, H., Michoux, N., Bandon, D., & Geissbuhler, A. (2009). A review of content-based image retrieval systems in medical applications - Clinical benefits and future directions. *International Journal of Medical Informatics*, 78(9), 638–638. doi:10.1016/j.ijmedinf.2009.05.001

Rahman, M. M., Bhattacharya, P., & Desai, B. C. (2009). A unified image retrieval framework on local visual and semantic concept-based feature spaces. *Journal of Visual Communication and Image Representation*, 20(7), 450–462. doi:10.1016/j.jvcir.2009.06.001

Roweis, S., & Saul, L. (2000). Nonlinear dimensionality reduction by locally linear embedding. *Science*, 290, 2323–2326. doi:10.1126/science.290.5500.2323

Saul, L. K., & Roweis, S. T. (2004). Think globally, fit locally: Unsupervised learning of low dimensional manifolds. *Journal of Machine Learning Research*, 4(2), 119–155. doi:10.1162/153244304322972667

Sha, F., & Saul, L. K. (2005). Analysis and extension of spectral methods for nonlinear dimensionality reduction. In *Proceedings of the 22nd International Conference on Machine Learning*, Bonn, Germany (pp. 785-792).

Verveer, P., & Duin, R. (1995). An evaluation of intrinsic dimensionality estimators. *IEEE Transactions on Pattern Analysis and Machine Intelligence*, 17(1), 81–86. doi:10.1109/34.368147

Wang, C., Zhao, J., He, X. F., Chen, C., & Bu, J. J. (2009). Image retrieval using nonlinear manifold embedding. *Neurocomputing*, 72(16-18), 3922–3929. doi:10.1016/j.neucom.2009.04.011

Weinberger, K. Q. Sha, & F., Saul, L. K. (2004). Learning a kernel matrix for nonlinear dimensionality reduction. In *Proceedings of the International Conference on Machine Learning*, Banff, AB, Canada (pp. 839-846).

Weinberger, K. Q., & Saul, L. K. (2006). Unsupervised learning of image manifolds by semidefinite programming. *International Journal of Computer Vision*, 70(1), 77–90. doi:10.1007/s11263-005-4939-z

Xiang, S. M., Nie, F. P., Zhang, C. S., & Zhang, C. X. (2006). Spline embedding for nonlinear dimensionality reduction. In *Proceedings of the European Conference on Machine Learning*, Berlin, Germany (pp. 825-832).

Xiao, Q. Z., Qin, K., Guan, Z. Q., & Wu, T. (2007). Image mining for robot vision based on concept analysis. In *Proceedings of the IEEE International Conference on Robotics and Biomimetics*, Yalong Bay, China (pp. 207-212).

Yang, C. W., & Shen, J. J. (2009). Recover the tampered image based on VQ indexing. *Signal Processing*, 90(1), 331–343. doi:10.1016/j.sigpro.2009.07.007

Yang, H. Y., Wu, J. F., Yu, Y. J., & Wang, X. Y. (2008). content based image retrieval using color edge histogram in HSV color space. *Journal of Image and Graphics, 13*(10), 2035–2038.

Zachary, J. M. (2000). *An information theoretic approach to content based image retrieval.* Unpublished doctoral dissertation, Louisiana State University and Agricultural and Mechanical College, Baton Rouge, LA.

Zhang, J., & Ye, L. (2009). Content based image retrieval using unclean positive examples. *IEEE Transactions on Image Processing, 18*(10), 2370–2375. doi:10.1109/TIP.2009.2026669

Zhang, Z., & Zha, H. Y. (2004). Principal manifolds and nonlinear dimensionality reduction via tangent space alignment. *Journal of Shanghai University, 8*(4), 406–424. doi:10.1007/s11741-004-0051-1

Zhao, L. W., Luo, S. W., Zhao, Y. C., & Liu, Y. H. (2005). Study on the low-dimensional embedding and the embedding dimensionality of manifold of high-dimensional data. *Journal of Software, 16*(8), 1423–1430. doi:10.1360/jos161423

Zhu, G. Y., Zheng, Y. F., Doermann, D., & Jaeger, S. (2009). Signature detection and matching for document image retrieval. *IEEE Transactions on Pattern Analysis and Machine Intelligence, 31*(11), 2015–2031. doi:10.1109/TPAMI.2008.237

This work was previously published in the International Journal of Cognitive Informatics and Natural Intelligence, Volume 5, Issue 2, edited by Yingxu Wang, pp. 97-112, copyright 2011 by IGI Publishing (an imprint of IGI Global).

Section 5
Applications of Cognitive Informatics and Cognitive Computing

Chapter 21
Equivalence between LDA/ QR and Direct LDA

Rong-Hua Li
The Hong Kong Polytechnic University, Hong Kong

Shuang Liang
The Hong Kong Polytechnic University, Hong Kong

George Baciu
The Hong Kong Polytechnic University, Hong Kong

Eddie Chan
The Hong Kong Polytechnic University, Hong Kong

ABSTRACT

Singularity problems of scatter matrices in Linear Discriminant Analysis (LDA) are challenging and have obtained attention during the last decade. Linear Discriminant Analysis via QR decomposition (LDA/QR) and Direct Linear Discriminant analysis (DLDA) are two popular algorithms to solve the singularity problem. This paper establishes the equivalent relationship between LDA/QR and DLDA. They can be regarded as special cases of pseudo-inverse LDA. Similar to LDA/QR algorithm, DLDA can also be considered as a two-stage LDA method. Interestingly, the first stage of DLDA can act as a dimension reduction algorithm. The experiment compares LDA/QR and DLDA algorithms in terms of classification accuracy, computational complexity on several benchmark datasets and compares their first stages. The results confirm the established equivalent relationship and verify their capabilities in dimension reduction.

DOI: 10.4018/978-1-4666-2476-4.ch021

1. INTRODUCTION

Classical Linear Discriminant Analysis (LDA) algorithm usually suffers from the singularity problem (Ye et al., 2004; Chen et al., 2000) of scatter matrices. Linear Discrimiant Analysis via QR decomposition (LDA/QR) (Ye & Qi, 2000; Ye & Li, 2004) and Direct Linear Discriminant Analysis (DLDA) (Yu & Yang, 2005) are two LDA algorithms to solve this singularity problem. This paper proves the equivalence relationship between these two LDA algorithms.

LDA/QR is a LDA extension for coping with the singularity problem and was developed in (Ye & Qi, 2000). It improves the efficiency by introducing QR decomposition on a small-sized matrix, while achieving competitive discriminant performance. There are two stages-of-process in LDA/QR: 1) In the first stage, the separation between different classes is maximized via applying QR decomposition to a small-sized matrix. Remind that this stage can be used independently as a dimension reduction algorithm and it is called Pre-LDA/QR (Ye & Li, 2004); 2) The second stage refines the result by applying LDA to the so-called reduced scatter matrices (Ye & Qi, 2000) resulting from the first stage. The solution by LDA/QR has been proved to be equivalent to that by generalized LDA named pseudo-inverse LDA (Ye et al., 2004; Zhang et al., 2005).

DLDA (Yu & Yang, 2005) bases on the observation that the null space of between-class scatter matrix carries little useful discriminative information (Chen et al., 2000). Accordingly, it first diagonalizes between-class scatter matrix S_b and then adopts traditional eigen-analysis to discard the zero eigenvalues together with their corresponding eigenvectors. Then, it diagonalizes within-class scatter matrix S_w but keeps the zero eigenvalues since its null space carries the most useful discriminant information. Hence DLDA can make use of all discriminant information. One advantage of DLDA is that it can be applied to high-dimensional input space directly without any intermediate dimension reduction stage, such as PCA (Martinez & Kak 2001; Turk & Pentland, 1996; Yang & Yang, 2006), which has been used in the preprocessing stage of a well-known face recognition algorithm, namely fisherfaces (Belhumeur et al., 1997; Jing et al., 2006).

In this paper, we study on the equivalence between LDA/QR and DLDA. Ye and Li (2004) and Ye and Qi (2000) have demonstrated the equivalence between LDA/QR and pseudo-inverse LDA. Here, we also demonstrate the equivalence of DLDA to pseudo-inverse LDA. Then we can bridge LDA/QR and DLDA by using pseudo-inverse LDA. This achieves the equivalence between LDA/QR and DLDA. Meanwhile, we show that DLDA is a special case of pseudo-inverse LDA where the inverse of the between-class matrix is replaced by pseudo-inverse. Furthermore, we point out that DLDA can also be regarded as a two-stage LDA method such as LDA/QR, in which the first stage can be used independently as a dimension reduction algorithm. But LDA/QR is more efficient than DLDA, since singular value decomposition (SVD) (Golub & Loan, 2001) is used to the first stage of DLDA to maximize the between-class distance, whereas QR decomposition (Golub & Loan, 2001) is applied to LDA/QR, which is faster than SVD.

We have conducted a series of experiments to verify our theoretical proof. We make use of the first stage of the two LDA methods independently as dimension reduction algorithms to perform recognition on different datasets, i.e. human facial images and text. Our experimental results confirm our theoretical analysis.

There are three main contributions of this paper. First, we prove the equivalence between LDA/QR and DLDA. Second, we show that DLDA is a special case of pseudo-inverse LDA (where pseudo-inverse is applied to between-class scatter matrix). Finally, we demonstrate that DLDA can also be regarded as a two-stage LDA method, in which the first stage can be used independently as a dimension reduction algorithm.

This paper is organized as follows. Section 2 reviews four LDA methods including classical LDA, pseudo-inverse LDA, LDA/QR and DLDA. In Section 3, we establish the equivalence relationship between LDA/QR and DLDA by pseudo-inverse LDA together with theoretical analysis. Experimental setup, results and discussions are presented in Section 4 and 5.

2. RELATED WORKS

In this section, we give a brief overview of classical LDA and its three extensions: pseudo-inverse LDA, LDA/QR, and DLDA. For convenience, we present in Table 1 the important notations used throughout this paper.

2.1 Classical Linear Discriminant Analysis (LDA)

Given a data matrix $A \in R^{m \times N}$, LDA aims to find a linear projection matrix $P \in \mathbb{R}$ that maps each column a_i of A, for $1 \leq i \leq N$, in the m-dimensional space to a vector $y_i \in \mathbb{R}^d$ in the d-dimensional lower space as follows:

$$P : y_i = P^T a_i \tag{1}$$

Assume that the original high-dimensional A is partitioned into k classes as $A = [\Pi_1, ..., \Pi_k]$, where $\Pi_i \in \mathbb{R}^{m \times N_i}$ contains data samples from the i-th class and $\sum_{i=1}^{k} N_i = N$. The optimal P is obtained such that the class structure of the original space is preserved in the low-dimensional space.

The so-called between-class scatter matrix S_b, within-class scatter matrix S_w, and total scatter matrix S_t in LDA can be defined as follows (Ye, etc. 2004):

Table 1. Notations

Notation	Description
N	number of training data samples
A	data matrix
k	number of classes
m	number of dimensions
S_w	within-class scatter matrix
S_b	between-class scatter matrix
S_t	total scatter matrix
H_w	precursor of within-class scatter
H_b	precursor of between-class scatter
P	transformation matrix
d	number of retained dimensions in LDA
c_i	centroid of the i-th class
c	global centroid of the training dataset
N_i	number of data samples in the i-th class
Π_i	data matrix of the i-th class
$tr(\cdot)$	trace
$\dagger$	pseudo-inverse

$$S_w = \frac{1}{N} \sum_{i=1}^{k} \sum_{j \in \Pi_i} (a_j - c_i)(a_j - c_i)^T \tag{2}$$

$$S_b = \frac{1}{N} \sum_{i=1}^{k} N_i (c_i - c)(c_i - c)^T \tag{3}$$

$$S_t = S_w + S_b = \frac{1}{N} \sum_{i=1}^{N} (a_i - c)(a_i - c)^T \tag{4}$$

The precursors (Chen et al., 2000) H_w, H_b, and H_t of the within-class, between-class, and total scatter matrices in Equations(2), (3), and (4) are:

$$H_t = \frac{1}{\sqrt{N}} [\Pi_1 - c_1 \cdot e_1^T, ..., \Pi_k - c_k \cdot e_k^T] \tag{5}$$

$$H_b = \frac{1}{\sqrt{N}}(\sqrt{N_1}(c_1 - c), ..., \sqrt{N_k}(c_k - c)) \qquad (6)$$

$$H_t = \frac{1}{\sqrt{N}}(A - ce^T) \qquad (7)$$

where $e_i = (1, ..., 1)^T \in \mathbb{R}^{N_i}$, A_i is the data matrix of the i-th class, e_i is the centroid of the i-th class, and c is the global centroid of the training dataset. It is easy to verify that: $S_w = H_w H_w^T$, $S_b = H_b H_b^T$ and $S_t = H_t H_t^T$.

However, classical LDA cannot be used directly due to the singularity problem (of scatter matrix S_b). Several extensions of LDA, including pseudo-inverse LDA (Turk & Pentland, 1996), LDA/QR, and DLDA were proposed in recent years to deal with this problem.

2.2 Pseudo-Inverse Linear Discriminant Analysis

Pseudo-inverse is commonly used to deal with the singularity of matrices (Ye et al., 2004; Zhang et al., 2005; Golub & Loan, 2001). Although the inverse of a matrix may not exist, the pseudo-inverse of any matrix is well defined. Moreover, when the matrix is invertible, its pseudo-inverse is consistent with its inverse.

The generalization of classical LDA is to replace the inverse of the scatter matrices by their pseudo-inverse, so that the matrix singularity can be taken out of consideration. This forms three new generalized optimization criterions:

$$F_1(P) = \arg\max_P \{tr((P^T S_b P)^\dagger P^T S_w P)\} \text{ or}$$

$$F_1(P) = \arg\max_P \{tr((P^T S_w P)^\dagger P^T S_b P)\} \text{ or}$$

$$F_1(P) = \arg\max_P \{tr((P^T S_t P)^\dagger P^T S_b P)\} \qquad (8)$$

LDA is then implemented by eigen-decomposition on matrix $S_b^\dagger S_w$, $S_w^\dagger S_b$, or $S_t^\dagger S_b$, where $\dagger$ denotes the pseudo-inverse.

The pseudo-inverse of a matrix can be computed by SVD (Golub & Loan, 2001) as $M^\dagger = V \sum^{-1} U^T$. The optimization criterions can be solved by a technique named simultaneous diagonalization of the three scatter matrices (Ye, 2005).

2.3 Linear Discriminant Analysis Via QR-Decomposition (LDA/QR)

LDA/QR is a two-stage LDA method (Ye & Li, 2004). The detailed procedure of LDA/QR is presented as follows:

Stage 1: Apply a skinny QR-decomposition to matrix H_b as $H_b = QR$, where $Q \in \mathbb{R}^{m \times k}$, $R \in \mathbb{R}^{k \times k}$. Let $Q = (Q_1, Q_2)$ be a column partition of Q, where $Q_1 \in \mathbb{R}^{m \times r}$, $Q_2 \in \mathbb{R}^{m \times (k-r)}$, then

$$H_b = (Q_1, Q_2)\begin{pmatrix} R_1 \\ 0 \end{pmatrix} = Q_1 R_1$$

where

$$R = \begin{pmatrix} R_1 \\ 0 \end{pmatrix}$$

is a row partition of R and $R_1 \in \mathbb{R}^{r \times k}$.

Stage 2: Compute the optimal transformation matrix by solving the following optimization problem:

$$G^* = \arg\min_G tr((G^T Q_1^T S_b Q_1 G)^{-1}(G^T Q_1^T S_w Q_1 G))$$

Then the optimal transformation matrix of LDA/QR is $P = Q_1 G^*$. Stage 1 of the LDA/QR method can be used independently as a dimension reduction algorithm, which is called Pre-LDA/QR in (Ye & Qi, 2000). It finds an optimal projection matrix W which solves the following optimization problems:

$$W^* = \arg \max_{W^T W = I_r} tr(W^T S_b W) \qquad (9)$$

As shown in (Ye & Qi, 2000), the matrix Q_1 is an optimal solution of Equation (9). The time complexity of LDA/QR algorithm is $O(mnk)$ and the detailed analysis can be found in (Ye & Qi, 2000).

2.4 Direct Linear Discriminant Analysis (DLDA)

Direct LDA (DLDA) was developed in (Yu & Yang, 2005) and can be applied directly in singularity problem. The goal is to look for a transformation matrix that simultaneously diagonalizes both between- and within-class scatter matrices S_b and S_w. The detailed procedure of DLDA is described as follows:

Step 1: Diagonalize S_b.

Find an orthogonal matrix V such that $V^T S_b V = \wedge$, where

$$\wedge = diag(\sigma_1, \sigma_2, ..., \sigma_m)$$

and

$$\sigma_1 \geq \sigma_2 \geq ... \geq \sigma_m \geq 0$$

Let $V = (Y, V_1)$ be a column partition of V as $Y \in \mathbb{R}^{m \times r}$ and $V_1 \in \mathbb{R}^{m \times (m-r)}$, where $r = rank(S_b)$. Based on the observation that the null space of S_b carries little useful discriminative information (Yu & Yang, 2005), we thus discard the zero-value diagonal elements of $\wedge$ so that we have $Y^T S_b Y = D_b$, where D_b is the $r \times r$ principle

submatrix of $\wedge$. Let $Z = YD_b^{-1/2}$, obviously, $Z^T S_b Z = I_b$ where I_b is an $r \times r$ unit matrix.

Step 2: Diagonalize the matrix $Z^T S_w Z$.

Perform eigen-analysis on matrix $Z^T S_w Z$:

$$U^T Z^T S_w Z U = D_w$$

where $U^T U = I_r$, D_w is a diagonal matrix with the diagonal entries sorted in ascending order. It is easy to get that the matrix $U^T Z^T$ diagonalizes both S_b and S_w simultaneously.

Step 3: Compute the transformation matrix P.

If the matrix D_w is nonsingular, the transformation matrix P is given by $P = ZUD_w^{-1/2}$, or $P = ZU$ otherwise.

To reduce the computational complexity, the diagonalization of matrix in step 1 and step 2 can be computed by SVD (Yu & Yang, 2005). The time complexity of DLDA algorithm is $O(mnk)$.

3. EQUIVALENCE BETWEEN LDA/QR AND DLDA

Since the solutions of LDA/QR and DLDA are closely related to the eigen-decomposition on $S_b^\dagger S_w$, it is reasonable to prove the equivalence between these two LDA methods. In this section, we establish a connection between them, i.e. LDA/QR and DLDA, by two steps. Section 3.1 demonstrates the equivalence between DLDA and pseudo-inverse LDA. Section 3.2 uses pseudo-inverse to bridge LDA/QR and DLDA.

3.1 Equivalence between DLDA and Pseudo-Inverse LDA

We first present a lemma, which is straightforward from linear algebra. Then we prove Theorem 1, based on which the equivalence between DLDA and pseudo-inverse LDA is established.

Lemma 1: *For any matrix* $M \in \mathbb{R}^{n \times n}$, *if* x *is an eigenvector of* M, *then* ax *is also an eigenvector of* M, *where* a *is an arbitrary nonzero scalar.*

Theorem 1: *Let* P *be the optimal projection matrix obtained by DLDA algorithm. Then the columns of* P *are eigenvectors of* $S_b^\dagger S_w$ *associated with the non-zero eigenvalues.*

Proof: Consider the following eigenequation (the pseudo-inverse is applied to the between-class scatter matrix):

$$S_b^\dagger S_w x = \lambda x \tag{10}$$

Let

$$S_b = V \wedge V^T = V \begin{pmatrix} D_b & 0 \\ 0 & 0 \end{pmatrix} V^T$$

be the SVD of S_b, where V and D_b are defined in Section 2.4. The matrix $S_b^\dagger$ can be computed as follows:

$$S_b^\dagger = V \begin{pmatrix} D_b^{-1} & 0 \\ 0 & 0 \end{pmatrix} V^T \tag{11}$$

as the real symmetric matrix S_b is positive semi-definite. Similar to the procedure of DLDA, let $V = (Y, V_1)$ be a column partition of V such that $Y \in \mathbb{R}^{m \times r}$ and $Y_1 \in \mathbb{R}^{m \times (m-r)}$, where $r = rank(S_b)$, with Equations (10) and (11) we then have

$$S_b^\dagger S_w x = (Y, V_1) \begin{pmatrix} D_b^{-1} & 0 \\ 0 & 0 \end{pmatrix} \begin{pmatrix} Y^T \\ V_1^T \end{pmatrix} S_w x \tag{12}$$

$$= \lambda x$$

Applying the same operator $V^T = (Y, V_1)^T$ on the left and the right of Equation (12) simultaneously, we have

$$\begin{pmatrix} D_b^{-1} & 0 \\ 0 & 0 \end{pmatrix} \begin{pmatrix} Y^T \\ V_1^T \end{pmatrix} S_w x = \lambda \begin{pmatrix} Y^T \\ V_1^T \end{pmatrix} x \tag{13}$$

Hence

$$\begin{pmatrix} D_b^{-1} Y^T S_w x \\ 0 \end{pmatrix} = \lambda \begin{pmatrix} Y^T x \\ V_1^T x \end{pmatrix}$$

Apparently, $V_1^T x = 0$. According to Equation (13), we also have

$$\begin{pmatrix} D_b^{-1} & 0 \\ 0 & 0 \end{pmatrix} \begin{pmatrix} Y^T \\ V_1^T \end{pmatrix} S_w (Y, V_1) \begin{pmatrix} Y^T \\ V_1^T \end{pmatrix} x = \lambda \begin{pmatrix} Y^T \\ V_1^T \end{pmatrix} x$$

With $V_1^T x = 0$, we can obtain

$$D_b^{-1} (Y^T S_w Y) Y^T x = \lambda Y^T x \tag{14}$$

Multiplying the both sides of Equation 14 by $D_b^{1/2}$, we can deduce

$$D_b^{-1/2} (Y^T S_w Y) D_b^{-1/2} (D_b^{1/2} Y^T x) = \lambda (D_b^{1/2} Y^T x)$$

So $D_b^{1/2} Y^T x$ is an eigenvector of $D_b^{-1/2} (Y^T S_w Y) D_b^{-1/2}$. Assume that each column of the matrix $X \in \mathbb{R}^{m \times r}$ is an eigenvector corresponding to the non-zero eigenvalue of $S_b^\dagger S_w$, and $r = rank(S_b)$. Obviously, the eigenvectors of

$$D_b^{-1/2} (Y^T S_w Y) D_b^{-1/2}$$

form the matrix $D_b^{1/2} Y^T X$.

In DLDA, the eigenvectors of $P = Y D_b^{-1/2} (Y^T S_w Y) D_b^{-1/2}$ form the matrix U (Yu & Yang, 2005). If D_w is singular, then $P = Y D_b^{-1/2} U$.

By applying some simple linear transformations, we have $U = D_b^{-1/2} Y^T P$. Hence each column of P is an eigenvector of $S_b^\dagger S_w$. If matrix D_w is nonsingular, we then have $U = D_b^{1/2} Y^T P D_w^{1/2}$. According to Lemma 1, each column of $D_b^{1/2} Y^T X D_w^{-1/2}$ is also an eigenvector of $D_b^{-1/2}(Y^T S_w Y) D_b^{-1/2}$, since $D_w^{-1/2}$ is a diagonal matrix with non-zero diagonal entries. Therefore, in this case, each column of P is an eigenvector of $S_b^\dagger S_w$ too. This completes the proof of Theorem 1.

Theorem 1 shows the equivalent relationship between DLDA and pseudo-inverse LDA, which implies that DLDA is a special case of pseudo-inverse LDA with the pseudo-inverse applied to matrix S_b.

3.2 Bridging LDA/QR and DLDA Via Pseudo-Inverse LDA

Pseudo-inverse LDA bridges LDA/QR and DLDA. As introduced in (Ye & Qi, 2000), the columns of the optimal transformation matrix solved by LDA/QR algorithm are eigenvectors of corresponding to the non-zero eigenvalues. According to Theorem 1, the eigenvectors of $S_b^\dagger S_w$ associate with the non-zero eigenvalues form the optimal projection matrix obtained from DLDA algorithm. Thus, the optimal transformation vectors of both LDA/QR and DLDA can be given by the eigenvectors of $S_b^\dagger S_w$ corresponding to the non-zero eigenvalues. Therefore, the subspace of LDA/QR spanned by its optimal projection vectors is equivalent to that of DLDA. In this sense, the LDA/QR is equivalent to DLDA. Moreover, both LDA/QR and DLDA can be considered as a special case of pseudo-inverse LDA, where the pseudo-inverse is applied to the between scatter matrix S_b.

3.3 Discussion

Similar to LDA/QR, DLDA can also be regarded as a two-stage method. In the first stage, it aims to find a matrix Y that diagonlizes the between-class scatter matrix S_b as $S_b = Y D_b Y^T$. In the second stage, DLDA solves the following generalized eigenvalue problem:

$$(Y^T S_w Y)x = \lambda(Y^T S_b Y)x \qquad (15)$$

The r eigenvectors which correspond to the smallest r eigenvalues of Equation (15) form the optimal transformation matrix.

For convenience, we call the first stage of DLDA as Pre-DLDA, like Pre-LDA/QR (the first stage of LDA/QR). Pre-DLDA can also be used independently as a dimension reduction algorithm. The following theorem states that Pre-DLDA solves the optimization problem in Equation (9).

Theorem 2: *For arbitrary orthogonal matrix $M \in \mathbb{R}^{r \times r}$, W=YM solves the optimization problems in Equation (9).*
Proof: It is easy to prove that $(YM)^T(YM) = I_r$. Consider

$$tr(W^T S_b W) = tr(W^T Y D_b Y^T W) \le tr(D_b)$$

if $Y^T W$ is an orthogonal matrix, the inequality becomes equality. Hence for an arbitrary orthogonal matrix $M \in \mathbb{R}^{r \times r}$, $W=YM$ solves the optimization problems in Equation (9). This completes the proof of Theorem 2.

Moreover, DLDA uses SVD to solve Equation (9), whereas LDA/QR uses QR decomposition to obtain the optimal solution. Since QR decomposition is generally considered as a sub-stage in SVD, it is more efficient than SVD numerically though the worst time complexity of these two methods are the same. Hence, LDA/QR can be regarded as an efficient implementation of DLDA.

4. EXPERIMENT SETUP

The purpose of the experiments is to verify the equivalence between LDA/QR and DLDA, as well as test the discriminant performance of LDA algorithms. In order to achieve this, three main steps are taken. First, we partition the dataset into training and testing sets. We perform LDA algorithm on the training sets to obtain the projection vectors. Second, we project both training and testing sets into the optimal LDA subspace, which helps to significantly reduce the dimensionality as mentioned before. Finally, we conduct classification task in the optimal LDA subspace of the data. We choose to use the 1-nearest neighbor classifier (1-NNC) and Euclidean distance metric in our experiments based on (Cover & Hart, 1967; Li & Yuan, 2000; Noushatha et al., 2000; Jing et al., 2006). Note that the selection of the classifier typically does not affect the performance of LDA algorithms. Other classifiers, such as Support Vector Machine (Gunn, 1998) and Bayesian classifier (Williams & Brber, 1998), can also be used.

In this section, we describe the datasets used in our experiments and the corresponding settings. We conduct two sets of experiments, i.e. face recognition and text classification. The facial image dataset is tested on a PC with Corel 2 Duo 2.33 GHz processor with 3.25GB RAM, while the text dataset is tested on another PC of Xeon 3.00 GHz processor with 32GB RAM. The reason that we separate the experiments on different machines is because the chosen text dataset is huge in storage and requires high computational capability and memory to process. The description of the two datasets we use is given in Section 4.1 and 4.2.

4.1 Experiment 1: Face Recognition

LDA methods have been widely tested on facial image datasets (Jing et al., 2006; Li & Yuan, 2000; Moghaddam et al., 2000; Noushatha et al., 2000; Xiong et al., 2006; Yang et al., 2000; Belhumeur et al., 1997; Price & Gee 2005; Yu & Yang, 2005; Jin et al., 2001). In our experiments, we choose three facial image datasets that are popularly used to verify our theoretical results, including: ORL (AT&T Laboratories, 2002), Yale (Yale University) and CMU PIE (Sim et al., 2003).

The ORL facial image dataset contains 400 facial images of 40 distinct individuals, in which each person has ten different images. Some images were captured at different times and vary in expressions (open or closed eyes, smiling or non-smiling) and facial details (glasses or no glasses). The images were taken with a tolerance for some face tilting and rotation up to 20 degrees and the size of each image is 92×92 pixels with 256 gray levels. We cropped the images to align eyes at the same position and resized them to 64×64 pixels.

The Yale facial image dataset consists of 165 facial images of 15 different individuals, so each person has 11 different images. The original size of each image is 243×320 pixels with 256 gray levels. We also resized them to 64×64 pixels each.

The CMU PIE facial image dataset contains 68 subjects with 41,368 facial images as a whole. The facial images were captured under varying poses, illumination and expressions. In our experiment, we choose five frontal views (C05, C07, C09, C27, C29) with varying lighting and illumination so that there are 170 images per subject. The size of each cropped facial image is 32×32 pixels with 256 grey levels per pixel.

We illustrate some sample facial images within the three datasets in Figure 1.

We conduct 20 random partitions of training and testing datasets. In particular, for the ORL and Yale datasets, $l=(2,3,...,8)$ facial images per individual are randomly selected as training set and the rest as testing set. For the CMU PIE dataset, $l=(5,10,20,30,40,50)$ percent of the images per individual are randomly selected as training and leave the rest for test. The final result is taken as an average over the 20 partitions.

Figure 1. Illustrations of facial images of the ORL, Yale, and CMU PIE datasets

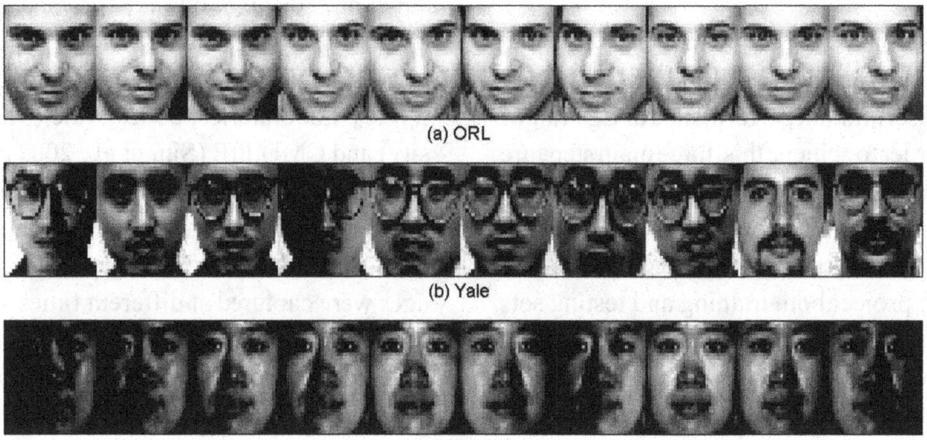

(a) ORL

(b) Yale

(c) CMU PIE

In each dataset, we test the classification accuracy and training time of both LDA/QR and DLDA. The classification accuracy is computed to verify whether LDA/QR is equivalent to DLDA. The same accuracy will indicate the same LDA solutions; The training time is used to measure the efficiency of LDA/QR and DLDA, which stands for the computational complexity for LDA/QR and DLDA to calculate the optimal projection vectors. According to our theoretical analysis, LDA/QR is equivalent to DLDA but is more efficient. Hence, we expect the same classification accuracy for both methods, and slightly less training time of LDA/QR than DLDA.

4.2 Experiment 2: Text Classification

LDA methods have also been tested on text datasets (Cai et al., 2003; Howland et al., 2003; Torkkola, 2003). In our text classification experiment, the test bed is the "bydate" version of the 20 Newsgroups (20NG) dataset (Lang, 1995). We remove the duplicates and newsgroup-identifying headers, and achieve 18,846 documents with the dimensionality of 26,214. This corpus contains 26,214 distinct terms after stemming and stop-word removal. Each document is then represented as a term-frequency vector and normalized. This dataset has a large number of features which might give rise to the high-dimension and singularity problem of scatter matrices (Ye et al., 2004).

Figure 2 shows the distribution of the documents in the 20NG dataset. The number of documents in each class mostly varies from 900 to 1000 with a little fluctuation, which approximates an even distribution over the 20 classes.

Similarly, we separate the dataset into training set and testing set. We randomly select $l=(10,20,30,40,50)$ percent of the documents as training samples and the rest for test. We average the results over the 5 random partitions. We also observe the classification accuracy and training time of LDA/QR and DLDA. Similar to face recognition, the classification accuracy of LDA/QR and DLDA should be the same, and less training time of LDA/QR is expected to be taken than DLDA.

5. RESULTS AND DISCUSSIONS

In this section we give the results of our two sets of experiments, i.e. face recognition and text classification. Section 5.1 and 5.2 reports the results

Figure 2. Illustration of the distribution over classes on the 20 Newsgroups dataset

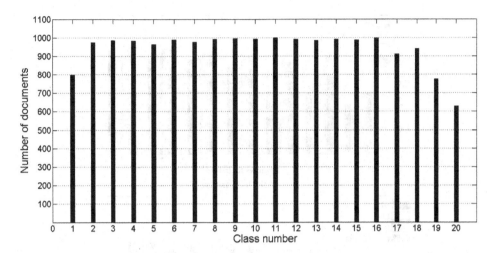

of the classification accuracy and training time of face recognition on the ORL, Yale and CMU PIE datasets respectively. Section 5.3 and 5.4 shows the results of accuracy and training time for text classification on the 20NG text dataset.

5.1 Face Recognition
Result 1: Accuracy

We compare the classification accuracy of LDA/QR and DLDA, Pre-LDA/QR and Pre-DLDA on the ORL, Yale and CMU PIE facial image datasets. The face recognition accuracy on ORL, Yale and CMU PIE datasets are reported in Figure 3(a), 3(b) and 3(c) respectively.

As can be observed in these figures, the LDA/QR and DLDA algorithms achieve the same classification accuracy on all datasets, which verifies our theoretical analysis. An interesting point to be stated here is that the Pre-LDA/QR and Pre-DLDA also obtain the same accuracy, which indicates these algorithms are equivalent too. The results also show that usually the Pre-LDA/QR and Pre-DLDA achieve lower accuracy than the LDA/QR and DLDA algorithms. This is because the Pre-LDA/QR and Pre-DLDA algorithms only use the between-class scatter information to com-

pute the projection vectors while ignoring the within-class information, and thus result in lower discriminant performance. Interestingly, on the Yale facial image dataset, the classification performance of Pre-LDA/QR and Pre-DLDA are competitive to LDA/QR and DLDA algorithms.

5.2 Face Recognition
Result 2: Training Time

Figure 4(a), 4(b), 4(c) shows the training time of face recognition on the ORL, Yale and CMU PIE datasets respectively.

Obviously, the curve of training time for LDA/QR is lower than that of DLDA on all the ORL, Yale and CMU PIE datasets. Similar results are also observed on Pre-LDA/QR and Pre-DLDA algorithms. This is consistent with our theoretical analysis that the QR decomposition used in (Pre-) LDA/QR is more efficient than SVD, which is applied in (Pre-)DLDA.

5.3 Text Classification
Result 1: Accuracy

Here we report our experiment results of text classification on the 20NG dataset. The illustration

Figure 3. Classification accuracy of face recognition on three different datasets

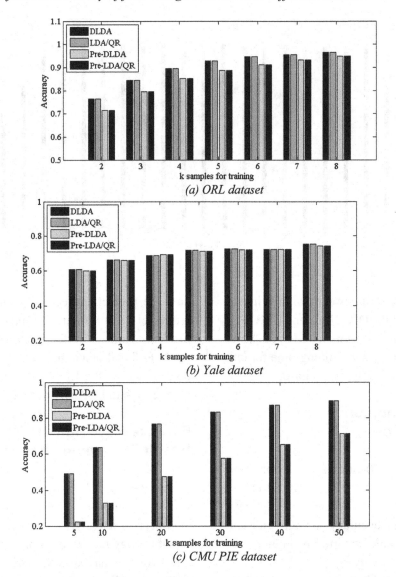

(a) ORL dataset

(b) Yale dataset

(c) CMU PIE dataset

of the classification accuracy of LDA/QR and DLDA, Pre-LDA/QR and Pre-DLDA are shown in Figure 5.

Similar to face recognition, the discriminant performance observed in Figure 5 for text classification is the same for both LDA/QR and DLDA, as well as Pre-LDA/QR and Pre-DLDA. These results further confirm our theoretical analysis.

5.4 Text Classification Result 2: Training Time

The training time of (Pre-)LDA/QR and (Pre-)DLDA algorithms are reported in Figure 6.

As we can see, the curve of the training time for DLDA algorithm is definitely higher than that of LDA/QR algorithm. This is also the case for Pre-LDA/QR and Pre-DLDA algorithms. These results are also consistent with our theoretical

Figure 4. Relationship between the training time and the number of facial samples using our method

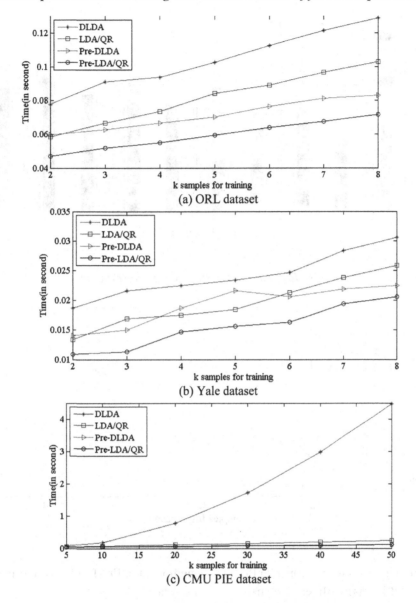

(a) ORL dataset

(b) Yale dataset

(c) CMU PIE dataset

analysis. This is to say, the QR decomposition-based LDA algorithms reduce the computational complexity for practical applications with respect to SVD-based algorithms.

5.5 Extension: Lite-LDA Methods

We discuss the above experimental results on classification accuracy here. Since Pre-LDA/QR and Pre-DLDA only incorporate the between-class information and ignore the within-class one, their average classification accuracy is typically lower than that of LDA/QR and DLDA's on all the tested ORL, Yale, CMU PIE, and 20NG datasets. However, as noticed with the Yale facial image dataset, the average accuracy of Pre-LDA/QR and Pre-DLDA is only lower than LDA/QR and DLDA by 0.0039 according to our statistic

Figure 5. Classification accuracy of text classification on the 20 Newsgroups dataset

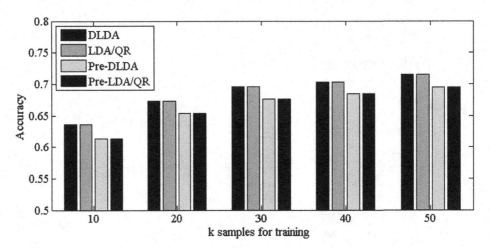

Figure 6. Training time of text classification on the 20 Newsgroups dataset

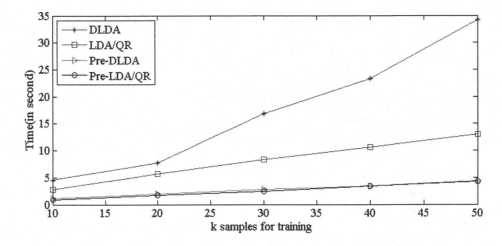

results. Hence, in some cases we can consider Pre-LDA/QR and Pre-DLDA as new linear dimension reduction methods. More specially, they are the lite-LDA algorithms because of their efficient computation without loss of much accuracy in some classification tasks.

Principal component analysis (PCA) is also a wildly used dimension reduction algorithms. However, LDA is generally considered better than PCA for classification (Martinez & Kak 2001; Yang & Yang, 2006) as it uses the label information which is important for encoding discriminant information. We also compare the two lite-LDA

methods, i.e. Pre-LDA/QR and Pre-DLDA with PCA here.

Same as before, the 1-NNC is used as the baseline. The ORL and Yale datasets are used in this experiment. l=(2,3,...,8) images for each subject are randomly selected as training samples. The results are averaged over random partitions and shown in Figure 7(a) and 7(b). We learn from the figures that the two lite-LDA methods perform better than PCA and the baseline. Moreover, the performance is better on the Yale dataset. This is to say, Pre-LDA/QR and Pre-DLDA gain better accuracy than PCA on these two popular facial

Figure 7. The relationship between the classification accuracy and the number of facial samples using lite-LDAs and PCA on the ORL and Yale datasets

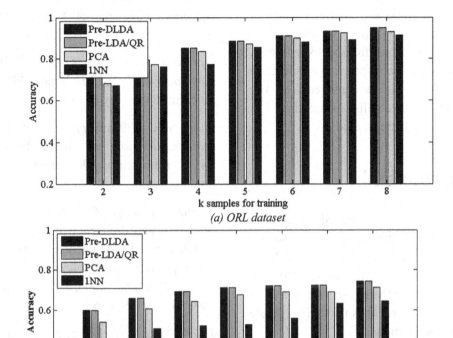

(a) ORL dataset

(b) Yale dataset

image datasets. In this sense, Pre-DLDA can be used independently as a dimension reduction algorithm.

6. CONCLUSION

In this paper, we establish the equivalent relationship between LDA/QR and DLDA. Similar to LDA/QR, we show that DLDA is also a two-stage LDA method. The first stage of DLDA, which is called Pre-DLDA, can be used independently as a dimension reduction algorithm. It is more interesting to find out that Pre-DLDA is equivalent to Pre-LDA/QR (the first stage of LDA/QR). We conduct extensive experiments on three famous

facial image datasets and one text dataset, and measure the classification accuracy and training time to evaluate our results. The experiments show that the classification accuracy of DLDA is equal to LDA/QR while the training time of DLDA is longer than LDA/QR on all the datasets. The same results are observed for Pre-LDA/QR and Pre-DLDA. In addition, in some applications Pre-LDA/QR and Pre-DLDA are competitive to LDA/QR and DLDA on classification accuracy with less training time. We consider Pre-LDA/QR and Pre-DLDA as lite-LDA algorithms, since they can achieve competitive discriminant performance while incorporating only the between-class information.

It is worth to explore two directions in our future work. First, our theoretical analysis is based on Euclidean space. It is also can be applied into Hilbert Reproduce Kernel Space, which produces nonlinear mapping. Second, from our experiments, the lite-LDA methods are shown to be effective in some applications with less time required. The theoretical analysis of these light-weighting LDA algorithms is worth to be studied.

ACKNOWLEDGMENT

The authors would like to acknowledge the partial funding support from the Hong Kong Research Grants Council (RGC) General Research Fund (GRF) grants PolyU 5101/09E, PolyU, 5101/08E and 5101/07E, The authors also would like to thank the anonymous reviews for their valuable comments and suggestions.

REFERENCES

AT&T Laboratories. (2002). *ORL face datasets.* Retrieved from http://www.cl.cam.ac.uk/research/dtg/attarchive/facedatabase.html

Belhumeur, P. N., Hespanha, J., & Kriegman, D. J. (1997). Eigenfaces vs. Fisherfaces: Recognition using class specific linear projection. *IEEE Transactions on Pattern Analysis and Machine Intelligence, 19*(7), 711–720. doi:10.1109/34.598228

Cai, D., He, X., & Han, J. (2005). Document clustering using locality preserving indexing. *IEEE Transactions on Knowledge and Data Engineering, 17*(12), 1624–1637. doi:10.1109/TKDE.2005.198

Chen, L., Liao, H., Ko, M., Lin, J., & Yu, G. (2000). A new LDA based face recognition system which can solve the small sample size problem. *Pattern Recognition, 33*(10), 1713–1726. doi:10.1016/S0031-3203(99)00139-9

Cover, T., & Hart, P. (1967). Nearest neighbor pattern classification. *IEEE Transactions on Information Theory, 13*(1), 21–27. doi:10.1109/TIT.1967.1053964

Golub, G. H., & Loan, C. F. V. (1996). *Matrix computations* (3rd ed.). Baltimore, MD: The Johns Hopkins University Press.

Gunn, S. R. (1998). *Support vector machines for classification and regression* (Tech. Rep. No. 6459). Hampshire, UK: University of Southampton.

Howland, P., Jeon, M., & Park, H. (2003). Structure preserving dimension reduction for clustered text data based on the generalized singular value decomposition. *SIAM Journal on Matrix Analysis and Applications, 25*(1), 165–179. doi:10.1137/S0895479801393666

Jin, Z., Yang, J. Y., Hu, Z. S., & Lou, A. (2001). Face recognition based on the uncorrelated discriminant transformation. *Pattern Recognition Letters, 34*, 1405–1416.

Jing, X.-Y., Wong, H.-S., & Zhang, D. (2006). Face recognition based on 2D Fisherface approach. *Pattern Recognition, 39*, 707–710. doi:10.1016/j.patcog.2005.10.020

Lang, K. (1995). Newsweeder: Learning to filter netnews. In *Proceedings of the International Conference on Machine Learning*, Tahoe City, CA. (pp. 331-339).

Li, M., & Yuan, B. (2005). 2D-LDA: A novel statistical linear discriminant analysis for image matrix. *Pattern Recognition Letters, 26*(5), 527–532. doi:10.1016/j.patrec.2004.09.007

Martinez, A. M., & Kak, A. C. (2001). PCA versus LDA. *IEEE Transactions on Pattern Analysis and Machine Intelligence, 23*(2), 228–233. doi:10.1109/34.908974

Moghaddam, B., Jebara, T., & Pentland, A. (2000). Bayesian face recognition. *Pattern Recognition Letters, 33*, 1771–1782.

Noushatha, S., Hemantha Kumar, G., & Shivakumara, P. (2006). (2D)2LDA: An efficient approach for face recognition. *Pattern Recognition Letters, 39*, 1396–1400.

Price, J. R., & Gee, T. F. (2005). Face recognition using direct, weighted linear discriminant analysis and modular subspaces. *Pattern Recognition Letters, 38*, 209–219.

Sim, T., Baker, S., & Bsat, M. (2003). The CMU pose, illumination, and expression database. *IEEE Transactions on Pattern Analysis and Machine Intelligence, 25*(12), 1615–1618. doi:10.1109/TPAMI.2003.1251154

Torkkola, K. (2003). Discriminative features for text document classification. *Pattern Analysis & Applications, 6*(4), 301–308.

Turk, M., & Pentland, A. (1991). Eigenfaces for recognition. *Journal of Cognitive Neuroscience, 3*(1), 71–86. doi:10.1162/jocn.1991.3.1.71

Williams, C. K. I., & Barber, D. (1998). Bayesian classification with gaussian processes. *IEEE Transactions on Pattern Analysis and Machine Intelligence, 20*(12), 1342–1351. doi:10.1109/34.735807

Xiong, H., Swamy, M. N. S., & Ahmad, M. O. (2005). Two-dimensional FLD for face recognition. *Pattern Recognition Letters, 38*, 112–1124.

Yale University. (1997). *Yale face datasets.* Retrieved from http://cvc.yale.edu/projects/yalefaces/yalefaces.html

Yang, J., & Yang, J. Y. (2003). Why can LDA be performed in PCA transformed space? *Pattern Recognition Letters, 36*(2), 563–566.

Yang, J., Zhang, D., Yong, X., & Yang, J.-Y. (2005). Two-dimensional discriminant transform for face recognition. *Pattern Recognition Letters, 38*, 1125–1129.

Ye, J. (2005). Characterization of a family of algorithms for generalized discriminant analysis on undersampled problems. *Journal of Machine Learning Research, 6*, 483–502.

Ye, J., Janardan, R., Park, C. H., & Park, H. (2004). An optimization criterion for generalized discriminant analysis on undersampled problems. *IEEE Transactions on Pattern Analysis and Machine Intelligence, 26*(8), 982–994. doi:10.1109/TPAMI.2004.37

Ye, J., & Li, Q. (2004). LDA/QR: An efficient and effective dimension reduction algorithm and its theoretical foundation. *Pattern Recognition Letters, 37*, 851–854.

Ye, J., & Li, Q. (2005). A two-stage linear discriminant analysis via QR-decomposition. *IEEE Transactions on Pattern Analysis and Machine Intelligence, 27*(6), 929–941. doi:10.1109/TPAMI.2005.110

Yu, H., & Yang, J. (2001). A direct LDA algorithm for high dimensional data with application to face recognition. *Pattern Recognition Letters, 34*(10), 2067–2070.

Zhang, P., Peng, J., & Riedel, N. (2005). Discriminant analysis: A least squares approximation view. In *Proceedings of the IEEE Conference on Computer Vision and Pattern Recognition* (p. 46).

This work was previously published in the International Journal of Cognitive Informatics and Natural Intelligence, Volume 5, Issue 1, edited by Yingxu Wang, pp. 94-112, copyright 2011 by IGI Publishing (an imprint of IGI Global).

Chapter 22
A Novel Algorithm for Block Encryption of Digital Image Based on Chaos

Jun Peng
Chongqing University of Science and Technology, China

Du Zhang
California State University, USA

Xiaofeng Liao
Chongqing University, China

ABSTRACT

This paper proposes a novel image block encryption algorithm based on three-dimensional Chen chaotic dynamical system. The algorithm works on 32-bit image blocks with a 192-bit secret key. The idea is that the key is employed to drive the Chen's system to generate a chaotic sequence that is inputed to a specially designed function G, in which we use new 8x8 S-boxes generated by chaotic maps (Tang, 2005). In order to improve the robustness against differential cryptanalysis and produce desirable avalanche effect, the function G is iteratively performed several times and its last outputs serve as the keystreams to encrypt the original image block. The design of the encryption algorithm is described along with security analyses. The results from key space analysis, differential attack analysis, and information entropy analysis, correlation analysis of two adjacent pixels prove that the proposed algorithm can resist cryptanalytic, statistical and brute force attacks, and achieve a higher level of security. The algorithm can be employed to realize the security cryptosystems over the Internet.

DOI: 10.4018/978-1-4666-2476-4.ch022

INTRODUCTION

In the last decades, extensive studies have been done in the theory of chaos in different fields such as physics, engineering, biology, and economics (Hao, 1993). Chaos theory consistently plays an active role in modern cryptography. As the basis for developing cryptosystems, the main advantage of the chaos-based approaches lies in the random behavior and sensitivity to the initial conditions and control parameters, hence the study on using chaos theory in information security has attracted great attentions (Chen, 2004; Li, 2005; Peng, 2008; Yang, 2004). The close relationship has been observed in (Álvarez, 2006; Fridrich, 1998; Kocarev, 2001) between chaotic maps and cryptosystems. In particular, the following connections between them can be established: (1) Ergodicity in chaos vs. confusion in cryptography; (2) Sensitive dependence on initial conditions and control parameters of chaotic maps vs. diffusion property of a good cryptosystem for a slight change in the plaintext and in the secret key; (3) Random-like behavior of deterministic chaotic-dynamics which can be used for generating pseudorandom sequences as key sequences in cryptography.

In recent years, the transmission of digital images over the Internet or personal digital mobile phones has been highly developed. Secure storage and transmission of digital images are becoming critically important. Most traditional ciphers, such as DES, IDEA, and AES are not suitable to conduct the digital image encryption in real time due to large data volume involved. Hence, the main purpose of this paper is to design a new image block encryption algorithm by using non-traditional methods such as chaos theory.

A brief review is in order on existing results of chaos-based encryption scheme. Yen et al. (2000) proposed a chaotic key-based algorithm (CKBA) in which a binary sequence as a key is generated based on a chaotic map. According to the binary sequence, image pixels are rearranged and then XORed with the key. However, as pointed out later in (Li, 2002), this algorithm has some drawbacks: it is vulnerable to the chosen or known-plain-text attack using only one plain image, and its security to brute-force attack is also questionable. Chen et al. (2004) used a 3D Arnolad's cat map for the purpose of substitution and employed Chen's chaotic system for generating key streaming and diffusion. A more complex system which combines discrete- and continuous-time chaotic systems has been proposed by Guan et al. (2005). Recently, Tong et al. (2009) presented a new compound two-dimensional chaotic functions design which acted as a chaotic sequence generator, and suggested a feedback image encryption scheme by using the new compound chaos and perturbation technology based on 3D Baker map. At the same time, in (Lian, 2009) the author employed a chaotic neural network (CNN) composed of a chaotic neuron layer and a linear neuron layer to construct a block cipher, in which the chaotic neuron layer realizes data diffusion and the linear neuron layer realizes data confusion, and the two layers are repeated for several times to strengthen the cipher. Obviously, these research results laid a good foundation to the subsequent studies on the chaos-based image encryption algorithm.

Motivated by the aforementioned results, a new image block encryption algorithm based on three-dimensional Chen dynamical system is proposed in this paper. Since cryptosystems play a pivotal role in a very important engineering application of cognitive informatics, i.e., information assurance and security, the subject topic of this paper is within the broad scope of cognitive informatics. The main novelty of the algorithm can be summarized as follows: (1) the 3D Chen system with complex dynamical behaviors is adapted; (2) within the algorithm, a 192-bit secret key is used to drive the Chen system and the keystreams are generated through iteratively performing a specially designed function G; (3) when designing the function G, we use new 8×8 S-boxes produced by chaotic maps in (Tang, 2005) in order to obtain desirable confusion and diffusion properties. Results of various types of analysis show that the new algorithm is highly secure for practical image encryption applications.

The remaining part of the paper is organized as follows. We first introduce the Chen chaotic system and show its chaotic attractors. The design of the encryption algorithm is then described in details, along with some experiments. Following that are the analyses that address the security performance of the proposed algorithm. Finally, conclusions and future work are drawn.

CHAOTIC SYSTEMS

When choosing a chaotic system, we mainly consider the following principles: (1) it can be easily implemented by software and hardware; (2) it should have complex dynamical behaviors; and (3) it is suitable for image encryption process. Research results in (Chen, 1998; Ueta, 2000; Yassen, 2003) indicate that Chen system is a three-order system which can be easily implemented by circuits, and has better three-dimensional dynamical properties in phase space than Lorenz system and Chua's system. Therefore, Chen chaotic system can be applied to designing a cryptosystem with higher security. Chen dynamical system is described by the following system of differential equations (Chen, 1999):

$$
\begin{cases}
\dot{x} = a(y - x), \\
\dot{y} = (c - a)x - xz + cy, \\
\dot{z} = xy - bz.
\end{cases}
\tag{1}
$$

where x, y and z are the state variables and a, b and c are three positive real constants. Chen dynamical system is chaotic when $a=35$, $b=3$ and $c \in [20, 28.4]$. Numerical experiments are carried out to integrate the system (1) by using fourth-order Runge-Kutta method with time step 0.001 and the chaotic attractors are shown in Figure 1. with parameters $a=35$, $b=3$ and $c=28$.

DESIGN OF THE IMAGE ENCRYPTION ALGORITHM

In this Section, we describe the proposed image encryption algorithm in details. Figure 2 is the diagrammatic depiction of the algorithm.

In general, we use a 192-bit key as the secret key of the encryption system, and the initial conditions of Chen system are determined through an initialization process. Four variables are generated from the output of Chen system and then used as input to a function G, which will be performed by several iterations. After that, the output keystreams are employed to encrypt plain-image block M_i through bitwise XOR operation. The encrypted image block would be E_i, which is also fed back to the Chen system for the encryption of the subsequent plain-image block. The introduction of this feedback mechanism would enhance the sensitivity of the ciphertext with respect to the plaintext and make their relationship more sophisticated, which in turn will result in the enhancement of the difficulties for cryptanalysis.

Description of Function G

Function G is similar to the round function of the DES-like iterated ciphers, where the ciphertext is calculated by recursively applying a round function to the plaintext. The block diagram of function G is displayed in Figure 3.

In Figure 3, k_1, k_2, k_3, k_4 are the four input variables, and l_1, l_2, l_3, l_4 are the corresponding four output variables of the function G. There are four identical S-boxes. As we know, the substitution boxes (S-boxes) have been extensively used in most of conventional cryptographic systems, such as DES, IDEA and AES. S-boxes are core component of the DES-like cryptosystems and the only nonlinear component of these ciphers. In block ciphers, they are typically used to obscure the relationship between the secret key and the

Figure 1. The chaotic attractor of Chen dynamical system in the (a) x-y plane, (b) x-z plane, and (c) y-z plane

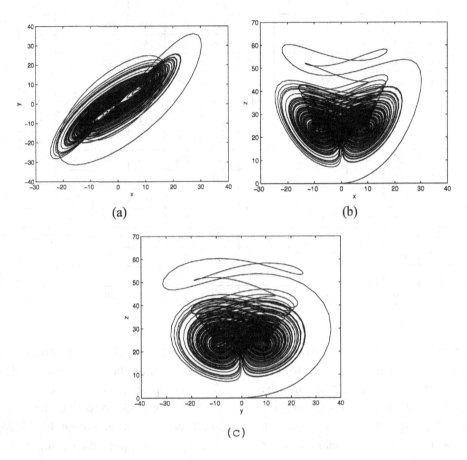

(a)

(b)

(c)

Figure 2. Diagram of proposed encryption algorithm

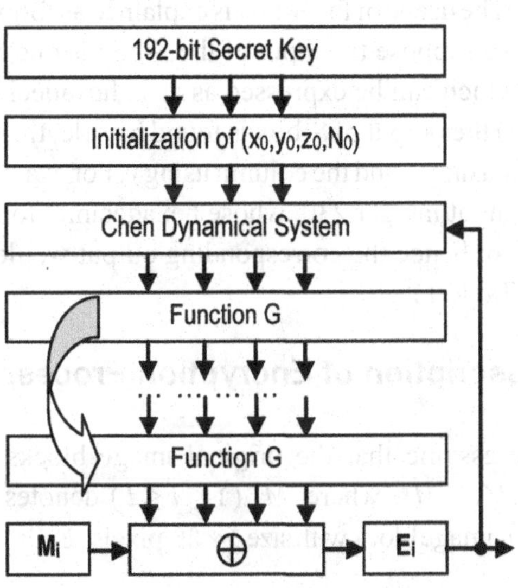

Figure 3. Diagram of function G

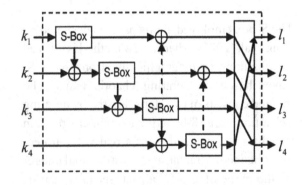

Table 1. An 8x8 S-Box produced

121	50	31	84	8	0	52	205	190	99	203	132	42	41	173	69
63	159	30	100	193	29	1	82	187	165	189	222	14	170	175	83
207	38	171	254	4	111	233	55	195	62	5	6	166	245	234	13
184	79	146	218	32	115	74	127	97	39	76	185	90	65	123	125
133	227	19	242	154	157	197	67	116	60	232	211	34	145	164	23
172	183	153	28	80	71	95	64	158	213	246	206	136	17	253	182
155	18	168	196	216	104	163	66	238	176	113	96	180	237	220	12
109	57	135	51	68	58	44	255	61	25	214	72	122	46	56	53
212	215	147	13	11	225	208	105	244	128	226	251	94	137	117	98
49	88	14	78	70	241	89	110	194	174	200	9	209	144	24	2
138	142	124	239	75	37	252	243	21	33	156	48	249	27	192	210
231	221	152	118	114	148	219	130	126	235	10	202	91	40	45	85
228	47	107	169	131	179	224	199	198	230	54	250	141	101	103	167
36	188	181	106	92	204	223	134	177	20	3	139	186	178	247	43
26	87	119	22	201	35	7	151	81	240	162	161	236	77	217	191
16	248	73	160	112	140	86	120	229	102	93	59	129	143	108	150

ciphertext - Shannon's property of confusion. In many cases, the S-boxes are carefully chosen to provide the cryptosystem with abilities of resisting cryptanalysis. Mathematically, an $n \times m$ S-box is a nonlinear mapping from V_n to V_m, where V_n and V_m represent the vector spaces of n and m tuples of elements from GF(2), respectively.

Description of S-Box

The S-box employed in this paper is designed by Tang et al. (2005), where a new method is utilized for obtaining cryptographically strong 8×8 S-boxes based on iterating chaotic maps. The method consists of two steps: Firstly, by iterating a chaotic map, an 8-bit sequence of binary random variables is generated from a real value trajectory obtained. After turning it to a decimal integer on the range of $0 - 2^n$, an integer table can be obtained then. Secondly, a key-dependent permuting is used to shuffle the table nonlinearly by applying a Baker map several times (Chen, 2004). The shuffled table is the desirable S-box, which is shown in Table 1.

The S-box produced in this new method satisfies many important properties (Adams, 1989) for cryptographically "good" S-boxes. These properties include: bijective, nonlinearity, strict avalanche criterion, output bits independence criterion and equiprobable input/output XOR distribution, besides the fact that it has high immunity to resist differential cryptanalysis. Therefore, it is a good candidate for use in the design image encryption algorithm.

The usage of the S-box is explained as follows: If we suppose the input of the S-box is one byte u, which can be expressed as xy in hexadecimal, then the output of S-box is found by selecting the row using x, and the column using y. For example, an input integer 230, whose hexadecimal format is E6; hence the corresponding output would be 7 (Table 1).

Description of Encryption Process

We assume that the original image blocks are $M_1 M_2 \cdots M_t$, where M_i ($1 \leq i \leq t$) denotes the i-th image block with size 1×32 pixels, $E_1 E_2 \cdots E_t$

are the corresponding encrypted image blocks, and t denotes the number of image blocks. Besides, we use a 192-bit secret key, $K_1 K_2 \cdots K_{24}$, to determine the initial conditions of Chen dynamical system (1), i.e., x_0, y_0, z_0, and its iteration times N_0. As shown in Figure 3, the detailed process for encrypting image is described as follows:

- **Input:** $M_1 M_2 \cdots M_t$ (original image blocks)
- **Output:** $E_1 E_2 \cdots E_t$ (encrypted image blocks)
- **Initialization:** Suppose v denotes the current image block, and let $v = 1$. The initial conditions of the Chen system are obtained through the following computational procedures:

$$S = (K_1 + K_2 + \cdots + K_{24}) \bmod 256, ,$$

$$P = K_1 \oplus K_2 \oplus \cdots \oplus K_{24},$$

$$Q = (\sum_{i=1}^{t} \sum_{j=1}^{4} M_{i,j}) \bmod 256,$$

$$x_0 = (K_1 \times K_5 \times K_9 \times K_{13} \times K_{17} \times S \times Q) / 256^7,$$

$$y_0 = (K_2 \times K_6 \times K_{10} \times K_{14} \times K_{18} \times S \times Q) / 256^7,$$

$$z_0 = (K_3 \times K_7 \times K_{11} \times K_{15} \times K_{19} \times S \times Q) / 256^7,$$

$$N_0 = (S^2 + P^2 + Q^2 + K_4 \times K_8 + K_{23} \times K_{24}) \bmod 256.$$

Step 1: Iterates Chen dynamical system $200 + N_0$ times from the initial conditions $x_0 + \delta$, $y_0 + \delta$ and $z_0 + \delta$ with parameters $a = 35$, $b = 3$ and $c = 28$, where $\delta = 0.01$. Assume (x_n, y_n, z_n) represents the n-th state. Calculate τ_n according to the following formula:

$$\tau_n = \lambda(x_n^2 + y_n^2 + z_n^2)^{1/2} \tag{2}$$

For the current image block, if $v = 1$ then $\lambda = 1$, else if $v > 1$ then

$$\lambda = \ln(E_{v-1,1} \oplus E_{v-1,2} \oplus E_{v-1,3} \oplus E_{v-1,4}) \tag{3}$$

Here, $E_{v-1,j} (1 \leq j \leq 4)$ denotes the j-th byte of the former encrypted image block whose size is four bytes (32 bits).

Step 2: Let $\zeta = \tau_n - floor(\tau_n)$, then $\zeta \in (0,1)$, and it can be expressed as

$$\zeta = 0.b_1(x)b_2(x) \cdots b_i(x) \cdots \tag{4}$$

in which the i-th bit $b_i(x)$ can be represented as:

$$b_i(x) = \sum_{r=1}^{2^i - 1} (-1)^{r-1} \Theta_{(r/2^i)}(x) \tag{5}$$

where $\Theta_t(x)$ is a threshold function which is defined as: if $x < t$, $\Theta_t(x) = 0$; otherwise $\Theta_t(x) = 1$.

Generate the following four byte variables:

$$k_i = (w_{8(i-1)+1} w_{8(i-1)+2} \cdots w_{8i}), 1 \leq i \leq 4 \tag{6}$$

where $w_j = b_j \oplus b_{1+(j+1) \bmod 32}$, $1 \leq j \leq 4$.

Step 3: Input the four variables k_1, k_2, k_3, k_4 to the function G, we get the four output variables l_1, l_2, l_3, l_4, which will be viewed as new input variables of function G in the subsequent iteration. After eight iterations, we obtain the keystream s_1, s_2, s_3, and s_4 Mathematically, $(s_1, s_2, s_3, s_4) = G^8(s_1, s_2, s_3, s_4)$.

Step 4: Encrypt the current v-th image block M_v with keystream $S = s_1 s_2 s_3 s_4$ as follows:

$$E_{v,j} = M_{v,j} \oplus s_j, 1 \leq j \leq 4 \tag{7}$$

Step 5: Assume

$p = (E_{v,1} + E_{v,2} + E_{v,3} + E_{v,4}) \bmod 256$,

$q = p^{1/2} / 256$, Let $x_0 = x_0 q$, $y_0 = y_0 q$,

$z_0 = z_0 q$, and $N_0 = N_0 \oplus p$. If all the original image blocks are encrypted (i.e., $v = t$) then encryption process **END** else goto **Step 1**, at the same time $v = v + 1$.

Remark 1: It is found that if the plaintext or secret key is all "0" bits then the variables S, P and Q are all zero, which will result in the initial conditions x_0, y_0 and z_0 being zero. As we know, (0, 0, 0) is a trivial solution of system (1). So to avoid this situation, we use a small positive increment δ to generate new initial conditions $x_0 + \delta$, $y_0 + \delta$ and $z_0 + \delta$, here $\delta = 0.01$.

Remark 2: To guarantee the sensitivity of the encryption algorithm with regard to the entire original image, we introduce one special parameter Q in the initialization process. It is clear that the generation of Q depends on the plaintext $M_{i,j}$, which implies that the ciphertext is sensitive to the plaintext. This can be verified in our experiments. If the encryption key is K, then the decryption key should be $K_E = (K, Q)$, here we call K_E an extend key.

Remark 3: As the presented encryption algorithm is a symmetrical one, thus the decryption process is similar to the encryption process. One can use the extend key K_E to regenerate the keystream s_1, s_2, s_3, s_4 and recover the original image according to the following formulae:

$$M_{v,j} = E_{v,j} \oplus s_j, \ 1 \le j \le 4 \tag{8}$$

Experimental Results

In this Section, we give some experimental results of the proposed encryption algorithm with regard to the 512×512 grey-scale "Lena" images. Figure 4 (a) is the original image and its corresponding

encrypted image with a randomly selected secret key K_1 is displayed in Figure 4 (b). Here, K_1 is selected as "JWe93F5a6&8hKw12Psb3#m7B". Furthermore, Figure 4(b) is processed for the 2nd time using the same algorithm and a completely difference image from Figure 4(b) is also obtained.

In the following step, we perform the histogram experiment. In statistics, an image histogram illustrates how pixels in an image are distributed by graphing the number of pixels at each grayscale value. Figure 4 (c) and Figure 4 (d) show the histograms for the images in Figure 4 (a) and Figure 4 (b), respectively. From the results, we found that the histogram of the encrypted image is fairly uniform and is significantly different from the respective histogram of the original image, indicating that this algorithm makes the encrypted image indeed look like a random one.

Security Analysis

From the cryptography point of view, an ideal encryption algorithm should have desirable features for withstanding most kinds of known attacks such as cryptanalytic attacks: statistical attacks, brute-force attacks, ciphertext only attack and known plaintext attack, etc. In this section, we will discuss some security analysis on the proposed algorithm in terms of key space, correlation analysis of two adjacent pixels, information entropy analysis, differential attack analysis, and diffusion and confusion analysis.

Key Space Analysis

A good encryption scheme should be sensitive to the secret keys, and the key space should be large enough to make the brute-force attacks infeasible. In this algorithm, since the size of the secret key is 192 bits, the key space size can reach up to 2^{192}, which is large enough to resist all kinds of brute-force attacks with the current computer technology.

Figure 4. Encryption experimental and histogram results. (a) and (c) are for "Lena" original image, (b) and (d) are for the encrypted image

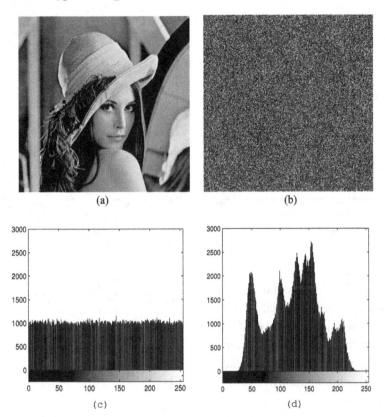

Correlation Analysis of Two Adjacent Pixels

To examine the correlation between two adjacent (in horizontal, vertical, and diagonal direction) pixels (Figure 5), we randomly select 2000 pairs of pixels from the original image and encrypted image, respectively, and compute their corresponding relevant coefficients by using the following formulae:

$$r_{xy} = \frac{\text{cov}(x,y)}{\sqrt{D(x)}\sqrt{D(y)}} \qquad (9)$$

$$\text{cov}(x,y) = \frac{1}{N}\sum_{i=1}^{N}(x_i - E(x))(y_i - E(y)) \qquad (10)$$

where $cov(x,y)$ is covariance, $D(x)$ is variance, x and y denote the gray-scale values of two adjacent pixels in the image. In numerical computation, the following discrete forms were used:

$$E(x) = \frac{1}{N}\sum_{i=1}^{N}x_i \qquad (11)$$

$$D(x) = \frac{1}{N}\sum_{i=1}^{N}(x_i - E(x))^2 \qquad (12)$$

Figure 5 illustrates the correlation distribution of two vertically adjacent pixels in the original "Lena" image and its corresponding encrypted image with K_1, and the correlation coefficients are 0.9730 and 0.0104, respectively. Similar results for vertical and diagonal directions were also obtained and shown in Table 2.

Figure 5. Correlations of two vertically adjacent pixels. (a) and (b) are correlation distribution analysis of the original 'Lena' image and the encrypted image, respectively

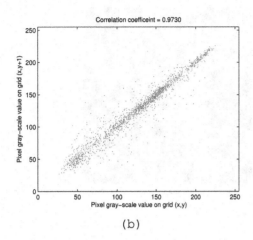

(b)

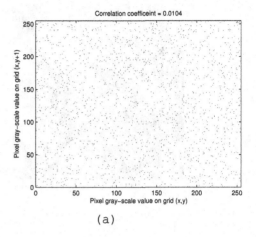

(a)

From the results of Table 2, we found that the correlation coefficients of the encrypted images are very small. These correlation analyses confirm that the chaotic encryption algorithm satisfies zero co-correlation, indicating that the attacker cannot obtain any valuable information by exploiting a statistic attack.

Differential Attack Analysis

As we know, in order to avoid the known-plaintext attack and the chosen-plaintext attack, a good encryption algorithm should have the desirable property which spreads the influence of a single plaintext bit over as much of the ciphertext as possible so as to hide the statistical structure of the plaintext (Schneier, 1996). This means the small difference of the plaintext should be diffused to the whole ciphertext. To evaluate the influence of one-pixel change on the whole encrypted image, two common measures (Chen, 2004) are used, i.e., number of pixels change rate (NPCR) and unified average changing intensity (UACI). The NPCR measures the different pixel numbers between two images, and the UACI measures the average intensity of differences between two images. These two measures are defined as follows:

$$NPCR = \frac{\sum_{i,j} D(i,j)}{W \times H} \times 100\% \quad (13)$$

$$UACI = \frac{1}{W \times H} \left[\sum_{i,j} \frac{\left| E^1(i,j) - E^2(i,j) \right|}{255} \right] \times 100\% \quad (14)$$

where E^1 and E^2 denote two encrypted images, respectively, W and H are the width and height of image, and the gray-scale values of the pixels at grid (i, j) of E^1 and E^2 are labeled as $E^1(i, j)$ or $E^2(i, j)$, respectively. $D(i,j)$ is related to $E^1(i, j)$ and $E^2(i, j)$. If $E^1(i, j) = E^2(i, j)$ then $D(i, j) = 1$ else $D(i, j) = 0$.

We randomly choose a pixel of the original image and make a slightly change on the gray-scale

Table 2. Correlation coefficients of two adjacent pixels in the original and encrypted image

Correlation coefficients	Original image	Encrypted image
Horizontal	0.9847	0.0189
Vertical	0.9730	0.0104
Diagonal	0.9640	-0.0076

value of this pixel. The encryption algorithm is performed on the modified original image and then the two measures NPCR and UACI are computed. This kind of experiment is carried out 512 times, and the average value of NPCR and UACI are calculated. We obtained NPCR $\approx$ 99.65%, and UACI $\approx$ 32.24%. The results show that a slightly change in the original image will result in a great change in the encrypted image, this implies that the proposed algorithm has an excellent capability to resist the differential attack.

Information Entropy Analysis

It is well known that the information entropy $H(m)$ of a plaintext message m can be calculated as

$$H(m) = \sum_{i=1}^{n} p(m_i) \log_2 \frac{1}{p(m_i)} \qquad (15)$$

where $p(m_i)$ represents the probability mass function of message m_i and $n = 256$ for image. For a 256-gray-scale image, if every gray value has an equal probability, then information entropy equals to 8, indicating that the image is a purely random one. When the information entropy of an image is less than 8, there exists a certain degree of predictability, which will threaten its security. Therefore, we strive for an entropy value of the encrypted image to be close to the ideal value of 8 so as to withstand the entropy attack effectively.

The information entropy of the original image "Lena" and its corresponding encrypted images are computed. They are 7.4455 and 7.9993, respectively. From the computation results we found that the entropy of the original images is smaller than the ideal one due to the fact that practical information sources seldom generate random messages. However the entropy of the encrypted one is very close to the theoretical value, showing the proposed encryption algorithm is sufficiently secure under the entropy attack.

Analysis of Diffusion and Confusion

Shannon (1949) has suggested two core techniques of "diffusion and confusion" for the design of practical ciphers. Diffusion means to spread the influence of a single plaintext or key bits over as much of the ciphertext as possible so as to hide the statistical structure of the plaintext. Confusion means to exploit some transformations to hide any relationship between the plaintext, the ciphertext and the key, thus making cryptanalysis more difficult (Schneier, 1996). Generally speaking, block ciphers' security depends on their computing security (Mollin, 2006) that is determined by the confusion and diffusion criteria.

As was pointed out earlier, several properties of chaotic systems such as ergodicity and sensitivity to the initial conditions are connected to the diffusion and confusion of the conventional ciphers. In this paper, the objective of introducing the Chen chaotic system with the unpredictable dynamical behaviors into the design of the image encryption algorithm is that we want to obtain the desirable features as conventional cryptosystem.

The term avalanche effect suggested by Feistel (1973) can be pertinently used to express the diffusion and confusion properties implicated in the encryption algorithm. An encryption algorithm is said to satisfy the strict avalanche criterion if, whenever a single input bit is complemented, each of the output bits should change with a probability of one half.

Let *HW* denote the Hamming weight (Figure 6) of a binary sequence, e_i^n denotes a unit vector whose length is n-th bit and only the i-th bit is "1". To evaluate the avalanche effect of this algorithm, we calculate the following four variables:

1. Suppose the plaintext block is e_i^{32}, and the Hamming weight of $f(e_i^{32}, z_j)$ is HW_i^1. Here, $f(e_i^{32}, z_j)$ denotes the obtained ciphertext when encrypting the plaintext e_i^{32} with a randomly selected secret key z_j, $1 \leq j \leq T$,

where T is the number of the secret keys. Calculate the following average Hamming weight HW^1 :

$$HW^1 = \frac{1}{32}\sum_{i=1}^{32}HW_i^1 \qquad (16)$$

2. Randomly select a non-zero plaintext block m_1, let $HW_{i,t}^2$ denote the Hamming weight of $f(m_1, z_t) \oplus f(m_1 \oplus e_i^{32}, z_t)$, where the meaning of function f is the same as in (i). Calculate the following average Hamming weight HW^2 :

$$HW^2 = \frac{1}{32}\sum_{i=1}^{32}(\frac{1}{T}\sum_{t=1}^{T}HW_{i,t}^2) \qquad (17)$$

3. Assume the plaintext m_1 is all-zero, the secret key z_1 also is all-zero. Let HW^3 denote the Hamming weight of $f(m_1, z_1) \oplus f(m_1, z_1 \oplus r_i^{192})$. Calculate the following average:

$$HW^3 = \frac{1}{192}\sum_{i=1}^{192}HW_i^3 \qquad (18)$$

4. Assume the plaintext m_1 is all-zero, but the secret key z_1 is non-zero. Let HW^4 denote the Hamming weight of $f(m_1, z_1) \oplus f(m_1, z_1 \oplus r_i^{192})$. Calculate the following average:

$$HW^4 = \frac{1}{192}\sum_{i=1}^{192}HW_i^4 \qquad (19)$$

In the experiments, we randomly select 50 secret keys, i.e., $T = 50$, and perform the aforementioned four calculations for the different iteration number of function G from 1 to 16, although we assign its value with 8 in Step 3. From the results shown in Figure 6, we found that the average of Hamming weight is very close to 16 (50%), especially HW^1=15.9781, HW^2=15.9456 for eight iterations of function G, and

Figure 6. Hamming weight results

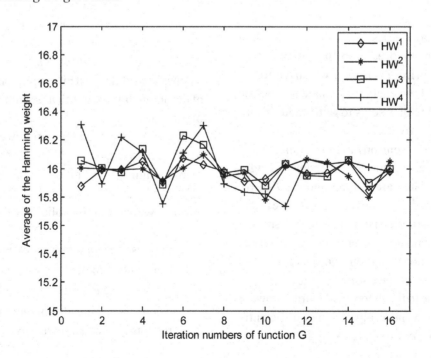

$\max_{1 \leq i \leq 4} \left| HW^i / 32 - 0.5 \right| = 0.0725$, which indicates that the proposed algorithm has the strict avalanche effect, implying the perfect diffusion and confusion features. This result also implies that if we want to improve the encryption speed, the function G could be iterated as few as two times while keeping the same security features.

CONCLUSION

In this paper, a novel image block encryption algorithm based on a three-dimensional Chen dynamical system is proposed. The main idea is to employ the output sequences of Chen's chaotic system to iteratively perform a specially designed function G, in which we use new 8×8 S-boxes generated by chaotic maps (Tang, 2005) so as to improve the robustness against difference cryptanalysis and produce desirable avalanche effect. The last output of function G worked as the keystream to encrypt the original image block. Security analyses indicate that the proposed algorithm has desirable properties such as excellent diffusion and confusion from cryptographic point of view and can be easily adopted to realize the security cryptosystems over the Internet.

The analysis of the capability of the algorithm withstanding some other attacks needs to be further investigated in the future work. Apart from the security considerations, some other issues on image encryption are also important, such as the running speed for real-time image encryption and decryption. Therefore, another future work is to consider and measure the speed performance with the proposed algorithm and to design a modified version for color image encryption.

ACKNOWLEDGMENT

The authors would like to thank the anonymous reviewers for their valuable suggestions. The work described in this paper was supported by the Natural Science Foundation Project of CQ CSTC under Grant No.2008BB2360, No.2008BB2199; the Science and Technology Research Project of Chongqing Municipal Education Commission of China under Grant No.KJ081406, No.KJ101413; and the First Batch of Supporting Program for University Excellent Talents in Chongqing.

REFERENCES

Adams, C., & Tavares, S. (1989). Good S-boxes are easy to find. In G. Brassard (Ed.), *Proceedings of Advances in Cryptology* (LNCS 435, pp. 612-615).

Álvarez, G., & Li, S. J. (2006). Some basic cryptographic requirements for chaos-based cryptosystems. *International Journal of Bifurcation and Chaos in Applied Sciences and Engineering, 16*(8), 2129–2151. doi:10.1142/S0218127406015970

Chen, G. R., & Dong, X. N. (1998). *From chaos to order methodologies, perspectives, and applications*. Singapore: World Scientific Press.

Chen, G. R., Mao, Y., & Chui, C. K. (2004). A symmetric image encryption based on 3D chaotic map. *Chaos, Solitons, and Fractals, 21*(3), 749–761. doi:10.1016/j.chaos.2003.12.022

Chen, G. R., & Ueta, T. (1999). Yet another chaotic attractor. *International Journal of Bifurcation and Chaos in Applied Sciences and Engineering, 9*(7), 1465–1466. doi:10.1142/S0218127499001024

Feistel, H. (1973). Cryptography and computer privacy. *Scientific American, 228*(5), 15–23. doi:10.1038/scientificamerican0573-15

Fridrich, J. (1998). Symmetric ciphers based on two-dimensional chaotic maps. *International Journal of Bifurcation and Chaos in Applied Sciences and Engineering, 8*(6), 1259–1284. doi:10.1142/S021812749800098X

Guan, Z. H., Huang, F., & Guan, W. (2005). Chaos based image encryption algorithm. *Physics Letters. [Part A]*, *346*(1-3), 153–157. doi:10.1016/j.physleta.2005.08.006

Hao, B. L. (1993). *Starting with parabolas: an introduction to chaotic dynamics*. Shanghai, China: Shanghai Scientific and Technological Education Publishing House.

Kocarev, L. (2001). Chaos-based cryptography: A brief overview. *IEEE Circuits and Systems Magazine*, *1*(3), 6–21. doi:10.1109/7384.963463

Li, S. J., Álvarez, G., & Chen, G. R. (2005). Breaking a chaos-based secure communication scheme designed by an improved modulation method. *Chaos, Solitons, and Fractals*, *25*, 109–120. doi:10.1016/j.chaos.2004.09.077

Li, S. J., & Zheng, X. (2002). Cryptanalysis of a chaotic image encryption method. In *Proceeding of IEEE International Symposium on Circuits and Systems*, Phoenix, AZ (Vol. 2, pp. 708-711). Washington, DC: IEEE Computer Society.

Lian, S. G. (2009). A block cipher based on chaotic neural networks. *Neurocomputing*, *72*, 1296–1301. doi:10.1016/j.neucom.2008.11.005

Mollin, R. A. (2006). *An introduction to cryptography*. Boca Raton, FL: CRC Press.

Peng, J., Zhang, D., Liu, Y. G., & Liao, X. F. (2008). A double-piped iterated hash function based on a hybrid of chaotic maps. In *Proceedings of the 7th IEEE International Conference on Cognitive Informatics* (pp. 358-365). Washington, DC: IEEE Computer Society Press.

Schneier, B. (1996). *Applied cryptography* (2nd ed.). New York, NY: John Wiley & Sons Press.

Shannon, C. E. (1949). Communication theory of secrecy system. *The Bell System Technical Journal*, *28*(4), 656–715.

Tang, G. P., Liao, X. F., & Chen, Y. (2005). A novel method for designing S-boxes based on chaotic maps. *Chaos, Solitons, and Fractals*, *23*(2), 413–419. doi:10.1016/j.chaos.2004.04.023

Tong, X. J., & Cui, M. G. (2009). Image encryption scheme based on 3D baker with dynamical compound chaotic sequence cipher generator. *Signal Processing*, *89*(4), 480–491. doi:10.1016/j.sigpro.2008.09.011

Ueta, T., & Chen, G. R. (2000). Bifurcation analysis of Chen's equation. *International Journal of Bifurcation and Chaos in Applied Sciences and Engineering*, *10*(8), 1917–1931. doi:10.1142/S0218127400001183

Yang, T. (2004). A survey of chaotic secure communication systems. *International Journal of Computational Cognition*, *2*(1), 81–130.

Yassen, M. T. (2003). Chaos control of Chen chaotic dynamical system. *Chaos, Solitons, and Fractals*, *15*(2), 271–283. doi:10.1016/S0960-0779(01)00251-X

Yen, J. C., & Guo, J. I. (2000). A new chaotic key-based design for image encryption and decryption. In *Proceedings of IEEE International Symposium on Circuits and Systems*, Geneva, Switzerland (Vol. 4, pp. 49-52). Washington, DC: IEEE Computer Society.

This work was previously published in the International Journal of Cognitive Informatics and Natural Intelligence, Volume 5, Issue 1, edited by Yingxu Wang, pp. 59-74, copyright 2011 by IGI Publishing (an imprint of IGI Global).

Chapter 23
Cognitive MIMO Radio:
Performance Analysis and Precoding Strategy

Mingming Li
Beijing University of Post and Telecommunication, China

Jiaru Lin
Beijing University of Post and Telecommunication, China

Fazhong Liu
College of Computing & Communication Engineering, GUCAS, China

Dongxu Wang
China Unicom, China

Li Guo
Beijing University of Post and Telecommunication, China

ABSTRACT

The authors consider a cognitive radio network in which a set of cognitive users make opportunistic spectrum access to one primary channel by time-division multiplexing technologies. Multiple Input Multiple Output techniques (MIMO) are similarly considered to enhance the stable throughput for cognitive links while they should guarantee co-channel interference constraints to the primary link. Here, two different cases are considered: one is that cognitive radio network is distributed; the other is centrally-controlled that cognitive radio network has a cognitive base station. In the first case, how to choose one fixed cognitive user and power control for each transmission antenna at the cognitive base station are considered to maximize the cognitive link's stable throughput. In the second case, a scheme to choose a group of cognitive users and a Zero-Forcing method to pre-white co-channel interference to the primary user, are also proposed in order to maximize cognitive base station's sum-rate. The algorithm can be employed to realize opportunistic spectrum transmission over the wireless fading channels.

DOI: 10.4018/978-1-4666-2476-4.ch023

INTRODUCTION

Cognitive informatics (CI) has been developed fast recently, Wang (2003) describes it as a transdisciplinary expansion of information science that studies computing and information processing problems by using cognitive science and neuropsychology theories, and studies the cognitive information processing mechanisms of the brain by using computing and informatics theories. In his following studies, he not only does many contributions on cognition, cognitive computing, and artificial intelligence, but also depicts the perspectives on cognitive informatics and cognitive computing. The authors of this paper are inspired by his famous architecture of contemporary cybernetics and cognitive informatics (Wang, 2009) and try to do some researches on cognition and automation for transmitting antennas in the wireless communication network.

Federal Communications Commission (FCC) had reported that the ever fixed spectrum allocation leads to vast spatial variations in the usage of allocated spectrum. This motivates the concepts of opportunistic spectrum access which guarantees the coexistence of primary licensed users and cognitive non-legitimate users in the same spectral resource opportunistically. Cognitive radio technology (CR) based on software radio concept has been recently proposed as such a smart and agile technology (Mitola, 1999; Haykin, 2005). It is proposed for wireless transmission in which either a base station or a cognitive node changes its transmission or reception parameters to communicate efficiently avoiding interference with licensed users. This alteration of parameters is obtained by radio frequency spectrum, user behaviors and network state. Research fields about CR technology include spectrum sensing, spectrum management, spectrum mobility and spectrum sharing. With the conception of CI, it is a paradigm of cognitive informatics and computational intelligence. In a cognitive radio network, cognitive users (CU) are the second users who must detect the licensed frequency band to find free time slots for transmitting without too much interference to the primary users. The authors in this paper focus on how to two different cases about CR network with intelligent antennas: one is that cognitive radio network is distributed structure; the other is centrally-controlled that cognitive radio network has a cognitive base station. Analysis about the first case is based on priority queuing framework model, which is characterized by the facts that primary link is oblivious to the cognitive activity and that the cognitive link required not to interference the primary link too much. The model was studied by Tsybakov and Mikhailov (1979), and Szpankowski et al. (1988) several decades ago. Recently, Sadek, Liu, and Ephremides (2007) and Simeone, Gambini, Bar-Ness, and Spagnolini (2007) had made attempts at analyzing cognitive multiple access protocols with cooperation, but Sadek et al. and Devroye (2006) also expounded the stability and performance in implementing the proposed multiple-access strategy. Simeone, Bar-Ness, and Spagnolini (2007) and their upper contributions mainly illustrated detection error occurring when primary link using one antenna simply. However, these prior works lacked some easy solution to the problem that both the primary and cognitive links communicate using multiple transmission antennas.

Multiple Input Multiple Output techniques (MIMO) is one of several forms of smart antenna technology which uses multiple antennas at both the transmitter and receiver to offer significant increases in data throughput and link range without additional bandwidth or transmit power (Gesbert, Shafi, Shiu, Smith, & Naguib, 2003). Nowadays, MIMO technology has attracted much attention in wireless communications and is one important part of modern wireless communication standards such as IEEE 802.11n. With the aims at extending the coverage, increasing connectivity and capacity, we try to study MIMO technology in cognitive radio network. Such motivation is attractive for three reasons:

1. Multiple antennas help the primary to diminish its transmission time, which leads to more transmission opportunities for the cognitive link, and multiple antennas help the cognitive achieve its diversity gain, and lessen its communicating time to reduce interference to primary;

2. It provides a benchmark on the performance of systems where the cognitive radio gains a partial understanding of the primary transmitter;

3. The ultimate limits on the cognitive transmitter are easily understood by giving it maximum information and allowing it to change its transmission and coding strategy based on all the information available at the primary user.

The representative contributions to cognitive MIMO system are mainly detailed by (Scutari, Palomar, & Barbarossa, 2008). Writers in this paper shed some light on game theory used to handle the resource allocating problem between cognitive users to achieve maximum rate. Hamdi, Zhang, and Ben Letaief (2007) had done some contributions on joint utilization between beam forming, scheduling, antenna selection in cognitive MIMO system while they forgot to consider the interference constraint for PU. Zhang, Liang, and Xin (2008) referred to an iterative power allocation scheme together with MMSE beam forming for the receiver is presented for the uplink channel of cognitive radio network to minimize the total transmitted power. In contrast, the object of MIMO technology in this paper has two parts: in the distributed cognitive radio network, both the primary and cognitive links communicate using multiple transmission antennas with the purpose is that maximize stable throughput and strengthening the system's stability; in the second case, we are going to obtain summation capacity of cognitive MU-MIMO broadcast channels with a novel block diagonalization in CR network considering coordinated transmit-receive processing.

In the distributed cognitive radio network like Figure 1, cognitive users (CU) will detect the primary channel separately and randomly access the primary spectrum. If without some scheme to choose a suitable cognitive user, the transmission model of cognitive users and the primary user is a collision one. Even when the CUs detect the primary channel, detection error probability p_e

Figure 1. Distributed CR network: K CU transmitter using M antennas, one PU transmitter using N antennas

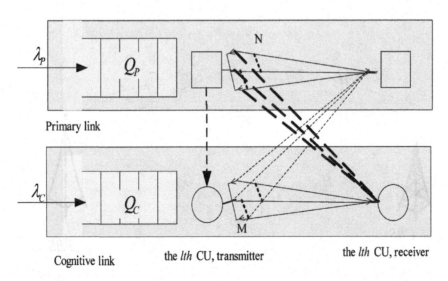

occurs with a missed detection probability p_f at the primary transmitter and random packet arrivals. The model depicted in the figure has two primary users (PU, marked with square) sharing their transmitting information to $L(L>2)$ cognitive users (CU, marked with circle), with the primary transmitter using N transmission antennas but cognitive transmitter using M transmission ones. To elaborate, transmitters are assumed to be equipped with one queue and times are slotted. At each queue, the packet arrival processes are independent and all packets have the same number of bits. They are also stationary with mean value λ_p (packets per slot) for the primary user and λ_{ck} (packets per slot) for the l th cognitive user. Such a model enlarges the possible schemes that can be implemented at the cognitive radio and it lends itself to information theoretic analysis. Here we suppose that the primary and cognitive users use the primary spectrum timely and cognitive users can derive the channels and system parameters by learning the primary link during the observation phase. Interference constraint requirement for primary link is focused on in terms of maximum average delays of primary packets.

Firstly, gamma distribution function is used to obtain accurate approximation to signal-to-interference-and-noise ratio (SINR) of the l th cognitive link. Based on these, by exploiting the concept of dominant systems, novel closed-form expressions of outage probability for both links are derived over Rayleigh fading channels in two different scenarios: one is detection error is omitted, then cognitive link's behaviors will bring no interference to primary link due to primary link idle; the other one is detection error occurs with a missed detection probability p_f due to misty and indetermination of wireless channels.

In the centrally-controlled cognitive radio network like Figure 2, cognitive base station will control all cognitive users to detect the primary channel corporately and access the primary spectrum. For maximizing the sum-rate of cognitive radio network, some Zero-Forcing methods should be utilized combined with constrains of interference to the primary user. Then we are going to give one scheme for a group cognitive user selection and try to obtain summation capacity of cognitive MU-MIMO broadcast channels with a novel block diagonalization in CR network (CR-

Figure 2. Centrally-controlled CR network: K CU transmitter using M antennas, one PU transmitter using N antennas

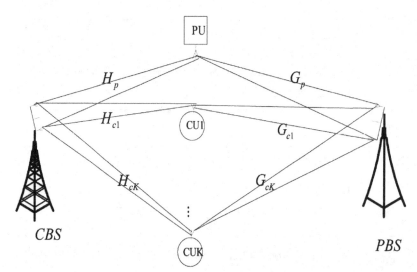

BD), meanwhile, the received interference power at PU caused by CUs in the same band should be under a tolerable limit and the total transmitted power for CUs is also restricted.

The remainder of this paper starts with a brief discussion of distributed system model and formulates the problem of the stable throughput. Next, user selection algorithm, stability, the maximum throughput of the cognitive MIMO system and power allocation schemes based on equivalent dominant system are elaborated in cognitive interference channels. Numerical results to complement the analysis and in the following part, we give centrally-controlled system model and formulate the problem of sum rate maximization. Next, the algorithm CR-BD and power allocation scheme are elaborated. Finally, we will show our simulation results to complement the sum-rate performance analysis. And the last part draws the conclusions of the paper.

DISTRIBUTED SYSTEM MODEL AND FORMULATION

Distributed System Model

In the distributed cognitive radio network shown like Figure 1, there has no cognitive base station to manage network resource and CUs can access the licensed spectrum with some promissory policies such as business priority, Quality of Service (QoS) guarantee and so on. In this model, each primary and cognitive transmitting node has a queue of unbounded capacity to store incoming packets, all packets have the same number of bits, and the transmission of each packet takes one slot. Queues are also stationary with an average throughput of primary link λ_p (packets per slot) for the primary user and λ_{ck} (packets per slot) for the l th cognitive user. The standard definition of stability with a queue size denoted $Q_j(t)$ is employed as in (Rao &

Ephremides, 1988), that is, a queue is stable if and only if it has positive probability of being empty $\lim_{t \to \infty} \Pr(Q_j(t) = 0) > 0$, where $Q_j(t)$ denotes the unfinished work (in packets) at time t, $j = p$ or c. At each node, the packet arrival processes are independent. Radio propagation between any pair of users is assumed to be affected by independent stationary time-variant Rayleigh fading channels. When the l th cognitive link is selected in Figure 1, the received signal can be expressed as $y = \sqrt{\gamma_i} h_i(t) x + n_i$, where x denotes the transmitted signal, $h_i(t)$ denote the channel fading coefficient with $E[|h_i(t)|^2] = 1$ (t denotes time and runs over time slots, $h_i(t)$ compose the transfer matrix $H_i(t)$ in Figure 1), γ_i is the average channel gain, n_i is additive white Gaussian noise, assumed to be with zero-mean and $E[|n_i|^2] = \sigma_i^2$, where i can be "pj" for the primary connection, "ck_s" for the secondary, "cpk_s" for the channel between secondary transmitter and primary receiver and "pcj" for the channel between primary transmitter and secondary receiver, and j refers to the jth antenna of primary links, $j = 1, 2 ... N$, k_s refers to the k_s th antenna of the l th cognitive links selected, $k = 1, 2 ... M$. In Figure 1, to keep the primary and the secondary queue stable, the total power allocated to transmission antennas should not exceed the maximum transmission power and interference from other systems is assumed to be negligible.

Formulating Problem of Stable Throughput

Stability is also an important network parameter that affects the tradeoff between rate and reliability of CR communication. In this paper, interference restrictions from cognitive node to primary node are defined here in terms of stability of the queue of the primary user. That is, as a result of the activity of the cognitive node, the primary node is guaranteed that its queue will remain stable. As has been stated, we adopt stochastic dominance technique to make ensure that the stationary distribution of the primary user. Buffer stability and packet delivery delay analysis for interacting queues were difficult problem that had been manifested in (Sidi & Segall, 1983; Nain, 1985; Naware, Mergen, & Tong, 2005) for ALOHA. Loynes' theorem (1962) is fit for acquiring the stability of a queue with non-independent interarrival and service times, which states that, if arrival rate λ_s is less than the average departure rate μ_s ($\lambda_s < \mu_s$), the queue is stable; if $\lambda_s > \mu_s$, the queue is unstable; finally if $\lambda_s = \mu_s$, the queue can be either stable or unstable, where $s = p$ or c. Whenever the Loynes' theorem is applicable, we define the average departure rate μ_s as the maximum stable throughput of the queue. Assuming that there is a given threshold β_p (β_c) at primary (cognitive) receiver, and the transmitted packet will be successfully received if the instantaneous received SINR is above the threshold. Then the outage probability on the primary or cognitive link is

$$P_{out,s} = \Pr[SINR_s < \beta_s] \qquad (1)$$

CU SELECTION ALGORITHMS AND AVERAGE DEPARTURE RATE ANALYSIS

In this section, the l th cognitive user selection algorithms is generated special for multiple CUs existing in the CR network. And the interacting queue theory is utilized to analyze average departure rates of both users.

CU Selection Algorithm

Our paper investigates the case of a machine to machine cognitive MIMO system with L cognitive users where user selection is needed. The scheme is learnt but different from the orthogonal or near-orthogonal multi-user selection algorithms for broadcast channels studied in the literatures (Hamdi, Zhang, & Ben Letaief, 2007; Yoo & Goldsmith, 2006; Bayesteh & Khandani, 2005).

Definition: $\Delta(H_p, H_{cl}) \triangleq \dfrac{\left|H_p H_{cl}^H\right|}{\left\|H_p\right\|\left\|H_{cl}^H\right\|}$, then, cognitive user l and primary user are near orthogonal if and only if $\Delta(H_p, H_{cl})$ is small enough. Now, the authors try to calculate the minimum value of $\Delta(H_p, H_{cl})$ decide how to choose cognitive user.

CU Selection Algorithm

1. Initializations: flag= 1, n = 1, $H_{cL} = [H_{c1}, H_{c2}, ..., H_{cL}]$;
2. Calculate $\Delta_l = \Delta(H_p, H_{cl})$;
3. If $\Delta_{l+1} < \Delta_l$, temp $= \Delta_{l+1}$; $\Delta_l =$ temp; flag $= l+1$; $l = l + 1$; end
4. Remember flag.

In upper algorithm, the last flag determines which CU is selected and $(\cdot)^H$ stands for the conjugate transpose. The idea of this algorithm is we try our best to find out the best near-orthogonal CU to PU. After this cognitive user selection, the interference of the selected cognitive user to the primary user is reduced. In MIMO channel, Dirty Paper Coding (DPC) is well known to be utilized on the optimal strategy with high complexity. Compared with DPC algorithm, time and space complexity of near-orthogonal user selection algorithm are much lower because computational complexity involving L matrixes but not all submatrixes of them. Then the proposed algorithm is able to lower the complexity compared to the exhaustive search and can be proposed as an efficient one. However, radio propagation between any pair of nodes is affected by independent stationary time-variant Rayleigh fading channels, sometimes the value of Δ_l is not little enough to decrease the interference between this CU and PU. In the following text, given the l th CU is our choice, we work out a new method to minimize interference brought by this CU to the PU.

Ideal System: Detection Error Not Considered

Without considering the probability of detection error, primary user and secondary user occupy spectrum hole alternately, then the queues of two links are pure. The signals received by the primary receiver contain only signals from its transmitter and the noise. Give no precoding schemes are done at both transmitters, Maximal Ratio Combing (MRC) are used as combing schemes and the weights are chosen like equation 9.2 in (Simon & Alouini, 2005) at the receiver.

PU's Stable Throughput: The SINR of primary link can be expressed as

$$SINR_{p,1} = \sum_{j=1}^{N} \frac{\gamma_{pj} |h_{pj}(t)|^2 P_{pj}}{n_p} \qquad (2)$$

where P_{pj} is the power allocated by the primary user for the j th antenna. The outage probability and $\mu_{P,ideal}$ of the primary user can be expressed as follows. Set $Y_j = \frac{\gamma_{pj} |h_{pj}(t)|^2 P_{pj}}{n_p}$ and $\theta_j = \frac{n_p}{\gamma_{pj} P_{pj}}$, then $SINR_{p,1} = \sum_{j=1}^{N} Y_j$. If $\theta = \theta_j$, Y_j is independent and identically distributed random variables. Due to the additivity of exponential distribution, outage probability $P_{out,P1}$ of primary link can be obtained by algorithms of multiple convolutions easily. From (1)-(2), given $\phi(\cdot)$ is the probability density function (PDF) of $SINR_{p,1}$, then $\phi(SINR_{p,1} = y) = \frac{\theta \exp(-\theta y)(\theta y)^{N-1}}{(N-1)!}$, and

$$\begin{aligned} P_{out,P1} &= \int_0^{\beta_p} \phi(SINR_{p,1} = y) dy = \int_0^{\beta_p} \frac{\theta \exp(-\theta y)(\theta y)^{N-1}}{(N-1)!} dy \\ &= 1 - \frac{\theta^N}{(N-1)!} \times \exp(-\theta \beta_p) \sum_{i=0}^{N-1} \frac{(N-1)! \beta_p^i}{i! \theta^{N-i}} = \frac{\gamma(N, \theta \beta_p)}{(N-1)!} \end{aligned}$$

$$(3)$$

If $\theta_m \neq \theta_n (m \neq n)$, Y_j is independent non-identically distributed random variables. So the distribution of $SINR_{p,1}$ is difficult to get. Because exponential distribution is a special case of Gamma distribution which can be proved by theories (Gradshteyn & Ryzhik, 2007), Gamma distribution is used to accurately approximate the distribution of $SINR_{p,1}$ in this paper. The following simulation results will demonstrate the accuracy of the theoretical approach. To the first order moment and the second order moment, we do the estimates as the following.

$$\begin{cases} E(SINR_{p,1}) = \sum_{j=1}^{N} \frac{1}{\theta_j} = \Gamma(\tilde{\alpha}+1)/(\Gamma(\tilde{\alpha})\tilde{\theta}^2) \\ E(SINR_{p,1}^2) = \sum_{j=1}^{N} \frac{2}{\theta_j^2} = \Gamma(\tilde{\alpha}+2)/(\Gamma(\tilde{\alpha})\tilde{\theta}^2) \end{cases} \text{,where}$$

$\tilde{\alpha} = \left\lfloor E^2(SINR_{p,1})/\text{Var}(SINR_{p,1}) \right\rfloor$, $\tilde{\theta} = \left\lfloor E(SINR_{p,1})/\text{Var}(SINR_{p,1}) \right\rfloor$. Then we can get the PDF of $SINR_{p,1}$, $\phi(SINR_{p,1} = y; \tilde{\alpha}, \tilde{\theta}) = \frac{\tilde{\theta}^{\tilde{\alpha}}}{\Gamma(\tilde{\alpha})} y^{\tilde{\alpha}-1} \exp(-\tilde{\theta} y)$. With integration method, the outage possibility is

$$\begin{aligned} P_{out,P1} &= \int_0^{\beta_p} \phi(SINR_{p,1} = y) dy = \int_0^{\beta_p} \frac{\tilde{\theta}^{\tilde{\alpha}}}{\Gamma(\tilde{\alpha})} y^{\tilde{\alpha}-1} \exp(-\tilde{\theta} y) dy \\ &= \frac{\tilde{\theta}^{\tilde{\alpha}}}{\Gamma(\tilde{\alpha})} \times \left[\frac{(\tilde{\alpha}-1)!}{\tilde{\theta}^{\tilde{\alpha}}} - \exp(-\tilde{\theta}\beta_p) \sum_{i=0}^{(\tilde{\alpha}-1)} \frac{(\tilde{\alpha}-1)! \beta_p^i}{i! \tilde{\theta}^{\tilde{\alpha}-i}} \right] = \frac{\gamma(\tilde{\alpha}, \tilde{\theta}\beta_p)}{\Gamma(\tilde{\alpha})} \end{aligned}$$

$$(4)$$

where $\Gamma(\cdot)$ and $\gamma(\cdot)$ are the incomplete gamma function and common gamma function (Gradshteyn & Ryzhik, 2007), $\lfloor \cdot \rfloor$ is the maximal integer but does not exceed the value $\cdot$. As mentioned in the upper, the queue of primary user is pure and the average departure rate of primary link is $\mu_{P,ideal} = 1 - P_{out,P1}$, substituting (3), (4) into this formula, we can obtain $\mu_{P,ideal}$.

CU's Stable Throughput: Similarly, the SINR of the l th cognitive link can be written as

$$SINR_{c,1} = \sum_{k=1}^{M} \frac{\gamma_{ck_s} \left| h_{ck_s}(t) \right|^2 P_{ck_s}}{n_c} \tag{5}$$

where P_{ck_s} is the power allocated by the cognitive link, $k_s = 1,2,...,$M. Set $Z_k = \frac{\gamma_{ck_s} \left| h_{ck_s}(t) \right|^2 P_{ck_s}}{n_c}$ and $\varsigma_k = \frac{n_c}{\gamma_{ck_s} P_{ck_s}}$, then $SINR_{c,1} = \sum_{k=1}^{M} Z_k$. If $\varsigma = \varsigma_k$, Z_k is independent and identically distributed random variables. Due to the additivity of exponential distribution, outage probability $P_{out,C1}$ of cognitive user can be obtained like (3).

$$P_{out,C1} = \int_0^{\beta_c} \phi'(SNR_{c,1} = y)dy = \int_0^{\beta_c} \frac{\varsigma \exp(-\varsigma y)(\varsigma y)^{M-1}}{(M-1)!} dy$$
$$= 1 - \frac{\varsigma^M}{(M-1)!} \times \exp(-\varsigma \beta_c) \sum_{i=0}^{M-1} \frac{(M-1)! \beta_c^i}{i! \varsigma^{N-i}} = \frac{\gamma(M, \varsigma \beta_c)}{(M-1)!} \tag{6}$$

where $\phi'(\cdot)$ is the PDF of $SINR_{c,1}$, $\gamma(\cdot)$ is referred as before. If $\varsigma_m \neq \varsigma_n (m \neq n)$, Z_k is independent non-identically distributed random variables. Gamma distribution is used to accurately approximate the distribution of $SINR_{c,1}$ as done like in (4).

$$P_{out,C1} = \int_0^{\beta_c} \phi(SINR_{c,1} = y)dy = \int_0^{\beta_c} \frac{\tilde{\varsigma}^{\tilde{\tau}}}{\Gamma(\tilde{\tau})} y^{\tilde{\tau}-1} \exp(-\tilde{\varsigma} y)dy$$
$$= \frac{\tilde{\varsigma}^{\tilde{\tau}}}{\Gamma(\tilde{\tau})} \times [\frac{(\tilde{\tau}-1)!}{\tilde{\varsigma}^{\tilde{\tau}}} - \exp(-\tilde{\varsigma} \beta_c) \sum_{i=0}^{(\tilde{\tau}-1)} \frac{(\tilde{\tau}-1)! \beta_c^i}{i! \tilde{\varsigma}^{\tilde{\tau}-i}}] = \frac{\gamma(\tilde{\tau}, \tilde{\varsigma} \beta_c)}{\Gamma(\tilde{\tau})} \tag{7}$$

where $\tilde{\tau} = \left| \frac{E^2(SINR_{c,1})}{Var(SINR_{c,1})} \right|$ and $\tilde{\varsigma} = \left| \frac{E(SINR_{c,1})}{Var(SINR_{c,1})} \right|$. Since the primary queue is not empty and primary user transmits with a probability of Since the primary queue is not empty and primary user transmits with a probability of $\frac{\lambda_p}{\mu_{P,ideal}}$, the maximum stable throughput for the cognitive user depends on the primary queue is empty or not and it can be expressed

$$\mu_{C,ideal} = (1 - \frac{\lambda_p}{\mu_{P,ideal}})(1 - P_{out,C1}) \tag{8}$$

Because the primary link and the cognitive link use different time slots, their transmitters can reach the maximum power $P_{P,\max}$ and $P_{C,\max}$, (8) is an optimal question with constraint conditions $\sum_{j=1}^{N} P_{pj} \leq P_{P,\max}$, $\sum_{k=1}^{M} P_{ck_s} \leq P_{C,\max}$, $P_{pj} \geq 0$ and $P_{ck_s} \geq 0$. Obviously, (8) is a convex problem which can be solved with Lagrange's interpolation to work out P_{pj} and P_{ck_s} for both links. In part four, to solve the problem (8), we set $P_{P,\max} = P_{C,\max} = 1$ for simplicity.

Real System: Detection Error Considered

We investigate the probability of slot existing probability and the detection error that occurs in real circumstances, the situation, in which the relationship of primary link and cognitive link is equivalent to interweave communications manifested in Haykin (2005) and Zhao and Sadler (2007). As the primary user selects its own arrival rate without ignoring the presence of the cognitive node, the cognitive link will transmit packets no matter the slot is idle or not, that will cause interference to the primary link. Actually, the queuing systems of primary user and cognitive user are interacting in this case and the stationary of the departure rates cannot be guaranteed. Then Loyne's theorem can no longer be used. While we have assumed the maximum stable throughput of the link equals the average departure rate in this case, the stability of the system can be analyzed as the following.

Throughput Analysis of PU: Since the primary user is unaware of the cognitive user, it will use the optimal power whenever there are packets left in its queue. The interferences to primary receiver include not only the noise but also interference from the cognitive user. Therefore, the SINR of primary user is

$$SINR_{p,2} = \sum_{j=1}^{N} \frac{\gamma_{pj} \left| h_{pj}(t) \right|^2 P_{pj}}{(n_p + \sum_{k=1}^{M} \gamma_{cpk_s} \left| h_{cpk_s}(t) \right|^2 P_{ck_s})} \qquad (9)$$

And the outage probability of the primary user is shown in Appendix-Derivations of Equations (10) and (11). Set $\theta_j = \dfrac{1}{\gamma_{pj} P_{pj}}$, for all m, n in (1, N), $\forall m \neq n$, if $\hat{\theta}_m = \hat{\theta}_n$. See Box 1 for Equations (10) and (11). Where $W(\cdot)$ is the Whittaker function (Gradshteyn & Ryzhik, 2007), and $_2F_1(\cdot)$ is the Gauss's Hypergeometric function (Magnus & Oberhettinger, 1949). Equations (6) and (7) (Chen & Chang, 1991) can be rewritten to get the average departure rate (the maximal throughput) of the primary user for this step, and we rewrite it

like considering its opposite side and detection error made by CR. See Box 2 for Equation (12).

Where the first item in (12) is $(1 - p_f)(1 - P_{out,P1})$ and the second item in upper formula is $p_f(1 - P_{out,P2})$. Substituting (3), (4), (10), (11) into (12), $\mu_{P,real}$ can be achieved.

Throughput Analysis of CU: If the probability of the successful detection $1 - p_e$ can be obtained by the cognitive user, they may transmit packets in an idle slot correctly. But if the probability of the detection error occurs with p_f, in a similar vein, the interferences to cognitive receiver contain both the noise from the cognitive link and the interference from the primary user. Therefore, the SINR of cognitive user is

Box 1.

$$P_{out,P2} = \Pr(\sum_{j=1}^{N} \frac{\gamma_{pj} \left| h_{pj}(t) \right|^2 P_{pj}}{(n_p + \sum_{k=1}^{M} \gamma_{cpk_s} \left| h_{cpk_s}(t) \right|^2 P_{ck_s})} < \beta_P)$$

$$= \frac{\Gamma(M)}{(M-1)!} - \frac{\exp(\dfrac{n_p \hat{\theta} - \beta_p n_p \hat{\theta}}{2}) \Gamma(n_p \hat{\theta})}{(M-1)!} \sum_{i=0}^{N-1} \frac{\beta_p^{\,i}}{i!} (\beta_p + 1)^{-\frac{M+i+1}{2}} (n_p \hat{\theta})^{\frac{M+i-1}{2}} W_{\frac{i+1-M}{2}, \frac{-i-M}{2}}(n_p \hat{\theta}(\beta_p + 1)) \qquad (10)$$

else

$$P_{out,P2} = \Pr(\sum_{j=1}^{N} \frac{\gamma_{pj} \left| h_{pj}(t) \right|^2 P_{pj}}{(n_p + \sum_{k=1}^{M} \gamma_{cpk_s} \left| h_{cpk_s}(t) \right|^2 P_{ck_s})} < \beta_P)$$

$$= \frac{(\tilde{\tau}-1)!(\tilde{\alpha}-1)!}{\Gamma(\tilde{\tau})\Gamma(\tilde{\alpha})} - \frac{\tilde{\theta}^{\tilde{\alpha}}}{\Gamma(\tilde{\tau})\Gamma(\tilde{\alpha})} \times \frac{\tilde{\varsigma}^{\tilde{\tau}}\Gamma(\tilde{\tau}+\tilde{\alpha})}{\tilde{\alpha}(\tilde{\varsigma}+\tilde{\theta})^{\tilde{\tau}+\tilde{\alpha}}} {}_2F_1(1, \tilde{\tau}+\tilde{\alpha}; \tilde{\tau}+1; \frac{\tilde{\varsigma}}{\tilde{\varsigma}+\tilde{\theta}}) \qquad (11)$$

Box 2.

$$\mu_{P,real} = p(\text{cognitive detection right}) \times p(\text{primary link transmit}|\text{with detection error})$$
$$+ p(\text{cognitive detection not right }) \times p(\text{primary link transmit}|\text{without detection error}) \qquad (12)$$

$$SINR_{c,2} = \sum_{k=1}^{M} \frac{\gamma_{ck_s} \left| h_{ck_s}(t) \right|^2 P_{ck_s}}{\left(n_c + \sum_{j=1}^{N} \gamma_{pcj} \left| h_{pcj}(t) \right|^2 P_{pj} \right)} \qquad (13)$$

We denote $\hat{\varsigma} = \dfrac{1}{\gamma_{ck_s} P_{ck_s}}$, for all m, n in (1, M),

$\forall m \neq n$, if $\hat{\varsigma}_m = \hat{\varsigma}_n$, as obtained in (10) and (11), the outage probability of the l th cognitive user can be expressed as Equations (14) and (15) in Box 3.where $\Gamma(\cdot)$, $W(\cdot)$, $_2F_1(\cdot)$, $\tilde{\varsigma}$, $\tilde{\tau}$, $\tilde{\alpha}$, $\tilde{\theta}$ are set as before.

Because the cognitive queue is the second queue in the system, cognitive user has to protect the purity of primary queue stated by CR principles, but the primary user transmits packets when there are still packets left in its queue. The l th cognitive user's maximum stable throughput should be considered with the probability of the

detection error occurs as well as false alarm probability. Formula (16), seen in Box 4, (Luo & Ephremides, 1999) can be write out for calculating average departure rate of the cognitive user like (10), but we could not ignore the probability of primary queue empty.

The first item represents cognitive link transmitting without false alarm and with the case of primary queue is empty, and second item means cognitive link transmitting when detection error occurs but primary queue is not empty. The formula is equal to

$$\mu_{C,real} = (1 - \frac{\lambda_p}{\mu_{P,real}})(1 - p_f)(1 - P_{out,C1})$$

$$+ \frac{\lambda_p}{\mu_{P,real}} p_e (1 - P_{out,C2}) + p_e \frac{\lambda_p}{\mu_{P,real}} (1 - p_{F \to B})(1 - P_{out,C2})$$

$$(17)$$

Box 3.

$$P_{out,C2} = \Pr\left(\sum_{k=1}^{N} \frac{\gamma_{ck_s} \left| h_{ck_s}(t) \right|^2 P_{ck_s}}{\left(n_c + \sum_{j=1}^{N} \gamma_{pcj} \left| h_{pcj}(t) \right|^2 P_{pj} \right)} < \beta_c \right)$$

$$= \frac{\Gamma(N)}{(N-1)!} - \frac{\exp(\frac{n_c \hat{\varsigma} - \beta_c n_c \hat{\varsigma}}{2})\Gamma(n_c \hat{\varsigma})}{(N-1)!} \sum_{i=0}^{M-1} \frac{\beta_c^{\ i}}{i!}(\beta_c + 1)^{-\frac{N+i+1}{2}} (n_c \hat{\varsigma})^{\frac{N+i-1}{2}} W_{\frac{i+1-N}{2}, \frac{-i-N}{2}}(n_c \hat{\varsigma}(\beta_c + 1)) \qquad (14)$$

else

$$P_{out,C2} = \Pr\left(\sum_{k=1}^{N} \frac{\gamma_{ck_s} \left| h_{ck_s}(t) \right|^2 P_{ck_s}}{\left(n_c + \sum_{j=1}^{N} \gamma_{pcj} \left| h_{pcj}(t) \right|^2 P_{pj} \right)} < \beta_c \right)$$

$$= \frac{(\tilde{\tau}-1)!(\tilde{\alpha}-1)!}{\Gamma(\tilde{\tau})\Gamma(\tilde{\alpha})} - \frac{\tilde{\varsigma}^{\tilde{\tau}}}{\Gamma(\tilde{\tau})\Gamma(\tilde{\alpha})} \times \frac{\tilde{\theta}^{\tilde{\alpha}}\Gamma(\tilde{\tau}+\tilde{\alpha})}{\tilde{\tau}(\tilde{\varsigma}+\tilde{\theta})^{\tilde{\tau}+\tilde{\alpha}}} \, _2F_1(1, \tilde{\tau}+\tilde{\alpha}; \tilde{\alpha}+1; \frac{\tilde{\theta}}{\tilde{\varsigma}+\tilde{\theta}}) \qquad (15)$$

Box 4.

$$\mu_{P,real} = p(\text{cognitive detection error}) \times p(\text{cognitive link transmit}|\text{with detection error})$$

$$+ p(\text{cognitive detection not error}) \times p(\text{cognitive link transmit}|\text{without detection error}) \qquad (16)$$

Then the maximum stability of cognitive MISO system can be achieved by solving optimization problem as $\max\limits_{P_{ck_s}} \mu_{C,real}(P_{c1_s}, P_{c2_s}, \ldots P_{cM_s})$, which the constraint conditions are $\lambda_p < \mu_{P,real}(P_{p1}, P_{p2}, \ldots P_{pN})$, $\sum\limits_{j=1}^{N} P_{pj} < P_{P,\max}$, and $\sum\limits_{k=1}^{M} P_{ck_s} < P_{c,\max}$. Though optimization problem is a little complex, multi-dimensional search can be used to solve it.

NUMERICAL RESULTS

In this section, various performance evaluation results have verified the mathematical analysis and show how much influence probability of detection error will bring to stability of the cognitive interference channel. We give the numerical and simulation results with $M = 2$, $N=2$ for simplicity. In this paper, to solve the problem $\max\limits_{P_{ck_s}} \mu_{C,real}(P_{c1_s}, P_{c2_s}, \ldots P_{cM_s})$, we use two dimensional search methods as the following.

Dimensional algorithm to find power allocation scheme for cognitive user:

1. Initializations: Set i=0, P_{p1}, P_{p2} initial values are in $0 \rightarrow P_{p,\max}$, P_{c1}, P_{c2} initial values are in $0 \rightarrow P_{c,\max}$ with step length $\triangle$=0.00001, $\mu_{C,real}$ and $\mu_{P,real}$ is zeros with length 100000;
2. Calculate $\mu_{P,real}$ with (12), when $\lambda_p < \mu_{P,real}$, remember $\mu_{P,real}$;
3. Calculate $\mu_{C,real}$ with (16), when $\lambda_C < \mu_{C,real}$, remember $\mu_{C,real}$, P_{c1}, P_{c2};
4. i=i+1, repeat step 2 and step 3 with 100000 times
5. Search $\mu_{C,real}$ to find the maximum, respectively, find P_{c1} and P_{c2}.

In the dimensional algorithm, the author try to insert interior point to split the solution space for the feasible solution, which jumps out of local optimization at a greater probability, the global optimization can be found. If one slot is assumed as one second, time complexity and space complexity of this algorithm will not exceed $o(\frac{1}{\triangle})^2$.

Figure 3 shows the parameter λ_p against the maximum stable throughput $\mu_{C,real}$ using different power allocation algorithm. It is evident from this figure that λ_p increases with a decrease of $\mu_{C,real}$. Furthermore, the optimal power allocation algorithm for the real scenario has the best performance when detection errors occur. The performance of the SISO system, in which the cognitive user only uses one transmission antenna, is also given for reference. We set the simulation parameters as follows:

$$\gamma_{p1} = \gamma_{c1} = \gamma_{pc2} = \gamma_{cp2} = -2dB,$$
$$\gamma_{p2} = \gamma_{c2} = \gamma_{pc1} = \gamma_{cp1} = -4dB, \beta_P = \beta_C = -5dB$$

Figure 4 is plotted as probability of detection error p_e versus the maximum stable throughput $\mu_{C,real}$ with $\lambda_p = 0.25, 0.35, 0.45$ and $p_f = 0.2$. The statement that probability of detection error can decrease $\mu_{C,real}$ is confirmed, and other parameters are the same as above in Figure 2 respectively.

In Figure 5, we plot the figure with missed detection probability of $p_f = 0.3, 0.4, 0.5$ and $\lambda_p = 0.35$, This figure can make a comparison with Figure 3 and the results show $\mu_{C,real}$ variation with p_f if we set λ_p with a fixed value.

CENTRALLY-CONTROLLED SYSTEM MODEL AND FORMULATION

In the distributed cognitive radio network with MIMO technology, the authors demonstrate the stable throughput with user selection method combined with power allocation scheme to show the system performance. In this section, we intend to learn sum-rate performance in centrally-controlled system to make a comparison with the upper one.

Figure 3. Maximum stable throughput $\mu_{C,real}$ allowed to the cognitive user versus the throughput selected by the primary user λ_P

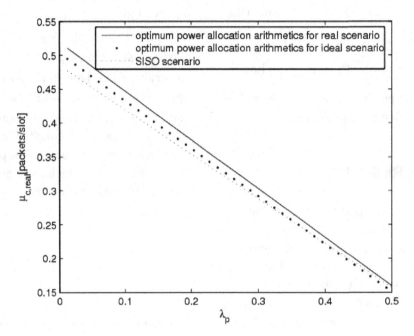

Figure 4. Maximum stable throughput $\mu_{C,real}$ allowed to the cognitive user versus probability of detection error p_e

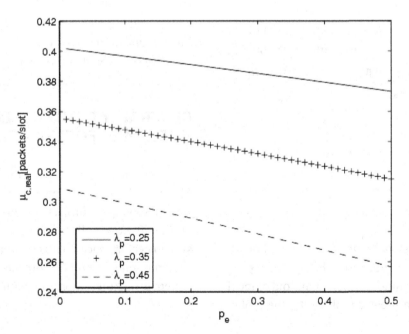

Figure 5. versus probability of detection error p_e and missed detection probability p_f

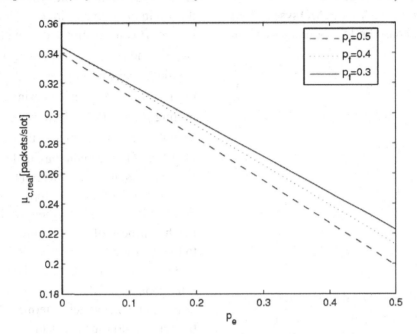

Centrally-Controlled System Model

In the centrally-controlled cognitive radio network like Figure 2, there is a central access point like cognitive base station will control network resources and designate all cognitive users to access the primary spectrum orderly. The cognitive base station is equipped with N_c transmit antennas, the j-th CU is uniformly spaced only with N_{rj} antenna ($j = 1,2,...K$, $K \gg N_c$), and the primary cognitive base station is equipped with N_p transmit antennas while the primary user is played with a single received antenna. The received signal vectors at the j-th cognitive receiver and the primary receivers, y_{cj} and y_p, are given respectively as $y_{cj} = \sum_{j=1}^{K} \sqrt{P_j} H_{cj} W_{cj} s_j + \sqrt{P_p} G_{cj} s_p + \omega_j$ and

$y_P = \sum_{j=1}^{K} \sqrt{P_j} H_P W_{ck} s_j + \sqrt{P_p} G_P s_p + \omega_P$, where $H_{cj} \in C^{N_{rj} \times N_c}$ is the channel between the cognitive base station and PU, $H_p \in C^{1 \times N_c}$ is the channel between the cognitive base station and the PU, $G_{cj} \in C^{N_{rj} \times N_p}$ is the channel between the primary

base station and the j-th CU, $G_p \in C^{1 \times N_p}$ is the channel between the primary base station and the PU. s_p and s_j denote transmitting signals, modulation matrix $W = \{W_{c1}, W_{c2}...W_{ck}\}$ is introduced to eliminate interference to the primary user caused by CUs and multiple user interference among CUs as well as ω_j (1×1) and ω_p (1×1) are Gaussian noise vector independent with zero mean and variance σ_c^2 and σ_p^2 respectively, for all j in 1 to k. To maximize the sum rate of CUs satisfying the condition that the co-channel interference received by the primary user under a tolerable threshold ς_p, the sum of transmitting power for all CUs at the cognitive base station is constrained with total power P_c, and the $SINR_{cj}$ is large than the value γ. With the recerver signal at the j-th CU, the $SINR_{cj}$ can be write as

$$SINR_{ck} = \frac{\left|H_{ck}W_{ck}\right|^2 P_{Ck}}{\sum_{j \neq k} \left|H_{ck}W_{ck}\right|^2 P_{cj} + \left|G_{ck}\right|^2 P_P + \sigma_c^2}$$

(18)

Then optimization problem in the downlink channel of the cognitive MU-MIMO system with the interference constrain to a primary user can be expressed as

$$\max_{P_{cj}, W, |S|=N_t} R = \sum_{j \subset S} \log(1 + SINR_{cj}) \tag{19}$$

s.t. $SINR_{cj} \geq \gamma$, $\forall j \in S$

$$\sum_{j \in S} \left| H_P W_{cj} \right|^2 P_{cj} < \beta_P$$

$$\sum_{j \in S} P_{cj} \leq P_c$$

In the upper equation, the first constrain is $SINR_{cj}$ constrain, the second one is interference constrain to the PU, and the last one is transmitting power constrain at the cognitive case station.

Cognitive User Selection Algorithm

The optimal problem (19) has shown that we have to maximize the sum rate of CUs while make the CUs are transparent to the primary user when it accesses the primary spectrum. Then a good choice of minimizing interference is providing a good zero forcing method to make the CU as a special user and pre-cancel its interference to the primary user firstly. Xu and Lin (2006) present game theory to handle the resource allocating problem between cognitive users to achieve maximum rate. Hamdi, Zhang, and Ben Letaief (2007) pointed out that the beamforming algorithm has been also exploited as a strategy that can serve many users at similar throughput but with lower complexity than dirty Dirty Paper Coding (DPC). Yet DPC is difficult to implement in practical systems due to its high complexity in successive encodings and decodings, especially when there

are too much users in the system. Now we check the optimal problem in Equation (19) again. Power P_p can be modulated with transmitting signals, and $SINR_{cj}$ can be enlarged by increasing cognitive transmitting power P_{cj} or decreasing primary power P_p. But increasing cognitive transmitting power P_{cj} or decreasing primary power P_p will be a contradiction to the second constrain in (19) and to the main maximal problem. There must be a balance point within this question and it is not so difficult to work out the maximal problem is non-convex. Then we choose a design for the purpose of deriving the feasible solution to the question.

Given that Δ_p is the threshold to make sure orthogonality of the cognitive channel and the primary channel which determined by the cognitive base station and the primary base station, and Δ_c the threshold to make sure orthogonality of the cognitive channels which only determined by the cognitive base station.

Δ_p Constrains to Make Sure Orthogonality Between Cognitive Channels and the Primary Channel

1. Initializations: $S_0 = \phi$,
2. For i from 1 to K, calculate $\Delta(H_P, H_{ci})$,

If $\Delta(H_P, H_{ci}) < \Delta_p$

$S_0 = S_0 \cup i$;

end

end

3. For i from 1 to K, calculate $\left| H_{ci} H_{ci}^H \right|$ and save it to the array $T[K]$; sort $T[K]$ from large to small.

Δ_c **Constrains to Make Sure Orthogonality of Cognitive Channels**

1.   Initializations: $i = 1$; $S = T_0$;

while $|S| < N_t$ ($|S|$,the number of element in set S)

$i = i + 1$;

If $\Delta(H_{T_0(i)}, H_{cj}) \leq \delta_C$, $\forall j \in \bar{S}$

$S = S \cup T_0(i)$

        end
end

CR-BD Algorithm to Diminish Co-Channel Interference Between CUs

In set S, define a matrix of
$\tilde{H}_{ci} = [H_{c1}^T ... H_{ci-1}^T, H_{ci+1}^T ... H_{cN_c}^T]^T$, which is the channel matrix for all other users except user i . Let $\tilde{L}_{ci} = rank(\tilde{H}_{ci})$, calculate singular value decomposition (SVD) of $\widetilde{H}_{ci}$:

$$\tilde{H}_{ci} = \tilde{U}_{ci} \tilde{\Sigma}_{ci} [\tilde{V}_{ci}^{(1)} \ \tilde{V}_{ci}^{(0)}]^H \qquad (20)$$

where the last right $N_t - \tilde{L}_{ci}$ singular vectors $\tilde{V}_{ci}^{(0)}$ forms an orthogonal basis for the null space of $\widetilde{H}_{ci}$.

Calculate $H_{ci} \tilde{V}_{ci}^{(0)}$ and let $L_{ci} = rank(H_{ci} \tilde{V}_{ci}^{(0)})$, then calculate the SVD:

$$H_{ci} \tilde{V}_{ci}^{(0)} = U_{ci} \begin{bmatrix} \Lambda_{ci}^{\frac{1}{2}} & 0 \\ 0 & 0 \end{bmatrix} [V_{ci}^{(1)} \ V_{ci}^{(0)}] \qquad (21)$$

where $V_{ci}^{(1)}$ holding the first L_{ci} , forms an orthogonal basis for the null space of $H_{ci} \tilde{V}_{ci}^{(0)}$.

When P_{cj} is the power loading matrix obtained through water-filling algorithm under total transmitted power constraint, we can get:

$$W = [\tilde{V}_{c1}^{(0)} V_{c1}^{(1)} \ \tilde{V}_{c2}^{(0)} V_{c2}^{(1)} ... \tilde{V}_{cN_t}^{(0)} V_{cN_t}^{(1)}] P^{\frac{1}{2}} \qquad (22)$$

where $P = [P_{c1}, P_{c2} ... P_{cN_c}]$.

Substituting (22) in (19), we can get the sum-rate of the selected cognitive users. One author of this paper Liu ever worked out one near orthogonality in cognitive mesh-network, whose achievements mainly focus on both the cognitive receivers and multiple primary receivers equipping with single antenna (Liu, Wang, Li, & Li, in press). Being different from the previous articles, we work out a new near orthogonality scheme for both the cognitive receivers and multiple primary receivers equipping with multiple antennas.

SIMULATION RESULTS

In this section, the simulation results of our proposed algorithm are presented. In order to evaluate the performance of cognitive user selection algorithm combined with CR-BD algorithm, the sum-rate performance is considered. In the simulation, we set $\gamma = 1$, $\varsigma_p = 8dB$.

In Figure 6, the sum rates of cognitive users under the proposed algorithm, averaged over the channel distributions, as a function of the number of users K . The conclusion from Figure 6 is that the sum rates of cgnitive users becomes larger via the number of cognitive users can be chosen with K and the number of cognitive users has been chosen with N_c . For a large number of cognitive users, the chance to have the best users selected gets higher so the sum-rate increases. And the sum-rate increases because the number of cognitive users exploits the multi-user diversity.

Figure 6. Sum rates of cognitive users under the number of cognitive users K

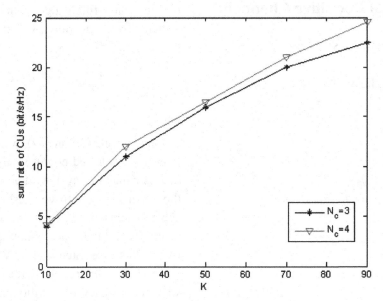

Figure 7. Sum rates of cognitive users via different power allocation schemes

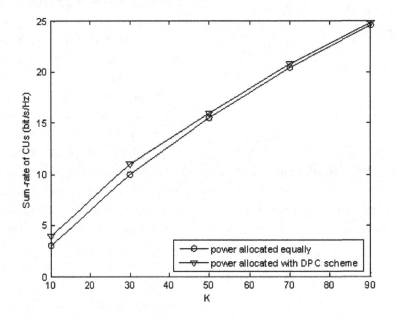

Figure 7 is plotted as sum rates via DPC power allocation method and using equal power allocation. The figure depicts simulation results using both schemes of power loading are similar in terms of throughput. Though DPC power allocation scheme has much better system performance than the equal power allocation, it is much complex to be employed in a real system.

CONCLUSION

In this paper, the authors investigate two cognitive MIMO systems with multiple cognitive transmitters using multiple transmission antennas: one is distributed and the other one is centrally-controlled. Each near orthogonal algorithm for proper cognitive user selection to the fixed PU is proposed in disparate network structure, which can reduce the co-channel interference from cognitive link to the primary link. To decrease co-channel interference further, the writers here have studied the possible schemes for power allocation to cognitive users per antenna in the aim of minishing the co-channel interference in the distributed cognitive radio network based on the analysis of the end-to-end ratio SINR studied in simple closed form and the outage probability are extracted in two scenarios. Based on these, the stable throughput, the influences of detection error and the probability of primary link idle are estimated while stability of the cognitive MIMO system is analyzed. And also, authors in this paper learn CR-BD algorithm in centrally-controlled cognitive radio network to achieve sum-rate performance and lightening Multi-User Interference (MUI) between cognitive users. These results can be used to study important performance criteria, and shed lights upon the application of MIMO techniques used in CR network.

ACKNOWLEDGMENT

This work is supported by National Basic Research Program of China (No.2009CB320401), National Natural Science Foundation of China (No. 60772108 and 61072055), National High Technology Research and Development Program of China (No.2009AA011501-2), and Major National Science and Technology Programs (No.2009ZX03007-001).

REFERENCES

Bayesteh, A., & Khandani, A. K. (2005). On the user selection for MIMO broadcast channels. In *Proceedings of the IEEE International Symposium on Information Theory* (pp. 2325-2329).

Chen, P., & Chang, J.-F. (1991). An improved transmission protocol for two interfering queues in packet radio networks. *IEEE Transactions on Wireless Communications*, *39*(3), 353–355.

Devroye, N. (2006). Achievable rates in cognitive radio channels. *IEEE Transactions on Information Theory*, *52*(5), 1813–1827. doi:10.1109/TIT.2006.872971

Gesbert, D., Shafi, M., Shiu, D., Smith, P. J., & Naguib, A. (2003). From theory to practice: An overview of MIMO space-time coded wireless systems. *IEEE Journal on Selected Areas in Communications*, *21*(3), 281–302. doi:10.1109/JSAC.2003.809458

Gradshteyn, I. S., & Ryzhik, I. M. (2007). *Table of integrals, series, and products*. New York, NY: Academic Press.

Hamdi, K., Zhang, W., & Ben Letaief, K. (2007). Joint beamforming and scheduling in cognitive radio networks. In *Proceedings of the IEEE Global Telecommunications Conference* (pp. 2977-2981).

Haykin, S. (2005). Cognitive radio: Brain-empowered wireless communications. *IEEE Journal on Selected Areas in Communications, 23*(2), 201–220. doi:10.1109/JSAC.2004.839380

Liu, F., Wang, D., Li, C., & Li, M. (in press). Zero forcing algorithms for a multi-user cognitive wireless communication system. In *International Conference on Future Computer and Communication*

Loynes, R. M. (1962). The stability of a queue with non-independent interarrival and service times. *Mathematical Proceedings of the Cambridge Philosophical Society, 58*, 494–520. doi:10.1017/S0305004100036781

Luo, W., & Ephremides, A. (1999). Stability of N interacting queues in random-access systems. *IEEE Transactions on Information Theory, 45*(5), 1579–1587. doi:10.1109/18.771161

Magnus, W., & Oberhettinger, F. (1949). Hypergeometric functions . In Abramowitz, M., & Stegun, I. A. (Eds.), *Handbook of mathematical functions with formulas, graphs, and mathematical tables: Hypergeometric functions*. New York, NY: Dover.

Mitola, J. III, & Maguire, G. Q. Jr. (1999). Cognitive radio: Making software radios more personal. *IEEE Personal Communications Magazine, 6*(4), 13–18. doi:10.1109/98.788210

Nain, P. (1985). Analysis of a two node ALOHA network with infinite capacity buffers. In *Proceedings of the International Conference of Computer Networking and Performance Evaluation* (pp. 49-63).

Naware, V., Mergen, G., & Tong, L. (2005). Stability and delay of finite-user slotted ALOHA with multi-packet reception. *IEEE Transactions on Information Theory, 51*(7), 2636–2656. doi:10.1109/TIT.2005.850060

Rao, R., & Ephremides, A. (1988). On the stability of interacting queues in multi-access system. *IEEE Transactions on Information Theory, 34*(5), 918–930. doi:10.1109/18.21216

Sadek, A. K., Liu, K. J. R., & Ephremides, A. (2007). Cognitive multiple access via cooperation: Protocol design and performance analysis. *IEEE Transactions on Information Theory, 53*(10), 3677–3696.

Scutari, G., Palomar, D. P., & Barbarossa, S. (2008). Cognitive MIMO radio. *IEEE Signal Processing Magazine, 25*(6), 46–59. doi:10.1109/MSP.2008.929297

Sidi, M., & Segall, A. (1983). Two interfering queues in packet-radio networks. *IEEE Transactions on Wireless Communications, 31*(1), 123–129.

Simeone, O., Bar-Ness, Y., & Spagnolini, U. (2007). Stable throughput of cognitive radios with and without relaying capability. *IEEE Transactions on Wireless Communications, 55*(12), 2351–2360.

Simeone, O., Gambini, J., Bar-Ness, Y., & Spagnolini, U. (2007). Cooperation and cognitive radio in Peter McLane. In *Proceedings of the IEEE International Conference on Communications* (pp. 6511-6515).

Simon, M. K., & Alouini, M.-S. (2005). *Digital communication over fading channels*. New York, NY: John Wiley & Sons.

Szpankowski, W. (1988). Stability conditions for multidimensional queueing systems with computer applications. *Operations Research, 36*(6), 944–957. doi:10.1287/opre.36.6.944

Tsybakov, B. S., & Mikhailov, V. A. (1979). Ergodicity of a slotted ALOHA system. *Problems of Information Transmission, 15*(4), 73–87.

Wang, Y. (2003). On cognitive informatics: Brain and mind. *Transdisciplinary Journal of Neuroscience and Neurophilosophy, 4*(2), 151–167.

Wang, Y. (2009). On cognitive computing. *International Journal of Software Science and Computational Intelligence, 1*(3), 1–15. doi:10.4018/jssci.2009070101

Xu, M. G., & Lin, D. (2006). Non-orthogonal precoding matrix design for MU-MIMO downlink channels. In *Proceedings of the IEEE Wireless Communications and Networking Conference* (pp. 1311-1315).

Yoo, T., & Goldsmith, A. (2006). On the optimality of multi-antenna broadcast scheduling using zero-forcing beamforming. *IEEE Journal on Selected Areas in Communications, 24*(3), 528–541. doi:10.1109/JSAC.2005.862421

Zhang, L., Liang, Y.-C., & Xin, Y. (2008). Joint beamforming and power allocation for multiple access channels in cognitive radio networks. *IEEE Journal on Selected Areas in Communications, 26*(1), 38–51. doi:10.1109/JSAC.2008.080105

Zhao, Q., & Sadler, B. (2007). A survey of dynamic spectrum access. *IEEE Signal Processing Magazine, 24*(3), 79–89. doi:10.1109/MSP.2007.361604

APPENDIX

Derivations of Equations (10) and (11)

From (9), set $\theta_j = \dfrac{n_p + \Omega}{\gamma_{pj} P_{pj}}$, if $\theta = \theta_m = \theta_n$, $\Omega = \sum_{j=1}^{N} \gamma_{cpj_s} \left| h_{cpj_s}(t) \right|^2 P_{cj_s}$, $\hat{\theta} = \dfrac{1}{\gamma_{pj} P_{pj}}$, $\forall m \neq n, m, n \in (1, N)$,

$$
P_{out,P2} = \Pr(\sum_{j=1}^{N} \frac{\gamma_{pj} \left| h_{pj}(t) \right|^2 P_{pj}}{(n_p + \Omega)} < \beta_p)
$$

$$
= \frac{1}{(M-1)!} \times \int_0^\infty [1 - \frac{\theta^N}{(N-1)!} \times \exp(-\theta \beta_p) \sum_{i=0}^{N-1} \frac{(N-1)! \beta_p^i}{i! \theta^{N-i}}] \hat{\theta} \exp(-\hat{\theta}\Omega)(\hat{\theta}\Omega)^{M-1} d\Omega
$$

$$
= \frac{\Gamma(M)}{(M-1)!} - \frac{\exp(\frac{n_p \hat{\theta} - \beta_p n_p \hat{\theta}}{2})\Gamma(n_p \hat{\theta})}{(M-1)!} \sum_{i=0}^{N-1} \frac{\beta_p^i}{i!} (\beta_p + 1)^{-\frac{M+i+1}{2}} (n_p \hat{\theta})^{\frac{M+i-1}{2}} W_{\frac{i+1-M}{2}, \frac{-i-M}{2}}(n_p \hat{\theta}(\beta_p + 1))
$$

$$
\tag{18}
$$

Otherwise, set $W = \sum_{k=1}^{M} \gamma_{cpk_s} \left| h_{cpk_s}(t) \right|^2 P_{ck_s}$, with integration by parts, we get

$$
P_{out,P2} = \Pr(\sum_{j=1}^{N} \frac{\gamma_{pj} \left| h_{pj}(t) \right|^2 P_{pj}}{(n_p + W)} < \beta_p)
$$

$$
= \frac{(\tilde{\tau}-1)!(\tilde{\alpha}-1)!}{\Gamma(\tilde{\tau})\Gamma(\tilde{\alpha})} - \int_0^\infty \frac{\gamma(\tilde{\tau}, \tilde{\varsigma}W)}{\Gamma(\tilde{\tau})} \frac{\tilde{\theta}^{\tilde{\alpha}}}{\Gamma(\tilde{\alpha})} W^{\tilde{\alpha}-1} \exp(-\tilde{\theta}W) dW \tag{19}
$$

$$
= \frac{(\tilde{\tau}-1)!(\tilde{\alpha}-1)!}{\Gamma(\tilde{\tau})\Gamma(\tilde{\alpha})} - \frac{\tilde{\theta}^{\tilde{\alpha}}}{\Gamma(\tilde{\tau})\Gamma(\tilde{\alpha})} \times \frac{\tilde{\varsigma}^{\tilde{\tau}}\Gamma(\tilde{\tau}+\tilde{\alpha})}{\tilde{\alpha}(\tilde{\varsigma}+\tilde{\theta})^{\tilde{\tau}+\tilde{\alpha}}} {}_2F_1(1, \tilde{\tau}+\tilde{\alpha}; \tilde{\tau}+1; \frac{\tilde{\varsigma}}{\tilde{\varsigma}+\tilde{\theta}})
$$

This work was previously published in the International Journal of Cognitive Informatics and Natural Intelligence, Volume 5, Issue 2, edited by Yingxu Wang, pp. 58-79, copyright 2011 by IGI Publishing (an imprint of IGI Global).

Chapter 24
Fuzzy Neural Network Control for Robot Manipulator Directly Driven by Switched Reluctance Motor

Baoming Ge
Beijing Jiaotong University, China, & Michigan State University, USA

Aníbal T. de Almeida
University of Coimbra, Portugal

ABSTRACT

Applications of switched reluctance motor (SRM) to direct drive robot are increasingly popular because of its valuable advantages. However, the greatest potential defect is its torque ripple owing to the significant nonlinearities. In this paper, a fuzzy neural network (FNN) is applied to control the SRM torque at the goal of the torque-ripple minimization. The desired current provided by FNN model compensates the nonlinearities and uncertainties of SRM. On the basis of FNN-based current closed-loop system, the trajectory tracking controller is designed by using the dynamic model of the manipulator, where the torque control method cancels the nonlinearities and cross-coupling terms. A single link robot manipulator directly driven by a four-phase 8/6-pole SRM operates in a sinusoidal trajectory tracking rotation. The simulated results verify the proposed control method and a fast convergence that the robot manipulator follows the desired trajectory in a 0.9-s time interval.

DOI: 10.4018/978-1-4666-2476-4.ch024

INTRODUCTION

There has recently been a considerable interest in developing efficient direct drive robot manipulator, because the elimination of gear boxes simplifies the construction of manipulators and removes sources of flexibility and nonlinearities, such as friction and backlash, which are known to cause difficulties in designing high-quality controls. Many motors, like conventional dc motor (Mohamed et al., 2004), brushless dc motor (Park et al., 2003), induction motor (Hsu et al., 2005), and switched reluctance motor (SRM) (Wallace et al., 1991; Spong et al., 1987; Chen et al., 2003; Amor et al., 1993; Bortoff et al., 1998; Milman et al., 1999), have been investigated in applications to direct drive manipulator, where the SRM is increasingly popular due to its simple structure, low cost and reliability in harsh environments. Moreover, the SRM can produce high torque at low speed, which is compatible with direct drive manipulator specifications.

SRM model, however, exhibits significant coupled nonlinear, multivariable, and uncertainty. Hence the classical linear control schemes cannot provide the required performances for high-precision position control.

The latest advances and engineering applications of cognitive informatics have gotten blooming achievements (Wang et al., 2010; Wang, 2009a, 2009b; Zhong, 2008). Artificial neural networks have been adopted extensively due to their abilities to achieve nonlinear mappings and fast autonomous learning, in a wide variety of domains, from modeling and simulation (Cai et al., 2011), rotor position estimation (Beno et al., 2011), classification of musical chords (Yaremchuk et al., 2008), sensorless control for a switched reluctance wind generator (Echenique et al., 2009), to robot control (Bu et al., 2009; Bugeja et al., 2009; Dierks et al., 2009, 2010; Ferreira et al., 2009; Hong et al., 2009; Hou et al., 2010; Tan et al., 2009; Wai et al., 2010; Wei et al., 2009; Zhao et al., 2009).

It can be seen that the neural network provides an effective approach to wide range of complex practical issues. Of course, the SRM control gets a big benefit of the neural network. Nevertheless, the conventional neural network may produce a degraded control performances when the existence of uncertainties. The fuzzy neural network (FNN) will overcome this disadvantage, because it possesses the merits of both fuzzy systems (e.g., humanlike rules thinking and ease of incorporating expert knowledge) and neural networks (e.g., learning and optimization abilities, and connectionist structures) (Wai, 2002). In this way, one can bring the low-level learning and computational power of neural networks into fuzzy systems and also high-level, humanlike rule thinking and reasoning of fuzzy systems into neural networks. Thus, the FNN-based control techniques have represented an alternative method to deal with uncertainties of the control system in recent years (Hu et al., 2006; Liu et al., 2006; Lachman et al., 2004). Reference Liu et al. (2006) employed a fuzzy logic-based neural network to establish the dynamic inversion of biped robots to compensate the complex nonlinearity. The neuro-fuzzy model of SRM in Lachman et al. (2004) used the adaptive neuro-fuzzy inference system techniques, which provided a method for the fuzzy modeling procedure to learn information about a data set, in order to compute the membership function parameters that best allow the associates fuzzy inference system to track the given input/output data. Reference Hu et al. (2006) combined the sliding mode control and the neural-network control with different weights, which were determined by a fuzzy supervisory controller, in its application to two-link manipulator with the presence of both structured and unstructured uncertainties.

The contribution of this paper is to present a FNN control method for robot manipulator directly driven by an SRM. The organization of the paper is as follows. In *Section Robot Manipulator*, we establish the model of robot manipulator directly driven by an SRM. In *Section FNN-based Torque*

Control, the torque control system based on FNN is designed, and the FNN model is proposed. In *Section Trajectory Tracking Control*, we design a trajectory tracking controller on the basis of the FNN torque control system. Finally, the simulated result presents that high precision trajectory tracking is fulfilled.

ROBOT MANIPULATOR

A single link robot manipulator is shown in Figure 1 (Spong et al., 1987), where Spong et al. controlled it by using a nonlinear state feedback linearizing control, and the accurate full state measurements (position, velocity, acceleration, stator currents) were required. Here we control it using FNN method, and the dynamics of the system can be modeled by

$$\frac{d\theta}{dt} = \omega, \tag{1}$$

$$\frac{d\omega}{dt} = \frac{1}{J}\left(T_e - T_l\right), \tag{2}$$

where θ denotes the angular position, ω is the angular speed, T_e is the torque supplied by SRM, J is called as the manipulator inertia, and the load torque written as

$$T_l = Mgl\sin\theta, \tag{3}$$

where M is the mass of manipulator, l is the distance from the joint to the center of the gravity of the link, and g is the acceleration of gravity.

For a typical m-phase SRM, its electric dynamic equations are given by

$$u_j = ri_j + \frac{d\psi_j}{dt}, \tag{4}$$

Figure 1. Single link direct drive manipulator

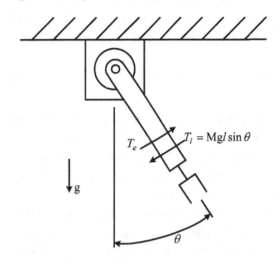

$$T_j(\theta, i_j) = \frac{d}{d\theta}\int_0^{i_j} \psi_j(\theta, i_j)di_j, \tag{5}$$

$$T_e = \sum_{j=1}^{m} T_j(\theta, i_j), \tag{6}$$

where u_j is the voltage applied to the stator terminals of phase j, i_j is the stator current of phase j, ψ_j is the flux linkage of phase j, r is the stator winding resistance, T_j is the torque produced by phase j. In general, ψ_j is a nonlinear function of θ and i_j due to magnetic saturation. As a result that T_j also is a nonlinear function of θ and i_j.

FNN-BASED TORQUE CONTROL

FNN-Based Inverse Model of Torque

From (6), the total torque of an SRM is

$$T_e = \sum_{j=1}^{m} f_{\text{phase}}(\theta_j, \ i_j), \tag{7}$$

$$\theta_j = \theta - (j-1)\frac{2\pi}{mN_r}, \tag{8}$$

where N_r is the number of rotor poles, $f_{\text{phase}}(\theta_j, i_j)$ represents an unknown nonlinear function of phase torque, then its inverse model is written as

$$i_j = f_{\text{phase}}^{-1}(\theta_j, T_j). \tag{9}$$

The proposed FNN used to determine the nonlinear inverse model of torque combines the learning ability of conventional neural network and fuzzy reasoning functions. As shown in Figure 2, the input vectors of two nodes are the rotor position and the phase torque. One node of output is the phase current. The signal propagation and the basic function in each layer of the FNN are introduced as follows:

*Input Layer*s 1 and 2: For every node i in this layer, the net input and the net output are represented as

$$net_i^k = x_i^1, \tag{10}$$

where x_i^1 is the ith input to the node of input layer.

Membership Layer: In this layer each node performs a membership function. The Gaussian function is adopted as the membership function. For the ith node, there are

$$net_{ij}^2 = -\frac{\left(y_i^1 - m_{ij}\right)^2}{\left(\sigma_{ij}\right)^2},$$

$$y_{ij}^2 = f_{ij}^2(net_{ij}^2) = \exp(net_{ij}^2), \; i = 1, \; 2,$$

$$j = 1, \; \cdots, \; p, \text{ or } j = 1, \; \cdots, \; q, \tag{11}$$

where m_{ij} and σ_{ij} are, respectively, the mean and the standard deviation of the Gaussian function in the jth term of the ith input variable y_i^1 to the node of membership layer, p is the number of the membership function with respect to the phase torque, and q corresponds to the rotor position.

Rule Layer: Each node i in this layer represents a rule in the rule base. The node itself performs the fuzzy AND operation, which is a minimum operation to obtain the degree of applicability of the rules. For the ith rule node

$$net_i^3 = \min\left\{y_{1j}^2, \; y_{2k}^2\right\}, \; j = 1, \; \cdots, \; p,$$
$$k = 1, \; \cdots, \; q,$$

$$y_i^3 = f_i^3\left(net_i^3\right) = net_i^3, \tag{12}$$

where m represents the number of rules, and $m = pq$.

Output Layer 1: Each node i of this layer performs the normalization computation of firing strengths, it respectively is the connecting weights w_{i1}^6 between the rule consequence layer and the output layer 2, namely

$$net_i^4 = \frac{y_i^3}{\sum_{j=1}^{m} y_j^3}, \; y_i^4 = f_i^4\left(net_i^4\right) = net_i^4,$$

$$w_{i1}^6 = y_i^4, \tag{13}$$

Rule Consequence Layer: Each node i in this layer represents a rule, finishes a calculation of consequent result of rules, which is given by

$$net_i^6 = w_{1i}^5 + w_{2i}^5 y_2^5 + w_{3i}^5 y_3^5,$$

$$y_i^6 = f_i^6\left(net_i^6\right) = net_i^6, \; i = 1, \; \cdots, \; m. \tag{14}$$

Output Layer 2: The single node in this layer is labeled as Sum, which computes the overall output as the summation of all input signals

$$net^o = \sum_{k=1}^{m} w_{k1}^6 y_k^6, \; I_{\text{ph}} = f^o\left(net^o\right) = net^o. \tag{15}$$

Figure 2. Proposed fuzzy-neural network to control SRM

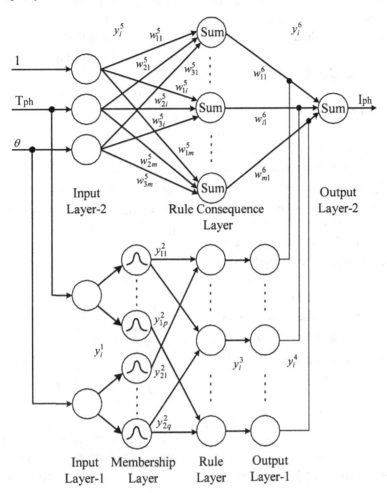

It can be seen that the proposed FNN realizes the Takagi-Sugeno fuzzy systems model, and the back-propagation algorithm is useful during the learning process due to its feed-forward model structure. The parameters m_{ij}, σ_{ij}, and the connecting weights w_{ij}^5 should be updated to minimize the cost function

$$J = 1/2\sum_{i=1}^{L}(d_i - \mathrm{I}_{\mathrm{ph}})^2, \qquad (16)$$

where d_i represents sample output, L is the node number of output layer, and I_{ph} is output of the output layer 2.

For the sake of demonstration, we present the learning algorithm of connection weights w_{ij}^5 by employing the gradient descent method as following

$$\Delta w_{ij}^5 = -\eta_1 \frac{\partial J}{\partial w_{ij}^5}, \qquad (17)$$

where η_1 is learning rates given by small positive constant.

The learning algorithm to update w_{ij}^5 is deduced as

$$w_{ij}^5(t+1) = w_{ij}^5(t) + \eta_1(d_i - \mathrm{I}_{\mathrm{ph}})w_{j1}^6 x_i^1. \qquad (18)$$

Without loss of generality, the learning algorithms of parameters m_{ij} and σ_{ij} can easily be obtained by using similar method.

FNN-Based Torque Control

FNN-based torque control system for SRM is shown in Figure 3, in which FNN model to determine the nonlinear inverse model of torque calculates the desired value of phase current that produces a desired phase torque. The torque share function (TSF) is elaborately designed (Sahoo et al., 2001), and the current closed-loop controller with hysteresis band is used to regulate the actual phase current tracking the desired current waveform.

The TSF is designed as Equation (19) presented in Box 1.where $\theta_{r0\text{-}j}$ is the initial position of phase-*j*, $\theta_{r0\text{-}j} \in [0, \quad 60°]$, and $\theta_{r0\text{-}(j+1)} = \theta_{r0\text{-}j} + 15°$. Figure 4 presents the waveforms of TSF in (19). If the total demand torque is T_d, there is

$$T_{d\text{-}j} = T_d \cdot T_{\text{TSF-}j}, \quad j = 1, \quad 2, \quad 3, \quad 4 \qquad (20)$$

TRAJECTORY TRACKING CONTROL

The control problem is to find appropriate torque to servo the single link manipulator in real time in order to track a desired time-based trajectory as closely as possible. The drive motor torque required to servo the robotic manipulator is based on a dynamic model of the manipulator. A useful control method that will cancel all nonlinearities and cross-coupling terms is described as

$$T_e = J\ddot{\theta}^* + Mgl\sin(\theta), \qquad (21)$$

$$\ddot{\theta}^* = \ddot{\theta}_d + K_d\dot{e} + K_pe + K_i\int_0^t edt, \qquad (22)$$

where $\ddot{\theta}_d$ denotes the reference of acceleration, we define θ_d as the desired trajectory, then the tracking error is

$$e = \theta_d - \theta. \qquad (23)$$

When the accurate model is available, the equation of the error satisfies the following linear equation

$$\ddot{e} + K_d\dot{e} + K_pe + K_i\int_0^t edt = 0, \qquad (24)$$

where the constant positive gains K_d, K_p, and K_i are selected according to well-known linear techniques, and $K_dK_p>K_i$ is necessary. The integral error feedback term is included to eliminate steady-state position error which is due to imperfect nonlinearity compensation. The block diagram of the proposed control system is shown in Figure 5.

Figure 3. FNN-based torque control

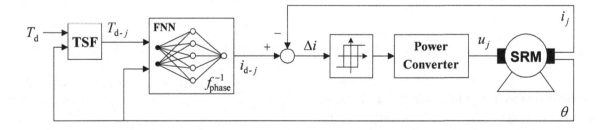

Box 1.

$$
T_{\text{TSF-j}} = \begin{cases}
C(\theta - \theta_{\text{r0-j}})^2 + D(\theta - \theta_{\text{r0-j}})^3, \theta \in [\theta_{\text{r0-j}}, \theta_{\text{r0-j}} + 10°] \\
1 \qquad\qquad\qquad\qquad\quad ,\theta \in (\theta_{\text{r0-j}} + 10°, \theta_{\text{r0-j}} + 15°) \\
A - C(\theta - 15° - \theta_{\text{r0-j}})^2 - D(\theta - 15° - \theta_{\text{r0-j}})^3, \theta \in [\theta_{\text{r0-j}} + 15°, \theta_{\text{r0-j}} + 25°] \\
0 \qquad , \quad \text{others}
\end{cases}
$$

$$
A = 1, C = \frac{3}{\left(\dfrac{\pi}{18}\right)^2}, D = \frac{-2}{\left(\dfrac{\pi}{18}\right)^3}, j = 1, \quad 2, \quad 3, \quad 4 . \tag{19}
$$

Figure 4. TSF waveforms

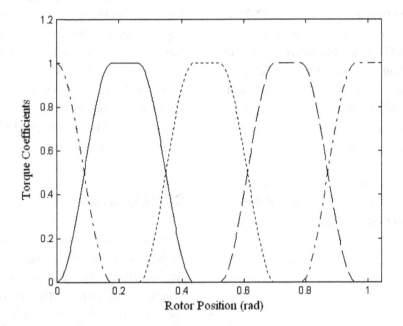

Figure 5. Trajectory tracking control

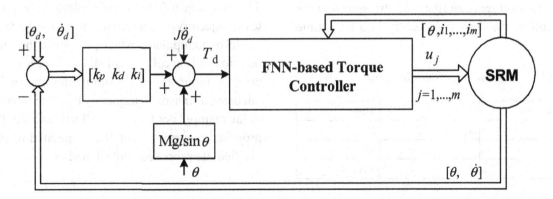

SIMULATED RESULTS

In order to evaluate the performances and feasibility of the proposed control technique presented in this paper, a single link robot manipulator directly driven by a four-phase 8/6- pole SRM is considered, as shown in Figure 1. The physical parameters of manipulator used in this simulation are listed in Table 1. The membership layer of FNN-based inverse model of torque includes 14 nodes, where the number of nodes related to position is 6, and six nodes are related to torque, that is, p=q=6. Therefore, the rule layer contains 36 nodes. Usually, some heuristics can be used to roughly initialize the parameters of the FNN for practical applications, e.g., the mean and standard deviation of the Gaussian functions can be determined according to the maximum variation of torque and position. The effect of the inaccurate selection of the initialized parameters can be retrieved by the on-line training methodology. The connecting weights are initialized to the uniform random number.

The control task used in this simulation is a sinusoidal trajectory tracking rotation of the manipulator, and the desired trajectory is expressed as

$$\theta_d = \frac{\pi}{2}\sin(\pi t), \qquad (25)$$

which shows a maximum range of 90° rotation. The simulated control results are shown in Figures 6 and 7, in which the proposed controller ensures a fast convergence and the robot manipulator tries to follow the desired trajectory in a 0.9-s time

Table 1. Parameters of manipulator

Parameters	Value	Unites
M	5	kg
l	0.5	m
J	1	Kgm^2

interval. A zero steady state error is obtained owing to the control integrator and the high-quality FNN-based torque controller.

Spong et al. (1987) published a nonlinear state feedback linearizing control method applied to control single link direct drive manipulator shown in Figure 1. It was based on an analytical SRM model, and required the accurate full state measurements (position, velocity, acceleration, stator currents). In fact, the SRM presents significant coupled nonlinear, multivariable, and uncertainty, and it is impossible to build an accurate SRM analytical model in a simple expression. For an un-accurate model, the nonlinear state feedback linearizing control method will greatly degrade its performances, for example, which causes high torque-ripple, even unstable operation. This paper built the SRM model using FNN, and an accurate model is easily available; also the built model can well deal with the uncertainty, which ensures the quality performance, and better position tracking than the previous method.

CONCLUSION

This paper presented a successful control method for robot manipulator directly driven by an SRM. An effective FNN-based inverse model of torque was proposed in application to the current closed-loop control system, and perfectly compensated the nonlinearities and uncertainties of SRM. The learning algorithms of the proposed FNN were obtained by employing the gradient descent method. The computed torque control method was applied to the trajectory tracking control using the dynamic model of manipulator, which performed the computation and cancellation of all nonlinearities and cross-coupling terms. The simulation was carried out using a sinusoidal trajectory tracking rotation of the manipulator to test the effectiveness of the proposed control system. The simulated results verified the proposed control method.

Figure 6. Trajectory tracking result

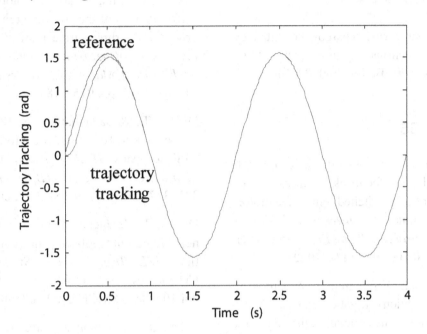

Figure 7. Position error

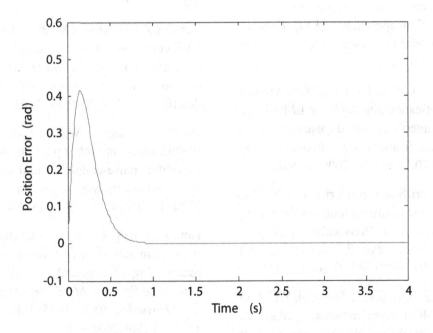

ACKNOWLEDGMENT

This work was supported in part by the State Key Lab. of Power System under grant SKLD09KM03, Tsinghua University, Beijing 100084, China.

REFERENCES

Amor, L., Dessaint, L., Akhrif, O., & Olivier, G. (1993). Adaptive feedback linearization for position control of a switched reluctance motor: analysis and simulation. *International Journal of Adaptive Control and Signal Processing*, 7(2), 117–136. doi:10.1002/acs.4480070205

Beno, M., Rajaji, L., Varatharaju, V., & Santos, A. (2011). Rotor position estimation of 6/4 switched reluctance motor using a novel neural network algorithm. In *Proceedings of the IEEE GCC Conference and Exhibition*, Dubai, United Arab Emirates (pp. 77-80).

Bortoff, S., Kohan, R., & Milman, R. (1998). Adaptive control of variable reluctance motors: a spline function approach. *IEEE Transactions on Industrial Electronics*, 45(3), 433–444. doi:10.1109/41.679001

Bu, N., Okamoto, M., & Tsuji, T. (2009). A hybrid motion classification approach for EMG-based human–robot interfaces using Bayesian and neural networks. *IEEE Transactions on Robotics*, 25(3), 502–511. doi:10.1109/TRO.2009.2019782

Bugeja, M., Fabri, S., & Camilleri, L. (2009). Dual adaptive dynamic control of mobile robots using neural networks. *IEEE Transactions on Systems, Man, and Cybernetics. Part B, Cybernetics*, 39(1), 129–141. doi:10.1109/TSMCB.2008.2002851

Cai, J., Deng, Z., Qi, R., Liu, Z., & Cai, Y. (2011). A novel BVC-RBF neural network based system simulation model for switched reluctance motor. *IEEE Transactions on Magnetics*, 47(4), 830–838. doi:10.1109/TMAG.2011.2105273

Chen, H., & Zhu, Y. (2003). Rotor position sensorless control of the switched reluctance motor drive for the direct-drive joint of the robot. In *Proceedings of the Sixth International Conference on Electrical Machines and Systems*, Beijing, China (Vol. 2, pp. 915-918).

Dierks, T., & Jagannathan, S. (2009). Neural network control of mobile robot formations using RISE feedback. *IEEE Transactions on Systems, Man, and Cybernetics. Part B, Cybernetics*, 39(2), 332–347. doi:10.1109/TSMCB.2008.2005122

Dierks, T., & Jagannathan, S. (2010). Neural network output feedback control of robot formations. *IEEE Transactions on Systems, Man, and Cybernetics. Part B, Cybernetics*, 40(2), 383–399. doi:10.1109/TSMCB.2009.2025508

Echenique, E., Dixon, J., Cárdenas, R., & Peña, R. (2009). Sensorless control for a switched reluctance wind generator based on current slopes and neural networks. *IEEE Transactions on Industrial Electronics*, 56(3), 817–825. doi:10.1109/TIE.2008.2005940

Ferreira, J., Crisóstomo, M., & Coimbra, A. (2009). SVR versus neural-fuzzy network controllers for the sagittal balance of a biped robot. *IEEE Transactions on Neural Networks*, 20(12), 1885–1897. doi:10.1109/TNN.2009.2032183

Hong, Q., Yang, S., Williams, A., & Zhang, Y. (2009). Real-time robot path planning based on a modified pulse-coupled neural network model. *IEEE Transactions on Neural Networks*, 20(11), 1724–1739. doi:10.1109/TNN.2009.2029858

Hou, Z., Cheng, L., & Tan, M. (2010). Multicriteria optimization for coordination of redundant robots using a dual neural network. *IEEE Transactions on Systems, Man, and Cybernetics. Part B, Cybernetics*, 40(4), 1075–1087. doi:10.1109/TSMCB.2009.2034073

Hsu, S., & Fu, L. (2005). Adaptive decentralized control of robot manipulators driven by current-fed induction motors. *IEEE/ASME Transactions on Mechatronics, 10*(4), 465–468. doi:10.1109/TMECH.2005.852453

Hu, H., & Woo, P. (2006). Fuzzy supervisory sliding-mode and neural-network control for robotic manipulators. *IEEE Transactions on Industrial Electronics, 53*(3), 929–940. doi:10.1109/TIE.2006.874261

Lachman, T., Mohamad, T., & Fong, C. (2004). Nonlinear modelling of switched reluctance motors using artificial intelligence techniques. *IEEE Electric Power Applications, 151*(1), 53–60. doi:10.1049/ip-epa:20040025

Liu, Z., & Zhang, Y. (2006). Fuzzy logic-based neural modeling and robust control for robot. In *Proceedings of the 6th World Congress on Intelligent Control and Automation*, Dalian, China (pp. 8976-8980).

Milman, R., & Bortoff, S. (1999). Observer-based adaptive control of a variable reluctance motor: experimental results. *IEEE Transactions on Control Systems Technology, 7*(5), 613–621. doi:10.1109/87.784425

Mohamed, M., Ashraf, S., & Aziz, I. (2004). Digital control of direct drive robot manipulators. In *Proceedings of the International Conference on Power System Technology*, Singapore (pp. 1810-1815).

Park, K., Kim, T., Ahn, S., & Hyun, D. (2003). Speed control of high-performance brushless DC motor drives by load torque estimation. In *Proceedings of the IEEE 34th Annual Power Electronics Specialist Conference* (Vol. 4, pp. 1677-1681).

Sahoo, N., Xu, J., & Panda, S. (2001). Low torque ripple control of switched reluctance motors using iterative learning. *IEEE Transactions on Energy Conversion, 16*(4), 318–326. doi:10.1109/60.969470

Spong, M., Marino, R., Peresada, S., & Taylor, D. (1987). Feedback linearizing control of switched reluctance motors. *IEEE Transactions on Automatic Control, 32*(5), 371–379. doi:10.1109/TAC.1987.1104616

Tan, X., Zhao, D., Yi, J., & Xu, D. (2009). Adaptive integrated control for omnidirectional mobile manipulators based on neural-network. *International Journal of Cognitive Informatics and Natural Intelligence, 3*(4), 34–53. doi:10.4018/jcini.2009062303

Wai, R. (2002). Hybrid fuzzy neural-network control for nonlinear motor-toggle servomechanism. *IEEE Transactions on Control Systems Technology, 10*(4), 519–532. doi:10.1109/TCST.2002.1014672

Wai, R., Huang, Y., Yang, Z., & Shih, C. (2010). Adaptive fuzzy-neural-network velocity sensorless control for robot manipulator position tracking. *IET Control Theory & Applications, 4*(6), 1079–1093. doi:10.1049/iet-cta.2009.0166

Wallace, R., & Taylor, D. (1991). Low-torque-ripple switched reluctance motors for direct-drive robotics. *IEEE Transactions on Robotics and Automation, 7*(6), 733–742. doi:10.1109/70.105382

Wang, Y. (2009a). On abstract intelligence: toward a unifying theory of natural, artificial, machinable, and computational intelligence. *International Journal of Software Science and Computational Intelligence, 1*(1), 1–17. doi:10.4018/jssci.2009010101

Wang, Y. (2009b). On cognitive computing. *International Journal of Software Science and Computational Intelligence*, *1*(3), 1–15. doi:10.4018/jssci.2009070101

Wang, Y., Baciu, G., Yao, Y., Kinsner, W., Chan, K., & Zhang, B. (2010). Perspectives on cognitive informatics and cognitive computing. *International Journal of Cognitive Informatics and Natural Intelligence*, *4*(1), 1–29. doi:10.4018/jcini.2010010101

Wei, R., & Liu, C. (2009). Design of dynamic petri recurrent fuzzy neural network and its application to path-tracking control of nonholonomic mobile robot. *IEEE Transactions on Industrial Electronics*, *56*(7), 2667–2683. doi:10.1109/TIE.2009.2020077

Yaremchuk, V., & Dawson, M. (2008). Artificial neural networks that classify musical chords. *International Journal of Cognitive Informatics and Natural Intelligence*, *2*(3), 22–30. doi:10.4018/jcini.2008070102

Zhao, Y., & Cheah, C. (2009). Neural network control of multifingered robot hands using visual feedback. *IEEE Transactions on Neural Networks*, *20*(5), 758–767. doi:10.1109/TNN.2008.2012127

Zhong, Y. (2008). A cognitive approach to the mechanism of intelligence. *International Journal of Cognitive Informatics and Natural Intelligence*, *2*(1), 1–16. doi:10.4018/jcini.2008010101

This work was previously published in the International Journal of Cognitive Informatics and Natural Intelligence, Volume 5, Issue 3, edited by Yingxu Wang, pp. 73-85, copyright 2011 by IGI Publishing (an imprint of IGI Global).

Compilation of References

Adams, C., & Tavares, S. (1989). Good S-boxes are easy to find. In G. Brassard (Ed.), *Proceedings of Advances in Cryptology* (LNCS 435, pp. 612-615).

Agrawal, R., & Srikant, R. (1994). Fast algorithms for mining association rules. In Proceedings of the International Conference on Very Large Data Bases, Santiago, Chile (pp. 487-499).

Ahuja, A. K., Behrend, M. R., Kuroda, M., Humayun, M. S., & Weiland, J. D. (2008). An in vitro model of a retinal prosthesis. *IEEE Transactions on Bio-Medical Engineering, 55*(6), 1744–1753. doi:10.1109/TBME.2008.919126

Al-Atabany, W. I., Memon, M. A., Downes, S. M., & Degenaar, P. A. (2010). Designing and testing scene enhancement algorithms for patients with retina degenerative disorders. *Biomedical Engineering Online, 9,* 25. doi:10.1186/1475-925X-9-27

Albus, J. S. (1991). Outline for a theory of intelligence. *IEEE Transactions on Systems, Man, and Cybernetics, 21*(3), 473–510. doi:10.1109/21.97471

Allan, R. (2009). *Survey of agent based modeling and simulation tools* (Tech. Rep. No. TR-2010-007). Warrington, UK: STFC Daresbury Laboratory.

Alpcan, T., & Başar, T. (2011). *Network security: A decision and game-theoretic approach.* Cambridge, UK: Cambridge University Press.

Álvarez, G., & Li, S. J. (2006). Some basic cryptographic requirements for chaos-based cryptosystems. *International Journal of Bifurcation and Chaos in Applied Sciences and Engineering, 16*(8), 2129–2151. doi:10.1142/S0218127406015970

Amor, L., Dessaint, L., Akhrif, O., & Olivier, G. (1993). Adaptive feedback linearization for position control of a switched reluctance motor: analysis and simulation. *International Journal of Adaptive Control and Signal Processing, 7*(2), 117–136. doi:10.1002/acs.4480070205

Angeline, P. (1998). Evolutionary optimization versus particle swarm optimization: Philosophy and performance differences. In V. W. Porto, N. Saravanan, D. Waagen, & A. E. Eiben (Eds.), *Proceedings of the 7th International Conference on Evolutionary Programming 4* (LNCS 1447, pp. 601-610).

Anghelescu, P., Ionita, S., & Sofron, E. (2008). FPGA implementation of hybrid additive programmable cellular automata encryption algorithm. In *Proceedings of the International Conference on Hybrid Intelligent Systems* (pp. 96-101).

Anghelescu, P., Sofron, E., & Ionita, S. (2007). VLSI implementation of high-speed cellular automata encryption algorithm. In *Proceedings of the International Conference on Semiconductor* (Vol. 2, pp. 509-512).

Anghelescu, P., Sofron, E., Rîncu, C. I., & Iana, V. G. (2008). Programmable cellular automata based encryption algorithm. In Proceedings of the International Conference on Semiconductor (Vol. 2, pp. 351-354).

Arasaratnan, I., & Haykin, S. (2009). Cubature Kalman Filters. *IEEE Transactions on Automatic Control, 54*(6), 1254–1269. doi:10.1109/TAC.2009.2019800

Aristotle,. (1989). *Prior analytics* (Smith, R., Trans.). Cambridge, MA: Hackett.

Arman, F., & Aggarwal, J. K. (1993). Model-based object recognition in dense-range images—a review. *ACM Computing Surveys*, *25*(1), 5–43. doi:10.1145/151254.151255

AT&T Laboratories. (2002). *ORL face datasets.* Retrieved from http://www.cl.cam.ac.uk/research/dtg/attarchive/facedatabase.html

Aziz-Zadeh, L., Wilson, S., Rizzolati, G., & Iacobani, M. (2006). Congruent embodied representations for visually presented actions and linguistic phrases describing actions. *Current Biology*, *16*, 1818–1823. doi:10.1016/j.cub.2006.07.060

Baciu, G., Yao, Y., Wang, Y., Zadeh, L. A., Chan, K., Kinsner, W., et al. (2009, June). Perspectives on cognitive informatics and cognitive computing: Summary of the panel. In *Proceedings of the 8th IEEE International Conference on Cognitive Informatics*, Kowloon, Hong Kong (pp. 9-27). Washington, DC: IEEE Computer Society.

Baddeley, A. (1990). *Human memory: Theory and practice*. Needham Heights, MA: Allyn and Bacon.

Bai, X., & Liu, W. J. (2009). Research of image retrieval based on color. In *Proceedings of the International Forum on Computer Science-Technology and Applications*, Chongqing, China (Vol. 2, pp. 283-286).

Bales, R. F. (2009). Interaction process analysis. K. Krippendorff & M. A. Bock (Eds.), *The content analysis reader* (pp. 75-83). London, UK: Sage.

Banerjee, M., Kundu, M. K., & Maji, P. (2009). Content-based image retrieval using visually significant point features. *Fuzzy Sets and Systems*, *160*(23), 3323–3341. doi:10.1016/j.fss.2009.02.024

Bansal, S., & Garg, R. (2010). A novel probabilistic approach for efficient information retrieval. *International Journal of Computers and Applications*, *9*(2), 44–48. doi:10.5120/1354-1827

Bargiela, A., & Pedrycz, W. (2003). *Granular computing: An introduction*. Dordrecht, The Netherlands: Kluwer Academic.

Bargiela, A., & Pedrycz, W. (2005). A model of granular data: a design problem with the Tchebyschev FCM. *Soft Computing*, *9*, 155–163. doi:10.1007/s00500-003-0339-2

Bargiela, A., & Pedrycz, W. (2005). Granular mappings. *IEEE Transactions on Systems, Man, and Cybernetics-Part A*, *35*(2), 292–297. doi:10.1109/TSMCA.2005.843381

Bargiela, A., & Pedrycz, W. (2008). Toward a theory of granular computing for human-centered information processing. *IEEE Transactions on Fuzzy Systems*, *16*(2), 320–330. doi:10.1109/TFUZZ.2007.905912

Bargiela, A., & Pedrycz, W. (Eds.). (2009). *Human-centric information processing through granular modelling*. Heidelberg, Germany: Springer-Verlag. doi:10.1007/978-3-540-92916-1

Barlett, M., & Sejnowski, T. (1997, May 17). Independent components of face images: A representation for face recognition. In *Proceedings of the 4th Annual Joint Symposium on Neural Computation*, Pasadena, CA.

Bayesteh, A., & Khandani, A. K. (2005). On the user selection for MIMO broadcast channels. In *Proceedings of the IEEE International Symposium on Information Theory* (pp. 2325-2329).

Beebee, H., Hitchcock, C., & Menzies, P. P. (Eds.). (2009). *The Oxford handbook of causation*. Oxford, UK: Oxford University Press.

Behrend, M. R., Ahuja, A. K., Humayun, M. S., Weiland, J. D., & Chowe, R. H. (2009). Selective labeling of retinal ganglion cells with calcium indicators by retrograde loading in vitro. *Journal of Neuroscience Methods*, *179*(2), 166–172. doi:10.1016/j.jneumeth.2009.01.019

Belhumeur, P. N., Hespanha, J., & Kriegman, D. J. (1997). Eigenfaces vs. Fisherfaces: Recognition using class specific linear projection. *IEEE Transactions on Pattern Analysis and Machine Intelligence*, *19*(7), 711–720. doi:10.1109/34.598228

Belkin, M., & Niyogi, P. (2003). Laplacia neigenmaps for dimensionality reduction and data representation. *Neural Computation*, *15*(6), 1373–1396. doi:10.1162/089976603321780317

Bell, D. A. (1953). *Information theory*. London, UK: Pitman.

Bender, E. A. (1996). *Mathematical methods in artificial intelligence*. Los Alamitos, CA: IEEE Press.

Bennett, C. H., & Brassard, G. (1984). Quantum cryptography: Public key distribution and coin tossing. In *Proceedings of the IEEE International Conference on Computers, Systems and Signal Processing*, Bangalore, India (pp. 175-179).

Beno, M., Rajaji, L., Varatharaju, V., & Santos, A. (2011). Rotor position estimation of 6/4 switched reluctance motor using a novel neural network algorithm. In *Proceedings of the IEEE GCC Conference and Exhibition*, Dubai, United Arab Emirates (pp. 77-80).

Bergh, F. (2001). *An analysis of particle swarm optimizers*. Unpublished doctoral dissertation, University of Pretoria, South Africa.

Bergh, F., & Engelbrecht, A. (2006). A study of particle swarm optimization particle trajectories. *Information Sciences, 176*(8), 937–971. doi:10.1016/j.ins.2005.02.003

Bertsekas, D. P. (2005). *Dynamic programming and optimal control* (3rd ed., Vol. 1). Hilliard, OH: Athena.

Berwick, R. C. (2011, August). Keynote: Songs to syntax: Cognition, combinatorial computation, and the origin of language. In *Proceedings of the 10th IEEE International Conference on Cognitive Informatics and Cognitive Computing*, Banff, AB, Canada (p. 1). Washington, DC: IEEE Computer Society.

Besl, P. J., & Jain, A. (1988). Segmentation through variable-order surface fitting. *IEEE Transactions on Pattern Analysis and Machine Intelligence, 10*(2), 167–192. doi:10.1109/34.3881

Besl, P. J., & Jain, R. C. (1985). Three-dimensional object recognition. *ACM Computing Surveys, 17*(1), 75–145. doi:10.1145/4078.4081

Besl, P. J., & McKay, N. (1992). A method for registration of 3D shapes. *IEEE Transactions on Pattern Analysis and Machine Intelligence, 14*(2), 239–256. doi:10.1109/34.121791

Beumier, C., & Acheroy, M. (2000). Automatic 3D face authentication. *Image and Vision Computing, 18*(4), 315–321. doi:10.1016/S0262-8856(99)00052-9

Beyer, H.-G., & Schwefel, H.-P. (2002). Evolution strategies - A comprehensive introduction. *Natural Computing, 1*, 3–52. doi:10.1023/A:1015059928466

Bezdek, J. C. (1981). *Pattern recognition with fuzzy objective function algorithms*. New York, NY: Plenum Press.

BISC. (2010). *Internal communications*. Berkeley, CA: University of California.

Bishop, C. M. (2006). *Pattern recognition and machine learning*. New York, NY: Springer.

Blasch, E., Kadar, I., Salerno, J., Kokar, M. M., Das, S., & Powell, G. M. (2006). Issues and challenges in situation assessment (level 2 fusion). *Journal of Advances in Information Fusion, 1*(2).

Bledsoe, W. W. (1961). *Lethally dependent genes using instant selection* (Tech. Rep. No. PRI1). Palo Alto, CA: Panoramic Research Inc.

Blei, D. M., & Lafferty, J. D. (2006). Correlated topic models. *Advances in Neural Information Processing Systems, 18*.

Blei, D. M., Ng, A. Y., & Jordan, M. I. (2003). Latent dirichlet allocation. *Journal of Machine Learning Research, 3*, 993–1022. doi:10.1162/jmlr.2003.3.4-5.993

Boden, M. A. (1987). *Artificial intelligence and natural man* (2nd ed.). New York, NY: Basic Books.

Boole, G. (2003). *The laws of thought, 1854*. Amherst, NY: Prometheus Books.

Boonyarattaphan, A., Bai, Y., & Chung, S. (2009). A security framework for e-Health service authentication and e-Health data transmission. In *Proceedings of the 9th International Symposium on Communications and Information Technology* (pp. 1213-1218).

Bortoff, S., Kohan, R., & Milman, R. (1998). Adaptive control of variable reluctance motors: a spline function approach. *IEEE Transactions on Industrial Electronics, 45*(3), 433–444. doi:10.1109/41.679001

Bosworth, J. (1972). *Comparison of genetic algorithms with conjugate gradient methods* (Tech. Rep. No. TR00312-1-T). Ann Arbor, MI: University of Michigan.

Bosworth, B. (1982). *Codes, ciphers, and computers: An introduction to information security*. Rochelle Park, NJ: Hayden.

Boyd, S., & Vandenberghe, L. (2004). *Convex optimization*. Cambridge, UK: Cambridge University Press.

Brand, M. (2003). Charting a manifold. *Advances in Neural Information Processing Systems, 15*, 961–968.

Bratton, D., & Kennedy, J. (2007, April). Defining a standard for particle swarm optimization. In *Proceedings of the IEEE Symposium on Swarm Intelligence*, Honolulu, HI (pp. 120-127). Washington, DC: IEEE Computer Society.

Breiman, L. (1996). Bagging predictors. *Machine Learning, 24*(2), 123–140. doi:10.1007/BF00058655

Breiman, L. (2001). Random forests. *Machine Learning, 45*(1), 5–32. doi:10.1023/A:1010933404324

Bremermann, H. J. (1962). Optimization through evolution and recombination. In Yovits, M. C., Jacobi, G. T., & Goldstein, G. D. (Eds.), *Self-organizing systems* (pp. 93–106). Washington, DC: Spartan Books.

Brooks, A. C., Zhao, X., & Pappas, T. N. (2008). Structural similarity quality metrics in a coding context: Exploring the space of realistic distortions. *IEEE Transactions on Image Processing*, 1261–1273. doi:10.1109/TIP.2008.926161

Brown, G., Wyatt, J., Harris, R., & Yao, X. (2005). Diversity creation methods: A survey and categorisation. *Journal of Information Fusion, 6*(1), 1–28.

Bruske, J., & Sommer, G. (1998). Intrinsic dimensionality estimation with optimally topology preserving maps. *IEEE Transactions on Pattern Analysis and Machine Intelligence, 20*(5), 572–575. doi:10.1109/34.682189

Budin, L., Golub, M., & Jakobović, D. (1998). Parallel adaptive genetic algorithm. In *Proceedings of the International ICSC/IFAC Symposium on Neural Computation*, Vienna, Austria (pp. 157-163).

Bugeja, M., Fabri, S., & Camilleri, L. (2009). Dual adaptive dynamic control of mobile robots using neural networks. *IEEE Transactions on Systems, Man, and Cybernetics. Part B, Cybernetics, 39*(1), 129–141. doi:10.1109/TSMCB.2008.2002851

Bu, N., Okamoto, M., & Tsuji, T. (2009). A hybrid motion classification approach for EMG-based human–robot interfaces using Bayesian and neural networks. *IEEE Transactions on Robotics, 25*(3), 502–511. doi:10.1109/TRO.2009.2019782

Cabri, G., Leonardi, L., & Zambonelli, F. (2003, January). Implementing role-based interactions for internet agents. In *Proceedings of the Symposium on Applications and the Internet* Orlando, FL (pp. 380-387).

Cai, Y., & Baciu, G. (2011, June). Detection of repetitive patterns in near regular texture images. In *Proceedings of the 10th IEEE Conference on Perception and Visual Signal Analysis*, Ithaca, NY (pp. 60-65). Washington, DC: IEEE Computer Society.

Cai, D., He, X., & Han, J. (2005). Document clustering using locality preserving indexing. *IEEE Transactions on Knowledge and Data Engineering, 17*(12), 1624–1637. doi:10.1109/TKDE.2005.198

Cai, J., Deng, Z., Qi, R., Liu, Z., & Cai, Y. (2011). A novel BVC-RBF neural network based system simulation model for switched reluctance motor. *IEEE Transactions on Magnetics, 47*(4), 830–838. doi:10.1109/TMAG.2011.2105273

Cangelosi, A., & Greco, S. H. (2002). Symbol grounding and the symbolic theft hypothesis. In Cangelosi, A., & Parisi, D. (Eds.), *Simulating the evolution of language* (pp. 91–210). New York, NY: Springer. doi:10.1007/978-1-4471-0663-0_9

Caspi, A., Dorn, J. D., McClure, K. H., Humayun, M. S., Greenberg, R. J., & McMahon, M. J. (2009). Feasibility study of a retinal prosthesis: Spatial vision with a 16-electrode implant. *Archives of Ophthalmology, 127*(4), 398–401. doi:10.1001/archophthalmol.2009.20

Casti, A., Hayot, F., Xiao, Y. P., & Kaplan, E. (2008). A simple model of retina-LGN transmission. *Journal of Computational Neuroscience, 24*(2), 235–252. doi:10.1007/s10827-007-0053-7

Chan, C., Kinsner, W., Wang, Y., & Miller, D. M. (Eds.). (2004, August). *Proceedings of the 3rd IEEE International Conference on Cognitive Informatics*, Victoria, BC, Canada Washington, DC: IEEE Computer Society.

Chen, H., & Zhu, Y. (2003). Rotor position sensorless control of the switched reluctance motor drive for the direct-drive joint of the robot. In *Proceedings of the Sixth International Conference on Electrical Machines and Systems*, Beijing, China (Vol. 2, pp. 915-918).

Chen, J., Pappas, T. N., Mojsilovic, A., & Rogowitz, B. E. (2002). Adaptive image segmentation based on color and texture. In *Proceedings of the IEEE International Conference on Information Processing* (Vol. 3, pp. 777-780). Washington, DC: IEEE Computer Society.

Chen, G. R., & Dong, X. N. (1998). *From chaos to order methodologies, perspectives, and applications.* Singapore: World Scientific Press.

Chen, G. R., Mao, Y., & Chui, C. K. (2004). A symmetric image encryption based on 3D chaotic map. *Chaos, Solitons, and Fractals, 21*(3), 749–761. doi:10.1016/j.chaos.2003.12.022

Chen, G. R., & Ueta, T. (1999). Yet another chaotic attractor. *International Journal of Bifurcation and Chaos in Applied Sciences and Engineering, 9*(7), 1465–1466. doi:10.1142/S0218127499001024

Chen, J., Pappas, T. N., Mojsilovic, A., & Rogowitz, B. E. (2005). Adaptive perceptual color-texture image segmentation. *IEEE Transactions on Image Processing,* 1524–1536. doi:10.1109/TIP.2005.852204

Chen, L., Liao, H., Ko, M., Lin, J., & Yu, G. (2000). A new LDA based face recognition system which can solve the small sample size problem. *Pattern Recognition, 33*(10), 1713–1726. doi:10.1016/S0031-3203(99)00139-9

Chen, P., & Chang, J.-F. (1991). An improved transmission protocol for two interfering queues in packet radio networks. *IEEE Transactions on Wireless Communications, 39*(3), 353–355.

Chen, X. Y. (2009). *Image registration and its MATLAB programming process.* Beijing, China: Electronic Industry Press.

Cherniak, C. (1983). Rationality and the structure of human memory. *Synthese, 57*(2), 163–186. doi:10.1007/BF01064000

Chitrakala, S., Shamini, P., & Manjula, D. (2009). Multiclass enhanced image mining of heterogeneous textual images using multiple image features. In *Proceedings of the IEEE International Advance Computing Conference,* Patiala, India (pp. 496-501).

Chuan, L., & Quanyuan, F. (2007, August). The standard particle swarm optimization algorithm convergence analysis and parameter selection. In *Proceedings of the Third International Conference on Natural Computation,* Hainan, China (Vol. 3, pp. 823-826). Washington, DC: IEEE Computer Society.

Clark & Parsia. (2004). *Pellet: OWL 2 reasoner for Java.* Retrieved from http://clarkparsia.com/pellet

Clerc, M., & Kennedy, J. (2002). The particle swarm - explosion, stability, and convergence in a multidimensional complex space. *IEEE Transactions on Evolutionary Computation, 6*(1), 58–73. doi:10.1109/4235.985692

Cocchiarella, N. (1996). Conceptual realism as a formal ontology. In Poli, R., & Simons, P. (Eds.), *Formal ontology* (pp. 27–60). London, UK: Kluwer Academic.

Codin, R., Missaoui, R., & Alaoui, H. (1995). Incremental concept formation algorithms based on Galois (concept) lattices. *Computational Intelligence, 11*(2), 246–267. doi:10.1111/j.1467-8640.1995.tb00031.x

Cover, T., & Hart, P. (1967). Nearest neighbor pattern classification. *IEEE Transactions on Information Theory, 13*(1), 21–27. doi:10.1109/TIT.1967.1053964

Craik, K. J. W. (1943). *The nature of explanation.* Cambridge, UK: Cambridge University Press.

Cuomo, K. M., & Oppenheim, A. V. (1993). Circuit implementation of synchronized chaos with applications to communications. *Transactions of the American Physical Society Journal,* 65-68.

Demri, S., & Orlowska, E. (1997). Logical analysis of indiscernibility. In Orlowska, E. (Ed.), *Incomplete information: Rough set analysis* (pp. 347–380). Heidelberg, Germany: Physica Verlag.

Devaney, R. L. (1992). *A first course in chaotic dynamical systems: Theory and experiment.* Reading, MA: Addison-Wesley.

Devroye, N. (2006). Achievable rates in cognitive radio channels. *IEEE Transactions on Information Theory, 52*(5), 1813–1827. doi:10.1109/TIT.2006.872971

Dierks, T., & Jagannathan, S. (2009). Neural network control of mobile robot formations using RISE feedback. *IEEE Transactions on Systems, Man, and Cybernetics. Part B, Cybernetics*, *39*(2), 332–347. doi:10.1109/TSMCB.2008.2005122

Dietterich, T. G. (2001). Ensemble methods in machine learning. In J. Kittler & F. Roli (Eds.), *Proceedings of the First International Workshop on Multiple Classifier Systems* (LNCS 1857, pp. 1-15).

Dietterich, T. G. (1997). Machine learning research: four current direction. *Artificial Intelligence Magazine*, *18*(4), 97–136.

Dietterich, T. G. (1998). An experimental comparison of three methods for constructing ensembles of decision trees: Bagging, boosting, and randomization. *Machine Learning*, *40*(2), 139–157. doi:10.1023/A:1007607513941

Dietterich, T. G., & Bakiri, G. (1995). Solving multi-class learning problem via error-correcting output codes. *Journal of Artificial Intelligence Research*, *2*, 263–286.

Diffie, W., & Hellman, M. E. (1976). New directions in cryptography. *IEEE Transactions on Information Theory*, *22*(6), 644–654. doi:10.1109/TIT.1976.1055638

Ding, Y., & Klein, K. (2010). Model-driven application-level encryption for the privacy of e-Health data. In *Proceedings of the International Conference on Availability, Reliability, and Security* (pp. 341-346).

Donoho, D. L., & Grimes, C. (2003). Hessian eigen-maps: Locally linear embedding techniques for high dimensional data. *Proceedings of the National Academy of Arts and Sciences*, *100*(10), 5591–5596. doi:10.1073/pnas.1031596100

Dorigo, M., Maniezzo, V., & Colorni, A. (1996). The ant system: Optimization by a colony of cooperating agents. *IEEE Transactions on Systems, Man, and Cybernetics B*, *26*(1), 29–41. doi:10.1109/3477.484436

Duda, R. O., & Hart, P. E. (1973). *Pattern classification and scene analysis*. New York, NY: John Wiley & Sons.

Dunn, F. A., Doan, T., Sampath, A. P., & Rieke, F. (2006). Controlling the gain of rod-mediated signals in the mammalian retina. *The Journal of Neuroscience*, *26*(15), 3959–3970. doi:10.1523/JNEUROSCI.5148-05.2006

Eberhart, R. C., & Kennedy, J. (1995). A new optimizer using particles warm theory. In *Proceedings of the 6th International Symposium on Micro Machine and Human Science* (pp. 39-43). Washington, DC: IEEE Computer Society.

Eberhart, R., & Shi, Y. (2000, August). Comparing inertia weights and constriction factors in particle swarm optimization. In *Proceedings of the Congress on Evolutionary Computation*, La Jolla, CA (Vol. 1, pp. 84-88). Washington, DC: IEEE Computer Society.

Eberhart, R., & Shi, Y. (2007). *Computational intelligence: Concepts to implementations* (p. 512). San Francisco, CA: Morgan Kaufmann.

Echenique, E., Dixon, J., Cárdenas, R., & Peña, R. (2009). Sensorless control for a switched reluctance wind generator based on current slopes and neural networks. *IEEE Transactions on Industrial Electronics*, *56*(3), 817–825. doi:10.1109/TIE.2008.2005940

Edelman, S. (1999). *Representation and recognition in vision*. Cambridge, MA: MIT Press.

Eiben, A., & Smith, J. (2003). *Introduction to evolutionary computing*. Berlin, Germany: Springer-Verlag.

Elbeltagi, E., Hegazy, T., & Grierson, D. (2005). Comparison among five evolutionary-based optimization algorithms. *Advanced Engineering Informatics*, *19*, 43–53. doi:10.1016/j.aei.2005.01.004

Enns, R. H. (2010). *It's a nonlinear world*. New York, NY: Springer.

Ester, M., Kriegel, H.-P., Sander, J., & Xu, X. (1996). A density-based algorithm for discovering clusters in large spatial databases with noise. In *Proceedings of the 2nd International Conference on Knowledge Discovery and Data Mining* (pp. 226-231).

Fabb, N., & Halle, M. (2008). *Meter in poetry*. Cambridge, UK: Cambridge University Press. doi:10.1017/CBO9780511755040

FaCT++ (2007). *OWL: FaCT++*. Retrieved from http://owl.man.ac.uk/factplusplus/

Fang, N., Luo, X. F., & Xu, W. M. (2009). Measuring textual context based on cognitive principles. *International Journal of Software Science and Computational Intelligence, 1*(4), 61–89. doi:10.4018/jssci.2009062504

Fang, S., McLaughlin, J., Fang, J., Huang, J., Foroud, T., & Autti-Rämö, I. (2008). Automated diagnosis of fetal alcohol syndrome using 3D facial image analysis. *Orthodontics & Craniofacial Research, 11*, 162–171. doi:10.1111/j.1601-6343.2008.00425.x

Fan, Z. Z., & Liu, E. G. (2008). New approach on image retrieval based on color information entropy. *Journal of Computer Applications, 25*(1), 281–282.

Feistel, H. (1973). Cryptography and computer privacy. *Scientific American, 228*(5), 15–23. doi:10.1038/scientificamerican0573-15

Feldman, J. (2006). An algebra of human concept learning. *Journal of Mathematical Psychology, 50*(4), 339–368. doi:10.1016/j.jmp.2006.03.002

Ferguson, N. (2010). *Cryptography engineering: Design principles and practical applications*. New York, NY: Wiley.

Ferreira, J., Crisóstomo, M., & Coimbra, A. (2009). SVR versus neural-fuzzy network controllers for the sagittal balance of a biped robot. *IEEE Transactions on Neural Networks, 20*(12), 1885–1897. doi:10.1109/TNN.2009.2032183

FIPA. (n. d.). *The foundation for intelligent physical agents*. Retrieved from http://www.fipa.org/

Floreano, D., & Mattiussi, C. (2008). *Bio-inspired artificial intelligence*. Cambridge, MA: MIT Press.

Freund, Y. (1995). Boosting a weak algorithm by majority. *Information and Computation, 121*(2), 256–285. doi:10.1006/inco.1995.1136

Fridrich, J. (1998). Symmetric ciphers based on two-dimensional chaotic maps. *International Journal of Bifurcation and Chaos in Applied Sciences and Engineering, 8*(6), 1259–1284. doi:10.1142/S021812749800098X

Friederici, A., Bahlmann, J., Friedrich, R., & Makuuchi, M. (2011). The neural basis of recursion and complex syntactic hierarchy. *Biolinguistics, 5*, 87–104.

Fukunaga, K., & Olsen, D. R. (1971). An algorithm for finding intrinsic dimensionality of data. *IEEE Transactions on Computers, 20*, 176–183. doi:10.1109/T-C.1971.223208

Gabrieli, J. D. E. (1998). Cognitive neuroscience of human memory. *Annual Review of Psychology, 49*, 87–115. doi:10.1146/annurev.psych.49.1.87

Gaines, H. F. (1956). *Cryptanalysis: a study of ciphers and their solution*. New York, NY: Dover.

Gait, J. (1978). Encryption standard: Validating hardware techniques. *Dimensions/NBS, 62*(7-8), 22–24.

Ganter, B., & Wille, R. (1999). *Formal concept analysis*. Berlin, Germany: Springer-Verlag.

Gavrilova, M. (2009). Adaptive computation paradigm in knowledge representation: traditional and emerging applications. *International Journal of Software Science and Computational Intelligence, 1*(1), 87–99. doi:10.4018/jssci.2009010106

Gentry, C. (2009). Fully homomorphic encryption using ideal lattices. In *Proceedings of the 41st Annual ACM Symposium on Theory of Computing*, Bethesda, MD.

Gerald, C., & Wheatley, P. (2004). *Applied numerical analysis* (7th ed.). Reading, MA: Addison-Wesley.

Gesbert, D., Shafi, M., Shiu, D., Smith, P. J., & Naguib, A. (2003). From theory to practice: An overview of MIMO space-time coded wireless systems. *IEEE Journal on Selected Areas in Communications, 21*(3), 281–302. doi:10.1109/JSAC.2003.809458

Giacinto, G., & Roli, F. (2001). Design of effective neural network ensembles for image classification purposes. *Image and Vision Computing, 19*(9), 699–707. doi:10.1016/S0262-8856(01)00045-2

Giorgio, C., Guido, T., & Francesco, P. (1997). Neural networks for region detection. In *Proceedings of the 9th International Conference on Image Analysis and Processing* (Vol. 2).

Giuliani, M., & Knoll, A. (2008). MultiML - a general purpose representation language for multimodal human utterances. In *Proceedings of the 10th International Conference on Multimodal Interfaces* (pp. 165-172).

Glover, N., & Dudley, T. (1991). *Practical error correction design for engineers* (2nd ed.). Broomfield, CO: Cirrus Logic-Colorado.

GOLD. (2010). *General ontology for linguistic description*. Retrieved from http://www.linguistics-ontology.org/gold/

Goldman, S. (1953). *Information theory*. Upper Saddle River, NJ: Prentice Hall.

Golub, G. H., & Loan, C. F. V. (1996). *Matrix computations* (3rd ed.). Baltimore, MD: The Johns Hopkins University Press.

Gordon, G. (1991). Face recognition based on depth maps and surface curvature. In. *Proceedings of the SPIE Conference on Geometric Method in Computer Vision, 1570*, 234–247.

Gotesky, R. (1968). The uses of inconsistency. *Philosophy and Phenomenological Research, 28*(4), 471–500. doi:10.2307/2105687

Gradshteyn, I. S., & Ryzhik, I. M. (2007). *Table of integrals, series, and products*. New York, NY: Academic Press.

Greco, S., Matarazzo, B., & Słowiński, R. (2009). Parameterized rough set model using rough membership and Bayesian confirmation measures. *International Journal of Approximate Reasoning, 49*, 285–300. doi:10.1016/j.ijar.2007.05.018

Grimson, W. E. L., & Grimson, W. (1981). *From images to surfaces: A computational study of the human early visual system* (*Vol. 4*). Cambridge, MA: MIT Press.

Grizzle, K. L., & Simms, M. D. (2005). (in Review). Early language development and language learning disabilities. *Pediatrics, 26*, 274–283.

Grundin, J. (2002). Group dynamics and ubiquitous computing. *Communications of the ACM, 45*(12), 74–78.

Guan, X. D., & Wei, H. (2009). Realistic simulation on retina photoreceptor layer. In *Proceedings of the First International Joint Conference on Artificial Intelligence* (pp. 179-184).

Guan, Z. H., Huang, F., & Guan, W. (2005). Chaos based image encryption algorithm. *Physics Letters. [Part A], 346*(1-3), 153–157. doi:10.1016/j.physleta.2005.08.006

Guarino, N. (1995). Formal ontology, conceptual analysis and knowledge representation. *Human-Computer Studies, 43*(5-6), 625–640. doi:10.1006/ijhc.1995.1066

Gunn, S. R. (1998). *Support vector machines for classification and regression* (Tech. Rep. No. 6459). Hampshire, UK: University of Southampton.

Hachtel, G. D., & Somenzi, F. (1996). *Logic synthesis and verification algorithms*. New York, NY: Springer Science & Business Media.

Haesler, S., Rochefort, C., Georgi, B., Licznerski, P., Osten, P., & Scharff, C. (2007). Incomplete and inaccurate vocal imitation after knockdown of FoxP2 in songbird basal ganglia nucleus Area X. *PLoS Biology, 5*, e321. doi:10.1371/journal.pbio.0050321

Hamdi, K., Zhang, W., & Ben Letaief, K. (2007). Joint beamforming and scheduling in cognitive radio networks. In *Proceedings of the IEEE Global Telecommunications Conference* (pp. 2977-2981).

Hancock, P. A., & Chignell, M. H. (1988). Mental workload dynamics in adaptive interface design. *IEEE Transactions on Systems, Man, and Cybernetics, 18*(4), 647–658. doi:10.1109/21.17382

Hansen, L. K., & Salamon, P. (1990). Neural network ensembles. *IEEE Transactions on Pattern Analysis and Machine Intelligence, 12*(10), 993–1001. doi:10.1109/34.58871

Hao, B. L. (1993). *Starting with parabolas: an introduction to chaotic dynamics*. Shanghai, China: Shanghai Scientific and Technological Education Publishing House.

Hardy, R. L. (1971). Multiquadric equations of topography and other irregular surfaces. *Journal of Geophysical Research, 76*, 1905–1915. doi:10.1029/JB076i008p01905

Harnad, S. (1987). *Categorical perception: the groundwork of cognition: Applied systems and cybernetics* (pp. 287–300). Cambridge, UK: Cambridge University Press.

Harnad, S. (1990). The symbol grounding problem. *Physica D. Nonlinear Phenomena, 42*, 335–346. doi:10.1016/0167-2789(90)90087-6

Harnad, S. (1994). Computation is just interpretable symbol manipulation: Cognition isn't. *Minds and Machines, 4,* 379–390. doi:10.1007/BF00974165

Harnad, S. (2007). Symbol grounding problem. *Scholarpedia, 2*(7), 2373. doi:10.4249/scholarpedia.2373

Hastie, T., Tibshirani, R., & Friedman, J. (2009). *The elements of statistical learning.* Berlin, Germany: Springer-Verlag.

Haykin, S. (2006b, November). Cognitive dynamic systems. In *Proceedings of the IEEE International Conference on Acoustics, Speech and Signal Processing* (pp. 1369-1372).

Haykin, S. (2011, August). Keynote: Cognitive dynamic systems: An integrative field that will be a hallmark of the 21st century. In *Proceedings of the 10th IEEE International Conference on Cognitive Informatics and Cognitive Computing,* Banff, AB, Canada (p. 2). Washington, DC: IEEE Computer Society.

Haykin, S., & Xue, Y. (n. d.). *Cognitive radar.* Manuscript submitted for publication.

Haykin, S. (2005). Cognitive radio: Brain-empowered wireless communication. *IEEE Journal on Selected Areas in Communications, 23*(2), 201–220. doi:10.1109/JSAC.2004.839380

Haykin, S. (2006). Cognitive Radar: A way of the future. *IEEE Signal Processing Magazine, 30,* 30–40. doi:10.1109/MSP.2006.1593335

Haykin, S. (2009). *Neural networks and learning machines* (3rd ed.). Upper Saddle River, NJ: Prentice Hall.

Haykin, S. (2011). *Cognitive dynamic systems: Perception-action cycle, radar, and radio.* Cambridge, UK: Cambridge University Press.

Haykin, S., Reed, J. H., Li, G. Y., & Shafi, M. (Eds.). (2009). Cognitive radio – Part 1: Practical perspectives. *Proceedings of the IEEE, 97*(4).

Haykin, S., Reed, J. H., Li, G. Y., & Shafi, M. (Eds.). (2009). Cognitive radio – Part 2: Fundamental issues. *Proceedings of the IEEE, 97*(5).

Haykin, S., Thomson, D., & Reed, J. (2009). Spectrum sensing for cognitive radio. *Proceedings of the IEEE, 97*(5), 849–877. doi:10.1109/JPROC.2009.2015711

Haykin, S., Zia, A., Xue, Y., & Arasaratnan, I. (2011). Control theoretic approach to tracking radar: First step towards cognition. *Digital Signal Processing, 21*(5). doi:10.1016/j.dsp.2011.01.004

He, L., & Pappas, T. N. (2010, September). An adaptive clustering and chrominance-based merging approach for image segmentation and abstraction. In *Proceedings of the IEEE 17th International Conference on Image Processing,* Hong Kong (pp. 241-244). Washington, DC: IEEE Computer Society.

Hebb, D. (1949). *Self-organization of behavior.* New York, NY: John Wiley & Sons.

Hecht-Nielssen, R. (1995). Replicator neural networks for universal optimal source coding. *Science, 269*(5232), 1860–1863. doi:10.1126/science.269.5232.1860

Heflin, J., Hendler, J., & Luke, S. (1999). SHOE: A knowledge representation language for internet applications (Tech. Rep. No. CS-TR-4078). Baltimore, MD: University of Maryland.

Hénon, M. (1976). A two-dimensional mapping with a strange attractor. *Communications in Mathematical Physics, 50,* 69–77. doi:10.1007/BF01608556

Herbert, J. P., & Yao, J. T. (2009). Game-theoretic rough sets. *Fundamenta Informaticae, 108*(3-4), 267–286.

Hermer-Vasquez, L., Spelke, E. S., & Katsnelson, A. S. (1999). Sources of flexibility in human cognition: Dual-task studies of space and language. *Cognitive Psychology, 39,* 3–36. doi:10.1006/cogp.1998.0713

Hirota, K. (1981). Concepts of probabilistic sets. *Fuzzy Sets and Systems, 5*(1), 31–46. doi:10.1016/0165-0114(81)90032-4

Hirota, K., & Pedrycz, W. (1984). Characterization of fuzzy clustering algorithms in terms of entropy of probabilistic sets. *Pattern Recognition Letters, 2*(4), 213–216. doi:10.1016/0167-8655(84)90027-8

Hoffman, T. (1999). Probabilistic latent semantic analysis. In Proceedings of the Conference on Uncertainty in Artificial Intelligence.

Hollnagel, E. (2000). Modeling the orderlineses of human actions. In Sarter, N. B., & Amalberti, R. (Eds.), *Cognitive engineering in the aviation domain* (pp. 65–98). Mahwah, NJ: Lawrence Erlbaum.

Hong, Q., Yang, S., Williams, A., & Zhang, Y. (2009). Real-time robot path planning based on a modified pulse-coupled neural network model. *IEEE Transactions on Neural Networks, 20*(11), 1724–1739. doi:10.1109/TNN.2009.2029858

Horrocks, I., Fensel, D., Broekstra, J., Decker, S., Erdmann, M., Goble, C., et al. (2000). OIL: The ontology inference layer (Tech. Rep. No. IR-479). Amsterdam, The Netherlands: Vrije Universiteit Amsterdam.

Hossain, E., & Bhargava, V. (Eds.). (2007). *Cognitive wireless communication networks* (1st ed.). New York, NY: Springer. doi:10.1007/978-0-387-68832-9

Ho, T. K. (1998). The random subspace method for constructing decision forests. *IEEE Transactions on Pattern Analysis and Machine Intelligence, 20*(8), 832–844. doi:10.1109/34.709601

Hou, M., Kobierski, R. D., & Brown, M. (2007). Intelligent adaptive interfaces for the control of multiple UAVs. *Journal of Cognitive Engineering and Decision Making, 1*(3), 327–362. doi:10.1518/155534307X255654

Hou, M., Zhu, H., Zhou, M., & Arrabito, G. R. (2011). Optimizing operator-agent interaction in intelligent adaptive interface design: A conceptual framework. *IEEE Transactions on Systems, Man and Cybernetics. Part C, Applications and Reviews, 41*(2), 161–178. doi:10.1109/TSMCC.2010.2052041

Hou, Z., Cheng, L., & Tan, M. (2010). Multicriteria optimization for coordination of redundant robots using a dual neural network. *IEEE Transactions on Systems, Man, and Cybernetics. Part B, Cybernetics, 40*(4), 1075–1087. doi:10.1109/TSMCB.2009.2034073

Howland, P., Jeon, M., & Park, H. (2003). Structure preserving dimension reduction for clustered text data based on the generalized singular value decomposition. *SIAM Journal on Matrix Analysis and Applications, 25*(1), 165–179. doi:10.1137/S0895479801393666

Hsu, S., & Fu, L. (2005). Adaptive decentralized control of robot manipulators driven by current-fed induction motors. *IEEE/ASME Transactions on Mechatronics, 10*(4), 465–468. doi:10.1109/TMECH.2005.852453

Hu, J., & Klein, A. (2009). A benchmark of transparent data encryption for migration of web applications in the cloud. In *Proceedings of the IEEE 8th International Conference on Dependable, Autonomic and Secure Computing* (pp. 735-740).

Hu, X. H. (2001). Using rough sets theory and database operations to construct a good ensemble of classifiers for data mining applications. In *Proceedings of the IEEE International Conference on Data Mining* (pp. 233-240).

Huang, J., Jain, A., Fang, S., & Riley, E. P. (2005). Using facial images to diagnose fetal alcohol syndrome (FAS). In *Proceedings of the IEEE International Conference on Information Technology: Coding and Computing* (pp. 66-71).

Hubpages. (2010). *Funny stories about computers.* Retrieved from http://hubpages.com/hub/about-funny-stories-2

Hu, H., & Woo, P. (2006). Fuzzy supervisory sliding-mode and neural-network control for robotic manipulators. *IEEE Transactions on Industrial Electronics, 53*(3), 929–940. doi:10.1109/TIE.2006.874261

Hu, K., Wang, Y., & Tian, Y. (2010). A web knowledge discovery engine based on concept algebra. *International Journal of Cognitive Informatics and Natural Intelligence, 4*(1), 80–97. doi:10.4018/jcini.2010010105

Hurley, P. J. (1997). *A concise introduction to logic* (6th ed.). London, UK: Wadsworth.

International Organization for Standardization (ISO). (2011). *ISO Standard 15836.* Geneva, Switzerland: ISO.

Ishii, N., Deguchi, T., & Sasaki, H. (2004). Parallel processing for movement detection in neural networks with nonlinear functions. In Z. R. Yang, H. Yin, & R. M. Everson (Eds.), *Proceedings of the 5th International Conference on Intelligent Data Engineering and Automated Learning* (LNCS 3177, pp. 626-633).

Jacobson, S. W., Stanton, M. E., Molteno, C. D., Burden, M. J., Fuller, D. S., & Hoyme, H. E. (2008). Impaired eyeblink conditioning in children with fetal alcohol syndrome. *Alcoholism, Clinical and Experimental Research*, *32*, 365–372. doi:10.1111/j.1530-0277.2007.00585.x

Jain, A. K., & Hoffman, R. (1988). Evidence based recognition of 3D objects. *IEEE Transactions on Pattern Analysis and Machine Intelligence*, *10*(6), 783–802. doi:10.1109/34.9102

JESS. (2008). *The rule engine for the Java platform.* Retrieved from http://www.jessrules.com/

Jin, S., & Peng, J. (2010). Access control for web services based on feedback and decay. In *Proceedings of the IEEE 9th International Conference on Cognitive Informatics* (pp. 501-505).

Jing, X.-Y., Wong, H.-S., & Zhang, D. (2006). Face recognition based on 2D Fisherface approach. *Pattern Recognition*, *39*, 707–710. doi:10.1016/j.patcog.2005.10.020

Jin, Z., Yang, J. Y., Hu, Z. S., & Lou, A. (2001). Face recognition based on the uncorrelated discriminant transformation. *Pattern Recognition Letters*, *34*, 1405–1416.

Johnson-Laird, P. N., Legrenzi, P., & Girotto, V. (2004). Reasoning from inconsistency to consistency. *Psychological Review*, *111*(3), 640–661. doi:10.1037/0033-295X.111.3.640

Johnson, R. A., & Bhattacharyya, G. K. (1996). *Statistics: Principles and methods* (3rd ed.). New York, NY: John Wiley & Sons.

Johnston, M. (2009). Building multimodal applications with EMMA. In *Proceedings of the International Conference on Multimodal Interfaces and Workshop on Machine Learning for Multimodal Interfaces*, Cambridge, MA (pp. 47-54).

Johnston, M., Baggia, P., Burnett, D. C., Carter, J., Dahl, D. A., MacCobb, G., et al. (2009). *EMMA: Extensible multimodal annotation markup language.* Retrieved from http://www.w3.org/TR/emma/

Jones, K. L., & Smith, D. W. (1973). Recognition of the fetal alcohol syndrome in early infancy. *Lancet*, *302*(7386), 999–1001. doi:10.1016/S0140-6736(73)91092-1

Kanade, T., Cohn, J. F., & Tian, Y. L. (2000). *The Cohn-Kanade AU-coded facial expression database.* Retrieved from http://vasc.ri.cmu.edu/idb/html/face/ facial_expression/index.html

Kantz, H., & Schreiber, T. (2004). *Nonlinear time series analysis* (2nd ed.). Cambridge, UK: Cambridge University Press.

KAON2. (n. d.). *Ontology management for the semantic web.* Retrieved from http://kaon2.semanticweb.org/

Katz, J., & Pesetsky, D. (2011). *The identity thesis for language and music.* Cambridge, MA: MIT. Retrieved from http://ling.auf.net/lingBuzz/000959

Katz, J., & Lindell, Y. (2007). *Introduction to modern cryptography: Principles and protocols.* Virginia Beach, VA: Chapman & Hall/CRC.

Kay, A. (1993). The early history of smalltalk. *ACM SIGPLAN Notice*, *28*, 69–95. doi:10.1145/155360.155364

Kennedy, J. (1997, April). The particle swarm: Social adaptation of knowledge. In *Proceedings of the IEEE International Conference on Evolutionary Computation*, Indianapolis, IN, (pp. 303-308). Washington, DC: IEEE Computer Society.

Kennedy, J. (2007). Some issues and practices for particle swarms. In *Proceedings of the Symposium on Swarm Intelligence*, Honolulu, HI (pp. 162-169). Washington, DC: IEEE Computer Society.

Kennedy, J., & Eberhart, R. (1995, June). Particle swarm optimization. In *Proceedings of the IEEE International Conference on Neural Networks*, Cambridge, UK, (Vol. 4, pp. 1942-1948). Washington, DC: IEEE Computer Society.

Kennedy, J., & Mendes, R. (2002, May). Population structure and particle swarm performance. In *Proceedings of the Congress on Evolutionary Computation*, Honolulu, HI (Vol. 2, pp. 1671-1676). Washington, DC: IEEE Computer Society.

Kennedy, J. (1999, July). Small worlds and mega-minds: effects of neighborhood topology on particle swarm performance. In []. Washington, DC: IEEE Computer Society.]. *Proceedings of the Congress on Evolutionary Computation*, *3*, 1931–1938.

Kennedy, J., & Eberhart, R. (2001). *Swarm intelligence.* San Francisco, CA: Morgan Kaufmann.

Kephart, J., & Chess, D. (2003). The vision of autonomic computing. *IEEE Computer, 26*(1), 41–50.

Khozemeih, F., & Haykin, S. (2010). Self-organizing dynamic spectrum management for cognitive radio networks. In *Proceedings of the 8th Annual Communication Networks and Services Research Conference*, Montreal, QC, Canada.

Kinsner, W. (2004, August). Is entropy suitable to characterize data and signals for cognitive informatics? In *Proceedings of the 3rd IEEE International Conference on Cognitive Informatics*, British Columbia, Canada (pp. 6-21). Washington, DC: IEEE Computer Society.

Kinsner, W. (2005, August). A unified approach to fractal dimensions. In *Proceedings of the 4th IEEE International Conference on Cognitive Informatics*, Irvine, CA (pp. 58-72). Washington, DC: IEEE Computer Society.

Kinsner, W. (2011, August). Keynote: It's time for multiscale analysis and synthesis in cognitive systems. In *Proceedings of the 10th IEEE International Conference on Cognitive Informatics and Cognitive Computing*, Banff, AB, Canada (pp. 7-10). Washington, DC: IEEE Computer Society.

Kinsner, W. (2011). It's time for polyscale analysis and synthesis in cognitive systems. In *Proceedings of the IEEE 10th International Conference on Cognitive Informatics and Cognitive Computing.*

Kinsner, W., & Grieder, W. (2008). Speech segmentation using multifractal measures and amplification of signal features. In *Proceedings of the IEEE 7th International Conference on Cognitive Informatics* (pp. 351-357).

Kinsner, W., Zhang, D., Wang, Y., & Tsai, J. (Eds.). (2005, August). *Proceedings of the 4th IEEE International Conference on Cognitive Informatics.* Washington, DC: IEEE Computer Society.

Kinsner, W. (2002). Towards cognitive machine: Multiscale measures and analysis. *International Journal of Cognitive Informatics and Natural Intelligence, 1*(1), 28–38. doi:10.4018/jcini.2007010102

Kinsner, W. (2007). A unified approach to fractal dimensions. *International Journal of Cognitive Informatics and Natural Intelligence, 1*(4), 26–46. doi:10.4018/jcini.2007100103

Kinsner, W. (2007). Towards cognitive machines: Multiscale measures and analysis. *International Journal of Cognitive Informatics and Natural Intelligence, 1*(1), 28–38. doi:10.4018/jcini.2007010102

Kinsner, W. (2009). Challenges in the design of adaptive, intelligent and cognitive systems. *International Journal of Software Science and Computational Intelligence, 1*(3), 16–35. doi:10.4018/jssci.2009070102

Kinsner, W. (2010). *Fractal and chaos engineering (course notes).* Winnipeg, MB, Canada: University of Manitoba.

Kinsner, W. (2010). System complexity and its measures: How complex is complex. In Wang, Y., Zhang, D., & Kinsner, W. (Eds.), *Advances in cognitive informatics and cognitive computing* (*Vol. 323*, pp. 265–295). Berlin, Germany: Springer-Verlag. doi:10.1007/978-3-642-16083-7_14

Kinsner, W., Cheung, V., Cannons, K., Pear, J., & Martin, T. (2006). Signal classification through multifractal analysis and complex domain neural networks. *IEEE Transactions on Systems, Man, and Cybernetics, 36*(2), 196–203. doi:10.1109/TSMCC.2006.871148

Kinsner, W., & Grieder, W. (2010). Amplification of signal features using variance fractal dimension trajectory. *International Journal of Cognitive Informatics and Natural Intelligence, 4*(4), 1–17. doi:10.4018/jcini.2010100101

Kline, M. (1972). *Mathematical thought: From ancient to modern times.* New York, NY: Oxford University Press.

Knowledge Interchange Format. (n. d.). *The KIF specification.* Retrieved from http://www.ksl.stanford.edu/knowledge-sharing/kif/

Koblitz, N. (1994). *A course in number theory and cryptography* (2nd ed.). New York, NY: Springer.

Kocarev, L. (2001). Chaos-based cryptography: A brief overview. *IEEE Circuits and Systems Magazine, 1*(3), 6–21. doi:10.1109/7384.963463

Kohonen, T. (1990). The self-organizing map. *Proceedings of the IEEE, 78*(9), 1464–1480. doi:10.1109/5.58325

Kranstedt, A., Kopp, S., & Wachsmuth, I. (2002). *Murml: A multimodal utterance representation markup language for conversational agents.* Paper presented at the First International Autonomous Agents and Multiagent Systems Workshop on Embodied Conversational Agents - Let's Specify and Evaluate Them, Bologna, Italy.

Kulczycki, M. (2008). Noncontinuous maps and Devaney's chaos. *Regular -and Chaotic Dynamics, 13*(2), 81-84.

Kuncheva, L. (2003). That elusive diversity in classifier ensembles. In F. J. Perales, A. J. C. Campilho, N. P. de la Blanca, & A. Sanfeliu (Eds.), *Proceedings of the First Iberian Conference on Pattern Recognition and Image Analysis* (LNCS 2652, pp. 1126-1138).

Kuncheva, L. I., & Whitaker, C. J. (2003). Measures of diversity in classifier ensembles and their relationship with the ensemble accuracy. *Machine Learning, 51*(2), 181–207. doi:10.1023/A:1022859003006

Kwak, J.-Y., Yoon, J. Y., & Shinn, R. H. (2006). An intelligent robot architecture based on robot mark-up languages. In *Proceedings of the IEEE International Conference on Engineering of Intelligent Systems* (pp. 1-6).

Lachman, T., Mohamad, T., & Fong, C. (2004). Nonlinear modelling of switched reluctance motors using artificial intelligence techniques. *IEEE Electric Power Applications, 151*(1), 53–60. doi:10.1049/ip-epa:20040025

Landauer, T. K., Foltz, P. W., & Laham, D. (1998). An introduction to latent semantic analysis. *Discourse Processes, 25*(2-3), 259–284. doi:10.1080/01638539809545028

Landragin, F., Denis, A., Ricci, A., & Romary, L. (2004). Multimodal meaning representation for generic dialogue systems architectures. In *Proceedings of the Fourth International Conference on Language Resources and Evaluation* (pp. 521-524).

Lang, K. (1995). Newsweeder: Learning to filter netnews. In *Proceedings of the International Conference on Machine Learning*, Tahoe City, CA. (pp. 331-339).

Leahey, T. H. (1997). *A history of psychology: Main currents in psychological thought* (4th ed.). Upper Saddle River, NJ: Prentice Hall.

Lebie, L., Rhoads, J. A., & McGrath, J. H. (1995). Interaction process in computer-mediated and face-to-face groups. *Computer Supported Cooperative Work, 4*(2-3), 127–152. doi:10.1007/BF00749744

Lee, D. J., Antani, S., Chang, Y., Gledhill, K., Rodney, L. L., & Christensen, P. (2009). CBIR of spine X-ray images on inter-vertebral disc space and shape profiles using feature ranking and voting consensus. *Data & Knowledge Engineering, 68*(12), 1359–1369. doi:10.1016/j.datak.2009.07.008

Lee, J. W., Chae, S. P., Kim, M. N., Kim, S. Y., & Cho, J. H. (2001). A moving detectable retina model considering the mechanism of an amacrine cell for vision. In. *Proceedings of the IEEE International Symposium on Industrial Electronics, 1-3*, 106–109.

Lee, J., Lendasse, A., & Verleysen, M. (2004). Nonlinear projection wit hcurvilinear distances: Isomap versus curvilinear distance analysis. *Neurocomputing, 57*(1), 49–76. doi:10.1016/j.neucom.2004.01.007

Lee, J., & Verleysen, M. (2005). Nonlinear dimensionality reduction of data manifolds with essential loops. *Neurocomputing, 67*(1-4), 2–53.

Lei, L., Wang, X., Yang, B., & Peng, J. (2010). Image dimensionality reduction based on the HSV feature. In *Proceedings of the 9th IEEE International Conference on Cognitive Informatics*, Beijing, China (pp. 127-131).

Levina, E., & Bickel, P. J. (2005). Maximum likelihood estimation of intrinsic dimension. In *Advances in neural information processing systems* (pp. 777–784). Cambridge, MA: MIT Press.

Lewis, D. (1973). Causation. *The Journal of Philosophy, 70*, 556–567. doi:10.2307/2025310

Li, S. J., & Zheng, X. (2002). Cryptanalysis of a chaotic image encryption method. In *Proceeding of IEEE International Symposium on Circuits and Systems*, Phoenix, AZ (Vol. 2, pp. 708-711). Washington, DC: IEEE Computer Society.

Liang, J. Y., & Qian, Y. H. (2008). Information granules and entropy theory in information systems. *Science in China F, 51*(9).

Liang, S., Chan, E., Baciu, G., & Li, R. (2010, July). Cognitive garment design interface using user behavior tree model. In *Proceedings of the 9th IEEE International Conference on Cognitive Informatics*, Beijing, China (pp. 496-500). Washington, DC: IEEE Computer Society.

Lian, S. G. (2009). A block cipher based on chaotic neural networks. *Neurocomputing, 72*, 1296–1301. doi:10.1016/j.neucom.2008.11.005

Licklider, J. C. R. (1960). Man-computer symbiosis. *IRE Transactions on Human Factors in Electronics, 1*, 4–11. doi:10.1109/THFE2.1960.4503259

Li, M., & Yuan, B. (2005). 2D-LDA: A novel statistical linear discriminant analysis for image matrix. *Pattern Recognition Letters, 26*(5), 527–532. doi:10.1016/j.patrec.2004.09.007

Lindblom, J., & Ziemke, T. (2003). Social situatedness of natural and artificial intelligence: Vygotsky and beyond. *Adaptive Behavior, 11*(2), 79–96. doi:10.1177/10597123030112002

Li, S. J., Álvarez, G., & Chen, G. R. (2005). Breaking a chaos-based secure communication scheme designed by an improved modulation method. *Chaos, Solitons, and Fractals, 25*, 109–120. doi:10.1016/j.chaos.2004.09.077

Liu, F., Wang, D., Li, C., & Li, M. (in press). Zero forcing algorithms for a multi-user cognitive wireless communication system. In *International Conference on Future Computer and Communication*

Liu, J. S., Zhang, D. Y., Liu, S. W., Fang, Y., & Zhang, M. (2008). The evaluation of wavelet and data driven feature selection for image understanding. In *Proceedings of the 1st International Conference on BioMedical Engineering and Informatics*, Sanya, China (Vol. 2, pp. 268-271).

Liu, Z., & Zhang, Y. (2006). Fuzzy logic-based neural modeling and robust control for robot. In *Proceedings of the 6th World Congress on Intelligent Control and Automation*, Dalian, China (pp. 8976-8980).

Liu, B., Wang, L., Jin, Y.-H., Tang, F., & Huang, D.-X. (2005). Improved particle swarm optimization combined with chaos. *Chaos, Solitons, and Fractals, 25*(5), 1261–1271. doi:10.1016/j.chaos.2004.11.095

Liu, H., & Singh, P. (2004). ConceptNet - A practical commonsense reasoning toolkit. *BT Technology Journal, 22*(4), 211–225. doi:10.1023/B:BTTJ.0000047600.45421.6d

Lorenz, E. N. (1963). Deterministic nonperiodic flow. *Journal of the Atmospheric Sciences, 20*, 130–141. doi:10.1175/1520-0469(1963)020<0130:DNF>2.0.CO;2

Loynes, R. M. (1962). The stability of a queue with non-independent interarrival and service times. *Mathematical Proceedings of the Cambridge Philosophical Society, 58*, 494–520. doi:10.1017/S0305004100036781

Lu, X., Colbry, D., & Jain, A. (2004, July). Matching 2.5D scans for face recognition. In *Proceedings of the International Conference on Biometric Authentication*, Hong Kong, China (pp. 30-36).

Luo, X. F., Xu, Z., Yu, J., & Liu, F. F. (2008). Discovery of associated topics for the intelligent browsing. In Proceedings of the 1st IEEE International Conference on Ubi-Media Computing, Lanzhou, China (pp.119-125).

Luo, W., & Ephremides, A. (1999). Stability of N interacting queues in random-access systems. *IEEE Transactions on Information Theory, 45*(5), 1579–1587. doi:10.1109/18.771161

Luo, X. F., Xu, Z., Yu, J., & Chen, X. (in press). Building association link network for semantic link on web resources. *IEEE Transactions on Automation Science and Engineering*.

Lyons, M., Akamatsu, S., Kamachi, M., & Gyoba, J. (1998). *The Japanese female facial expression (JAFFE) database*. Retrieved from http://www.kasrl.org/jaffe.html

Macal, C. M., & North, M. J. (2006). Tutorial on agent-based modeling and simulation part 2: How to model with agents. In *Proceedings of the 38th Winter Simulation Conference* (pp. 73-83).

Magnus, W., & Oberhettinger, F. (1949). Hypergeometric functions. In Abramowitz, M., & Stegun, I. A. (Eds.), *Handbook of mathematical functions with formulas, graphs, and mathematical tables: Hypergeometric functions*. New York, NY: Dover.

Marks, J., Andalman, B., Beardsley, P. A., Freeman, W., Gibson, S., Hodgins, J., et al. (1997). Design galleries: A general approach to setting parameters for computer graphics and animation. In *Proceedings of the 24th Annual Conference on Computer Graphics and Interactive Techniques* (pp. 389-400).

Martinez, A. M., & Kak, A. C. (2001). PCA versus LDA. *IEEE Transactions on Pattern Analysis and Machine Intelligence, 23*(2), 228–233. doi:10.1109/34.908974

Matlin, M. W. (1998). *Cognition* (4th ed.). New York, NY: Harcourt Brace College.

Mayo, M. J. (2003). Symbol grounding and its implications for artificial intelligence. In *Proceedings of the Twenty-Sixth Australian Computer Science Conference* (pp. 55-60).

May, P. A., & Gossage, J. P. (2001). Estimating the prevalence of fetal alcohol syndrome. A summary. *Alcohol Research & Health, 25*(3), 159–167.

McCallum, A., Corrada-Emmanuel, A., & Wang, X. (2004). *The author-recipient-topic model for topic and role discovery in social networks: Experiments with Enron and academic email*. Amherst, MA: University of Massachusetts.

McCarthy, J. (2010). *What is artificial intelligence?* Retrieved from http://www-formal.stanford.edu/jmc/whatisai/whatisai.html

McDermid, J. (Ed.). (1991). *Software engineer's reference book*. Oxford, UK: Butterworth Heinemann.

McGuinness, D. L., & Harmelen, F. (2004). OWL web ontology language. Retrieved from http://www.w3.org/TR/owl-features/

McGuire, W. J. (1960). A syllogistic analysis of cognitive relationships. In Hovland, C., & Rosenberg, M. (Eds.), *Attitude organization and change*. New Haven, CT: Yale University Press.

Medin, D. L., & Shoben, E. J. (1988). Context and structure in conceptual combination. *Cognitive Psychology, 20*, 158–190. doi:10.1016/0010-0285(88)90018-7

Mei, S. Y., Liu, Y., Wu, G. F., & Zhang, B. F. (2005). Rough reducts based SVM ensemble. In *Proceedings of the IEEE International Conference on Granular Computing* (pp. 571-574).

Mellor, D. H. (1995). *The facts of causation*. London, UK: Routledge. doi:10.4324/9780203302682

Mendes, R. (2004). *Population topologies and their infuence in particle swarm performance*. Unpublished doctoral dissertation, Universidade do Minho, Portugal.

Mendes, R., Kennedy, J., & Neves, J. (2004). The fully informed particle swarm: Simpler, maybe better. *IEEE Transactions on Evolutionary Computation, 8*(3), 204–210. doi:10.1109/TEVC.2004.826074

Menezes, A. J. (1993). *Elliptic curve public key cryptosystems*. New York, NY: Springer.

Meyer, M., Desbrun, M., Schroder, P., & Barr, A. (2002). *Discrete differential geometry operators for triangulated 2-manifolds*. Paper presented at the Visual Mathematics Workshop, Berlin, Germany.

Micciancio, D., & Regev, O. (2008). Lattice-based cryptography. In Bernstein, D. J., Buchmann, J., & Dahmen, E. (Eds.), *Post quantum cryptography* (pp. 147–191). Berlin, Germany: Springer-Verlag.

Michal, R. Z., Thomas, G., Mark, S., & Padhraic, S. (2004). The author-topic model for authors and documents. In Proceedings of the 20th Conference on Uncertainty in Artificial Intelligence, Banff, AB, Canada (pp. 487-494).

Michalski, R. S., Carbonell, J. G., & Mitchell, T. M. (Eds.). (1983). *Machine learning, an artificial intelligence approach*. San Francisco, CA: Morgan Kaufmann.

Miller, G. A. (1956). The magical number seven, plus or minus two: Some limits of our capacity for processing information. *Psychological Review, 63*, 81–97. doi:10.1037/h0043158

Miller, G. A. (1995). WordNet: A lexical database for English. *Communications of the ACM, 38*(11), 39–41. doi:10.1145/219717.219748

Miller, G. A., Beckwith, R., Fellbaum, C. D., Gross, D., & Miller, K. (1990). WordNet: An online lexical database. *International Journal of Lexicograph, 3*(4), 235–244. doi:10.1093/ijl/3.4.235

Mill, J. S. (1874). *A system of logic*. New York, NY: Harper & Brothers.

Milman, R., & Bortoff, S. (1999). Observer-based adaptive control of a variable reluctance motor: experimental results. *IEEE Transactions on Control Systems Technology*, 7(5), 613–621. doi:10.1109/87.784425

Minsky, M. (1965). Matter, mind and models. In *Proceedings of the International Federation for Information Processing Congress*, New York, NY (pp. 45-49).

Mirza, M., Sommers, J., Barford, P., & Zhu, X. J. (2010). A machine learning approach to TCP throughput prediction. *IEEE/ACM Transactions on Networking*, 18(4), 1026–1039. doi:10.1109/TNET.2009.2037812

Mitchell, T. M. (1982). Generalization as search. *Artificial Intelligence*, 18, 203–226. doi:10.1016/0004-3702(82)90040-6

Mitchell, T. M. (1997). *Machine learning*. New York, NY: McGraw-Hill.

Mitola, J. III, & Maguire, G. Q. Jr. (1999). Cognitive radio: Making software radios more personal. *IEEE Personal Communications Magazine*, 6(4), 13–18. doi:10.1109/98.788210

Modha, D. S., Ananthanarayanan, R., Esser, S. K., Ndirango, A., Sherbondy, A. J., & Singh, R. (2011). Cognitive computing. *Communications of the ACM*, 54(8), 62–71. doi:10.1145/1978542.1978559

Moghaddam, B., Jebara, T., & Pentland, A. (2000). Bayesian face recognition. *Pattern Recognition Letters*, 33, 1771–1782.

Mohamed, M., Ashraf, S., & Aziz, I. (2004). Digital control of direct drive robot manipulators. In *Proceedings of the International Conference on Power System Technology*, Singapore (pp. 1810-1815).

Mollin, R. A. (2006). *An introduction to cryptography*. Boca Raton, FL: CRC Press.

Moore, E. S., Ward, R. E., Jamison, P. L., Morris, C. A., Bader, P. I., & Hall, B. D. (2002). New perspectives on the face in fetal alcohol syndrome: What anthropometry tells us. *American Journal of Medical Genetics*, 109(4), 249–260. doi:10.1002/ajmg.10197

Morillas, C. A., Romero, S. F., Martinez, A., Pelayo, F. J., Ros, E., & Fernandez, E. (2007). A design framework to model retinas. *Bio Systems*, 87(2-3), 156–163. doi:10.1016/j.biosystems.2006.09.009

Moscato, P. (1989). *On evolution, search, optimization, genetic algorithms and martial arts: Towards memetic algorithms* (Tech. Rep. No. C3P826). Pasadena, CA: California Institute of Technology.

Moulin, C., & Sbodio, M. L. (2010). Improving the accessibility and efficiency of e-Government processes. In *Proceedings of the IEEE 9th International Conference on Cognitive Informatics* (pp. 603-610).

Moyer, M. J., & Ahamad, M. (2001, April). Generalized role-based access control. In *Proceedings of the 21st IEEE International Conference on Distributed Computing Systems*, Mesa, AZ (pp. 391-398).

Müller, H., Michoux, N., Bandon, D., & Geissbuhler, A. (2009). A review of content-based image retrieval systems in medical applications - Clinical benefits and future directions. *International Journal of Medical Informatics*, 78(9), 638–638. doi:10.1016/j.ijmedinf.2009.05.001

Murphy, G. L. (1993). Theories and concept formation. In Mechelen, I. V. (Ed.), *Categories and concepts, theoretical views and inductive data analysis* (pp. 173–200). New York, NY: Academic Press.

Myers, B. A. (1998). A brief history of human computer interaction technology. *Interactions (New York, N.Y.)*, 5(2), 44–54. doi:10.1145/274430.274436

Nain, P. (1985). Analysis of a two node ALOHA network with infinite capacity buffers. In *Proceedings of the International Conference of Computer Networking and Performance Evaluation* (pp. 49-63).

National Bureau of Standards. (1977). *Data encryption standard*. Washington, DC: NTIS.

Naware, V., Mergen, G., & Tong, L. (2005). Stability and delay of finite-user slotted ALOHA with multi-packet reception. *IEEE Transactions on Information Theory*, 51(7), 2636–2656. doi:10.1109/TIT.2005.850060

Nenadic, A., Zhang, N., Shi, Q., & Goble, C. (2005). DSA-based verifiable and recoverable encryption of signatures and its application in certified e-Goods delivery. In *Proceedings of the IEEE International Conference e-Technology, e-Commerce and e-Service* (pp. 94-99).

Newton, I. (1687). *The mathematical principles of natural philosophy* (1st ed.). Berkeley, CA: University of California Press.

Nichols, R. N. (1998). *ICSA guide to cryptography*. New York, NY: McGraw-Hill.

Niu, W. Q., & Yuan, J. Q. (2007). Recurrent network simulations of two types of non-concentric retinal ganglion cells. *Neurocomputing, 70*(13-15), 2576–2580. doi:10.1016/j.neucom.2007.01.008

Niu, W. Q., & Yuan, J. Q. (2008). A multi-subunit spatiotemporal model of local edge detector cells in the cat retina. *Neurocomputing, 72*(1-3), 302–312. doi:10.1016/j.neucom.2008.01.012

Norman, D. A. (1986). Cognitive engineering. In Norman, D. A., & Draper, S. W. (Eds.), *User centered system design: New perspectives on human-computer interaction* (pp. 32–65). Mahwah, NJ: Lawrence Erlbaum.

Noushatha, S., Hemantha Kumar, G., & Shivakumara, P. (2006). (2D)2LDA: An efficient approach for face recognition. *Pattern Recognition Letters, 39*, 1396–1400.

Obrst, L. (2003). Ontologies for semantically interoperable systems. In *Proceedings of the Twelfth International Conference on Information and Knowledge Management*, New Orleans, LA (pp. 366-369).

Olson, J. R., & Olson, G. M. (1990). The growth of cognitive modeling in human-computer interaction since GOMS. *Human-Computer Interaction, 5*(2), 221–265. doi:10.1207/s15327051hci0502&3_4

Opitz, D. (1999). Feature selection for ensembles. In *Proceedings of the 16th National Conference on Artificial Intelligence* (pp. 379-384).

Palmeri, T. J., & Gauthier, I. (2004). Visual object understanding. *Nature Reviews. Neuroscience, 5*(4), 291–303. doi:10.1038/nrn1364

Panayiotou, C., & Bennett, B. (2009). Critical thinking attitudes for reasoning with points of view. In *Proceedings of the IEEE 8th International Conference on Cognitive Informatics* (pp. 371-377).

Pappas, T. N., Tartter, V. C., Seward, A. G., Genzer, B., Gourgey, K., & Kretzshmar, I. (2009). Perceptual dimensions for a dynamic tactile display, human vision and electronic imaging. *Proceedings of the Society for Photo-Instrumentation Engineers, 7240*.

Park, K., Kim, T., Ahn, S., & Hyun, D. (2003). Speed control of high-performance brushless DC motor drives by load torque estimation. In *Proceedings of the IEEE 34th Annual Power Electronics Specialist Conference* (Vol. 4, pp. 1677-1681).

Parzen, E. (1962). On the estimation of a probability density function and mode. *Annals of Mathematical Statistics, 33*(3). doi:10.1214/aoms/1177704472

Patel, D., Patel, S., & Wang, Y. (Eds.). (2003, August). *Proceedings of the 2nd IEEE International Conference on Cognitive Informatics*. Washington, DC: IEEE Computer Society.

Pawlak, Z., & Skowron, A. (2007). Rudiments of rough sets. *Information Sciences, 177*(1), 1 3-27.

Pawlak, Z. (1984). On rough sets. *Bulletin of the EATCS, 24*, 94–108.

Pawlak, Z. (1984). Rough classification. *International Journal of Man-Machine Studies, 20*(5), 469–483. doi:10.1016/S0020-7373(84)80022-X

Pawlak, Z. (1985). Rough sets and fuzzy sets. *Fuzzy Sets and Systems, 17*(1), 99–102. doi:10.1016/S0165-0114(85)80029-4

Pawlak, Z. (1991). *Rough sets: Theoretical aspects of reasoning about data, system theory*. Dordrecht, The Netherlands: Kluwer Academic.

Pawlak, Z., Wong, S. K. M., & Ziarko, W. (1988). Rough sets: probabilistic versus deterministic approach. *International Journal of Man-Machine Studies, 29*, 81–95. doi:10.1016/S0020-7373(88)80032-4

Pearl, J. (2009). *Causality: Models, reasoning, and inference*. Berkeley, CA: California University Press.

Pedrycz, W. (2011, August). Keynote: Human centricity and perception-based perspective of architectures of granular computing. In *Proceedings of the 10th IEEE International Conference on Cognitive Informatics and Cognitive Computing*, Banff, AB, Canada (p. 3). Washington, DC: IEEE Computer Society.

Pedrycz, W. (1998). Shadowed sets: representing and processing fuzzy sets. *IEEE Transactions on Systems, Man, and Cybernetics. Part B, 28*, 103–109.

Pedrycz, W. (1999). Shadowed sets: bridging fuzzy and rough sets. In Pal, S. K., & Skowron, A. (Eds.), *Rough fuzzy hybridization: A new trend in decision-making* (pp. 179–199). Berlin, Germany: Springer-Verlag.

Pedrycz, W. (2005). Interpretation of clusters in the framework of shadowed sets. *Pattern Recognition Letters, 26*(15), 2439–2449. doi:10.1016/j.patrec.2005.05.001

Pedrycz, W., & Gomide, F. (2007). *Fuzzy systems engineering: Toward human-centric computing*. Hoboken, NJ: John Wiley & Sons.

Pedrycz, W., & Hirota, K. (2008). A consensus-driven clustering. *Pattern Recognition Letters, 29*, 1333–1343. doi:10.1016/j.patrec.2008.02.015

Pedrycz, W., & Rai, P. (2008). Collaborative clustering with the use of Fuzzy C-Means and its quantification. *Fuzzy Sets and Systems, 159*(18), 2399–2427. doi:10.1016/j.fss.2007.12.030

Pedrycz, W., & Song, M. (2011). Analytic Hierarchy Process (AHP) in group decision making and its optimization with an allocation of information granularity. *IEEE Transactions on Fuzzy Systems, 19*(3), 527–539. doi:10.1109/TFUZZ.2011.2116029

Peitgen, H. O., Jürgens, H., & Saupe, D. (2004). *Chaos and fractals: New frontiers of science*. New York, NY: Springer Science & Business Media.

Peng, J., Zhang, D., & Liao, X. (2009). Design of a novel image block encryption algorithm based on chaotic systems. In *Proceedings of the IEEE 8th International Conference on Cognitive Informatics* (pp. 215-221).

Peng, J., Zhang, D., Liu, Y. G., & Liao, X. F. (2008). A double-piped iterated hash function based on a hybrid of chaotic maps. In *Proceedings of the 7th IEEE International Conference on Cognitive Informatics* (pp. 358-365). Washington, DC: IEEE Computer Society Press.

Peng, F., Tang, K., Chen, G., & Yao, X. (2010). Population-based algorithm portfolios for numerical optimization. *IEEE Transactions on Evolutionary Computation, 14*(5), 782–800. doi:10.1109/TEVC.2010.2040183

Pescovitz, D. (2002). Autonomic computing: Helping computers help themselves. *IEEE Spectrum, 39*(9), 49–53. doi:10.1109/MSPEC.2002.1030968

Peterson, W. W. (1962). *Error-correcting codes*. Cambridge, MA: MIT Press.

Petitto, L. (2005). How the brain begets language. In McGilvray, J. (Ed.), *The Cambridge Companion to Chomsky* (pp. 84–101). Cambridge, UK: Cambridge University Press. doi:10.1017/CCOL0521780136.005

Picard, R. W. (1997). *Affective computing*. Cambridge, MA: MIT Press.

Picard, R. W. (2003). Affective computing: Challenges. *International Journal of Human-Computer Studies, 59*(1), 55–64. doi:10.1016/S1071-5819(03)00052-1

Picard, R. W., Vyzas, E., & Healey, J. (2001). Toward machine emotional intelligence: Analysis of affective physiological state. *IEEE Transactions on Pattern Analysis and Machine Intelligence, 23*(10), 1175–1191. doi:10.1109/34.954607

Pinel, J. P. J. (1997). *Biopsychology* (3rd ed.). Needham Heights, MA: Allyn and Bacon.

Post, F. J., van Walsum, T., Post, F. H., & Silver, D. (1995). Iconic techniques for feature visualization. *IEEE Visualization*, 288-295.

Powell, M. (1964). An efficient method for finding the minimum of a function of several variables without calculating derivatives. *The Computer Journal, 7*(2), 155–162. doi:10.1093/comjnl/7.2.155

Price, J. R., & Gee, T. F. (2005). Face recognition using direct, weighted linear discriminant analysis and modular subspaces. *Pattern Recognition Letters, 38*, 209–219.

Publio, R., Oliveira, R. F., & Roquea, A. C. (2006). A realistic model of rod photoreceptor for use in a retina network model. *Neurocomputing, 69*(10-12), 1020–1024. doi:10.1016/j.neucom.2005.12.037

Pulvermüller, F., & Fadiga, L. (2010). Active perception: sensorimotor circuits as a cortical basis for language. *Nature Reviews. Neuroscience, 11,* 351–360. doi:10.1038/nrn2811

Qian, Y. H., Liang, J. Y., & Dang, C. Y. (2009). Knowledge structure, knowledge granulation and knowledge distance in a knowledge base. *International Journal of Approximate Reasoning, 50*(1), 174–188. doi:10.1016/j.ijar.2008.08.004

Qu, Z., Ma, T., & Zhang, Y. (2008). Application of parameter modulation in e-Commerce security based on chaotic encryption. In *Proceedings of the International Symposium in Electronic Commerce and Security* (pp. 390-393).

Quillian, R. (1968). Semantic memory. In Minsky, M. (Ed.), *Semantic Information Processing.* Cambridge, MA: MIT Press.

Quinlan, J. R. (1983). Learning efficient classification procedures and their application to chess endgames. In Michalski, J. S., Carbonell, J. G., & Michell, T. M. (Eds.), *Machine learning: An artificial intelligence approach* (Vol. 1, pp. 463–482). San Francisco, CA: Morgan Kaufmann.

Rahman, M. M., Bhattacharya, P., & Desai, B. C. (2009). A unified image retrieval framework on local visual and semantic concept-based feature spaces. *Journal of Visual Communication and Image Representation, 20*(7), 450–462. doi:10.1016/j.jvcir.2009.06.001

Rajarshi, R. (2007). Synchronization, chaos and consistency. In *Proceedings of the Quantum Electronics and Laser Science Conference* (pp. 1-2).

Rangarajan, G., & Ding, M. (Eds.). (2003). *Processes with long-range correlations: theory and applications.* Berlin, Germany: Springer-Verlag. doi:10.1007/3-540-44832-2

Rao, R., & Ephremides, A. (1988). On the stability of interacting queues in multi-access system. *IEEE Transactions on Information Theory, 34*(5), 918–930. doi:10.1109/18.21216

Reddick, W. E., Glass, J. O., Cook, E. N., Elkin, T. D., & Deaton, R. J. (1997). Automated segmentation and classification of multispectral magnetic resonance images of brain using artificial neural networks. *IEEE Transactions on Medical Imaging, 16*(6), 911–918. doi:10.1109/42.650887

Regev, O. (2006). Lattice-based cryptography. In C. Dwork (Ed.), *Proceedings of the 26th Annual International Cryptology Conference on Advances in Cryptography* (LNCS 4117, pp. 131-141).

Rhee, M. Y. (1989). *Error correcting coding theory.* New York, NY: McGraw-Hill.

Riesenhuber, M., & Poggio, T. (2000). Models of object recognition. *Nature Neuroscience, 3,* 1199–1204. doi:10.1038/81479

RinkWorks. (2010). *Computer stupidities.* Retrieved from http://www.rinkworks.com/stupid/

Risinger, L., & Kaikhah, K. (2008). Motion detection and object tracking with discrete leaky integrate-and-fire neurons. *Applied Intelligence, 29*(3), 248–262. doi:10.1007/s10489-007-0092-9

Ritter, F. E., Baxter, G. D., Jones, G., & Young, R. M. (2000). Supporting cognitive models as users. *ACM Transactions on Computer-Human Interaction, 7*(2), 141–173. doi:10.1145/353485.353486

Rivest, R., Shamir, A., & Adelman, L. (1978). A method of obtaining digital signatures and public-key cryptosystems. *Communications of the ACM, 21*(2), 120–126. doi:10.1145/359340.359342

Rohde, G. K., Nichols, J. M., & Bucholtz, F. (2008). Chaotic signal detection and estimation based on attractor sets: application to secure communications. *Chaos (Woodbury, N.Y.), 18*(1). doi:10.1063/1.2838853

Rolls, E. T. (2005). *Emotion explained.* Oxford, UK: Oxford University Press. doi:10.1093/acprof:oso/9780198570035.001.0001

Rolls, E. T. (2008). *Memory, attention, and decision-making: A unifying computational neuroscience approach.* Oxford, UK: Oxford University Press.

Rolls, E. T. (2008). The affective and cognitive processing of touch, oral texture, and temperature. *Journal of Neuroscience and Biobehavioral Reviews, 34*, 237–245. doi:10.1016/j.neubiorev.2008.03.010

Rong, L., Yi, S., & Qiang, W. (2004, May 18-20). Edge detection based on early vision model incorporating improved directional median filtering. In *Proceedings of the 21st IEEE Conference on Instrumentation and Measurement Technology.*

Rosing, M. (1998). *Implementing elliptic curve cryptography.* Greenwich, CT: Manning.

Roska, B., Molnar, A., & Werblin, F. S. (2006). Parallel processing in retinal ganglion cells: How integration of space-time patterns of excitation and inhibition form the spiking output. *Journal of Neurophysiology, 95*(6), 3810–3822. doi:10.1152/jn.00113.2006

Rössler, O. (1977). Continuous chaos. In *Proceedings of the International Workshop on Synergetics at Schloss Elmau.*

Ross, T. J. (1995). *Fuzzy logic with engineering applications.* New York, NY: McGraw-Hill.

Roweis, S., & Saul, L. (2000). Nonlinear dimensionality reduction by locally linear embedding. *Science, 290*, 2323–2326. doi:10.1126/science.290.5500.2323

Rule, M. L. (2004). *The rule markup initiative.* Retrieved from http://ruleml.org/

Russel, B. (1903). *The principles of mathematics.* London, UK: George Allen & Unwin.

Russell, B. (1996). *The principles of mathematics, 1903.* New York, NY: W.W. Norton.

Russell, S. J., & Norvig, P. (2003). *Artificial intelligence: A modern approach* (2nd ed.). Upper Saddle River, NJ: Pearson Education.

Russell, S., & Norvig, P. (2010). *Artificial intelligence: A modern approach.* Upper Saddle River, NJ: Prentice Hall.

Sadek, A. K., Liu, K. J. R., & Ephremides, A. (2007). Cognitive multiple access via cooperation: Protocol design and performance analysis. *IEEE Transactions on Information Theory, 53*(10), 3677–3696.

Sahoo, N., Xu, J., & Panda, S. (2001). Low torque ripple control of switched reluctance motors using iterative learning. *IEEE Transactions on Energy Conversion, 16*(4), 318–326. doi:10.1109/60.969470

Salcedo, C. G., & Whitley, D. (1999). Feature selection mechanisms for ensemble creation: A genetic search perspective. In Freitas, A. A. (Ed.), *Data mining with evolutionary algorithms: Research directions.* Menlo Park, CA: AAAI Press.

Salton, G., Wong, A., & Yang, C. S. (1975). A vector space model for automatic indexing. *Communications of the ACM, 18*(11), 613–620. doi:10.1145/361219.361220

Saul, L. K., & Roweis, S. T. (2004). Think globally, fit locally: Unsupervised learning of low dimensional manifolds. *Journal of Machine Learning Research, 4*(2), 119–155. doi:10.1162/153244304322972667

Schiller, P. H., Slocum, W. M., & Weiner, V. S. (2007). How the parallel channels of the retina contribute to depth processing. *The European Journal of Neuroscience, 26*(5), 1307–1321. doi:10.1111/j.1460-9568.2007.05740.x

Schneier, B. (1995). *Applied cryptography: Protocols, algorithms and source code in C* (2nd ed.). New York, NY: John Wiley & Sons.

Schneier, B. (1996). *Applied cryptography* (2nd ed.). New York, NY: John Wiley & Sons Press.

Schoning, U. (1989). *Logic for computer scientists.* Boston, MA: Birkhauser.

Schor, D., & Kinsner, W. (2010, July). A study of particle swarm optimization for cognitive machines. In *Proceedings of the 9th International Conference on Cognitive Informatics*, Beijing, China (pp. 26-33). Washington, DC: IEEE Computer Society.

Schor, D., Kinsner, W., & Anderson, J. (2010, May). A study of optimal topologies in swarm intelligence. In *Proceedings of the IEEE Canadian Conference on Electrical and Computer Engineering*, Calgary, AB, Canada (pp. 1-8). Washington, DC: IEEE Computer Society.

Schroder, P., & Sweldens, W. (2001). Digital geometry processing. In *Proceedings of the Sixth Annual Symposium on Frontiers of Engineering* (pp. 41-44).

Scutari, G., Palomar, D. P., & Barbarossa, S. (2008). Cognitive MIMO radio. *IEEE Signal Processing Magazine*, *25*(6), 46–59. doi:10.1109/MSP.2008.929297

Searle, J. R. (1980). Minds, brains, and programs. *The Behavioral and Brain Sciences*, *3*, 417–457. doi:10.1017/S0140525X00005756

Setoodeh, P., & Haykin, S. (2009). Robust transmit power control for cognitive radio. *Proceedings of the IEEE*, *97*(5), 915–939. doi:10.1109/JPROC.2009.2015718

Sha, F., & Saul, L. K. (2005). Analysis and extension of spectral methods for nonlinear dimensionality reduction. In *Proceedings of the 22nd International Conference on Machine Learning*, Bonn, Germany (pp. 785-792).

Shah, S., & Levine, M. D. (1996). Visual information processing in primate cone pathways. 1. A model. *IEEE Transactions on Systems, Man, and Cybernetics. Part B, Cybernetics*, *26*(2), 259–274. doi:10.1109/3477.485837

Shah, S., & Levine, M. D. (1996). Visual information processing in primate cone pathways. 2. Experiments. *IEEE Transactions on Systems, Man, and Cybernetics. Part B, Cybernetics*, *26*(2), 275–289. doi:10.1109/3477.485878

Shannon, C. E. (1948). A mathematical theory of communication. *The Bell System Technical Journal*, *27*, 379–423, 623–656.

Shannon, C. E. (1949). Communication theory of secrecy system. *The Bell System Technical Journal*, *28*(4), 656–715.

Shannon, C. E., & Weaver, W. (1949). *The mathematical theory of communication*. Urbana, IL: Illinois University Press.

Shastri, L., & Grannes, D. J. (1996). A connectionist treatment of negation and inconsistency. In *Proceedings of the 18th Annual Conference of the Cognitive Science Society*, San Diego, CA (pp. 142-147).

Shaw, D., & Kinsner, W. (1996, May). Chaotic simulated annealing in multilayer feedforward networks. In *Proceedings of the IEEE Canadian Conference on Electrical and Computer Engineering*, Calgary, AB, Canada (Vol. 1, pp. 265-269). Washington, DC: IEEE Computer Society.

Shimonomura, K., Kushima, T., & Yagi, T. (2008). Binocular robot vision emulating disparity computation in the primary visual cortex. *Neural Networks*, *21*(2-3), 331–340. doi:10.1016/j.neunet.2007.12.033

Shipp, C. A., & Kuncheva, L. I. (2002). Relationships between combination methods and measures of diversity in combining classifiers. *Information Fusion*, *3*(2), 135–148. doi:10.1016/S1566-2535(02)00051-9

Shi, Y., & Eberhart, R. C. (1999). Empirical study of particle swarm optimization. In []. Washington, DC: IEEE Computer Society.]. *Proceedings of the Congress on Evolutionary Computation*, *3*, 1945–1950.

Sidi, M., & Segall, A. (1983). Two interfering queues in packet-radio networks. *IEEE Transactions on Wireless Communications*, *31*(1), 123–129.

Silver, D., & Wang, X. (1997). Tracking and visualizing turbulent 3D features. *IEEE Transactions on Visualization and Computer Graphics*, *3*(2), 129–141. doi:10.1109/2945.597796

Simeone, O., Gambini, J., Bar-Ness, Y., & Spagnolini, U. (2007). Cooperation and cognitive radio in Peter McLane. In *Proceedings of the IEEE International Conference on Communications* (pp. 6511-6515).

Simeone, O., Bar-Ness, Y., & Spagnolini, U. (2007). Stable throughput of cognitive radios with and without relaying capability. *IEEE Transactions on Wireless Communications*, *55*(12), 2351–2360.

Simon, H. A. (1982). *Models of bounded rationality*. Cambridge, MA: MIT Press.

Simon, M. K., & Alouini, M.-S. (2005). *Digital communication over fading channels*. New York, NY: John Wiley & Sons.

Sim, T., Baker, S., & Bsat, M. (2003). The CMU pose, illumination, and expression database. *IEEE Transactions on Pattern Analysis and Machine Intelligence*, *25*(12), 1615–1618. doi:10.1109/TPAMI.2003.1251154

Sinkov, A. (1968). *Elementary cryptoanalysis: A mathematical approach*. New York, NY: Random House.

Skowron, A., & Pal, S. K. (2003). Rough sets, pattern recognition, and data mining. *Pattern Recognition Letters*, *24*(6), 829–933.

Slezak, D., & Ziarko, W. (2002, December 9). Bayesian rough set model. In *Proceedings of the Conference on the Foundation of Data Mining*, Maebashi, Japan (pp. 131-135).

Slezak, D., & Ziarko, W. (2005). The investigation of the Bayesian rough set model. *International Journal of Approximate Reasoning*, *40*, 81–91. doi:10.1016/j.ijar.2004.11.004

Smith, S. F. (1980). *A learning system based on genetic adaptive algorithms.* Unpublished doctoral dissertation, University of Pittsburgh, PA.

Smith, E. E. (1989). Concepts and induction. In Posner, M. I. (Ed.), *Foundations of cognitive science* (pp. 501–526). Cambridge, MA: MIT Press.

Smith, E. E., & Medin, D. L. (1981). *Categories and concepts.* Cambridge, MA: Harvard University Press.

Socha, K. (2008). *Ant colony optimization for continuous and mixed-variable domains.* Unpublished doctoral dissertation, Universite Libre de Bruxelles, Belgium.

Sowa, J. F. (1984). *Conceptual structures, information processing in mind and machine.* Reading, MA: Addison-Wesley.

Sperschneider, V., & Antoniou, G. (1991). *Logic: A foundation for computer science.* Reading, MA: Addison-Wesley.

Spong, M., Marino, R., Peresada, S., & Taylor, D. (1987). Feedback linearizing control of switched reluctance motors. *IEEE Transactions on Automatic Control*, *32*(5), 371–379. doi:10.1109/TAC.1987.1104616

Squire, L. R., Knowlton, B., & Musen, G. (1993). The structure and organization of memory. *Annual Review of Psychology*, *44*, 453–459. doi:10.1146/annurev.ps.44.020193.002321

Steedman, M., & Baldridge, J. (in press). Combinatory categorial grammar. In Borsley, R., & Borjars, K. (Eds.), *Non-transformational syntax: Formal and explicit models of grammar.* New York, NY: John Wiley & Sons.

Sternberg, R. J. (1998). *In search of the human mind* (2nd ed.). Orlando, FL: Harcourt Brace & Co.

Stinson, D. R. (2006). *Cryptography theory and practice* (3rd ed.). Boca Raton, FL: Chapman & Hall/CRC.

Strogatz, S. H. (2000). *Nonlinear dynamics and chaos.* Cambridge, MA: Westview Press, Perseus Books.

Sun, F., Wang, Y., Lu, J., Zhang, B., Kinsner, W., & Zadeh, L. A. (Eds.). (2010, July). *Proceedings of the 9th IEEE International Conference on Cognitive Informatics*, Beijing, China. Washington, DC: IEEE Computer Society.

Sung, W. T., Hsu, Y. C., & Chen, K. Y. (2010). Enhance information acquired efficiency for wireless sensors networks via multi-bit decision fusion. In *Proceedings of the IEEE 9th International Conference on Cognitive Informatics* (pp. 154-159).

Sun, R. (2000). Symbol grounding: A new look at an old idea. *Philosophical Psychology*, *13*, 149–172. doi:10.1080/09515080050075663

Sun, R., & Peterson, T. (1998). Some experiments with a hybrid model for learning sequential decision making. *Information Sciences*, *111*, 83–107. doi:10.1016/S0020-0255(98)00007-3

Sutton, R. S., & Barto, A. (1998). *Reinforcement learning.* Cambridge, MA: MIT Press.

Swenson, C. (2008). *Modern cryptanalysis: Techniques for advanced code breaking.* New York, NY: John Wiley & Sons.

Szpankowski, W. (1988). Stability conditions for multidimensional queueing systems with computer applications. *Operations Research*, *36*(6), 944–957. doi:10.1287/opre.36.6.944

Taddeo, M., & Floridi, L. (2005). Solving the symbol grounding problem: a critical review of fifteen years of research. *Journal of Experimental & Theoretical Artificial Intelligence*, *17*(4), 419–445. doi:10.1080/09528130500284053

Tanaka, H. T., Ikeda, M., & Chiaki, H. (1998). Curvature-base face surface recognition using spherical correlation – principal directions for curved object recognition. In *Proceedings of the International Conference on Automatic Face Gesture Recognition* (pp. 372-377).

Tang, Q. L., Sang, N., & Zhang, T. X. (2005). A neural network model for extraction of salient contours. In J. Wang, X.-F. Liao, & Z. Yi (Eds.), *Proceedings of the Second International Symposium on Advances in Neural Networks* (LNCS 3497, pp. 316-320).

Tang, Y., & Zhang, L. (2005). Adaptive bucket formation in encrypted databases. In *Proceedings of the IEEE International Conference in e-Technology, e-Commerce and e-Service* (pp. 116-119).

Tang, G. P., Liao, X. F., & Chen, Y. (2005). A novel method for designing S-boxes based on chaotic maps. *Chaos, Solitons, and Fractals, 23*(2), 413–419. doi:10.1016/j.chaos.2004.04.023

Tan, X., Zhao, D., Yi, J., & Xu, D. (2009). Adaptive integrated control for omnidirectional mobile manipulators based on neural-network. *International Journal of Cognitive Informatics and Natural Intelligence, 3*(4), 34–53. doi:10.4018/jcini.2009062303

Tattersall, I. (1998). *The origin of the human capacity (68th James Arthur Lecture on the Evolution of the Human Brain)*. New York, NY: American Museum of Natural History.

Tattersall, I. (2008). An evolutionary framework for the acquisition of symbolic cognition by *Homo sapiens. Comparative Cognition Behavior Reviews, 3*, 99–114.

Tattersall, I. (2010). Human evolution and cognition. *Theory in Biosciences, 129*, 193–201. doi:10.1007/s12064-010-0093-9

Taubin, G. (2000). *Geometric signal processing on polygonal meshes.* Paper presented at the Eurographics Workshop State of the Art Report (STAR), Interlaken, Switzerland.

Tian, Y., Wang, Y., Gavrilova, M. L., & Ruhe, G. (in press). A formal knowledge representation system for the cognitive learning engine. In *Proceedings of the 10th IEEE International Conferences on Cognitive Informatics and Cognitive Computing.* Washington, DC: IEEE Computer Society.

Tian, Y., Wang, Y., Gavrilova, M., & Rehe, G. (2011). A formal knowledge representation system for the cognitive learning engine. *International Journal of Software Science and Computational Intelligence, 3*(4).

Tian, Y., Wang, Y., & Hu, K. (2009). A knowledge representation tool for autonomous machine learning based on concept algebra. *Transactions of Computational Science, 5*, 143–160. doi:10.1007/978-3-642-02097-1_8

Tomassi, P. (1999). *Logic.* London, UK: Routledge. doi:10.4324/9780203197035

Tong, X. J., & Cui, M. G. (2009). Image encryption scheme based on 3D baker with dynamical compound chaotic sequence cipher generator. *Signal Processing, 89*(4), 480–491. doi:10.1016/j.sigpro.2008.09.011

Torkkola, K. (2003). Discriminative features for text document classification. *Pattern Analysis & Applications, 6*(4), 301–308.

Trelea, I. C. (2003). The particle swarm optimization algorithm: Convergence analysis and parameter selection. *Information Processing Letters, 85*(6), 317–325. doi:10.1016/S0020-0190(02)00447-7

Tsui, C. (400BC). Autumn water (Ch. 17). In *Outer chapters.*

Tsybakov, B. S., & Mikhailov, V. A. (1979). Ergodicity of a slotted ALOHA system. *Problems of Information Transmission, 15*(4), 73–87.

Tsymbal, A., Pechenizkiy, M., & Cunningham, P. (2005). Diversity in search strategies for ensemble feature selection. *Information Fusion, 6*(1), 83–98. doi:10.1016/j.inffus.2004.04.003

Turk, M., & Pentland, A. (1991). Eigenfaces for recognition. *Journal of Cognitive Neuroscience, 3*(1), 71–86. doi:10.1162/jocn.1991.3.1.71

Turner, S. (2010). *Causality.* Thousand Oaks, CA: Sage.

Ueta, T., & Chen, G. R. (2000). Bifurcation analysis of Chen's equation. *International Journal of Bifurcation and Chaos in Applied Sciences and Engineering, 10*(8), 1917–1931. doi:10.1142/S0218127400001183

Van Assche, G. (2006). *Quantum cryptography and secret-key distilation.* Cambridge, UK: Cambridge University Press. doi:10.1017/CBO9780511617744

van Mechelen, I., Hampton, J., Michalski, R. S., & Theuns, P. (Eds.). (1993). *Categories and concepts, theoretical views and inductive data analysis.* New York, NY: Academic Press.

Vanstone, S. A., & van Oorshot, P. C. (1989). *An introduction to error correcting codes with applications.* Boston, MA: Kluwer Academic.

Vapnik, V. N. (1998). *The statistical learning theory*. New York, NY: Springer.

Verveer, P., & Duin, R. (1995). An evaluation of intrinsic dimensionality estimators. *IEEE Transactions on Pattern Analysis and Machine Intelligence, 17*(1), 81–86. doi:10.1109/34.368147

Vossen, P. (Ed.). (1998). *EuroWordNet: A multilingual database with lexical semantic networks*. Dordrecht, The Netherlands: Kluwer Academic.

W3C. (2004). *SWRL: A semantic web rule language combining OWL and RuleML*. Retrieved from http://www.w3.org/Submission/SWRL/

W3C. (2007). *SOAP: Version 1.2 part 1: Messaging framework*. Retrieved from http://www.w3.org/TR/soap12-part1/

W3C. (2009). *OWL: Web ontology language*. Retrieved from http://www.w3.org/TR/owl-features/

Wai, R. (2002). Hybrid fuzzy neural-network control for nonlinear motor-toggle servomechanism. *IEEE Transactions on Control Systems Technology, 10*(4), 519–532. doi:10.1109/TCST.2002.1014672

Wai, R., Huang, Y., Yang, Z., & Shih, C. (2010). Adaptive fuzzy-neural-network velocity sensorless control for robot manipulator position tracking. *IET Control Theory & Applications, 4*(6), 1079–1093. doi:10.1049/iet-cta.2009.0166

Wakerley, J. (1978). *Error correcting codes, self-checking circuits and applications*. New York, NY: North-Holland.

Wallace, A. (1871). *Contributions to the theory of natural selection*. New York, NY: Macmillan.

Wallace, R., & Taylor, D. (1991). Low-torque-ripple switched reluctance motors for direct-drive robotics. *IEEE Transactions on Robotics and Automation, 7*(6), 733–742. doi:10.1109/70.105382

Wang, Q. D., Wang, X. J., & Huang, H. (2004). Rough set based feature ensemble learning. In *Proceedings of the 5th World Congress on Intelligent Control and Automation* (pp. 1890-1894).

Wang, Y. (2002). Keynote: On cognitive informatics. In Proceedings of the 1st IEEE International Conference on Cognitive Informatics, Calgary, AB, Canada (pp. 34-42).

Wang, Y. (2002). On cognitive informatics. In *Proceedings of the International Conference on Cognitive Informatics* (pp. 34-42).

Wang, Y. (2003). On cognitive informatics. *Brain and Mind: A Transdisciplinary Journal of Neuroscience and Neurophilisophy, 4*(3), 151-167.

Wang, Y. (2006). On concept algebra and knowledge representation. In Proceedings of the 5th IEEE International Conference on Cognitive Informatics, Beijing, China (pp. 320-331).

Wang, Y. (2006, July). Keynote: Cognitive informatics - towards the future generation computers that think and feel. In *Proceedings of the 5th IEEE International Conference on Cognitive Informatics*, Beijing, China (pp. 3-7). Washington, DC: IEEE Computer Society.

Wang, Y. (2007). *Software engineering foundations: A software science perspective* (CRC Series in Software Engineering, Vol. 2). Boca Raton, FL: CRC Press.

Wang, Y. (2007a, August). Cognitive information foundations of nature and machine intelligence. In *Proceedings of the 6th IEEE International Conference on Cognitive Informatics*, Lake Tahoe, CA (pp. 3-12).

Wang, Y. (2007d, August). The theoretical framework and cognitive process of learning. In *Proceedings of the 6th IEEE International Conference on Cognitive Informatics* (pp. 470-479).

Wang, Y. (2009c, June). Fuzzy inferences methodologies for cognitive informatics and computational intelligence. In *Proceedings of the 8th IEEE International Conference on Cognitive Informatics*, Hong Kong (pp. 241-248).

Wang, Y. (2009e, June). Granular algebra for modeling granular systems and granular computing. In *Proceedings of the 8th IEEE International Conference on Cognitive Informatics*, Hong Kong (pp. 145-154). Washington, DC: IEEE Computer Society.

Wang, Y. (2010a, July). Keynote: Cognitive computing and World Wide Wisdom (WWW+). In *Proceedings of the 9th IEEE International Conference on Cognitive Informatics*, Beijing, China. Washington, DC: IEEE Computer Society.

Wang, Y. (2010b, August). Keynote: Cognitive informatics and denotational mathematics means for brain informatics. In *Proceedings of the 1st International Conference on Brain Informatics*, Toronto, ON, Canada.

Wang, Y. (2011b, August 18-20). Keynote: On inference algebra: A formal means for machine reasoning and cognitive computing. In *Proceedings of the 10th IEEE International Conference on Cognitive Informatics and Cognitive Computing*, Banff, AB, Canada (pp. 4-6).

Wang, Y. (2012). Inference Algebra (IA): A denotational mathematics for cognitive computing and machine reasoning (II). *International Journal of Cognitive Informatics and Natural Intelligence, 6*(1).

Wang, Y., Celikyilmaz, A., Kinsner, W., Pedrycz, W., Leung, H., & Zadeh, L. A. (Eds.). (2011, August). *Proceedings of the 10th IEEE International Conference on Cognitive Informatics and Cognitive Computing*, Banff, AB, Canada. Washington, DC: IEEE Computer Society.

Wang, Y., Liu, D., & Wang, Y. (2003). Discovering the capacity of human memory. *Brain and Mind: A Transdisciplinary Journal of Neuroscience and Neurophilosophy, 4*(2), 189-198.

Wang, Y., Tian, Y., & Hu, K. (2011, August). The operational semantics of concept algebra for cognitive computing and machine learning. *Proceedings of the 10th IEEE International Conference on Cognitive Informatics and Cognitive Computing*, Banff, AB, Canada. Washington, DC: IEEE Computer Society.

Wang, Y., Zhang, D., Latombe, J.-C., & Kinsner, W. (Eds.). (2008, August). *Proceedings of the 7th IEEE International Conference on Cognitive Informatics*, Stanford, CA. Washington, DC: IEEE Computer Society.

Wang, Y. (in press). On cognitive models of causal inferences and causation networks. International Journal of Software Science and Computational Intelligence, 3*(1)*.

Wang, C., Zhao, J., He, X. F., Chen, C., & Bu, J. J. (2009). Image retrieval using nonlinear manifold embedding. *Neurocomputing, 72*(16-18), 3922–3929. doi:10.1016/j.neucom.2009.04.011

Wang, G. Y. (2001). *Rough set theory and knowledge acquisition.* Shaanxi, China: Xi'an Jiaotong University Press.

Wang, G. Y. (2003). Rough reduction in algebra view and information view. *International Journal of Intelligent Systems, 18*(6), 679–688. doi:10.1002/int.10109

Wang, Y. (2002). The Real-Time Process Algebra (RTPA). *Annals of Software Engineering,* (14): 235–274. doi:10.1023/A:1020561826073

Wang, Y. (2003). On cognitive informatics: Brain and mind. *Transdisciplinary Journal of Neuroscience and Neurophilosophy, 4*(2), 151–167.

Wang, Y. (2003). Cognitive informatics: A new transdisciplinary research filed. *Brain and Mind, 4*(2), 115–127. doi:10.1023/A:1025419826662

Wang, Y. (2003). Using process algebra to describe human and software behavior. *Brain and Mind, 4*(2), 199–213. doi:10.1023/A:1025457612549

Wang, Y. (2007). Cognitive informatics: exploring theoretical foundations for natural intelligence, neural Informatics, autonomic computing, and agent systems. *International Journal of Cognitive Informatics and Natural Intelligence, 1*, 1–10. doi:10.4018/jcini.2007040101

Wang, Y. (2007). The theoretical framework of cognitive informatics. *International Journal of Cognitive Informatics and Natural Intelligence, 1*(1), 1–27. doi:10.4018/jcini.2007010101

Wang, Y. (2007). *Software engineering foundations: A software science perspective* (*Vol. 2*). Boca Raton, FL: Auerbach.

Wang, Y. (2007). The cognitive processes of formal inferences. *International Journal of Cognitive Informatics and Natural Intelligence, 1*(4), 75–86. doi:10.4018/jcini.2007100106

Wang, Y. (2007). The theoretical framework of cognitive informatics. *International Journal of Cognitive Informatics and Natural Intelligence, 1*(1), 1–27. doi:10.4018/jcini.2007010101

Wang, Y. (2007). On the cognitive processes of perception with emotions, motivations, and attitudes. *International Journal of Cognitive Informatics and Natural Intelligence*, *1*(4), 1–13. doi:10.4018/jcini.2007100101

Wang, Y. (2007). The OAR model of neural informatics for internal knowledge representation in the brain. *International Journal of Cognitive Informatics and Natural Intelligence*, *1*(3), 64–75. doi:10.4018/jcini.2007070105

Wang, Y. (2007). Towards theoretical foundations of autonomic computing. *International Journal of Cognitive Informatics and Natural Intelligence*, *1*(3), 1–16. doi:10.4018/jcini.2007070101

Wang, Y. (2008). On concept algebra: A denotational mathematical structure for knowledge and software modeling. *International Journal of Cognitive Informatics and Natural Intelligence*, *2*(2), 1–19. doi:10.4018/jcini.2008040101

Wang, Y. (2008). On contemporary denotational mathematics for computational intelligence. *Transactions of Computational Science*, *2*, 6–29. doi:10.1007/978-3-540-87563-5_2

Wang, Y. (2008). RTPA: A denotational mathematics for manipulating intelligent and computational behaviors. *International Journal of Cognitive Informatics and Natural Intelligence*, *2*(2), 44–62. doi:10.4018/jcini.2008040103

Wang, Y. (2008). Deductive semantics of RTPA. *International Journal of Cognitive Informatics and Natural Intelligence*, *2*(2), 95–121. doi:10.4018/jcini.2008040106

Wang, Y. (2008). On contemporary denotational mathematics for computational intelligence. *Transactions of Computational Science*, *2*, 6–29. doi:10.1007/978-3-540-87563-5_2

Wang, Y. (2008). On system algebra: A denotational mathematical structure for abstract system modeling. *International Journal of Cognitive Informatics and Natural Intelligence*, *2*(2), 20–42. doi:10.4018/jcini.2008040102

Wang, Y. (2009). Toward a formal knowledge system theory and its cognitive informatics foundations. *Transactions of Computational Science*, *5*, 1–19. doi:10.1007/978-3-642-02097-1_1

Wang, Y. (2009). A formal syntax of natural languages and the deductive grammar. *Fundamenta Informaticae*, *90*(4), 353–368.

Wang, Y. (2009). Cognitive computing. *International Journal of Software Science and Computational Intelligence*, *1*(3). doi:10.4018/jssci.2009070101

Wang, Y. (2009). Formal description of the cognitive process of memorization. *Transactions of Computational Science*, *5*, 81–98. doi:10.1007/978-3-642-02097-1_5

Wang, Y. (2009). On abstract intelligence: Toward a unifying theory of natural, artificial, machinable, and computational intelligence. *International Journal of Software Science and Computational Intelligence*, *3*(1), 1–17. doi:10.4018/jssci.2009010101

Wang, Y. (2009). On abstract intelligence: Toward a unified theory of natural, artificial, machinable, and computational intelligence. *International Journal of Software Science and Computational Intelligence*, *1*(1), 1–18. doi:10.4018/jssci.2009010101

Wang, Y. (2009). Paradigms of denotational mathematics for cognitive informatics and cognitive computing. *Fundamenta Informaticae*, *90*(3), 282–303.

Wang, Y. (2009). On Visual Semantic Algebra (VSA): A denotational mathematical structure for modeling and manipulating visual objects and patterns. *International Journal of Software Science and Computational Intelligence*, *1*(4), 1–15. doi:10.4018/jssci.2009062501

Wang, Y. (2009). A cognitive informatics reference model of Autonomous Agent Systems (AAS). *International Journal of Cognitive Informatics and Natural Intelligence*, *3*(1), 1–16. doi:10.4018/jcini.2009010101

Wang, Y. (2009). Paradigms of denotational mathematics for cognitive informatics and cognitive computing. *Fundamenta Informaticae*, *90*(3), 282–303.

Wang, Y. (2010). A sociopsychological perspective on collective intelligence in metaheuristic computing. *International Journal of Applied Metaheuristic Computing*, *1*(1), 110–128. doi:10.4018/jamc.2010102606

Wang, Y. (2010). Cognitive robots: A reference model towards intelligent authentication. *IEEE Robotics and Automation*, *17*(4), 54–62. doi:10.1109/MRA.2010.938842

Wang, Y. (2010). A sociopsychological perspective on collective intelligence in metaheuristic computing. *International Journal of Applied Metaheuristic Computing*, *1*(1), 110–128. doi:10.4018/jamc.2010102606

Wang, Y. (2010). Cognitive robots: A reference model towards intelligent authentication. *IEEE Robotics and Automation*, *17*(4), 54–62. doi:10.1109/MRA.2010.938842

Wang, Y. (2010). On concept algebra for computing with words (CWW). *International Journal of Semantic Computing*, *4*(3), 331–356. doi:10.1142/S1793351X10001061

Wang, Y. (2010). On formal and cognitive semantics for semantic computing. *International Journal of Semantic Computing*, *4*(2), 203–237. doi:10.1142/S1793351X10000833

Wang, Y. (2011). Inference Algebra (IA): A denotational mathematics for cognitive computing and machine reasoning (I). *International Journal of Cognitive Informatics and Natural Intelligence*, *5*(4).

Wang, Y. (2011). On cognitive models of causal inferences and causation networks. *International Journal of Software Science and Computational Intelligence*, *3*(1), 50–60.

Wang, Y. (2011). On concept algebra for Computing with Words (CWW). *International Journal of Semantic Computing*, *4*(3), 331–356. doi:10.1142/S1793351X10001061

Wang, Y. (Ed.). (2009). *International journal of software science and computational intelligence* (*Vol. 1*). Hershey, PA: IGI Global.

Wang, Y., Baciu, G., Yao, Y., Kinsner, W., Chan, K., & Zhang, B. (2010). Perspectives on cognitive informatics and cognitive computing. *International Journal of Cognitive Informatics and Natural Intelligence*, *4*(1), 1–29. doi:10.4018/jcini.2010010101

Wang, Y., Carley, K., Zeng, D., & Mao, W. (2007). Social computing: From social informatics to social intelligence. *IEEE Intelligent Systems*, *22*(2), 79–83. doi:10.1109/MIS.2007.41

Wang, Y., & Chiew, V. (2010). On the cognitive process of human problem solving. *Cognitive Systems Research: An International Journal*, *11*(1), 81–92. doi:10.1016/j.cogsys.2008.08.003

Wang, Y., & Kinsner, W. (2006). Recent advances in cognitive informatics. [C]. *IEEE Transactions on Systems, Man, and Cybernetics*, *36*(2), 121–123. doi:10.1109/TSMCC.2006.871120

Wang, Y., Kinsner, W., Anderson, J. A., Zhang, D., Yao, Y., & Sheu, P. (2009). A doctrine of cognitive informatics. *Fundamenta Informatica*, *90*(3), 203–228.

Wang, Y., Kinsner, W., & Zhang, D. (2009). Contemporary cybernetics and its faces of cognitive informatics and computational intelligence. *IEEE Transactions on System, Man, and Cybernetics. Part B*, *39*(4), 823–833.

Wang, Y., & Wang, Y. (2006). Cognitive informatics models of the brain. *IEEE Transactions on Systems, Man and Cybernetics. Part C, Applications and Reviews*, *36*(2), 203–207. doi:10.1109/TSMCC.2006.871151

Wang, Y., & Wang, Y. (2006). Cognitive informatics models of the brain. *IEEE Transactions on Systems, Man and Cybernetics. Part C, Applications and Reviews*, *36*(2), 203–207. doi:10.1109/TSMCC.2006.871151

Wang, Y., Wang, Y., Patel, S., & Patel, D. (2006). A layered reference model of the brain (LRMB). *IEEE Transactions on Systems, Man, and Cybernetics. Part C*, *36*(2), 124–133.

Wang, Y., Widrow, B. C., Zhang, B., Kinsner, W., Sugawara, K., & Sun, F. C. (2011). Perspectives on the field of cognitive informatics and its future development. *International Journal of Cognitive Informatics and Natural Intelligence*, *5*(1), 1–17.

Wang, Y., Zadeh, L. A., & Yao, Y. (2009). On the system algebra foundations for granular computing. *International Journal of Software Science and Computational Intelligence*, *1*(1), 1–17. doi:10.4018/jssci.2009010101

Wang, Y., Zhang, D., & Kinsner, W. (Eds.). (2010). *Advances in cognitive informatics and cognitive computing*. Berlin, Germany: Springer-Verlag.

Wang, Y., Zhang, D., & Tsumoto, S. (2009). Cognitive informatics, cognitive computing, and their denotational mathematical foundations (1). *Fundamenta Informaticae*, *90*(3), 1–7.

Washington, L. C. (2008). *Elliptic curves: Number theory and cryptography* (2nd ed.). Boca Raton, FL: Chapman and Hall/CRC.

Weiler, M., Botchen, R., Huang, J., Jang, Y., Stegmaier, S., & Gaither, K. P. (2005). Hardware-assisted feature analysis and visualization of procedurally encoded milti-level volumetric data. *IEEE Computer Graphics and Applications*, *25*(5), 72–81. doi:10.1109/MCG.2005.106

Weinberger, K. Q. Sha, & F., Saul, L. K. (2004). Learning a kernel matrix for nonlinear dimensionality reduction. In *Proceedings of the International Conference on Machine Learning*, Banff, AB, Canada (pp. 839-846).

Weinberger, K. Q., & Saul, L. K. (2006). Unsupervised learning of image manifolds by semidefinite programming. *International Journal of Computer Vision*, *70*(1), 77–90. doi:10.1007/s11263-005-4939-z

Weinberg, R. A. (1989). Intelligence and IQ: Landmark issues and great debates. *The American Psychologist*, *44*(2), 98–104. doi:10.1037/0003-066X.44.2.98

Wei, R., & Liu, C. (2009). Design of dynamic petri recurrent fuzzy neural network and its application to path-tracking control of nonholonomic mobile robot. *IEEE Transactions on Industrial Electronics*, *56*(7), 2667–2683. doi:10.1109/TIE.2009.2020077

Weise, T. (2009). *Global optimization algorithms – theory and application*. Retrieved from http://www.it-weise.de/

Weise, T., Podlich, A., Reinhard, K., Gorldt, C., & Geihs, K. (2009). Evolutionary freight transportation planning. In M. Giacobini, A. Brabazon, S. Cagnoni, G. Di Caro, A. Ekart, A. Esparcia et al. (Eds.), *Proceedings of the EvoWorkshops on Applications of Evolutionary Computing* (LNCS 5484, pp.768-777).

Widrow, B., & Aragon, J. C. (2010). Cognitive memory: Human like memory. *International Journal of Software Science and Computational Intelligence*, *2*(4), 1–15. doi:10.4018/jssci.2010100101

Widrow, B., & Hoff, M. E. (1960). Adaptive switching circuits. *IRE WESCON Convention Record*, *4*(1), 96–104.

Wiese, E. E., & Lee, J. D. (2007). Attention grounding: A new approach to in-vehicle information system implementation. *Theoretical Issues in Ergonomics Science*, *8*(3), 255–276. doi:10.1080/14639220601129269

Wille, R. (1982). Restructuring lattice theory: An approach based on hierarchies of concepts. In Rival, I. (Ed.), *Ordered sets* (pp. 445–470). Dordrecht, The Netherlands: Reidel.

Wille, R. (1992). Concept lattices and conceptual knowledge systems. *Computers & Mathematics with Applications (Oxford, England)*, *23*, 493–515. doi:10.1016/0898-1221(92)90120-7

Williams, C. K. I., & Barber, D. (1998). Bayesian classification with gaussian processes. *IEEE Transactions on Pattern Analysis and Machine Intelligence*, *20*(12), 1342–1351. doi:10.1109/34.735807

Wilson, R. A., & Keil, F. C. (2001). *The MIT encyclopaedia of the cognitive sciences*. Cambridge, MA: MIT Press.

Wolfram, S. (2002). *A new kind of science*. Champaign, IL: Wolfram Media.

Wolpert, D., & Macready, W. (1997). No free lunch theorems for optimization. *IEEE Transactions on Evolutionary Computation*, *1*(1), 67–82. doi:10.1109/4235.585893

Wroblewski, J. (2001). Ensemble of classifiers based on approximate reducts. *Fundamenta Informaticae*, *47*(3), 351–360.

Wu, L. S., Yang, Z., Basham, E., & Liu, W. T. (2008). An efficient wireless power link for high voltage retinal implant. In *Proceedings of the IEEE Conference on Biomedical Circuits and Systems Conference - Intelligent Biomedical Systems* (pp. 101-104).

Wu, Q. X., McGinnity, T. M., Maguire, L., Valderrama-Gonzalez, G. D., & Dempster, P. (2010). Colour image segmentation based on a spiking neural network model inspired by the visual system. In D. S. Huang, Z. M. Zhao, V. Bevilacqua, & J. C. Figueroa (Eds.), *Proceedings of the International Conference on Advanced Intelligent Computing Theories and Applications* (LNCS 6215, pp. 49-57).

Wu, X., Liu, W., Zhao, L., & Fu, J. S. (2001). Chaotic phase code for radar pulse compression. In *Proceedings of the IEEE National Radar Conference* (pp. 279-283).

Xiang, S. M., Nie, F. P., Zhang, C. S., & Zhang, C. X. (2006). Spline embedding for nonlinear dimensionality reduction. In *Proceedings of the European Conference on Machine Learning*, Berlin, Germany (pp. 825-832).

Xiao, Q. Z., Qin, K., Guan, Z. Q., & Wu, T. (2007). Image mining for robot vision based on concept analysis. In *Proceedings of the IEEE International Conference on Robotics and Biomimetics*, Yalong Bay, China (pp. 207-212).

Xiong, H., Swamy, M. N. S., & Ahmad, M. O. (2005). Two-dimensional FLD for face recognition. *Pattern Recognition Letters, 38*, 112–1124.

Xu, M. G., & Lin, D. (2006). Non-orthogonal precoding matrix design for MU-MIMO downlink channels. In *Proceedings of the IEEE Wireless Communications and Networking Conference* (pp. 1311-1315).

Yale University. (1997). *Yale face datasets.* Retrieved from http://cvc.yale.edu/projects/yalefaces/yalefaces.html

Yanco, H. A., Drury, J. L., & Scholtz, J. (2004). Beyond usability evaluation: Analysis of human-robot interaction at a major robotics competition. *Human-Computer Interaction, 19*(1), 117–149. doi:10.1207/s15327051hci1901&2_6

Yang, Y. (2008). *Chongqing University of Posts and Telecommunications Emotional database (CQUPTE).* Retrieved from http://cs.cqupt.edu.cn/users/ 904/docs/9317-1.rar

Yang, Y., Wang, G. Y., & He, K. (2007). An approach for selective ensemble feature selection based on rough set theory. In *Proceedings of the Second International Conference on Rough Sets and Knowledge Technology* (pp. 518-525).

Yang, C. W., & Shen, J. J. (2009). Recover the tampered image based on VQ indexing. *Signal Processing, 90*(1), 331–343. doi:10.1016/j.sigpro.2009.07.007

Yang, H. Y., Wu, J. F., Yu, Y. J., & Wang, X. Y. (2008). content based image retrieval using color edge histogram in HSV color space. *Journal of Image and Graphics, 13*(10), 2035–2038.

Yang, J., & Yang, J. Y. (2003). Why can LDA be performed in PCA transformed space? *Pattern Recognition Letters, 36*(2), 563–566.

Yang, J., Zhang, D., Yong, X., & Yang, J.-Y. (2005). Two-dimensional discriminant transform for face recognition. *Pattern Recognition Letters, 38*, 1125–1129.

Yang, T. (2004). A survey of chaotic secure communication systems. *International Journal of Computational Cognition, 2*(1), 81–130.

Yang, T., Wu, C. W., & Chua, L. O. (1997). Cryptography based on chaotic systems. *IEEE Transactions on Circuits and Systems. I, Fundamental Theory and Applications, 44*(5), 469–472. doi:10.1109/81.572346

Yao, Y. Y. (2004). A partition model of granular computing. In J. F. Peters, A. Skowron, J. W. Grzymala-Busse, B. Kostek, R. W. Swiniarski, & M. S. Szczuka (Eds.), *Transactions on Rough Sets I* (LNCS 3100, pp. 232-253).

Yao, Y. Y. (2007). Decision-theoretic rough set models. In J. T. Yao, P. Lingras, W.-Z. Wu, M. Szczuka, N. J. Cercone, & D. Slezak (Eds.), *Proceedings of the Second International Conference on Rough Sets and Knowledge Technology* (LNCS 4481, pp. 1-12).

Yao, Y. Y. (2007). The art of granular computing. In M. Kryszkiewicz, J. F. Peters, H. Rybinski, & A. Skowron (Eds.), *Proceeding of the International Conference on Rough Sets and Emerging Intelligent Systems Paradigms* (LNCS 4585, pp. 101-112).

Yao, Y. Y., Shi, Z., Wang, Y., & Kinsner, W. (Eds.). (2006, July). *Proceedings of the 5th IEEE International Conference on Cognitive Informatics*, Beijing, China. Washington, DC: IEEE Computer Society.

Yao, Y. Y. (2004). Granular computing. *Computer Science, 31*, 1–5.

Yao, Y. Y. (2009). Interpreting concept learning in cognitive informatics and granular computing. *IEEE Transactions on Systems, Man, and Cybernetics. Part B, Cybernetics, 39*(4), 855–866. doi:10.1109/TSMCB.2009.2013334

Yao, Y. Y. (2010). Three-way decisions with probabilistic rough sets. *Information Sciences, 180*(3), 341–353. doi:10.1016/j.ins.2009.09.021

Yao, Y. Y., & Wong, S. K. M. (1992). A decision theoretic framework for approximating concepts. *International Journal of Man-Machine Studies, 37*, 793–809. doi:10.1016/0020-7373(92)90069-W

Yao, Y. Y., Wong, S. K. M., & Lingras, P. (1990). A decisiontheoretic rough set model. In Ras, Z. W., Zemankova, M., & Emrich, M. L. (Eds.), *Methodologies for intelligent systems* (Vol. 5, pp. 17–24). New York, NY: North-Holland.

Yaremchuk, V., & Dawson, M. (2008). Artificial neural networks that classify musical chords. *International Journal of Cognitive Informatics and Natural Intelligence*, 2(3), 22–30. doi:10.4018/jcini.2008070102

Yassen, M. T. (2003). Chaos control of Chen chaotic dynamical system. *Chaos, Solitons, and Fractals*, 15(2), 271–283. doi:10.1016/S0960-0779(01)00251-X

Ye, J. (2005). Characterization of a family of algorithms for generalized discriminant analysis on undersampled problems. *Journal of Machine Learning Research*, 6, 483–502.

Ye, J., Janardan, R., Park, C. H., & Park, H. (2004). An optimization criterion for generalized discriminant analysis on undersampled problems. *IEEE Transactions on Pattern Analysis and Machine Intelligence*, 26(8), 982–994. doi:10.1109/TPAMI.2004.37

Ye, J., & Li, Q. (2004). LDA/QR: An efficient and effective dimension reduction algorithm and its theoretical foundation. *Pattern Recognition Letters*, 37, 851–854.

Ye, J., & Li, Q. (2005). A two-stage linear discriminant analysis via QR-decomposition. *IEEE Transactions on Pattern Analysis and Machine Intelligence*, 27(6), 929–941. doi:10.1109/TPAMI.2005.110

Yen, J. C., & Guo, J. I. (2000). A new chaotic key-based design for image encryption and decryption. In *Proceedings of IEEE International Symposium on Circuits and Systems*, Geneva, Switzerland (Vol. 4, pp. 49-52). Washington, DC: IEEE Computer Society.

Yifang, W., Rong, Z., & Yi, C. (2009). A self-synchronous stream cipher based on composite discrete chaos. In *Proceedings of the IEEE 8th International Conference on Cognitive Informatics* (pp. 210-214).

Yoo, T., & Goldsmith, A. (2006). On the optimality of multi-antenna broadcast scheduling using zero-forcing beamforming. *IEEE Journal on Selected Areas in Communications*, 24(3), 528–541. doi:10.1109/JSAC.2005.862421

Yu, H., & Yang, J. (2001). A direct LDA algorithm for high dimensional data with application to face recognition. *Pattern Recognition Letters*, 34(10), 2067–2070.

Zachary, J. M. (2000). *An information theoretic approach to content based image retrieval*. Unpublished doctoral dissertation, Louisiana State University and Agricultural and Mechanical College, Baton Rouge, LA.

Zadeh, L. A. (1999). From computing with numbers to computing with words – from manipulation of measurements to manipulation of perception. *IEEE Transactions on Circuits and Systems 1, 45*(1), 105-119.

Zadeh, L. A. (2008, August). Toward human level machine intelligence – Is it achievable? In *Proceedings of the 7th IEEE International Conference on Cognitive Informatics*, Stanford, CA (p. 1). Washington, DC: IEEE Computer Society.

Zadeh, L. A. (1965). Fuzzy sets and systems. In Fox, J. (Ed.), *Systems theory* (pp. 29–37). Brooklyn, NY: Polytechnic Press.

Zadeh, L. A. (1975). Fuzzy logic and approximate reasoning. *Syntheses*, 30, 407–428. doi:10.1007/BF00485052

Zadeh, L. A. (1975). Fuzzy logic and approximate reasoning. *Syntheses*, 30, 407–428. doi:10.1007/BF00485052

Zadeh, L. A. (1997). Towards a theory of fuzzy information granulation and its centrality in human reasoning and fuzzy logic. *Fuzzy Sets and Systems*, 19, 111–127. doi:10.1016/S0165-0114(97)00077-8

Zadeh, L. A. (1998). Some reflections on soft computing, granular computing and their roles in the conception, design and utilization of information/intelligent systems. *Soft Computing*, (2): 23–25. doi:10.1007/s005000050030

Zadeh, L. A. (1999). From computing with numbers to computing with words – from manipulation of measurements to manipulation of perception. *IEEE Transactions on Circuits and Systems I, 45*(1), 105–119. doi:10.1109/81.739259

Zadeh, L. A. (2004). Precisiated Natural Language (PNL). *AI Magazine*, 25(3), 74–91.

Zadeh, L. A. (2008). Is there a need for fuzzy logic? *Information Sciences*, 178, 2751–2779. doi:10.1016/j.ins.2008.02.012

Zarri, G. (2009). *Representation and processing of complex events.* Paper presented at the Association for the Advancement of Artificial Intelligence Spring Symposium, Stanford, CA.

Zhang, B. (2010, July). Computer vision vs. human vision. In *Proceedings of the 9ᵗʰ IEEE International Conference on Cognitive Informatics*, Beijing, China (p. 3). Washington, DC: IEEE Computer Society.

Zhang, D. (2005). Fixpoint semantics for rule base anomalies. In *Proceedings of the 4ᵗʰ IEEE International Conference on Cognitive Informatics*, Irvine, CA, (pp.10-17). Washington, DC: IEEE Computer Society.

Zhang, D. (2008). Quantifying knowledge base inconsistency via fixpoint semantics. In M. Gavrilova, C. J. K. Tan, Y. Wang, Y. Yao, & G. Wang (Eds.), *Transactions on Computational Science II* (LNCS 5150, pp. 145-160).

Zhang, D. (2009). Taming inconsistency in value-based software development. In *Proceedings of the 21ˢᵗ International Conference on Software Engineering and Knowledge Engineering*, Boston, MA (pp.450-455).

Zhang, D. (2011). Inconsistency-induced heuristics for problem solving. In *Proceedings of the 23ʳᵈ International Conference on Software Engineering and Knowledge Engineering*, Miami, FL (pp. 137-142).

Zhang, D., & Lu, M. (2011, August). Inconsistency-induced learning: A step toward perpetual learners. *Proceedings of the 10ᵗʰ IEEE International Conference on Cognitive Informatics*, Banff, AB, Canada. Washington, DC: IEEE Computer Society.

Zhang, D., Wang, Y., & Kinsner, W. (Eds.). (2007, August). *Proceedings of the 6th IEEE International Conference on Cognitive Informatics*, Lake Tahoe, CA. Washington, DC: IEEE Computer Society.

Zhang, J., Baciu, G., Liang, S., & Liang, C. (2010, November). A creative try: Composing weaving patterns by playing on a multi-input device. In *Proceedings of the 17th ACM Symposium on Virtual Reality Software and Technology* (pp. 127-130).

Zhang, P., Peng, J., & Riedel, N. (2005). Discriminant analysis: A least squares approximation view. In *Proceedings of the IEEE Conference on Computer Vision and Pattern Recognition* (p. 46).

Zhang, D. (2007). Fixpoint semantics for rule base anomalies. *International Journal of Cognitive Informatics and Natural Intelligence, 1*(4), 14–25. doi:10.4018/jcini.2007100102

Zhang, J., & Ye, L. (2009). Content based image retrieval using unclean positive examples. *IEEE Transactions on Image Processing, 18*(10), 2370–2375. doi:10.1109/TIP.2009.2026669

Zhang, L., Liang, Y.-C., & Xin, Y. (2008). Joint beamforming and power allocation for multiple access channels in cognitive radio networks. *IEEE Journal on Selected Areas in Communications, 26*(1), 38–51. doi:10.1109/JSAC.2008.080105

Zhang, Z., & Zha, H. Y. (2004). Principal manifolds and nonlinear dimensionality reduction via tangent space alignment. *Journal of Shanghai University, 8*(4), 406–424. doi:10.1007/s11741-004-0051-1

Zhao, C., Zhao, D., & Chen, Y. (1996). Simplified Gaussian and mean curvatures to range image segmentation. In *Proceedings of the International Conference on Pattern Recognition* (pp. 427-431).

Zhao, F. (2010). Sensors meet the cloud: Planetary-scale distributed sensing and decision making. In *Proceedings of the IEEE 9th International Conference on Cognitive Informatics* (p. 998).

Zhao, L. W., Luo, S. W., Zhao, Y. C., & Liu, Y. H. (2005). Study on the low-dimensional embedding and the embedding dimensionality of manifold of high-dimensional data. *Journal of Software, 16*(8), 1423–1430. doi:10.1360/jos161423

Zhao, Q., & Sadler, B. (2007). A survey of dynamic spectrum access. *IEEE Signal Processing Magazine, 24*(3), 79–89. doi:10.1109/MSP.2007.361604

Zhao, Y., & Cheah, C. (2009). Neural network control of multifingered robot hands using visual feedback. *IEEE Transactions on Neural Networks, 20*(5), 758–767. doi:10.1109/TNN.2008.2012127

Zhao, Y., Chen, Y., & Yao, Y. (2008). User-centered interactive data mining. *International Journal of Cognitive Informatics and Natural Intelligence, 2*(1), 58–72. doi:10.4018/jcini.2008010105

Zheng, D., Baciu, G., & Hu, J. (2010). Weave pattern accurate indexing and classification using entropy-based computing. *International Journal of Cognitive Informatics and Natural Intelligence, 4*(4), 76–92. doi:10.4018/jcini.2010100106

Zheng, H. T., Kang, B. Y., & Kim, H. G. (2009). Exploiting noun phrases and semantic relationships for text document clustering. *Information Science, 179*(13), 2249–2262. doi:10.1016/j.ins.2009.02.019

Zhong, N. (2009, July). A unified study on human and web granular reasoning. In *Proceedings of the 8th International Conference on Cognitive Informatics*, Kowloon, Hong Kong (pp. 3-4). Washington, DC: IEEE Computer Society.

Zhong, N., Dong, J. Z., & Ohsuga, S. (2001). Using rough sets with heuristics for feature selection. *Journal of Intelligent Information Systems, 16*(3), 199–214. doi:10.1023/A:1011219601502

Zhong, Y. (2008). A cognitive approach to the mechanism of intelligence. *International Journal of Cognitive Informatics and Natural Intelligence, 2*(1), 1–16. doi:10.4018/jcini.2008010101

Zhou, B., & Yao, Y. Y. (2008). A logic approach to granular computing. *International Journal of Cognitive Informatics and Natural Intelligence, 2*(2), 63–79. doi:10.4018/jcini.2008040104

Zhou, Z. H., Wu, J. X., & Tang, W. (2002). Ensembling neural networks: Many could be better than all. *Artificial Intelligence, 137*(1-2), 239–263. doi:10.1016/S0004-3702(02)00190-X

Zhu, H. (2008, May). Fundamental issues in the design of a role engine. In *Proceedings of the International Symposium on Collaborative Technologies and Systems*, Irvine, CA (pp. 399-407).

Zhu, H., & Grenier, M. (2009, July). Agent evaluation for role assignment. In *Proceedings of the IEEE 8th International Conference on Cognitive Informatics*, Hong Kong, China (pp. 405-411).

Zhu, H., & Hou, M. (2009, September). Restrain mental workload with roles in HCI. In *Proceedings of the IEEE Toronto International Conference – Science and Technology for Humanity*, Toronto, ON, Canada (pp. 387-392).

Zhu, H., & Zhou, M. C. (2003, November). Methodology first and language second: A way to teach object-oriented programming. In *Proceedings of the Educator's Symposium on Object-Oriented Programming, Systems, Languages and Applications*, Anaheim, CA (pp. 140-147).

Zhu, G. Y., Zheng, Y. F., Doermann, D., & Jaeger, S. (2009). Signature detection and matching for document image retrieval. *IEEE Transactions on Pattern Analysis and Machine Intelligence, 31*(11), 2015–2031. doi:10.1109/TPAMI.2008.237

Zhuge, H., & Sun, Y. C. (2010). The schema theory for semantic link network. *Future Generation Computer Systems, 26*(3), 408–420. doi:10.1016/j.future.2009.08.012

Zhu, H. (2007). Role as dynamics of agents in multi-agent systems. *System and Informatics Science Notes, 1*(2), 165–171.

Zhu, H. (2010). Role-based autonomic systems. *International Journal of Software Science and Computational Intelligence, 2*(3), 32–51. doi:10.4018/jssci.2010070103

Zhu, H., & Zhou, M. C. (2006). *Object-oriented programming with C++: A project-based approach*. Beijing, China: Tsinghua University Press.

Zhu, H., & Zhou, M. C. (2006). Role-based collaboration and its kernel mechanisms. *IEEE Transactions on Systems, Man and Cybernetics. Part C, 36*(4), 578–589.

Zhu, H., & Zhou, M. C. (2008). Role mechanisms in information systems - a survey. *IEEE Transactions on Systems, Man and Cybernetics. Part C, 38*(6), 377–396.

Zhu, H., & Zhou, M. C. (2008). Role transfer problems and their algorithms. *IEEE Transactions on Systems, Man, and Cybernetics. Part A, Systems and Humans, 38*(6), 1442–1450. doi:10.1109/TSMCA.2008.2003965

Zhu, H., & Zhou, M. C. (2009). M–M role-transfer problems and their solutions. *IEEE Transactions on Systems, Man, and Cybernetics. Part A, Systems and Humans, 39*(2), 448–459. doi:10.1109/TSMCA.2008.2009924

Zhu, H., & Zhou, M. C. (in press). Efficient role transfer. *IEEE Transactions on Systems, Man, and Cybernetics. Part A, Systems and Humans.*

Ziarko, W. (1993). Variable precision rough sets model. *Journal of Computer and System Sciences, 46,* 39–59. doi:10.1016/0022-0000(93)90048-2

Zielinski, K., Peters, D., & Laur, R. (2005, December). Stopping criteria for single-objective optimization. In *Proceedings of the 3rd International Conference on Computational Intelligence, Robotics, and Autonomous Systems* (pp. 1-6).

Ziemke, T. (2004). Embodied AI as science: Models of embodied cognition, embodied models of cognition, or both? In F. Iida, R. Pfeifer, L. Steels, & Y. Kuniyoshi (Eds.), *Proceedings of the International Seminar on Embodied Artificial Intelligence* (LNCS 3139, pp. 27-36).

About the Contributors

Yingxu Wang is Professor of Cognitive Informatics and Software Science, President of International Institute of Cognitive Informatics and Cognitive Computing (ICIC), Director of Laboratory for Cognitive Informatics and Cognitive Computing, and Director of Laboratory for Denotational Mathematics and Software Science at the University of Calgary. He is a Fellow of WIF (UK), a Fellow of ICIC, a P.Eng of Canada, a Senior Member of IEEE and ACM. He received a PhD in Software Engineering from the Nottingham Trent University, UK, and a BSc in Electrical Engineering from Shanghai Tiedao University. He has industrial experience since 1972 and has been a full Professor since 1994. He was a visiting Professor on sabbatical leaves in the Computing Laboratory at Oxford University in 1995, Dept. of Computer Science at Stanford University in 2008, the Berkeley Initiative in Soft Computing (BISC) Lab at University of California, Berkeley in 2008, and MIT (2012), respectively. He is the founder and steering committee chair of the annual IEEE International Conference on Cognitive Informatics and Cognitive Computing (ICCI*CC). He is founding Editor-in-Chief of *International Journal of Cognitive Informatics and Natural Intelligence* (IJCINI), founding Editor-in-Chief of *International Journal of Software Science and Computational Intelligence* (IJSSCI), Associate Editor of *IEEE Trans on System, Man, and Cybernetics* (Part A), and Editor-in-Chief of *Journal of Advanced Mathematics and Applications* (JAMA). Dr. Wang is the initiator of a few cutting-edge research fields or subject areas such as *Cognitive Informatics* (CI, the theoretical framework of CI, neuroinformatics, the logical model of the brain (LMB), the layered reference model of the brain (LRMB), the cognitive model of brain informatics (CMBI), the mathematical model of consciousness, and the cognitive learning engine); Abstract Intelligence (αI); Cognitive Computing (such as cognitive computers, cognitive robots, cognitive agents, and cognitive Internet); Denotational Mathematics (i.e., concept algebra, inference algebra, semantic algebra, real-time process algebra, system algebra, granular algebra, and visual semantic algebra); Software Science (on unified mathematical models and laws of software, cognitive complexity of software, and automatic code generators, the coordinative work organization theory, and built-in tests (BITs)); basic studies in Cognitive Linguistics (such as the cognitive linguistic framework, the deductive semantics of languages, deductive grammar of English, and the cognitive complexity of online text comprehension). He has published over 130 peer reviewed journal papers, 220+ peer reviewed conference papers, and 25 books in cognitive informatics, cognitive computing, software science, denotational mathematics, and computational intelligence. He is the recipient of dozens international awards on academic leadership, outstanding contributions, research achievement, best papers, and teaching in the last three decades.

* * *

Aníbal T. de Almeida (IEEE Member'85, SM'03) received the Ph.D. degree in electrical engineering from the Imperial College, University of London, London, U.K., 1977. He is currently a Professor with the Department of Electrical and Computer Engineering, University of Coimbra, Coimbra, Portugal, where he has been the Director of the Institute of Systems and Robotics, since 1993. He is a consultant with the European Commission Framework Programmes. He is the coauthor of six books and more than 150 papers in international journals, meetings, and conferences. He has coordinated several European and national research projects.

George Baciu holds a PhD degree in Engineering and a B.Math degree in Computer Science and Applied Mathematics from the University of Waterloo. Currently, he is a professor in the Department of Computing at The Hong Kong Polytechnic University. His research interests are computer graphics, image processing, user interfaces, motion tracking and information retrieval.

Virendrakumar C. Bhavsar is professor of computer science, Director of Advanced Computational Research Lab, former dean of the faculty of computer science at University of New Brunswick, Canada. Dr. Bhavsar is expert in intelligent systems, bioinformatics, and parallel/distributed computing.

Jeff Bancroft is an MSc candidate in the Department of Electrical and Computer Engineering at the University of Calgary, Canada. He received a BSc in computer engineering from the University of Calgary in 2008. His research interests are cognitive computing and computational intelligence.

Robert C. Berwick is Professor of Computational Linguistics in the Department of Electrical Engineering and Computer Science and the Department of Brain and Cognitive Sciences at the Massachusetts Institute of Technology. Professor Berwick received his A.B. degree from Harvard University in Applied Mathematics and his S.M. and Ph.D. degrees from the Massachusetts Institute of Technology in Computer Science in Artificial Intelligence. Since then he has been a member of the MIT faculty, and is currently co-Director of the MIT Center for Biological and Computational Learning. He is the recipient of a Guggenheim Award, and the author of 7 books and many articles in the area of natural language processing, complexity theory, language acquisition, and the biology and evolution of language. His latest book, to be published by Oxford University Press, is *Rich Grammars from Poor Inputs*.

Chuanliang Cai received the bachelor degree in Wuhan University of Science and Technology in 2008. He is currently a graduate student at School of Computer Engineering and Science in Shanghai University and his main research interests include cognitive informatics.

Eddie Chan received his BSc Degree in Computing and MSc Degree from The Hong Kong Polytechnic University in 2005 and 2007. Currently, he is a PhD student in the same university. His research interests include wireless communication, localization, and agent technology.

Sébastien Dourlens is IT and R&D Director in software development company and also works as researcher in Artificial intelligence & Human Robot Interaction in Laboratoire d'ingénierie des systèmes de Versailles (LISV), Versailles University, in France.

Shiaofen Fang is a Professor of Computer Science and the Chairman of the Department of Computer and Information Science at Indiana University Purdue University Indianapolis (IUPUI). He also served as director of the Signature Center for Bio-computing (CBC) and director of the Center for Visual Information Sensing and Computing (VISC). Prof. Fang received his Ph.D in Computer Science from the University of Utah, and his BS and MS in Mathematics from Zhejiang University. Prof. Fang's research interest is in Scientific and Information Visualization, Medical Imaging, Volume Graphics, and Geometric Modeling. He has published over 70 papers in peer reviewed international journals and conferences. His research has been funded by the National Science Foundation (NSF), Nation Institutes of Health (NIH), and the National Institute of Justice (NIJ). He is a regular panelists and reviewers for NSF and NIH, and has chaired or served in program committees in many international conferences and workshops.

Tatiana Foroud, Ph.D., was trained as in population genetics at Indiana University. She is a Professor in the Department of Medical and Molecular Genetics at Indiana University School of Medicine. Her research in the area of fetal alcohol syndrome spectrum disorders is focused on the utility of facial imaging to improve our understanding of the biological effects of prenatal alcohol exposure. She is coupling these data along with DNA from the mother and the child to also understand the role that genetics may play in the risk of fetal alcohol syndrome spectrum disorders following prenatal alcohol exposure. This work is funded by the National Institute on Alcohol Abuse and Alcoholism (NIAAA).

Marina L. Gavrilova is an Associate Professor and associate head of the Department of Computer Science, University of Calgary. Dr. Gavrilova's research interests include computational geometry, image processing, optimization, exact computation and computer modeling. Dr. Gavrilova is a founder of two innovative research labs, the SPARCS Laboratory for Spatial Analysis in Computational Sciences and the Biometric Technologies Laboratory. Her publication list includes over 80 research papers, books and book chapters. Dr. Gavrilova is an Editor-in-Chief of Transactions on Computational Science, Springer and serves on the Editorial Board for International Journal of Computational Sciences and Engineering and Computer Graphics and CAD/CAM Journal. She is an ACM, IEEE and Computer Society member.

Baoming Ge (IEEE Member'11) received the Ph.D. degree in electrical engineering from Zhejiang University, Hangzhou, China, in 2000. He was a Postdoctoral Researcher with the Department of Electrical Engineering, Tsinghua University, Beijing, China, from 2000 to 2002, was a Visiting Scholar with the Department of Electrical and Computer Engineering, University of Coimbra, Coimbra, Portugal, from 2004 to 2005, and was a Visiting Professor with the Department of Electrical and Computer Engineering, Michigan State University, East Lansing, from 2007 to 2008, and from 2010 to present. He is currently with the School of Electrical Engineering, Beijing Jiaotong University, Beijing, China, as a Professor. His research interests include nonlinear control theories, artificial intelligence and applications to electric drives.

Jesus D. Terrazas Gonzalez completed a B.Eng. in Mechatronic Engineering at the Technological Institute of Cuauhtémoc City, México, in 2005. He is now pursuing a M.Sc. in Electrical and Computer Engineering at the University of Manitoba, Canada. His professional experience involves mechatronic applications and mining incursions in Scotland, Germany, and México. His research interest includes advanced signal processing, cellular automata, cryptography, cryptology, dynamical systems, and fractals and chaos. He is a member of the Institute of Electrical and Electronic Engineers (IEEE).

Xudong Guan received his B.Sc. degree in software engineering from East China Normal University, China, in 2006, and the M.Sc. degree in Computer Science from Fudan University in 2009. His research interests include artificial intelligence, pattern recognition and computer vision.

Li Guo is a professor in Information and Communication Engineering, BUPT. Her research interests include communication theory, cognitive MIMO system and mobile communication. Currently, she has published about 20 papers which can be indexed by EI.

Simon Haykin is professor of electrical and computer engineering, FIEEE. FRSC, Distinguished University Professor. He is director of the Cognitive Systems Lab in Dept. of Electrical and Computer Engineering at McMaster University, Canada. Prof. Haykin is a pioneer of cognitive radio/radar and adaptive signal processing, as well as an expert in cognitive systems and communication/information systems.

Xiang He received the bachelor degree in Agricultural University of Hebei in 2007. He is currently a graduate student at School of Computer Engineering and Science in Shanghai University and his main research interests include text semantic analysis.

Ming Hou is the Head of the Advanced Interface Group, Defence Research and Development Canada (DRDC) - Toronto, where he is responsible for providing informed decisions to the Canadian Forces on investment in and application of advanced technologies for operator machine interface requirements. Dr. Hou is the Canadian National Leader of Human Systems Integration Technical Panel for the Air in The Technical Cooperation Program (TTCP). He is a Human Factors Specialist in two NATO Tasking Groups. Dr. Hou is the Human Factors executive member in DRDC Working Group on Uninhabited Autonomous Systems. Dr. Hou received the Ph.D. degree in Human Factors Engineering from the University of Toronto, Canada, in 2002. He is a senior member of IEEE, a member of the Human Factors and Ergonomics Society (HFES), and a member of the Association of Computing Machinery (ACM). He was the Chair of the Symposium on Human Factors and Ergonomics at the 2009 IEEE Toronto International Conference Science and Technology for Humanity. He has been the Co-Chair of the International Symposium on Mixed and Virtual Reality since 2004.

H. Eugene Hoyme is Professor and Chair of the Department of Pediatrics at the Sanford School of Medicine of The University of South Dakota, Chief Medical Officer of Sanford Children's Hospital and Senior Vice President for Children's Services at Sanford Clinic in Sioux Falls. Until 2007, Dr. Hoyme served as Professor and Chief of the Division of Medical Genetics and Associate Chair of the Department of Pediatrics at Stanford University School of Medicine in Palo Alto, CA. Dr. Hoyme received his BA from Augustana College in Sioux Falls and his MD from the University of Chicago Pritzker School of Medicine. He completed his pediatric residency and clinical genetics fellowship training at the University of California, San Diego. Among his professional affiliations, he is a member of the American Academy of Pediatrics, the American Pediatric Society, the American Society of Human Genetics and the American College of Medical Genetics. In 2007-2008, he served as the President of the Western Society for Pediatric Research. In 2011, he received the Joseph St. Geme, Jr., Education Award from the WSPR for outstanding contributions to pediatric education. Dr. Hoyme has served numerous leadership roles in regional and national clinical and research organizations in pediatrics and medical genetics. His research focuses on the delineation of genetic and malformation syndromes, with a specific emphasis on fetal alcohol spectrum disorders. He has authored over 300 original articles, monographs, book chapters and research abstracts.

Kendal K. Hu received an MSc degree in the Cognitive Informatics and Cognitive Computing Lab and the Theoretical and Empirical Software Engineering Center (TESERC) in Dept. of Electrical and Computer Engineering at the University of Calgary in 2009. He received a MEng from Zhejiang University, China, in Electrical Engineering in 1986. He is a software engineer working at ITSportsnet Inc., Calgary. His research interest is in cognitive informatics, cognitive computing, software engineering, cognitive search engines, and machine learning.

Sandra W. Jacobson, Ph.D., was trained as a clinical and developmental psychologist at Harvard University. She is a Professor in the Department of Psychiatry and Behavioral Neurosciences at Wayne State University School of Medicine and Honorary Professor in the Departments of Human Biology and of Psychiatry at University of Cape Town Faculty of Health Sciences. Her research focuses on the effects of prenatal exposure to alcohol and environmental contaminants on cognitive and behavioral development in infancy and childhood. She is currently conducting a prospective longitudinal study of fetal alcohol spectrum disorders in the Cape Coloured (mixed ancestry) community in Cape Town, South Africa, that is being funded by the NIH/National Institute on Alcohol Abuse and Alcoholism.

Witold Kinsner is Professor and Associate Head at the Department of Electrical and Computer Engineering, University of Manitoba, Winnipeg, Canada. He is also Affiliate Professor at the Institute of Industrial Mathematical Sciences, Winnipeg, and Adjunct Scientist at the Telecommunications Research Laboratories (TRLabs), Winnipeg. He obtained his PhD in electrical and computer engineering from McMaster University, Hamilton in 1974. He has authored and co-authored over 630 publications in his research areas. Dr. Kinsner is a life member of the Institute of Electrical & Electronics Engineers (IEEE), a member of the Association of Computing Machinery (ACM), a member the Mathematical and Computer Modelling Society, a member of Sigma Xi, and a life member of the Radio Amateur of Canada (RAC).

Oliver Kramer is Assistant Professor at the Bauhaus-University in Weimar, Germany. His main research interests are stochastic optimization, machine learning, and cognitive systems. In 2008 he has finished his PhD at the Graduate School of Dynamic Intelligent Systems at the University of Paderborn, Germany. After a postdoc stay at the University of Dortmund in 2009 he was researcher at the International Computer Science Institute in Berkeley, California, in 2010. At the Bauhaus University he is working in various projects related to optimization and stochastic modeling in civil engineering, simulation data mining, and optimization in sustainable energy projects since 2011. He is the author of the book "Self-Adaptive Heuristics for Evolutionary Computation" in the Springer Series Studies in Computational Intelligence. He is the author of the German textbook "Computational Intelligence" with a focus on machine learning and optimization.

Liang Lei was born in Sichuan, China in March 1973. He is a PhD candidate of Chongqing University. And he received a master degree in control theory and control engineering from YanShan University in 2005, a bachelor's degree in Computer Application from Liaoning University of Science and Technology (the original Anshan University of iron and steel). He works at Chongqing iron and steel group company in 1998, where he is a computer administrator. From 1999 to present he works at Chongqing University of Science and Technology, where he is currently an associate professor in School of Electrical and Information Engineering. He has published more than 20 international journal and conference papers and published more than 10 books used in teaching. His current research interests are on Image retrieval, pattern recognition and database technology.

Mingming Li is a Ph.D. candidate in Information and Communication Engineering, Beijing University of Post and Telecommunication. Her research includes wireless communication, cognitive MIMO system and sensor networks, who has published six papers in international journals and conferences and searched by EI Compendex.

Rong-Hua Li received his BSE degree and MPhil degree in Computer Science from Jiangxi Normal University, 2007 and South China University of Technology, 2010. Currently, he is a PhD student at the Chinese University of Hong Kong. His recent research focuses on machine learning, data mining and online social network analysis.

Shuang Liang received her BSE in computer science from Zhejiang University in 2003 and her PhD in computer science from Nanjing university in 2008. Currently she is a postdoc fellow at The Hong Kong Polytechnic University. Her research interests include computer graphics, machine learning, pattern recognition, multimedia and information retrieval.

Xiaofeng Liao received the BS and MS degrees in mathematics from Sichuan University, Chengdu, China, in 1986 and 1992, respectively, and the PhD degree in circuits and systems from the University of Electronic Science and Technology of China in 1997. From 1999 to 2001, he was involved in post-doctoral research at Chongqing University, where he is currently a professor. From November 1997 to April 1998, he was a research associate at the Chinese University of Hong Kong. From October 1999 to October 2000, he was a research associate at the City University of Hong Kong. From March 2001 to June 2001 and March 2002 to June 2002, he was a senior research associate at the City University of Hong Kong. From March 2006 to April 2007, he was a research fellow at the City University of Hong Kong. He has published more than 150 international journal and conference papers. His current research interests include neural networks, nonlinear dynamical systems, bifurcation and chaos, and cryptography.

Jiaru Lin is a professor in Information and Communication Engineering, Beijing University of Post and Telecommunication. His major research interests include information theory, communication theory and mobile communication. Currently, Lin is taking charge of a National Basic Research Program of China and has published about 50 papers in IEEE or others, 20 of which can be indexed by either SCI or EI.

Fazhong Liu is a Ph.D. candidate of College of Computing & Communication Engineering Graduate University of Chinese Academy of Sciences. He is a Senior Engineer working at China Unicom Shandong branch and interested in mobile value-added services, Internet and telecommunications business operation.

Jianhua Lu is a professor of electronic engineering and aerospace information systems. He is the Associate Dean in the School of Aerospace at Tsinghua University, Beijing, China. He was the Program Committee Co-Chair of the 9th IEEE International Conference on Cognitive Informatics (ICCI 2010).

Xiangfeng Luo received the master's and PhD degrees from the Hefei University of Technology in 2000 and 2003, respectively. He was a postdoctoral researcher with the China Knowledge Grid Research Group, Institute of Computing Technology (ICT), Chinese Academy of Sciences (CAS), from 2003 to 2005. He is currently an associate professor in the School of Computer Engineering and Science, Shanghai University. His main research interests include web content analysis, Semantic Networks, web knowledge flow, and Semantic Grid. His publications have appeared in IEEE Trans. on Automation Science and Engineering, IEEE Trans. on Learning Technology, Concurrency and Computation: Practice and Experience, and New Generation Computing.

Jason McLaughlin is a PhD student in the Purdue Department of Computer and Information Science at Indiana University Purdue University Indianapolis (IUPUI). He received his MS in CSCI from IUPUI and his BS in Computer Science at the US Military Academy at West Point. He is a recipient of the IUPUI University Fellowship, and his research interests include Visual Analytics, Information Visualization, and Polygon Mesh Processing. His primary advisor is Dr. Shiaofen Fang.

Witold Pedrycz is Professor and Canada Research Chair (CRC - Computational Intelligence) in the Department of Electrical and Computer Engineering, University of Alberta, Edmonton, Canada. He is also with the Systems Research Institute of the Polish Academy of Sciences, Warsaw, Poland. In 2009 Dr. Pedrycz was elected a foreign member of the Polish Academy of Sciences. He main research directions involve Computational Intelligence, fuzzy modeling and Granular Computing, knowledge discovery and data mining, fuzzy control, pattern recognition, knowledge-based neural networks, relational computing, and Software Engineering. He has published numerous papers in this area. He is also an author of 14 research monographs covering various aspects of Computational Intelligence and Software Engineering. Witold Pedrycz has been a member of numerous program committees of IEEE conferences in the area of fuzzy sets and neurocomputing. Dr. Pedrycz is intensively involved in editorial activities. He is an Editor-in-Chief of *Information Sciences* and Editor-in-Chief of *IEEE Transactions on Systems, Man, and Cybernetics - part A*. He currently serves as an Associate Editor of *IEEE Transactions on Fuzzy Systems* and is a member of a number of editorial boards of other international journals. In 2007 he received a prestigious Norbert Wiener award from the IEEE Systems, Man, and Cybernetics Council. He is a recipient of the IEEE Canada Computer Engineering Medal 2008. In 2009 he has received a Cajastur Prize for Soft Computing from the European Centre for Soft Computing for *"pioneering and multifaceted contributions to Granular Computing"*.

Jun Peng was born in Chongqing, China in July 1970. He received a Ph.D. in Computer Software and Theory from Chongqing University in 2003, a MA in computer system architecture from Chongqing University in 2000, and a BSc in Applied Mathematics from the Northeast University in 1992. From 1992 to present he works at Chongqing University of Science and Technology, where he is currently a Professor in school of Electrical and Information Engineering. He was a visiting scholar in the Laboratory of Cryptography and Information Security at Tsukuba University, Japan in 2004, and Department of Computer Science at California State University, Sacramento in 2007, respectively. He has authored or coauthored over 40 peer reviewed journal or conference papers. He has served as the program committee member or session co-chair for over 10 international conferences such as SEKE'08, SEKE'10, ICCI'09, ICCI'10, ICTAI'10, WCICA'08, ICNC-FSKD'08, SocialNet'09, SEA'08, ICCIT'08, and NISS'10. His current research interests are on cryptography, chaos and network security.

Amar Ramdane-Cherif received his Ph.D. degree from Université Pierre et Marie Curie, Paris in 1998 in Neural networks and AI optimization for robotic applications. Then, he has occupied an Assistant Professor position at the PRISM Laboratory, Versailles University in France. His is now a full professor in LISV Laboratory, Versailles University in France. His research interests include Software Architecture, Dynamic Architecture, Multimodal Interaction, Human-Robot Interaction, styles and quality attributes.

Luther K. Robinson, MD is a Professor of Pediatrics at the State University of New York School of Medicine and Biomedical Sciences. He is Director of Dysmorphology and Clinical Genetics in the Division of Genetics of the Women and Children's Hospital of Buffalo at Kaleida Health. Dr Robinson graduated from Oberlin College and earned his medical degree at the University of Cincinnati College of Medicine. Following a four-year commitment in the US Public Health Service Dr Robinson undertook postgraduate training in the Department of Pediatrics at the University of California at San Diego. There he met Kenneth Lyons Jones, MD who would become his mentor in Dysmorphology (altered structural development). Dr Robinson is former medical director of the New York Pregnancy Risk Network, a program that provides information concerning exposures in pregnancy to pregnant women and their physicians. Dr. Robinson was an inaugural member of the National Task Force on Fetal Alcohol Syndrome/Fetal Alcohol Effects (NTFFASE) and is involved in studies on fetal alcohol spectrum disorder in the United States, Russia, Europe and South Africa.

Dario Schor completed a B.Sc. in Computer Engineering in 2008 and is currently pursuing a M.Sc. in Computer Engineering both at the University of Manitoba. His research interest includes evolutionary algorithms for cognitive machines, and particularly the design of hardware and software for space applications. He is a member of the Institute of Electrical and Electronic Engineers and the Association of Computing Machinery.

Kenji Sugawara is a professor of the Department of Information and Network Science, and the dean of Faculty of Information and Computer Science, Chiba Institute of Technology, Chiba, Japan. He received a doctoral degree in engineering (1983) from Tohoku University, Japan. His research interests include Multi-agent System, Artificial Intelligence, Ubiquitous Computing and Symbiotic Computing. He is a fellow of IEICE Japan and a member of IEEE, ACM and IPSJ.

Fuchun Sun is a professor of computer science with the Tsinghua National Laboratory for Information Science and Technology, Department of Computer Science and Technology, Tsinghua University, Beijing, China. He was the Program Committee Co-Chair of the 9th IEEE International Conference on Cognitive Informatics (ICCI 2010).

Yousheng Tian is a PhD candidate in cognitive computing and software engineering with the Cognitive Informatics and Cognitive Computing Lab and the Theoretical and Empirical Software Engineering Center (TESERC) in Dept. of Electrical and Computer Engineering at the University of Calgary, Canada. He received a PhD from Xian Jiantong University, China, in Computer Science in 2002. His research interests are in cognitive informatics, cognitive computing, software engineering, machine learning, and denotational mathematics.

Dongxu Wang is an Intermediate Engineer at China Unicom headquarters, Master's Degree. He has long been engaged in cognitive radio research work and mobile value-added services.

Guoyin Wang was born in Chongqing, China, in 1970. He received the B.Sc. degrees, the M.Sc. degrees and Ph. D. degree in Computer Science from Xi'an Jiaotong University, Xi'an, China, in 1992, 1994, and 1996 respectively. He is Steering Committee chair of International Rough Set Society, he is Chairman of the Rough Set and Soft Computation Society, Chinese Association for Artificial Intelligence. His research interestes include rough set theory, granular computing, data mining.

TongQing Wang was born in Chongqing, China in November 1949. Today, he works at Chongqing University, where he is currently a Professor and PhD. supervisor at College of Optoelectronic Engineering. He won four progress prizes in science and technology of country, and in science and technology of ministry of education, 10 national patents. The main technical of "AV - 100 form automatic reading machine" indexes in the world advanced level, and won the third national scientific and technological progress, American invention and new products at the 7th international exposition award and computer gold prize, the results are regarding as the high value by the United Nations statistical agency with America, Germany, Italy and other developed countries. And he has published more than 60 international journal and conference papers. His current research interests in optic-mechanical integration technology, the research and application of computer automatic identification technology.

Hui Wei received a Ph.D. degree in the Department of Computer Science, Beijing University of Aeronautics and Astronautics, China, in 1998. From September 1998 to November 2000, he was a Post-doctoral Fellow in the Department of Computer Science and Institute of Artificial Intelligence, Zhejiang University. In November 2000, he joined the Department of Computer Science and Engineering at Fudan University. Currently, he serves as a professor of the center for Information Intelligence. His research interests include artificial intelligence and cognitive science.

Thomas Weise received the Diplom Informatiker (equivalent to M.Sc.) degree from the Department of Computer Science, Chemnitz University of Technology, Chemnitz, Germany, in 2005, and the Ph.D. degree at the Distributed Systems Group of the Fachbereich Elektrotechnik and Informatik, University of Kassel, Kassel, Germany in 2009. Since 2009, he has been researcher at the Nature Inspired Computation and Applications Laboratory, School of Computer Science and Technology, University of Science and Technology of China, Hefei, Anhui, China where he became Associate Professor in 2011. His major research interests include Evolutionary Computation, Genetic Programming (GP), and real-world applications of optimization algorithms. His experience ranges from applying GP to distributed systems and multi-agent systems, efficient web service composition for Service Oriented Architectures, to solving large-scale real-world vehicle routing problems for multimodal logistics and transportation. Besides being the author/co-author of over 50 refereed publications, Dr. Weise also authors the electronic book "Global Optimization Algorithms -- Theory and Application" which is freely available at his website http://www.it-weise.de/.

Bernard Widrow received the S.B., S.M., and Sc.D. degrees in Electrical Engineering from the Massachusetts Institute of Technology in 1951, 1953, and 1956, respectively. He joined the MIT faculty and taught there from 1956 to 1959. In 1959, he joined the faculty of Stanford University, where he is currently Professor of Electrical Engineering. Dr. Widrow began research on adaptive filters, learning processes, and artificial neural models in 1957. Together with M.E. Hoff, Jr., his first doctoral student at Stanford, he invented the LMS algorithm in the autumn of 1959. Today, this is the most widely used learning algorithm, used in every MODEM in the world. He has continued working on adaptive signal processing, adaptive controls, and neural networks since that time. Dr. Widrow is a Life Fellow of the IEEE and a Fellow of AAAS. He received the IEEE Centennial Medal in 1984, the IEEE Alexander Graham Bell Medal in 1986, the IEEE Signal Processing Society Medal in 1986, the IEEE Neural Networks Pioneer Medal in 1991, the IEEE Millennium Medal in 2000, and the Benjamin Franklin Medal for Engineering from the Franklin Institute of Philadelphia in 2001. He was inducted into the National Academy of Engineering in 1995 and into the Silicon Valley Engineering Council Hall of Fame in 1999. Dr. Widrow is a past president and member of the Governing Board of the International Neural Network Society. He is associate editor of several journals and is the author of over 125 technical papers and 21 patents. He is co-author of Adaptive Signal Processing and Adaptive Inverse Control, both Prentice-Hall books. A new book, Quantization Noise, was published by Cambridge University Press in June 2008.

Bo Yang was born in Chongqing, China in 1973. He received a Ph.D. in Instrument science and technology from Chongqing University in 2002. From 2003 to 2006, he was involved in postdoctoral research at machine Manufacture Company of Changan of China. From 2008 to present he works at Chongqing University of Science and Technology, where he is currently a associate Professor in School of Electrical and Information Engineering. He has published more than 40 international journal and conference papers. His current research interests include pattern recognition and intelligent system; machine vision; clean energy development and utilization.

Yong Yang was born in Heqing, China, in 1976. He received the B.Sc. degrees in Telecommunication Engineering from Chongqing University of Posts and Telecommunications, Chongqing, China, in 1997, and the M.Sc. degrees in Computer Science from Chongqing University of Posts and Telecommunications, Chongqing, China, in 2003, and the Ph.D. degree in Computer Science from the Southwest Jiaotong University, Chengdu, China, in 2009. His research interestes include affective computing, human-compuer interaction, cloud computing, and parallel data mining.

Yiyu Yao is a Professor with the Department of Computer Science, University of Regina, Canada. He received his B.Eng. (1983) in Computer Science from Xi'an Jiaotong University, China, M.Sc. (1988) and Ph.D. (1991) in Computer Science from University of Regina, Canada. From 1992 to 1998, he was an Assistant Professor and an Associate Professor with the Department of Mathematical Science, Lakehead University, Canada. Dr. Yao's research interests include Web Intelligence, information retrieval, uncertainty management (fuzzy sets, rough sets, interval computing, and granular computing), data mining, and intelligent information systems. He was invited to give talks at many international conferences and universities. Dr. Yao is the chair of steering committee of the International Rough Set Society, a member Technical Committee of Web Intelligence Consortium. He has served, and is serving, as Conference Chair, Program Committee Chair, member of Advisory and/or Program Committee of many international conferences. He is serving as an editorial board member of several international journals.

Bo Zhang is professor in the Department of Computer Science and Technology, Tsinghua University and a Fellow of Chinese Academy of Sciences, Beijing, China. He graduated from the Tsinghua University, Beijing, China, in 1958. His main research interests include artificial intelligence, neural networks and pattern recognition. He has published about 150 papers and three books in these fields.

Du Zhang received his Ph.D. degree in Computer Science from the University of Illinois. He is a Professor of the Computer Science Department at California State University, Sacramento. His current research interests include: knowledge base inconsistency, machine learning in software engineering, and knowledge-based systems and multi-agent systems. He has authored or coauthored over 150 publications in journals, conference proceedings, and book chapters, in these and other areas. In addition, he has edited or co-edited eleven books and conference proceedings. He has served as the conference general chair, the program committee chair, a program committee co-chair, or a program vice chair/area chair for 22 international conferences, most of which are IEEE sponsored international conferences. Currently, he is an Associate Editor for International Journal on Artificial Intelligence Tools, a member of editorial board for International Journal of Cognitive Informatics and Natural Intelligence, and a member of editorial board for International Journal of Software Science and Computational Intelligence. In addition, he has served as a guest editor for special issues of International Journal of Software Engineering and Knowledge Engineering, Software Quality Journal, IEEE Transactions on SMC-Part B, EATCS Fundamenta Informaticae, and International Journal of Computer Applications in Technology. Du Zhang is a senior member of IEEE and a senior member of ACM.

Jun Zhang received the bachelor degree in Shanghai University in 2008 and then was recommended for admission to graduate school. Currently, he is a PhD student at School of Computer Engineering and Science in Shanghai University and his main research interests include online word relation discovery and topic detection and tracking.

Bing Zhou is a PhD candidate in the Department of Computer Science, University of Regina, Canada. She received her MSc in Computer Science from the University of Regina in 2008. Her research interests include machine learning, data mining, granular computing and formal languages.

Haibin Zhu is Full Professor of the Department of Computer Science and Mathematics, Director and Founder of Collaborative Systems Laboratory, Nipissing University, Canada. He received his Ph.D. degree in computer science from the National University of Defense Technology, Changsha, China, in 1997. He has published more than 100 research papers, four books and two book chapters. He is a senior member of IEEE, and a member of the Association of Computing Machinery (ACM). He is serving and served as co-chair of the technical committee of Distributed Intelligent Systems of IEEE SMC Society; guest (co-)editor for 3 special issues of prestigious journals; PC chair, PC vice chair, and other co-chairs for many IEEE conferences. He is the recipient of the 2006-2007 research award from Nipissing University, the 2004 and 2005 IBM Eclipse Innovation Grant Awards, the Best Paper Award from the 11th ISPE Int'l Conf. on Concurrent Engineering (ISPE/CE2004), the Educator's Fellowship of OOPSLA'03, a 2nd Class National Award of Excellent Textbook from the Ministry of Education of China (2002), a 2nd class National Award for Education Achievement(1997), and three 1st Class Ministerial Research Achievement Awards from The Commission of Science Technology and Industry for National Defense of China (1997, 1994, and 1991).

Qingsong Zuo received his B.Sc. degree in software engineering from Nanjing University of Science & Technology, China, in 2007. From 2007 to 2008, he was a software engineer in Rigol Technologies, INC., Beijing. Since 2009, he has been a Master candidate in the Department of Computer Science and Engineering, Fudan University, China. His research interests include artificial intelligence, cognitive model, computer vision and image processing, with emphasis on models inspired by human visual system.

Index

G

Ganglion cell 56, 58-66
General Intelligence Model (GIM) 38
General Research Fund (GRF) 352
genetic algorithm (GA) 130
global feedback loop 252, 259
GOLD 195, 218
granular algebra 8, 23, 144, 155
granular computing 19, 26, 33, 138, 146-148, 152, 154-156, 177-179, 192, 219-220, 238, 240, 246-248
granular logic descriptor 189, 191-192
graphical representor 111, 113, 118
Group Agent (GA) 27

H

Hessian LLE (HLLE) 325
hidden Markov model (HMM) 129
Hominidae 72
Homo heidelbergensis 72
horizontal cell layer 57
human cognition 25, 77, 80, 146, 216, 250-251, 285
human-computer interaction (HCI) 35
human-human interactions (HHI) 36
Human-Machine Interaction (HMI) 39
Human-Robot Interaction (HRI) 35
hype-structure 195

I

image encryption 272, 354-356, 358, 363, 365-366
Independent Component Analysis (ICA) 100, 324
Individual Intelligences Interaction (I3) 39
inductive learning 240-241
inference algebra 144, 156, 159, 162, 170-171, 175, 177, 196, 219
information granularity 178-179, 183-187, 189, 192-193
Information-Matter-Energy (IME) 3, 22
Information-Matter-Energy-Intelligence (IME-I) 142-143
Information Retrieval (IR) 83
integral error feedback 392
Intelligence 36
    ability 4, 37, 45, 62, 71, 74, 77-79, 82-84, 112, 120, 130-131, 183, 194-195, 222, 230, 233, 291-292, 327, 390
    Individual Intelligence (I2) 37
    Intelligence Quotient (IQ) 37
    Known Intelligence (KI) 37

object 7, 14, 25, 27, 37-39, 44, 46-50, 55, 62-63, 68, 74-76, 78, 83, 86, 88, 100, 109-110, 114, 117-118, 120, 141, 145, 198, 206-207, 210-215, 222, 224, 238-241, 268, 326, 369
intension of a concept 116-117, 241
interface optimization 311, 313-315, 317-318
internal thought 77
International Conferences on Cognitive Informatics and Cognitive Computing (ICCI*CC) 141
International Conferences on Cognitive Informatics (ICCI) 21
in-vehicle information systems (IVIS) 46
isometric mapping (Isomap) 325
Iterative Closest Point (ICP) 101

J

Japanese female facial expression (JAFFE) 134, 138

K

Knowledge Management (KM) 83
Knowledge Representation Languages (KRL) 83

L

Laplacian eigenmaps(LE) 325
latent dirichlet allocation (LDA) 233
latent semantic analysis (LSA) 222
Layered Reference Model of the Brain (LRMB) 3, 11-12, 18, 23, 26, 33, 127, 142, 144, 156, 177
leaf node 237-238, 242-246
least-mean-square (LMS) 253
least squares (LS) 286
linear combiner 251
Linear Discriminant Analysis (LDA) 338-340
Linear Discriminant Analysis via QR decomposition (LDA/QR) 338
lingua franca 71
Local Linear Projection (LLP) 324
locally linear embedding (LLE) 325
local tangent space alignment(LTSA) 325
Long-Term Memory (LTM) 6, 29, 113, 145, 199
loop inference 165

M

machinable intelligence 5, 15, 143
machinable intelligence (MI) 5
manipulator 118, 387-389, 392, 394, 397
Maximal Ratio Combing (MRC) 373
maximum likelihood estimator (MLE) 326
membership layer 390, 394